Classics in Mathematics

J.W.S. Cassels An Introduction to the Geometry of Numbers

Springer
*Berlin*
*Heidelberg*
*New York*
*Barcelona*
*Budapest*
*Hong Kong*
*London*
*Milan*
*Paris*
*Santa Clara*
*Singapore*
*Tokyo*

J. W. S. Cassels (known to his friends by the Gaelic form "Ian" of his first name) was born of mixed English-Scottish parentage on 11 July 1922 in the picturesque cathedral city of Durham. With a first degree from Edinburgh, he commenced research in Cambridge in 1946 under L. J. Mordell, who had just succeeded G. H. Hardy in the Sadleirian Chair of Pure Mathematics. He obtained his doctorate and was elected a Fellow of Trinity College in 1949. After a year in Manchester, he returned to Cambridge and in 1967 became Sadleirian Professor. He was Head of the Department of Pure Mathematics and Mathematical Statistics from 1969 until he retired in 1984.

Cassels has contributed to several areas of number theory and written a number of other expository books:

- *An introduction to diophantine approximations*
- *Rational quadratic forms*
- *Economics for mathematicians*
- *Local fields*
- *Lectures on elliptic curves*
- *Prolegomena to a middlebrow arithmetic of curves of genus 2* (with E. V. Flynn).

J.W.S. Cassels

# An Introduction to the Geometry of Numbers

Reprint of the 1971 Edition

J.W.S. Cassels
University of Cambridge
Department of Pure Mathematics
and Mathematical Statistics
16, Mill Lane
CB2 1SB Cambridge
United Kingdom

Originally published as Vol. 99 of the
*Die Grundlehren der mathematischen Wissenschaften in Einzeldarstellungen*

Mathematics Subject Classification (1991): 10Exx

CIP data applied for

Die Deutsche Bibliothek – CIP-Einheitsaufnahme

**Cassels, John W.S.:**
An introduction to the geometry of numbers / J.W.S. Cassels - Reprint of the 1971 ed. - Berlin; Heidelberg; New York; Barcelona; Budapest; Hong Kong; London; Milan; Paris; Santa Clara; Singapore; Tokyo: Springer, 1997
(Classics in mathematics)
ISBN 3-540-61788-4

ISBN 3-540-61788-4 Springer-Verlag Berlin Heidelberg New York


Printed in Germany



SPIN 10554506 41/3143-5 4 3 2 1 0 – Printed on acid-free paper

J. W. S. Cassels

# An Introduction to the Geometry of Numbers

Second Printing, Corrected

Springer-Verlag Berlin · Heidelberg · New York 1971

---

AMS Subject Classifications (1970): 10 E xx

---

ISBN 3-540-02397-6 Springer-Verlag Berlin · Heidelberg · New York
ISBN 0-387-02397-6 Springer-Verlag New York · Heidelberg · Berlin

 Library of Congress Catalog Card Number 75-154801. Printed in Germany.

Printing: Fotocop Wilhelm Weihert, Darmstadt.
Binding: Universitätsdruckerei Stürtz, Würzburg.

# Preface

> **Of making many bookes there is no end, and much studie is a wearinesse of the flesh.**
>
> *Ecclesiastes* XII, 12.

When I first took an interest in the Geometry of Numbers, I was struck by the absence of any book which gave the essential skeleton of the subject as it was known to the experienced workers in the subject. Since then the subject has developed, as will be clear from the dates of the papers cited in the bibliography, but the need for a book remains. This is an attempt to fill the gap. It aspires to acquaint the reader with the main lines of development, so that he may with ease and pleasure follow up the things which interest him in the periodical literature. I have attempted to make the account as self-contained as possible.

References are usually given to the more recent papers dealing with a particular topic, or to those with a good bibliography. They are given only to enable the reader to amplify the account in the text and are not intended to give a historical picture. To give anything like a reasonable account of the history of the subject would have involved much additional research.

I owe a particular debt of gratitude to Professor L. J. MORDELL, who first introduced me to the Geometry of Numbers.

The proof-sheets have been read by Professors K. MAHLER, L. J. MORDELL and C. A. ROGERS. It is a pleasure to acknowledge their valuable help and advice both in detecting errors and obscurities and in suggesting improvements. Dr. V. ENNOLA has drawn my attention to several slips which survived into the second proofs.

I should also like to take the opportunity to thank Professor F. K. SCHMIDT and the Springer-Verlag for accepting this book for their celebrated yellow series and the Springer-Verlag for its readiness to meet my typographical whims.

Cambridge, June, 1959 J. W. S. CASSELS

# Contents

# Notation

An effort has been made to distinguish different types of mathematical object by the use of different alphabets. It is not necessary to describe the scheme in full since an acquaintance with it is not presupposed. However the following conventions are made throughout the book without explicit mention.

Bold Latin letters (large and small) always denote vectors. The dimensions is $n$, unless the contrary is explicitly stated: and the letter $n$ is not used otherwise, except in one or two places where there can be no fear of ambiguity. The co-ordinates of a vector are denoted by the corresponding italic letter with a suffix 1, 2, ..., $n$. If the bold letter denoting the vector already has a suffix, then that is put after the co-ordinate suffix. Thus:

$$\boldsymbol{a} = (a_1, \ldots, a_n)$$
$$\boldsymbol{b}_r = (b_{1r}, \ldots, b_{nr})$$
$$\boldsymbol{X}'_\varepsilon = (X'_{1\varepsilon}, \ldots, X'_{n\varepsilon}).$$

The origin is always denoted by $\boldsymbol{o}$. The length of $\boldsymbol{x}$ is

$$|\boldsymbol{x}| = (x_1^2 + \cdots + x_n^2)^{\frac{1}{2}}.$$

Sanserif Greek capitals, in particular $\mathsf{\Lambda}$, $\mathsf{M}$, $\mathsf{N}$, $\mathsf{\Gamma}$, denote lattices.

The notation $d(\mathsf{\Lambda})$, $\Delta(\mathscr{S})$, $V(\mathscr{S})$ for respectively the determinant of the lattice $\mathsf{\Lambda}$ and for the lattice-constant and volume of a set $\mathscr{S}$ will be standard, once the corresponding concepts have been introduced.

Chapters are divided into sections with titles. These sections are subdivided, for convenience, into subsections, which are indicated by a decimal notation. The numbering of displayed formulae starts afresh in each subsection. The prologue is just subdivided into sections without titles, and it was convenient to number the displayed formulae consecutively throughout.

# Prologue

P1. We owe to Minkowski the fertile observation that certain results which can be made almost intuitive by the consideration of figures in $n$-dimensional euclidean space have far-reaching consequences in diverse branches of number theory. For example, he simplified the theory of units in algebraic number fields and both simplified and extended the theory of the approximation of irrational numbers by rational ones (Diophantine Approximation). This new branch of number theory, which Minkowski christened "The Geometry of Numbers", has developed into an independent branch of number-theory which, indeed, has many applications elsewhere but which is well worth studying for its own sake.

In this prologue we first discuss some of the concepts and results which will play a leading rôle. The arguments we shall use are sometimes rather different from those in the main body of the text: since here we wish to make the geometrical situation intuitive in simple cases without necessarily giving complete proofs, while later we may need to sacrifice picturesqueness for precision. The proofs in the text are independent of this prologue, which may be omitted if desired.

P2. A fundamental and typical problem in the geometry of numbers is as follows:

Let $f(x_1, \ldots, x_n)$ be a real-valued function of the real variables $x_1, \ldots, x_n$. How small can $|f(u_1, \ldots, u_n)|$ be made by suitable choice of the integers $u_1, \ldots, u_n$? It may well be that one has trivially $f(0, \ldots, 0) = 0$, for example when $f(x_1, \ldots, x_n)$ is a homogeneous form; and then one excludes the set of values $u_1 = u_2 = \cdots = u_n = 0$. (The "homogeneous problem".)

In general one requires estimates which are valid not merely for individual functions $f$ but for whole classes of functions. Thus a typical result is that if

$$f(x_1, x_2) = a_{11} x_1^2 + 2 a_{12} x_1 x_2 + a_{22} x_2^2 \tag{1}$$

is a positive definite quadratic form, then there are integers $u_1, u_2$ not both 0 such that

$$f(u_1, u_2) \leq (4D/3)^{\frac{1}{2}}, \tag{2}$$

where

$$D = a_{11} a_{22} - a_{12}^2$$

is the discriminant of the form. It is trivial that if the result is true then it is the best possible of its kind, since

$$u_1^2 + u_1 u_2 + u_2^2 \geqq 1$$

for all pairs of integers $u_1, u_2$ not both zero; and here $D = \frac{3}{4}$.

Of course the positive definite binary quadratic forms are a particularly simple case. The result above was known well before the birth of the Geometry of Numbers; and indeed we shall give a proof substantially independent of the Geometry of Numbers in Chapter II, § 3. But positive definite binary quadratic forms display a number of arguments in a particularly simple way so we shall continue to use them as examples.

P3. The result just stated could be represented graphically. An inequality of the type

$$f(x_1, x_2) \leqq k,$$

where $f(x_1, x_2)$ is given by (1) and $k$ is some positive number, represents the region $\mathscr{R}$ bounded by an ellipse in the $(x_1, x_2)$-plane. Thus our result above states that $\mathscr{R}$ contains a point $(u_1, u_2)$, other than the origin, with integer coordinates provided that $k \geqq (4D/3)^{\frac{1}{2}}$.

A result of this kind but not so precise follows at once from a fundamental theorem of MINKOWSKI. The 2-dimensional case of this states that a region $\mathscr{R}$ always contains a point $(u_1, u_2)$ with integral co-ordinates other than the origin provided that it satisfies the following three conditions.

(i) $\mathscr{R}$ is symmetric about the origin, that is if $(x_1, x_2)$ is in $\mathscr{R}$ then so is $(-x_1, -x_2)$.

(ii) $\mathscr{R}$ is convex, that is if $(x_1, x_2)$ and $(y_1, y_2)$ are two points of $\mathscr{R}$ then the whole line segment

$$\{\lambda x_1 + (1-\lambda) y_1, \lambda x_2 + (1-\lambda) y_2\} \qquad (0 \leqq \lambda \leqq 1)$$

joining them is also in $\mathscr{R}$.

(iii) $\mathscr{R}$ has area greater than 4.

Any ellipse $f(x_1, x_2) \leqq k$ satisfies (i) and (ii). Since its area is

$$\frac{k\pi}{(a_{11}a_{22} - a_{12}^2)^{\frac{1}{2}}} = \frac{k\pi}{D^{\frac{1}{2}}},$$

it also satisfies (iii), provided that $k\pi > 4D^{\frac{1}{2}}$. We thus have a result similar to (2), except that the constant $(\frac{4}{3})^{\frac{1}{2}}$ is replaced by any number greater than $4/\pi$.

P4. It is useful to consider briefly the basic ideas behind the proof of MINKOWSKI's theorem, since in the formal proofs in Chapter 3 they may be obscured by the need to obtain powerful theorems which are as widely applicable as possible. Instead of the region $\mathscr{R}$, MINKOWSKI works with the region $\mathscr{S}=\frac{1}{2}\mathscr{R}$ of points $(\frac{1}{2}x_1, \frac{1}{2}x_2)$, where $(x_1, x_2)$ is in $\mathscr{R}$. Thus $\mathscr{S}$ is symmetric about the origin and convex: its area is $\frac{1}{4}$ that of $\mathscr{R}$ and so is greater than 1. More generally, MINKOWSKI considers the set of bodies $\mathscr{S}(u_1, u_2)$ similar and similarly situated to $\mathscr{S}$ but with centres at the points $(u_1, u_2)$ with integer co-ordinates.

We note first that if $\mathscr{S}$ and $\mathscr{S}(u_1, u_2)$ overlap then[1] $(u_1, u_2)$ is in $\mathscr{R}$. For let a point of overlap be $(\xi_1, \xi_2)$. Since $(\xi_1, \xi_2)$ is in $\mathscr{S}(u_1, u_2)$ the point $(\xi_1-u_1, \xi_2-u_2)$ must be in $\mathscr{S}$. Hence, by the symmetry of $\mathscr{S}$, the point $(u_1-\xi_1, u_2-\xi_2)$ is in $\mathscr{S}$. Finally, the mid-point of $(u_1-\xi_1, u_2-\xi_2)$ and $(\xi_1, \xi_2)$ is in $\mathscr{S}$ because of convexity, that is $(\frac{1}{2}u_1, \frac{1}{2}u_2)$ is in $\mathscr{S}$, and $(u_1, u_2)$ is in $\mathscr{R}$, as required. It is clear that $\mathscr{S}(u_1, u_2)$ overlaps $\mathscr{S}(u_1', u_2')$ when and only when $\mathscr{S}$ overlaps $\mathscr{S}(u_1-u_1', u_2-u_2')$.

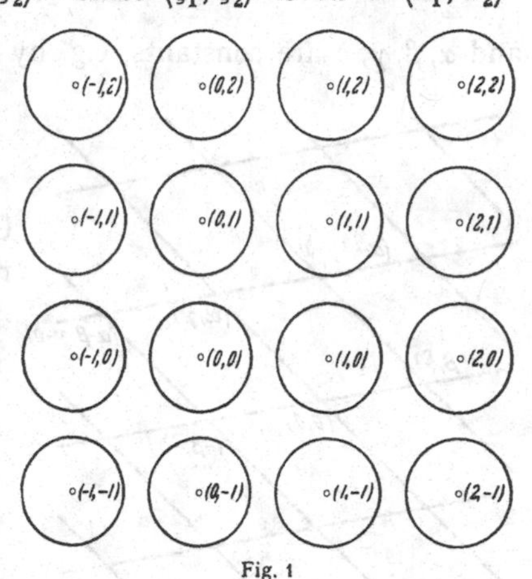

Fig. 1

To prove MINKOWSKI's theorem, it is thus enough to show that when the $\mathscr{S}(u_1, u_2)$ do not overlap then the area of each is at most 1. A little reflection convinces one that this must be so. A formal proof is given in Chapter 3. Another argument, which is perhaps more intuitive is as follows, where we suppose that $\mathscr{S}$ is entirely contained in a square

$$|x_1| \leqq X, \quad |x_2| \leqq X.$$

Let $U$ be a large integer. There are $(2U+1)^2$ regions $\mathscr{S}(u_1, u_2)$ whose centres $(u_1, u_2)$ satisfy

$$|u_1| \leqq U, \quad |u_2| \leqq U.$$

These $\mathscr{S}(u_1, u_2)$ are all entirely contained in the square

$$|x_1| \leqq U+X, \quad |x_2| \leqq U+X$$

of area

$$4(U+X)^2.$$

Since the $\mathscr{S}(u_1, u_2)$ are supposed not to overlap, we have

$$(2U+1)^2 V \leqq 4(U+X)^2,$$

[1] The converse statement is trivially true. If $(u_1, u_2)$ is in $\mathscr{R}$ then $(\frac{1}{2}u_1, \frac{1}{2}u_2)$ is in both $\mathscr{S}$ and $\mathscr{S}(u_1, u_2)$.

where $V$ is the area of $\mathscr{S}$; and so of each $\mathscr{S}(u_1, u_2)$. On letting $U$ tend to infinity we have $V \leqq 1$, as required.

P5. A change in the co-ordinate system in our example of a definite binary quadratic form $f(x_1, x_2)$ leads to another point of view. We may represent $f(x_1, x_2)$ as the sum of the squares of two linear forms:

$$f(x_1, x_2) = X_1^2 + X_2^2, \tag{3}$$

where

$$X_1 = \alpha x_1 + \beta x_2, \quad X_2 = \gamma x_1 + \delta x_2 \tag{4}$$

and $\alpha, \beta, \gamma, \delta$ are constants, e.g. by putting

$$\alpha = a_{11}^{\frac{1}{2}}, \quad \beta = a_{11}^{-\frac{1}{2}} a_{12},$$
$$\gamma = 0, \quad \delta = a_{11}^{-\frac{1}{2}} D^{\frac{1}{2}}.$$

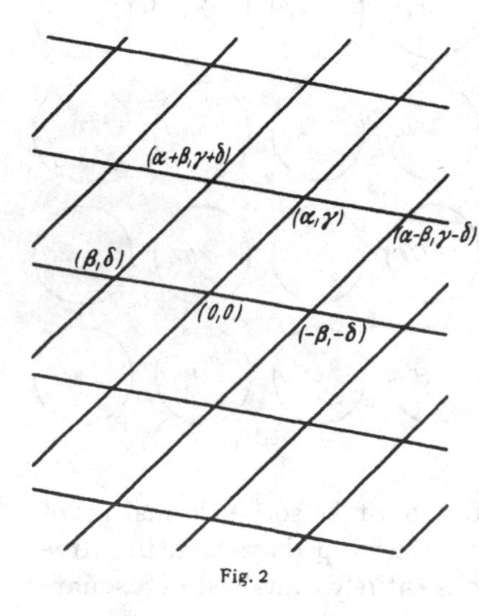

Fig. 2

Conversely if $\alpha, \beta, \gamma, \delta$ are any real numbers with $\alpha\delta - \beta\gamma \neq 0$ and $X_1, X_2$ are given by (4), then

$$X_1^2 + X_2^2 = a_{11} x_1^2 + 2 a_{12} x_1 x_2 + a_{22} x_2^2,$$

with

$$\left.\begin{aligned} a_{11} &= \alpha^2 + \gamma^2, \\ a_{12} &= \alpha\delta + \beta\gamma, \\ a_{22} &= \beta^2 + \delta^2, \end{aligned}\right\} \tag{5}$$

is a positive definite quadratic form with

$$D = a_{11} a_{22} - a_{12}^2 = (\alpha\delta - \beta\gamma)^2. \tag{6}$$

We now consider $X_1, X_2$ as a system of rectangular cartesian coordinates. The points $X_1, X_2$ corresponding to integers $x_1, x_2$ in (4) are then said to form a (2-dimensional) lattice $\Lambda$. In vector notation $\Lambda$ is the set of points

$$(X_1, X_2) = u_1(\alpha, \gamma) + u_2(\beta, \delta), \tag{7}$$

where $u_1, u_2$ run through all integer values.

We must now examine the properties of lattices more closely. Since we consider $\Lambda$ merely as a set of points, it can be expressed in terms of more than one basis. For example

$$(\alpha - \beta, \gamma - \delta), \quad (-\beta, -\delta)$$

is another basis for $\Lambda$. A fixed basis $(\alpha, \beta)$, $(\gamma, \delta)$ for $\Lambda$ determines a subdivision of the plane by two families of equidistant parallel lines, the first family consisting of those points $(X_1, X_2)$ which can be expressed in the form (7) with $u_2$ integral and $u_1$ only real, while for the lines of the second family the rôles of $u_1$ and $u_2$ are interchanged. In this way the plane is subdivided into parallelograms whose vertices are just the points of $\Lambda$. Of course the subdivision into parallelograms depends on the choice of basis, but we show that the area of each parallelogram, namely

$$|\alpha\delta-\beta\gamma|,$$

is independent of the particular basis. We can do this by showing that the number $N(X)$ of points of $\Lambda$ in a large square

$$\mathcal{Q}(X):\quad |X_1|\leqq X,\ |X_2|\leqq X$$

satisfies

$$\frac{N(X)}{4X^2}\to\frac{1}{|\alpha\delta-\beta\gamma|}\qquad (X\to\infty).$$

Indeed a consideration along the lines of the proof of MINKOWSKI's convex body theorem sketched above shows that the number of points of $\Lambda$ in $\mathcal{Q}(X)$ is roughly equal to the number of parallelograms contained in $\mathcal{Q}(X)$, which again is roughly equal to the area of $\mathcal{Q}(X)$ divided by the area $|\alpha\delta-\beta\gamma|$ of an individual parallelogram. The strictly positive number

$$d(\Lambda)=|\alpha\delta-\beta\gamma| \tag{8}$$

is called the determinant of $\Lambda$. As we have seen, it is independent of the choice of basis.

P6. In terms of the new concepts we see that the statement that there is always an integer solution of $f(x_1,x_2)\leqq(4D/3)^{\frac{1}{2}}$ is equivalent to the statement that every lattice $\Lambda$ has a point, other than the origin, in

$$X_1^2+X_2^2\leqq(\tfrac{4}{3})^{\frac{1}{2}}d(\Lambda). \tag{9}$$

On grounds of homogeneity this is again equivalent to the statement that the open circular disc

$$\mathcal{D}:\quad X_1^2+X_2^2<1 \tag{10}$$

contains a point of every lattice $\Lambda$ with $d(\Lambda)<(\frac{3}{4})^{\frac{1}{2}}$, and the fact that there are forms such that equality is necessary in (2) is equivalent to the existence of a lattice $\Lambda_c$ with determinant $d(\Lambda_c)=(\frac{3}{4})^{\frac{1}{2}}$ having no point in $\mathcal{D}$. So our problem about all definite binary quadratic forms is equivalent to one about the single region $\mathcal{D}$ and all lattices. Similarly consideration of the lattices with points in

$$|X_1X_2|<1$$

gives us information about the minima of indefinite binary quadratic forms:

$$\inf_{\substack{u_1, u_2 \text{ integers} \\ \text{not both } 0}} |f(u_1, u_2)|:$$

and so on.

These considerations prompt the following definitions. A lattice $\Lambda$ is said to be admissible for a region (point-set) $\mathscr{R}$ in the $(X_1, X_2)$-plane if it contains no point of $\mathscr{R}$ other than perhaps the origin, if that is a point of $\mathscr{R}$. We may say then that $\Lambda$ is $\mathscr{R}$-admissible. The lower bound $\Delta(\mathscr{R})$ of $d(\Lambda)$ over all $\mathscr{R}$-admissible lattices is the lattice-constant of $\mathscr{R}$: if there are no $\mathscr{R}$-admissible lattices we put $\Delta(R)=\infty$. Then any lattice $\Lambda$ with $d(\Lambda)<\Delta(R)$ certainly contains a point of $\mathscr{R}$ other than the origin. An $\mathscr{R}$-admissible lattice $\Lambda$ with $d(\Lambda)=\Delta(\mathscr{R})$ is called critical (for $\mathscr{R}$): of course critical lattices need not exist in general.

The importance of critical lattices was already recognized by MINKOWSKI. If $\Lambda_c$ is critical for $\mathscr{R}$ and $\Lambda$ is obtained from $\Lambda_c$ by a slight distortion (i.e. by making small changes in a pair of base-points) then either $\Lambda$ has a point in $\mathscr{R}$ other than the origin or $d(\Lambda)\geqq d(\Lambda_c)$ (or both).

As an example, let us again consider the open circular disc

$$\mathscr{D}: \quad X_1^2+X_2^2<1.$$

Suppose that $\Lambda_c$ is a critical lattice for $\mathscr{D}$. We outline a proof that a critical lattice, if it exists, must have three pairs of points $\pm(A_1, A_2)$, $\pm(B_1, B_2)$, $\pm(C_1, C_2)$ on the boundary $X_1^2+X_2^2=1$ of $\mathscr{D}$. For if $\Lambda_c$ had no points on $X_1^2+X_2^2=1$, we could obtain an $\mathscr{D}$-admissible lattice with smaller determinant from $\Lambda_c$ by shrinking it about the origin, that is by considering the lattice $\Lambda=t\Lambda_c$ of points $(tX_1, tX_2)$, where $(X_1, X_2)\in\Lambda$ and $0<t<1$ is fixed. Then $d(\Lambda)=t^2 d(\Lambda_c)<d(\Lambda_c)$, and clearly $\Lambda$ would be also $\mathscr{D}$-admissible if $t$ is near enough to 1. Hence $\Lambda_c$ contains a pair of points on $X_1^2+X_2^2=1$, which, after a suitable rotation of the co-ordinate system, we may suppose to be $\pm(1, 0)$. If there were no further points of $\Lambda_c$ on $X_1^2+X_2^2=1$ then we could obtain a $\mathscr{D}$-admissible lattice $\Lambda$ of smaller determinant by shrinking $\Lambda_c$ perpendicular to the $X_1$-axis, that is by taking $\Lambda$ to be the lattice of $(X_1, tX_2)$, $(X_1, X_2)\in\Lambda_c$, where $t$ is near enough to 1. Finally, if $\Lambda_c$ had only two pairs of points $\pm(1, 0)$, $\pm(B_1, B_2)$ on the boundary, then it is not difficult to see that it could be slightly distorted so that $(1, 0)$ remains fixed but $(B_1, B_2)$ moves along $X_1^2+X_2^2=1$ nearer to the $X_1$-axis, cf. Fig. 3.

This can be verified to decrease the determinant of the lattice [indeed $(1, 0)$ and $(B_1, B_2)$ can be shown to be a basis for $\Lambda_c$], and for small distortions the distorted lattice $\Lambda$ will still be $\mathscr{D}$-admissible. Hence a critical lattice $\Lambda_c$ (if it exists) must have three pairs of points on $X_1^2+X_2^2=1$: and it is easy to verify that the only lattice with three pairs of points on $X_1^2+X_2^2=1$, one of them being $\pm(1,0)$, is the lattice $\Lambda'$ with basis

$$(1,0), \quad \left(\tfrac{1}{2}, \sqrt{\tfrac{3}{4}}\right).$$

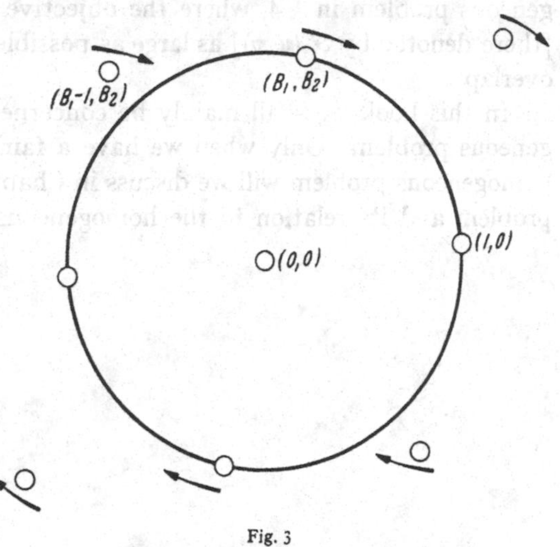

Fig. 3

This has the vertices of a regular hexagon

$$\pm(1,0),$$
$$\pm\left(\tfrac{1}{2}, \sqrt{\tfrac{3}{4}}\right),$$
$$\pm\left(-\tfrac{1}{2}, \sqrt{\tfrac{3}{4}}\right)$$

on $X_1^2+X_2^2=1$, but no points in $X_1^2+X_2^2<1$. We have thus shown that $\Delta(D)=d(\Lambda')=\left(\tfrac{3}{4}\right)^{\frac{1}{2}}$ provided that $\mathscr{D}$ has a critical lattice. MINKOWSKI showed that critical lattices exist for a fairly wide set of regions $\mathscr{R}$ by, roughly speaking, showing that any $\mathscr{R}$-admissible lattice $\Lambda$ can be gradually shrunk and distorted until it becomes critical. In the text we give a more general proof of the existence of critical lattices using concepts due to MAHLER which turn out to have much wider significance.

P7. Another general type of problem is the typical "inhomogeneous problem": Let $f(x_1, \ldots, x_n)$ be some real-valued function of the real variables $x_1, \ldots, x_n$. It is required to find a constant $k$ with the following property: If $\xi_1, \ldots, \xi_n$ are any real numbers there are integers $u_1, \ldots, u_n$ such that

$$|f(\xi_1-u_1, \ldots, \xi_n-u_n)| \leqq k.$$

Questions of this sort turn up naturally, for example in the theory of algebraic numbers. Again there is a simple geometric picture. For simplicity let $n=2$. Let $\mathscr{R}$ be the set of points $(x_1, x_2)$ in the 2-dimensional euclidean plane with

$$|f(x_1, x_2)| \leqq k.$$

Denote by $\mathscr{R}(u_1, u_2)$, where $u_1, u_2$ are any integers, the region similar to $\mathscr{R}$ but with the displacement $u_1, u_2$; that is $\mathscr{R}(u_1, u_2)$ is the set of points $x_1, x_2$ such that

$$|f(x_1-u_1, x_2-u_2)| \leqq k.$$

Then the inhomogeneous problem is clearly to choose $k$ so that the regions $\mathscr{R}(u_1, u_2)$ cover the whole plane. In general one will wish to choose $k$, and so $\mathscr{R}$, as small as possible so that it still has this covering property. Here we have a contrast with the treatment of the homogeneous problem in § 4, where the objective was to make the regions [there denoted by $\mathscr{S}(u, v)$] as large as possible but so that they did not overlap.

In this book we shall mainly be concerned at first with the homogeneous problem. Only when we have a fairly complete theory of the homogeneous problem will we discuss in Chapter XI the inhomogeneous problem and its relation to the homogeneous one.

## Chapter I

# Lattices

**I.1. Introduction.** In this chapter we introduce the most important concept in the geometry of numbers, that of a lattice, and develop some of its basic properties. The contents of this chapter, except § 2.4 and § 5, are fundamental for almost everything that follows.

In this book we shall be concerned only with lattices over the ring of rational integers. A certain amount of work has been done on lattices over complex quadratic fields, see e.g. MULLENDER (1945a) and K. ROGERS (1955a). Many of the concepts should carry over practically unaltered. Again, work on approximation to complex numbers by integers of a complex quadratic field [e.g. MULLENDER (1945a), CASSELS, LEDERMANN and MAHLER (1951a), POITOU (1953a)] and on the minima of hermitian forms when the variables are integers in a quadratic field [e.g. OPPENHEIM (1932a, 1936a, 1953f) and K. ROGERS (1956a)] may be regarded as a generalization of the geometry of numbers to lattices over complex quadratic fields. We shall not have occasion to mention lattices over complex quadratic fields again in this book; we mention them here only for completeness. For lattices over general algebraic number fields see ROGERS and SWINNERTON-DYER (1958a).

**I.2. Bases and sublattices.** Let $\boldsymbol{a}_1, \ldots, \boldsymbol{a}_n$ be linearly independent real vectors in $n$-dimensional real euclidean space, so that the only set of numbers $t_1, \ldots, t_n$ for which $t_1\boldsymbol{a}_1 + \cdots + t_n\boldsymbol{a}_n = \boldsymbol{o}$ is $t_1 = t_2 = \cdots = t_n = 0$. The set of all points

$$\boldsymbol{x} = u_1\boldsymbol{a}_1 + \cdots + u_n\boldsymbol{a}_n \tag{1}$$

with integral $u_1, \ldots, u_n$ is called the lattice with basis $\boldsymbol{a}_1, \ldots, \boldsymbol{a}_n$. We note that, since $\boldsymbol{a}_1, \ldots, \boldsymbol{a}_n$ are linearly independent, the expression of any vector $\boldsymbol{x}$ in the shape (1) with real $u_1, \ldots, u_n$ is unique. Hence if $\boldsymbol{x}$ is in $\Lambda$ and (1) is any expression for $\boldsymbol{x}$ with real $u_1, \ldots, u_n$, then $u_1, \ldots, u_n$ are integers. We shall make use of these remarks frequently, often without explicit reference.

The basis is not uniquely determined by the lattice. For let $\boldsymbol{a}_i'$ be the points

$$\boldsymbol{a}_i' = \sum_j v_{ij}\boldsymbol{a}_j \qquad (1 \leqq i, j \leqq n), \tag{2}$$

where $v_{ij}$ are any integers with

$$\det(v_{ij}) = \pm 1. \tag{3}$$

Then

$$\boldsymbol{a}_i = \sum_j w_{ij} \boldsymbol{a}'_j \tag{4}$$

with integral $w_{ij}$. It follows easily that the set of points (1) is precisely the set of points

$$u'_1 \boldsymbol{a}'_1 + \cdots + u'_n \boldsymbol{a}'_n$$

where $u'_1, \ldots, u'_n$ run through all integers; that is $\boldsymbol{a}_1, \ldots, \boldsymbol{a}_n$ and $\boldsymbol{a}'_1, \ldots, \boldsymbol{a}'_n$ are bases of the same lattice. We show now that every basis $\boldsymbol{a}'_i$ of a lattice $\Lambda$ may be obtained from a given basis $\boldsymbol{a}_i$ in this way. For since $\boldsymbol{a}'_i$ belongs to the lattice with basis $\boldsymbol{a}_1, \ldots, \boldsymbol{a}_n$ there are integers $v_{ij}$ such that (2) holds: and since $\boldsymbol{a}_i$ belongs to the lattice with basis $\boldsymbol{a}'_1, \ldots, \boldsymbol{a}'_n$ there are integers $w_{ij}$ such that (4) holds. On substituting (2) in (4) and making use of the linear independence of the $\boldsymbol{a}_i$, we have

$$\sum w_{ij} v_{jl} = \begin{cases} 1 & \text{if } i = l \\ 0 & \text{otherwise.} \end{cases}$$

Hence

$$\det(w_{ij}) \det(v_{jl}) = 1$$

and so each of the integers $\det(w_{ij})$ and $\det(v_{jl})$ must be $\pm 1$; that is (3) holds as required.

We denote lattices by capital sanserif Greek letters, and in particular by $\Lambda$, $\mathsf{M}$, $\mathsf{N}$, $\Gamma$.

If $\boldsymbol{a}_1, \ldots, \boldsymbol{a}_n$ and $\boldsymbol{a}'_1, \ldots, \boldsymbol{a}'_n$ are bases of the same lattice, so that they are related by (2) and (3), then we have

$$\det(\boldsymbol{a}'_1, \ldots, \boldsymbol{a}'_n) = \det(v_{ij}) \det(\boldsymbol{a}_1, \ldots, \boldsymbol{a}_n) = \pm \det(\boldsymbol{a}_1, \ldots, \boldsymbol{a}_n),$$

where, for example, $\det(\boldsymbol{a}_1, \ldots, \boldsymbol{a}_n)$ denotes the determinant of the $n \times n$ array whose $j$-th row is the vector $\boldsymbol{a}_j$. Hence

$$d(\Lambda) = |\det(\boldsymbol{a}_1, \ldots, \boldsymbol{a}_n)|$$

is independent of the particular choice of basis for $\Lambda$. Because of the linear independence of $\boldsymbol{a}_1, \ldots, \boldsymbol{a}_n$ we have

$$d(\Lambda) > 0.$$

We call $d(\Lambda)$ the determinant of $\Lambda$.

An example of a lattice is the set $\Lambda_0$ of all vectors with integral coordinates. A basis for $\Lambda_0$ is clearly the set of vectors

$$\boldsymbol{e}_j = (\overbrace{0, \ldots, 0}^{j-1 \text{ zeros}}, 1, \overbrace{0, \ldots, 0}^{n-j \text{ zeros}}) \qquad (1 \leqq j \leqq n);$$

and so

$$d(\Lambda_0) = 1.$$

We note that the vectors of a lattice $\Lambda$ form a group under addition: if $\boldsymbol{a}\in\Lambda$ then $-\boldsymbol{a}\in\Lambda$; and if $\boldsymbol{a}, \boldsymbol{b}\in\Lambda$ then $\boldsymbol{a}\pm\boldsymbol{b}\in\Lambda$. We shall see later (Chapter III, § 4) that a lattice is the most general group of vectors in $n$-dimensional space which contains $n$ linearly independent vectors and which satisfies the further property that there is some sphere about the origin $\boldsymbol{o}$ which contains no other vector of the group except $\boldsymbol{o}$.

**I.2.2.** Let $\boldsymbol{a}_1, \ldots, \boldsymbol{a}_n$ be vectors of a lattice $\mathsf{M}$ with basis $\boldsymbol{b}_1, \ldots, \boldsymbol{b}_n$, so that

$$\boldsymbol{a}_i = \sum_j v_{ij} \boldsymbol{b}_j \tag{1}$$

with integers $v_{ij}$. The integer

$$I = |\det(v_{ij})| = \frac{|\det(\boldsymbol{a}_1, \ldots, \boldsymbol{a}_n)|}{|\det(\boldsymbol{b}_1, \ldots, \boldsymbol{b}_n)|} = \frac{|\det(\boldsymbol{a}_1, \ldots, \boldsymbol{a}_n)|}{d(\mathsf{M})}$$

is called the index of the vectors $\boldsymbol{a}_1, \ldots, \boldsymbol{a}_n$ in $\mathsf{M}$. From the last expression it is independent of the particular choice of basis for $\mathsf{M}$. By definition, $I \geqq 0$; and $I=0$ only if $\boldsymbol{a}_1, \ldots, \boldsymbol{a}_n$ are linearly dependent.

If every point of the lattice $\Lambda$ is also a point of the lattice $\mathsf{M}$ then we say that $\Lambda$ is a sublattice of $\mathsf{M}$. Let $\boldsymbol{a}_1, \ldots, \boldsymbol{a}_n$ and $\boldsymbol{b}_1, \ldots, \boldsymbol{b}_n$ be bases of $\Lambda$ and $\mathsf{M}$ respectively. Then there are integers $v_{ij}$ such that (1) holds, since $\boldsymbol{a}_i\in\mathsf{M}$. The index of $\boldsymbol{a}_1, \ldots, \boldsymbol{a}_n$ in $\mathsf{M}$, namely

$$D = |\det(v_{ij})| = \frac{|\det(\boldsymbol{a}_1, \ldots, \boldsymbol{a}_n)|}{|\det(\boldsymbol{b}_1, \ldots, \boldsymbol{b}_n)|} = \frac{d(\Lambda)}{d(\mathsf{M})} \tag{2}$$

is called the index of $\Lambda$ in $\mathsf{M}$. From the last expression the index depends only on $\Lambda$ and $\mathsf{M}$, not on the choice of bases. Since $\boldsymbol{a}_1, \ldots, \boldsymbol{a}_n$ are linearly independent, we have $D>0$. On solving (1) for the $\boldsymbol{b}_i$ and using (2), we have

$$D\boldsymbol{b}_i = \sum_j w_{ij} \boldsymbol{a}_j,$$

where the $w_{ij}$ are integers. Hence

$$D\mathsf{M} \subset \Lambda \subset \mathsf{M}, \tag{3}$$

where $D\mathsf{M}$ is the lattice of $D\boldsymbol{b}$, $\boldsymbol{b}\in\mathsf{M}$.

It is often convenient to choose particular bases for $\Lambda$ and $\mathsf{M}$ so that (1) takes a particularly simple shape.

THEOREM I. *Let $\Lambda$ be a sublattice of $\mathsf{M}$.*

A. *To every base $\boldsymbol{b}_1, \ldots, \boldsymbol{b}_n$ of $\mathsf{M}$ there can be found a base $\boldsymbol{a}_1, \ldots, \boldsymbol{a}_n$ of $\Lambda$ of the shape*

$$\left.\begin{aligned}
\boldsymbol{a}_1 &= v_{11}\boldsymbol{b}_1 \\
\boldsymbol{a}_2 &= v_{21}\boldsymbol{b}_1 + v_{22}\boldsymbol{b}_2 \\
&\cdots\cdots\cdots\cdots \\
\boldsymbol{a}_n &= v_{n1}\boldsymbol{b}_1 + \cdots + v_{nn}\boldsymbol{b}_n,
\end{aligned}\right\} \tag{4}$$

*where the $v_{ij}$ are integers and $v_{ii} \neq 0$ for all $i$.*

B. *Conversely, to every basis* $\boldsymbol{a}_1, \ldots, \boldsymbol{a}_n$ *of* $\Lambda$ *there exists a basis* $\boldsymbol{b}_1, \ldots, \boldsymbol{b}_n$ *of* $\mathsf{M}$ *such that* (4) *holds.*

Proof of A. For each $i$ $(1 \leqq i \leqq n)$ there certainly exist points $\boldsymbol{a}_i$ in $\Lambda$ of the shape

$$\boldsymbol{a}_i = v_{i1}\boldsymbol{b}_1 + \cdots + v_{ii}\boldsymbol{b}_i$$

where $v_{i1}, \ldots, v_{ii}$ are integers and $v_{ii} \neq 0$, since, as we have seen, $D\boldsymbol{b}_i \in \Lambda$. We choose for $\boldsymbol{a}_i$ such an element of $\Lambda$ for which the positive integer $|v_{ii}|$ is as small as possible (but not 0), and will show that $\boldsymbol{a}_1, \ldots, \boldsymbol{a}_n$ are in fact a basis for $\Lambda$. Since $\boldsymbol{a}_1, \ldots, \boldsymbol{a}_n$ are in $\Lambda$, by construction, so is every vector

$$w_1\boldsymbol{a}_1 + \cdots + w_n\boldsymbol{a}_n, \tag{5}$$

where $w_1, \ldots, w_n$ are integers. Suppose, if possible, that $\boldsymbol{c}$ is a vector of $\Lambda$ not of the shape (5). Since $\boldsymbol{c}$ is in $\mathsf{M}$, it certainly can be expressed in terms of $\boldsymbol{b}_1, \ldots, \boldsymbol{b}_n$, and so can be written in the shape

$$\boldsymbol{c} = t_1\boldsymbol{b}_1 + \cdots + t_k\boldsymbol{b}_k,$$

where $1 \leqq k \leqq n$, $t_k \neq 0$ and $t_1, \ldots, t_k$ are integers. If there are several such $\boldsymbol{c}$, then we choose one for which the integer $k$ is as small as possible. Now, since $v_{kk} \neq 0$, we may choose an integer $s$ such that

$$|t_k - s v_{kk}| < |v_{kk}|. \tag{6}$$

The vector

$$\boldsymbol{c} - s\boldsymbol{a}_k = (t_1 - s v_{11})\boldsymbol{b}_1 + \cdots + (t_k - s v_{kk})\boldsymbol{b}_k$$

is in $\Lambda$ since $\boldsymbol{c}$ and $\boldsymbol{a}_k$ are; but it is not of the shape (5) since $\boldsymbol{c}$ is not. Hence $t_k - s v_{kk} \neq 0$ by the assumption that $k$ was chosen as small as possible. But then (6) contradicts the assumption that the non-zero integer $v_{kk}$ was chosen as small as possible. The contradiction shows that there are no $\boldsymbol{c}$ in $\Lambda$ which cannot be put in the form (5), and so proves part A of the theorem.

Proof of B. Let $\boldsymbol{a}_1, \ldots, \boldsymbol{a}_n$ be some fixed basis of $\Lambda$. Since $D\mathsf{M}$ is a sublattice of $\Lambda$ by (3), where $D$ is the index of $\Lambda$ in $\mathsf{M}$, there exists by Part A a basis $D\boldsymbol{b}_1, \ldots, D\boldsymbol{b}_n$ of $D\mathsf{M}$ of the type

$$\left.\begin{aligned}
D\boldsymbol{b}_1 &= w_{11}\boldsymbol{a}_1 \\
D\boldsymbol{b}_2 &= w_{21}\boldsymbol{a}_1 + w_{22}\boldsymbol{a}_2 \\
&\cdots\cdots\cdots\cdots \\
D\boldsymbol{b}_n &= w_{n1}\boldsymbol{a}_1 + \cdots + w_{nn}\boldsymbol{a}_n,
\end{aligned}\right\} \tag{7}$$

with integral $w_{ij}$ and $w_{ii} \neq 0$ $(1 \leqq i \leqq n)$. On solving (7) for $\boldsymbol{a}_1, \ldots, \boldsymbol{a}_n$ in succession we obtain a series of equations of the type (4) but where at first we know only that the $v_{ij}$ are rational. But clearly $\boldsymbol{b}_1, \ldots, \boldsymbol{b}_n$ are a basis for M and so the $v_{ij}$ are in fact integers, since the $\boldsymbol{a}_i$ are in M, and since the representation of any vector $\boldsymbol{a}$ in the shape

$$\boldsymbol{a} = t_1 \boldsymbol{b}_1 + \cdots + t_n \boldsymbol{b}_n \qquad (t_1, \ldots, t_n, \text{ real numbers})$$

is unique by the independence of $\boldsymbol{b}_1, \ldots, \boldsymbol{b}_n$.

From this theorem we have a number of simple but useful corollaries.

COROLLARY 1. *In theorem I we may suppose further that*

$$v_{ii} > 0 \tag{8}$$

*and that*

$$0 \leqq v_{ij} < v_{jj} \quad \textit{in case } A, \tag{9}$$

$$0 \leqq v_{ij} < v_{ii} \quad \textit{in case } B. \tag{10}$$

Proof of A. To obtain (8) it is necessary only to replace $\boldsymbol{a}_i$ or $\boldsymbol{b}_i$ by $-\boldsymbol{a}_i$, $-\boldsymbol{b}_i$ respectively if originally $v_{ii} < 0$. To obtain (9) we replace the $\boldsymbol{a}_i$ by

$$\boldsymbol{a}_i' = t_{i1} \boldsymbol{a}_1 + \cdots + t_{i,i-1} \boldsymbol{a}_{i-1} + \boldsymbol{a}_i,$$

where the $t_{ij}$ are integers to be determined. For any choice of the $t_{ij}$ the $\boldsymbol{a}_i'$ are a basis for $\Lambda$. We have

$$\boldsymbol{a}_i' = v_{i1}' \boldsymbol{b}_1 + \cdots + v_{ii}' \boldsymbol{b}_i,$$

where

$$v_{ii}' = v_{ii};$$

and, for $j < i$, we have

$$v_{ij}' = t_{ij} v_{jj} + t_{i,j+1} v_{j+1,j} + \cdots + t_{i,i-1} v_{i-1,j} + v_{ij}.$$

For each $i$ we may now choose $t_{i,i-1}, t_{i,i-2}, \ldots, t_{i1}$ in that order so that

$$0 \leqq v_{ij}' < v_{jj} = v_{jj}',$$

as was required.

Proof of B. Similar.

COROLLARY 2. *Let $\boldsymbol{a}_1, \ldots, \boldsymbol{a}_m$ be linearly independent vectors of a lattice* M. *Then there is a basis $\boldsymbol{b}_1, \ldots, \boldsymbol{b}_n$ of* M *such that*

$$\begin{aligned}
\boldsymbol{a}_1 &= v_{11} \boldsymbol{b}_1 \\
\boldsymbol{a}_2 &= v_{21} \boldsymbol{b}_1 + v_{22} \boldsymbol{b}_2 \\
&\cdots\cdots\cdots\cdots \\
\boldsymbol{a}_m &= v_{m1} \boldsymbol{b}_1 + \cdots + v_{mm} \boldsymbol{b}_m,
\end{aligned}$$

*with integers $v_{ij}$ such that*

$$v_{jj} > 0 \quad 0 \leqq v_{ij} < v_{ii} \qquad (1 \leqq j < i \leqq m). \tag{11}$$

We can choose vectors $\boldsymbol{a}_{m+1}, \ldots, \boldsymbol{a}_n$ in $\mathsf{M}$ such that $\boldsymbol{a}_1, \ldots, \boldsymbol{a}_n$ are linearly independent. Corollary 2 follows now on applying Corollary 1 to the lattice $\Lambda$ with basis $\boldsymbol{a}_1, \ldots, \boldsymbol{a}_n$.

COROLLARY 3. *Let $\boldsymbol{a}_1, \ldots, \boldsymbol{a}_m$ $(m<n)$ be linearly independent vectors of a lattice $\mathsf{M}$. A necessary and sufficient condition for the existence of vectors $\boldsymbol{a}_{m+1}, \ldots, \boldsymbol{a}_n$ in $\mathsf{M}$ such that $\boldsymbol{a}_1, \ldots, \boldsymbol{a}_n$ is a basis is the following: every vector $\mathbf{c} \in \mathsf{M}$ which is of the shape*

$$\mathbf{c} = u_1 \boldsymbol{a}_1 + \cdots + u_m \boldsymbol{a}_m \tag{12}$$

*with real $u_1, \ldots, u_m$ necessarily has $u_1, \ldots, u_m$ integral.*

If $\boldsymbol{a}_1, \ldots, \boldsymbol{a}_m$ is part of a basis $\boldsymbol{a}_1, \ldots, \boldsymbol{a}_n$ the condition is clearly satisfied. Conversely if $\boldsymbol{a}_1, \ldots, \boldsymbol{a}_m$ satisfy the condition, let $\boldsymbol{b}_1, \ldots, \boldsymbol{b}_n$ be the basis of $\mathsf{M}$ given by Corollary 2 and let $v_{ij}$ be the corresponding integers. Then $\mathbf{c} = \boldsymbol{b}_1, \ldots, \boldsymbol{b}_m$ are of the shape (12) and indeed the coefficient of $\boldsymbol{a}_i$ in the expression for $\boldsymbol{b}_i$ is $v_{ii}^{-1}$. Hence $v_{ii} = 1$ and so $v_{ij} = 0$ for $i \neq j$, that is $\boldsymbol{a}_i = \boldsymbol{b}_i$ $(1 \leqq i \leqq m)$ and we may put $\boldsymbol{a}_i = \boldsymbol{b}_i$ $(m+1 \leqq i \leqq n)$.

In some contexts we shall need the following more specialized corollary which follows at once from Corollary 3.

COROLLARY 4. *Let $\boldsymbol{b}_1, \ldots, \boldsymbol{b}_n$ be a basis for a lattice $\mathsf{M}$ and let*

$$\mathbf{c} = u_1 \boldsymbol{b}_1 + \cdots + u_n \boldsymbol{b}_n \in \mathsf{M}.$$

*A necessary and sufficient condition that*

$$\boldsymbol{b}_1, \ldots, \boldsymbol{b}_{m-1}, \mathbf{c}$$

*be part of some basis*

$$\boldsymbol{b}_1, \ldots, \boldsymbol{b}_{m-1}, \mathbf{c}, \mathbf{c}_{m+1}, \ldots, \mathbf{c}_n$$

*of $\mathsf{M}$ is that $u_m, u_{m+1}, \ldots, u_n$ have no common factors $\neq \pm 1$.*

Proof. Clear.

The following characterisation of the index of a sublattice $\Lambda$ of a lattice $\mathsf{M}$ is sometimes useful. We say that two vectors $\mathbf{c}, \boldsymbol{d}$ of $\mathsf{M}$ are in the same class with respect to $\Lambda$ if $\mathbf{c} - \boldsymbol{d}$ is in $\Lambda$. Clearly this is a subdivision into classes: if $\mathbf{c} - \boldsymbol{d}$ and $\boldsymbol{d} - \boldsymbol{e}$ are in $\Lambda$, then $\mathbf{c} - \boldsymbol{e}$ is in $\Lambda$.

LEMMA 1. *The index of the sublattice $\Lambda$ of $\mathsf{M}$ is the number of classes in $\mathsf{M}$ with respect to $\Lambda$.*

For let $\boldsymbol{a}_j, \boldsymbol{b}_j$ be bases for $\Lambda$ and $\mathsf{M}$ respectively in the shape (4) given by Theorem I. Then clearly the index $D$ of $\Lambda$ in $\mathsf{M}$ is

$$D = \prod_i |v_{ii}|.$$

But now every $\boldsymbol{c} \in \mathsf{M}$ is in the same class as precisely one of the vectors

$$q_1 \boldsymbol{b}_1 + \cdots + q_n \boldsymbol{b}_n \qquad (0 \leqq q_j < v_{jj}),$$

as is readily verified (cf. proof of Theorem I, Corollary 1).

**I.2.3.** There is a useful transformation of the criterion of Theorem I, Corollary 3, for deciding whether or not a set of vectors $\boldsymbol{a}_1, \ldots, \boldsymbol{a}_m$ $(m<n)$ of a lattice $\Lambda$ can be extended to a basis for $\Lambda$.

LEMMA 2. *Let $\boldsymbol{b}_1, \ldots, \boldsymbol{b}_n$ be a basis for a lattice $\Lambda$ and let*

$$\boldsymbol{a}_i = \sum_{1 \leqq j \leqq n} v_{ij} \boldsymbol{b}_j \qquad (1 \leqq i \leqq m) \tag{1}$$

*be vectors of $\Lambda$. A necessary and sufficient condition that $\boldsymbol{a}_1, \ldots, \boldsymbol{a}_m$ be extendable to a basis $\boldsymbol{a}_1, \ldots, \boldsymbol{a}_n$ of $\Lambda$ is that the $m \times m$ determinants formed by taking $m$ columns of the array*

$$(v_{ij}) \qquad (1 \leqq i \leqq m,\ 1 \leqq j \leqq n) \tag{2}$$

*shall not have a common factor.*

The condition is certainly necessary. For let $\boldsymbol{a}_{m+1}, \ldots, \boldsymbol{a}_n$ form a basis with $\boldsymbol{a}_1, \ldots, \boldsymbol{a}_m$, so that

$$\boldsymbol{a}_i = \sum_{1 \leqq j \leqq n} v_{ij} \boldsymbol{b}_j \qquad (m+1 \leqq i \leqq n) \tag{3}$$

for some integers $v_{ij}$. Since $\boldsymbol{a}_i$ $(1 \leqq i \leqq n)$ and $\boldsymbol{b}_i$ $(1 \leqq i \leqq n)$ are bases of the same lattice, we have

$$\det(v_{ij}) = \pm 1. \tag{4}$$

We may expand the determinant (4) by the first $m$ and last $(n-m)$ rows [Laplace-development] and obtain

$$\sum_{1 \leqq r \leqq R} V_r W_r = \det(v_{ij}), \tag{5}$$

where the $V_r$ are the $\binom{n}{m}$ determinants formed from columns of (2) and $W_r$ is the $(n-m) \times (n-m)$ determinant formed from the remaining $(n-m)$ columns and the $(n-m)$ rows,

$$v_{ij} \qquad (m < i \leqq n,\ 1 \leqq j \leqq n),$$

taken with an appropriate sign. Since the $W_r$ are integers, it follows from (4) and (5) that the $V_r$ have no common factor.

The condition is also sufficient. For let $\boldsymbol{c}$ be a vector of $\Lambda$ of the shape

$$\boldsymbol{c} = u_1 \boldsymbol{a}_1 + \cdots + u_m \boldsymbol{a}_m \tag{6}$$

for real numbers $u_1, \ldots, u_m$. On inserting (1) in (6) we have

$$\sum_{1 \leq i \leq m} u_i v_{ij} = \text{integer} = l_j \quad \text{(say)} \qquad (1 \leqq j \leqq n), \tag{7}$$

since $\boldsymbol{b}_1, \ldots, \boldsymbol{b}_n$ is a basis for $\Lambda$. We may solve (7) for the $u_i$, and indeed in a multitude of ways. For example let $\tilde{v}_j$ be the cofactor of $v_{1j}$ in the expansion of the determinant

$$V_1 = \det(v_{ij}) \qquad (1 \leqq i \leqq m,\ 1 \leqq j \leqq m).$$

Then

$$\sum_{1 \leq j \leq m} \tilde{v}_j l_j = V_1 u_1,$$

so $V_1 u_1$ is an integer. Similarly

$$V_r u_i = \text{integer} \qquad (1 \leqq i \leqq m),$$

where $V_r$ is any $m \times m$ determinant formed from (2). Since, by hypothesis, the $V_r$ are integers without common divisor, the $u_i$ must be integers. Hence by Theorem I, Corollary 3 it is possible to extend $\boldsymbol{a}_1, \ldots, \boldsymbol{a}_m$ to a basis $\boldsymbol{a}_1, \ldots, \boldsymbol{a}_n$.

**I.2.4**[1]. We shall now apply Lemma 2 to obtain a result of DAVENPORT (1955a) about the way in which a basis for a lattice may be chosen. This will be used only in Chapter V, § 10 and then only to prove a result on Diophantine Approximation rather aside from the main theme of the book.

THEOREM II. *Let $\Lambda$ be an n-dimensional lattice, let*

$$\mathbf{c}_i \qquad (1 \leqq i \leqq n-1)$$

*be $(n-1)$ arbitrary real vectors and let $\varepsilon > 0$ be an arbitrarily small real number. Then for all real numbers $N$ greater than a number $N_0$ depending only on $\Lambda$, $\varepsilon$ and the $\mathbf{c}_i$, there exists a basis $\boldsymbol{a}_1, \ldots, \boldsymbol{a}_n$ of $\Lambda$ such that*

$$|\boldsymbol{a}_i - N\mathbf{c}_i| < N^{\varepsilon} \qquad (1 \leqq i \leqq n-1). \tag{1}$$

*Here*

$$|\boldsymbol{x}| = (x_1^2 + \cdots + x_n^2)^{\frac{1}{2}} \tag{2}$$

*denotes the usual euclidean distance.*

To prove Theorem II we shall need a result about the distribution of integers prime to a given integer. We prove this first, and then Theorem II.

LEMMA 3. *For each $\delta > 0$ there is a number $k(\delta)$ with the following property: Every interval of length $k(\delta)\, q^{\delta}$, where $q$ is a positive integer, contains an integer prime to $q$.*

[1] § 2.4 may well be omitted at a first reading

Let

$$q = \prod_{1 \leq j \leq J} p_j^{\alpha_j}, \tag{3}$$

where the $p_j$ are distinct primes and the $\alpha_j > 0$ are integers. An integer is prime to $q$ if and only if it is not divisible by any of $p_1, \dots, p_J$. Consider some interval

$$V < u \leq V + U \tag{4}$$

of length $U$, where $U$, $V$ are fixed integers. For $j_1 < j_2 < \dots < j_s$, where $s \leq J$, let

$$M(j_1, \dots, j_s)$$

be the number of integers $u$ in the interval (4) which are divisible by $p_{j_1}, p_{j_2}, \dots, p_{j_s}$ (and perhaps also by other primes from $p_1, \dots, p_J$). We show next that

$$W = U + \sum_{\substack{s>0 \\ j_1 < j_2 < \dots < j_s}} (-1)^s M(j_1, \dots, j_s) \tag{5}$$

gives the number of integers $u$ in (4) prime to $q$, where $U$ is the number of integers $u$ in (4). For let the integer $u$ be divisible by precisely $r$ primes $p_j$, where $r \geq 1$: say by $p_1, \dots, p_r$, but not by $p_{r+1}, \dots, p_J$. Then $u$ is one of the integers counted in $M(j_1, \dots, j_s)$ if and only if $s \leq r$ and $j_1, \dots, j_s$ is one of the $\binom{r}{s}$ combinations of $s$ out of the numbers $1, 2, \dots, r$. Also $u$ contributes 1 to $U$ regarded as giving the number of integers in (4). Hence the total contribution of $u$ to (5) is

$$1 - \binom{r}{1} + \binom{r}{2} \cdots = (1-1)^r = 0.$$

If, however, $u$ is prime to $q$, then it contributes 1 to $U$ but does not contribute to the $M(j_1, \dots, j_s)$; so $W$ is the number of integers in (4) prime to $q$, as asserted. But

$$\left| M(j_1, \dots, j_s) - \frac{U}{p_{j_1} \cdots p_{j_s}} \right| < 1,$$

since $M(j_1, \dots, j_s)$ is the number of integers

$$u = p_{j_1} \cdots p_{j_s} u',$$

where $u'$ is an integer and

$$\frac{V}{p_{j_1} \cdots p_{j_s}} < u' \leq \frac{U+V}{p_{j_1} \cdots p_{j_s}}.$$

Since (5) contains $2^J$ summands, we have

$$W > U \left\{ 1 + \sum_{\substack{s>0 \\ j_1 < \dots < j_s}} \frac{(-1)^s}{p_{j_1} \cdots p_{j_s}} \right\} - 2^J = U \prod_j \left(1 - \frac{1}{p_j}\right) - 2^J \geq 2^{-J} U - 2^J.$$

Hence there is an integer prime to $q$ in the required interval provided that

$$U \geqq U_0(q) = 4^J.$$

If $\delta$ is the arbitrarily small number given in the lemma, we now have

$$\frac{U_0(q)}{q^\delta} \leqq \prod_j \left(\frac{4}{p_j^\delta}\right) \leqq \prod_{p \leqq 4^{1/\delta}} \left(\frac{4}{p^\delta}\right) = k(\delta) \quad \text{(say)},$$

where the second product is taken now over all primes less than $4^{1/\delta}$. This proves the lemma.

We shall use the lemma in the following apparently more general shape.

COROLLARY. *Let $q$, $\delta$, $k(\delta)$ be as in the lemma and let $s$, $t$ be integers of which $t$ is prime to $q$. Then an interval of length greater than $k(\delta)\,q^\delta$ contains an integer $u$ such that $tu+s$ is prime to $q$.*

For since $t$ is prime to $q$ we may write

$$s = s_1 t + s_2 q$$

for integers $s_1$ and $s_2$. Then

$$tu + s = t(u + s_1) + s_2 q.$$

Since $t$ is prime to $q$ we need only choose $u$ so that $u+s_1$ is prime to $q$; and this is possible by the lemma.

We now revert to the proof of Theorem II. Let $\boldsymbol{b}_1, \ldots, \boldsymbol{b}_n$ be any basis for $\Lambda$, and let the given vectors $\boldsymbol{c}_i$ be

$$\boldsymbol{c}_i = \sum_{1 \leqq j \leqq n} \gamma_{ij} \boldsymbol{b}_j \qquad (1 \leqq i \leqq n-1) \tag{6}$$

for real numbers $\gamma_{ij}$. We shall choose a basis

$$\boldsymbol{a}_i = \sum_j v_{ij} \boldsymbol{b}_j \qquad (1 \leqq i \leqq n) \tag{7}$$

for $\Lambda$ such that

$$v_{ij} = N\gamma_{ij} + O(N^{i\delta}), \tag{8}$$

where $N>1$ is the given positive number, $\delta>0$ is arbitrarily small, and the constant implied by the $O$ symbol may depend on $n$, $\delta$ and the $\gamma_{ij}$. We shall choose the $v_{ij}$ so that for each $I<n$ the two integers

$$R_I = \det(v_{ij}) \qquad (1 \leqq i \leqq I,\ 1 \leqq j \leqq I)$$

and

$$S_I = \det(v_{ij}) \qquad (1 \leqq i \leqq I,\ 2 \leqq j \leqq I+1)$$

are non-zero and without common factor.

Suppose, first, that $I=1$. We take for $v_{11}$ one of the integers nearest to $N\gamma_{11}$ which is not 0. Next we choose for $v_{12}$ the integer nearest to

$N\gamma_{12}$ which is not 0 and prime to $v_{11}$. For $j>2$ we choose for $v_{1j}$ the integer nearest to $N\gamma_{1j}$. Then (8) holds for $i=1$ and $j\neq 2$ trivially and for $i=1$, $j=2$ by Lemma 3, and since clearly $v_{11}=O(N)$. The integers $R_1=v_{11}$ and $S_1=v_{12}$ have the required properties.

Now suppose that $I>1$, and that the $v_{ij}$ with $i<I$ have already been constructed. For $j\neq I$, $I+1$ we take for $v_{Ij}$ just the nearest integer to $N\gamma_{Ij}$. On expanding $R_I$ and $S_I$ by their last rows, we now have

$$R_I=\pm v_{II}R_{I-1}+A,$$
$$S_I=\pm v_{I,I+1}S_{I-1}+v_{II}B+C,$$

where $A$, $B$, $C$ are integers which have already been determined. Since $R_{I-1}$ is prime to $S_{I-1}$, we may choose the integer $v_{II}$ so that $R_I$ is not 0 and prime to $S_{I-1}$. We choose for $v_{II}$ the integer nearest to $N\gamma_{II}$ for which this is true, so that, by the corollary to Lemma 3,

$$v_{II}-N\gamma_{II}=O(S_{I-1}^{\delta})=O(N^{(I-1)\delta}),$$

since $S_{I-1}=O(N^{I-1})$, being a sum of products of $I-1$ numbers $v_{ij}$ each of order $N$. Having determined $v_{II}$ we now take for $v_{I,I+1}$ the integer nearest to $N\gamma_{I,I+1}$ such that $S_I$ is not 0 and prime to $R_I$, so that similarly

$$v_{I,I+1}-N\gamma_{I,I+1}=O(S_I^{\delta})=O(N^{I\delta}).$$

This completes one stage of the induction. We have thus shown the existence of integers $v_{ij}$ satisfying (8).

From (7) and (8) we have

$$|\boldsymbol{a}_i-N\boldsymbol{c}_i|=O(N^{(n-1)\delta})\qquad(1\leq i\leq n-1).$$

The truth of the statement of the theorem now follows on taking $\delta=\varepsilon/n$.

**I.3. Lattices under linear transformation.** It is convenient here to consider briefly the effect of a non-singular affine transformation $\boldsymbol{x}\rightarrow\boldsymbol{X}=\boldsymbol{\alpha x}$ of $n$-dimensional space into itself. Let the transformation $\boldsymbol{X}=\boldsymbol{\alpha x}$ be given by

$$X_i=\sum_{1\leq j\leq n}\alpha_{ij}x_j\qquad(1\leq i\leq n),\tag{1}$$

where

$$\boldsymbol{X}=(X_1,\ldots,X_n),\quad \boldsymbol{x}=(x_1,\ldots,x_n)$$

are corresponding points in the transformation and $\alpha_{ij}$ are real numbers such that

$$\det(\boldsymbol{\alpha})=\det(\alpha_{ij})\neq 0.$$

Let $\Lambda$ be a lattice and denote by $\boldsymbol{\alpha}\Lambda$ the set of points $\boldsymbol{\alpha x}$, $\boldsymbol{x}\in\Lambda$. If $\boldsymbol{b}_1,\ldots,\boldsymbol{b}_n$ is a basis for $\Lambda$, then the general point $\boldsymbol{b}=u_1\boldsymbol{b}_1+\cdots+u_n\boldsymbol{b}_n$

($u_1, \ldots, u_n$ integers) of $\Lambda$ has the transform

$$\boldsymbol{\alpha b} = \boldsymbol{\alpha}(u_1 \boldsymbol{b}_1 + \cdots + u_n \boldsymbol{b}_n) = u_1 \boldsymbol{\alpha b}_1 + \cdots + u_n \boldsymbol{\alpha b}_n.$$

Hence $\boldsymbol{\alpha}\Lambda$ is a lattice with basis $\boldsymbol{\alpha b}_1, \ldots, \boldsymbol{\alpha b}_n$, and

$$d(\boldsymbol{\alpha}\Lambda) = |\det(\boldsymbol{\alpha b}_1, \ldots, \boldsymbol{\alpha b}_n)| = |\det(\boldsymbol{\alpha})|\,|\det(\boldsymbol{b}_1, \ldots, \boldsymbol{b}_n)| = |\det(\boldsymbol{\alpha})|\, d(\Lambda).$$

We note two particular cases. First, if $t \neq 0$ is a real number, then the set of $t\boldsymbol{b}$, $\boldsymbol{b} \in \Lambda$ is a lattice of determinant $|t|^n d(\Lambda)$ which we shall denote by $t\Lambda$. Secondly every lattice M can be put in the shape $\mathsf{M} = \boldsymbol{\alpha}\Lambda_0$, where $\boldsymbol{\alpha}$ is of the type (1) and $\Lambda_0$ is the lattice of integer vectors. For if $\boldsymbol{a}_1, \ldots, \boldsymbol{a}_n$ is any basis for $\Lambda$, we may define $\alpha_{ij}$ by

$$\boldsymbol{a}_j = (\alpha_{1j}, \ldots, \alpha_{nj}).$$

**I.4. Forms and lattices.** We consider first quadratic forms. Let

$$f(\boldsymbol{x}) = \sum_{i,j=1}^{n} f_{ij} x_i x_j \qquad (f_{ij} = f_{ji}), \tag{1}$$

where

$$\boldsymbol{x} = (x_1, \ldots, x_n), \tag{2}$$

be a non-singular quadratic form of signature[1] $(r, n-r)$; that is, there exist independent real linear forms

$$X_i = \sum_{1 \leq j \leq n} d_{ij} x_j \qquad (1 \leq i \leq n) \tag{3}$$

such that identically

$$f(\boldsymbol{x}) = \varphi(\boldsymbol{X}), \tag{4}$$

where

$$\boldsymbol{X} = (X_1, \ldots, X_n) \tag{5}$$

and

$$\varphi(\boldsymbol{X}) = X_1^2 + \cdots + X_r^2 - X_{r+1}^2 - \cdots - X_n^2 \tag{6}$$

(for $r = 0, n$ there are no positive or negative squares respectively). We have clearly

$$\det(f_{ij}) = \pm \{\det(d_{ij})\}^2. \tag{7}$$

Conversely, if $d_{ij}$ is any set of real numbers with $\det(d_{ij}) \neq 0$, then (3), (4) and (6) determine a quadratic form (1) of signature $(r, n-r)$ and (7) holds. We shall be concerned a great deal with the values which $f(\boldsymbol{x})$ takes when $x_1, \ldots, x_n$ are integers. By (3), these are precisely the

---

[1] Many writers define the signature of a form to be the number of positive squares less the number of negative squares in (6), i.e. $2r - n$. But it is more convenient to give explicitly the number of each kind of square than to do the arithmetic every time.

values which $\varphi(\boldsymbol{X})$ takes when $\boldsymbol{X}$ runs through the vectors of the lattice $\Lambda$ with basis

$$\boldsymbol{d}_j = (d_{1j}, \ldots, d_{nj}).$$

Then, by (7), we have

$$\{d(\Lambda)\}^2 = |\det(f_{ij})|. \tag{8}$$

In this way statements about different quadratic forms of signature $(r, n-r)$ at integral values are equivalent to statements about the single form $\varphi(\boldsymbol{X})$ and different lattices. For later reference we formulate a typical result as a Lemma.

LEMMA 4. *The following four statements about a number $\varkappa$ are equivalent, where*

$$\varphi(\boldsymbol{X}) = X_1^2 + \cdots + X_r^2 - X_{r+1}^2 - \cdots - X_n^2.$$

*(i) In every lattice $\Lambda$ there is a vector $\boldsymbol{A} \neq \boldsymbol{o}$ with*

$$|\varphi(\boldsymbol{A})| \leqq \varkappa \{d(\Lambda)\}^{2/n}.$$

*(ii) In every lattice $\Lambda$ of determinant 1 there is a vector $\boldsymbol{A} \neq \boldsymbol{o}$ with*

$$|\varphi(\boldsymbol{A})| \leqq \varkappa.$$

*(iii) In every lattice $\Lambda$ of determinant $d(\Lambda) \leqq \varkappa^{-n/2}$ there is a vector $\boldsymbol{A} \neq \boldsymbol{o}$ in*

$$|\varphi(\boldsymbol{A})| \leqq 1.$$

*(iv) For every quadratic form $\sum f_{ij} x_i x_j$ of signature $(r, n-r)$ there is an integer vector $\boldsymbol{a} \neq \boldsymbol{o}$ such that*

$$|f(\boldsymbol{a})| \leqq \varkappa |\det(f_{ij})|^{1/n}.$$

That (i), (ii) and (iii) are equivalent follows from homogeneity, since $\varphi(t\boldsymbol{X}) = t^2 \varphi(\boldsymbol{X})$ and since the set $t\Lambda$ of all $t\boldsymbol{X}$ $(\boldsymbol{X} \in \Lambda)$ is a lattice $t\Lambda$ of determinant $|t|^n d(\Lambda)$; and we may choose $t$ so that $t^n d(\Lambda) = 1$. That (iii) and (iv) are equivalent follows at once from the earlier discussion and, in particular, from (8).

The foregoing argument is quite general. For example the behaviour for integer values of the variables of any form $f(\boldsymbol{x})$ of degree $n$ which can be expressed as the product of $n$ real linear forms:

$$f(\boldsymbol{x}) = \prod_{1 \leqq j \leqq n} (d_{j1} x_1 + \cdots + d_{jn} x_n)$$

is equivalent to the behaviour of the function

$$\varphi(\boldsymbol{X}) = X_1 \ldots X_n$$

at the points of an appropriate lattice $\Lambda$. A single function $\varphi(\boldsymbol{X})$ corresponds to the set of all functions $f(\boldsymbol{x})$ that can be deduced from it by a real non-singular affine transformation

$$X_i = \sum d_{ij} x_j \qquad (d_{ij} \text{ real, } \det(d_{ij}) \neq 0).$$

**I.4.2.** Of course the form $\varphi(\boldsymbol{x})$ and the lattice $\Lambda$ do not determine the function $f(\boldsymbol{x})$ uniquely, since $f(\boldsymbol{x})$ depends on the choice of a particular basis for $\Lambda$; and we shall discuss this ambiguity here. The transformation

$$X_i = \sum_j d_{ij} x_j$$

of § 4.1 is just of the type

$$\boldsymbol{X} = \boldsymbol{\alpha} \boldsymbol{x}$$

discussed in § 3. Identifying these transformations we see that

$$\Lambda = \boldsymbol{\alpha} \Lambda_0,$$

where $\Lambda_0$ is the lattice of all integer vectors; the particular basis

$$\boldsymbol{d}_1, \ldots, \boldsymbol{d}_n$$

of $\Lambda$ corresponding to the basis

$$\boldsymbol{e}_j = \Big(\overbrace{0, \ldots, 0}^{j-1}, 1, \overbrace{0, \ldots, 0}^{n-j}\Big) \qquad (1 \leq j \leq n)$$

of $\Lambda_0$. Hence any other basis $\boldsymbol{d}'_1, \ldots, \boldsymbol{d}'_n$ of $\Lambda$ is of the shape

$$\boldsymbol{d}'_j = \boldsymbol{\alpha} \boldsymbol{e}'_j,$$

where $\boldsymbol{e}'_j$ is some other basis for $\Lambda_0$. Let $f'$ be the form corresponding to the basis $\boldsymbol{d}'_j$ as $f$ does to $\boldsymbol{d}_j$. Then clearly there is the identical relation

$$f'(\boldsymbol{x}') = f'(x'_1, \ldots, x'_n) = \varphi(x'_1 \boldsymbol{d}'_1 + \cdots + x'_n \boldsymbol{d}'_n) = f(x'_1 \boldsymbol{e}'_1 + \cdots + x'_n \boldsymbol{e}'_n).$$

But now since $\boldsymbol{e}'_j$ is a basis for $\Lambda_0$ we have

$$\boldsymbol{e}'_j = (v_{1j}, \ldots, v_{nj}),$$

where the $v_{ij}$ are integers and

$$\det(v_{ij}) = \pm 1: \tag{1}$$

so that identically

$$f'(\boldsymbol{x}') = f(\boldsymbol{x}) \tag{2}$$

if

$$x_i = \sum_j v_{ij} x'_j. \tag{3}$$

Conversely, if the $v_{ij}$ are integers such that (1), (2), (3), hold then $f'$ and $f$ correspond to the same lattice $\Lambda$. Two forms in this relationship are said to be equivalent; they take the same set of values as the variables run through all integral values, since, by (1) and (3), integral $\boldsymbol{x}'$ correspond to integral $\boldsymbol{x}$ and vice versa.

It is sometimes useful to distinguish between $\det(v_{ij}) = +1$ (proper equivalence) and $\det(v_{ij}) = -1$ (improper equivalence) in (1). We shall not do this, however, since it does not correspond to anything intrinsic in the corresponding lattices.

**I.4.3.** The forms $f(\boldsymbol{x})$ and $\varphi(\boldsymbol{X})$ do not in general determine the lattice uniquely, since for example a quadratic form $f(\boldsymbol{x})$ of signature $(r, s)$ with $r+s=n$ may be expressed in the shape

$$X_1^2 + \cdots + X_r^2 - X_{r+1}^2 - \cdots - X_{r+s}^2$$

in many different ways. Let $\boldsymbol{a}_1, \ldots, \boldsymbol{a}_n$ and $\boldsymbol{b}_1, \ldots, \boldsymbol{b}_n$ be bases of lattices $\Lambda$ and $M$ respectively and suppose that

$$\varphi\left(\sum_j u_j \boldsymbol{a}_j\right) = \varphi\left(\sum_j u_j \boldsymbol{b}_j\right) \tag{1}$$

for all integral $u = (u_1, \ldots, u_n)$. Since $\varphi(\boldsymbol{X})$ is a form, (1) is an identity in the variables $u_1, \ldots, u_n$. Let $\boldsymbol{\omega}$ be the uniquely determined homogeneous transformation such that

$$\boldsymbol{\omega} \boldsymbol{a}_j = \boldsymbol{b}_j \qquad (1 \leqq j \leqq n).$$

Then

$$\boldsymbol{\omega}\left(\sum_j u_j \boldsymbol{a}_j\right) = \sum_j u_j \boldsymbol{b}_j$$

for all $u$, and so

$$\varphi(\boldsymbol{X}) = \varphi(\boldsymbol{\omega} \boldsymbol{X}) \tag{2}$$

for all $\boldsymbol{X}$, by (1) and since every vector is of the shape $\boldsymbol{X} = \sum u_j \boldsymbol{a}_j$ for some real numbers $u_j$. If the homogeneous transformation $\boldsymbol{\omega}$ satisfies (2) we call it an automorph of $\varphi$. We have just shown that if (1) holds there is an automorph $\boldsymbol{\omega}$ of $\varphi$ such that $\boldsymbol{\omega} \boldsymbol{a}_j = \boldsymbol{b}_j$. The converse is, of course trivial that if $\boldsymbol{\omega}$ is an automorph of $\varphi$ and $\boldsymbol{\omega} \boldsymbol{a}_j = \boldsymbol{b}_j$ then (1) holds.

We shall study the automorphs of forms intensively in Chapter X.

**I.5. The polar lattice**[1]. We denote the scalar product of two $n$-dimensional vectors $\boldsymbol{x}, \boldsymbol{y}$ by

$$\boldsymbol{x}\boldsymbol{y} = x_1 y_1 + \cdots + x_n y_n. \tag{1}$$

Let $\boldsymbol{b}_1, \ldots, \boldsymbol{b}_n$ be a basis of a lattice $\Lambda$. Since the $\boldsymbol{b}_j$ are linearly independent, there exist vectors $\boldsymbol{b}_j^*$ such that

$$\boldsymbol{b}_j^* \boldsymbol{b}_i = \begin{cases} 1 & \text{if } i = j \\ 0 & \text{otherwise.} \end{cases} \tag{2}$$

[1] This section will not be referred to until Chapter VIII and will not be of importance until Chapter X and XI.

The lattice $\Lambda^*$ with basis $\boldsymbol{b}_j^*$ is called the polar (or dual or reciprocal) lattice of $\Lambda$, and $\boldsymbol{b}_j^*$ is the polar basis to $\boldsymbol{b}_j$. The polar lattice $\Lambda^*$ of $\Lambda$ is independent of the choice of the particular basis, as we now show.

LEMMA 5. *The polar lattice $\Lambda^*$ of $\Lambda$ consists of all vectors $\boldsymbol{a}^*$ such that $\boldsymbol{a}^*\boldsymbol{a}$ is an integer for all $\boldsymbol{a}$ in $\Lambda$. Then $\Lambda$ is conversely the polar lattice of $\Lambda^*$. Further,*

$$d(\Lambda)\, d(\Lambda^*) = 1.$$

Suppose, first, that

$$\boldsymbol{a}^* = \sum u_j \boldsymbol{b}_j^*, \quad \boldsymbol{a} = \sum v_j \boldsymbol{b}_j$$

are in $\Lambda^*$ and $\Lambda$ respectively, so that the $u_j$, $v_j$ are integers. Then

$$\boldsymbol{a}^*\boldsymbol{a} = \sum u_j v_j$$

is an integer. Now let $\boldsymbol{c}$ be any vector such that $\boldsymbol{c}\boldsymbol{a}$ is an integer for all $\boldsymbol{a}$ in $\Lambda$. In particular

$$\boldsymbol{c}\boldsymbol{b}_j = u_j \qquad (1 \leqq j \leqq n)$$

is an integer. Put $\boldsymbol{a}^* = \sum u_j \boldsymbol{b}_j^*$. Then

$$(\boldsymbol{c} - \boldsymbol{a}^*)\, \boldsymbol{b}_j = 0 \qquad (1 \leqq j \leqq n);$$

and so $\boldsymbol{c} = \boldsymbol{a}^*$ since the $\boldsymbol{b}_j$ are linearly independent. This proves the first sentence of the theorem. The second sentence follows immediately from the first and also from (2). Finally, (2) implies that

$$\det(\boldsymbol{b}_1^*, \ldots, \boldsymbol{b}_n^*) \det(\boldsymbol{b}_1, \ldots, \boldsymbol{b}_n) = 1,$$

and so $d(\Lambda^*)\, d(\Lambda) = 1$. This concludes the proof of the lemma.

**I.5.2.** When $\boldsymbol{y} \neq \boldsymbol{o}$ is fixed, the points $\boldsymbol{x}$ such that $\boldsymbol{y}\boldsymbol{x} = 0$ lie in a hyperplane through $\boldsymbol{o}$.

LEMMA 6. *A necessary and sufficient condition that there be $n-1$ linearly independent points $\boldsymbol{a}_1, \ldots, \boldsymbol{a}_{n-1}$ in $\Lambda$ with $\boldsymbol{y}\boldsymbol{a}_i = 0$ $(1 \leqq i \leqq n-1)$ is that $\boldsymbol{y} = t\boldsymbol{a}^*$ for some real $t$ and some $\boldsymbol{a}^*$ in $\Lambda^*$.*

Suppose first that $\boldsymbol{y}\boldsymbol{a}_i = 0$ $(1 \leqq i \leqq n-1)$. Then by Theorem I Corollary 2 there is a basis $\boldsymbol{b}_j$ $(1 \leqq j \leqq n)$ for $\Lambda$ such that

$$\boldsymbol{a}_i = v_{i1}\boldsymbol{b}_1 + \cdots + v_{ii}\boldsymbol{b}_i \qquad (v_{ii} \neq 0)$$

for integers $v_{ij}$. Hence $\boldsymbol{y}\boldsymbol{b}_i = 0$ $(1 \leqq i \leqq n-1)$. Let $\boldsymbol{y}\boldsymbol{b}_n = t$. Then clearly $\boldsymbol{y} = t\boldsymbol{b}_n^*$ where $\boldsymbol{b}_j^*$ $(1 \leqq j \leqq n)$ is the polar basis to $\boldsymbol{b}_j$. This proves half the lemma.

Suppose now that $\boldsymbol{y} = t\boldsymbol{a}^*$, where $\boldsymbol{a}^* \in \Lambda^*$. If $\boldsymbol{a}^* = 0$ there is nothing to prove. Otherwise, $\boldsymbol{a}^* = m\boldsymbol{b}_1^*$, where $m$ is an integer and $\boldsymbol{b}_1^*$ is primitive[1].

[1] That is, not of the shape $u\boldsymbol{c}^*$, $\boldsymbol{c}^* \in \Lambda^*$ for an integer $u > 1$.

Then $\boldsymbol{b}_1^*$ can be extend to a basis $\boldsymbol{b}_j^*$ for $\Lambda^*$. Let $\boldsymbol{b}_j$ be the polar basis. Then

$$\boldsymbol{y}\boldsymbol{b}_j = m t \boldsymbol{b}_1^* \boldsymbol{b}_j = 0 \qquad (2 \leqq j \leqq n).$$

This concludes the proof of the lemma.

Let $\Lambda(\boldsymbol{a}^*)$ be the set of $\boldsymbol{a}$ in $\Lambda$ such that $\boldsymbol{a}^*\boldsymbol{a} = 0$. Clearly if $\boldsymbol{a}_1$ and $\boldsymbol{a}_2$ are in $\Lambda(\boldsymbol{a}^*)$, so is $u_1\boldsymbol{a}_1 + u_2\boldsymbol{a}_2$ for any integers $u_1, u_2$. By Lemma 6 if $\boldsymbol{a}^* \in \Lambda^*$ there are $n-1$ linearly independent points of $\Lambda(\boldsymbol{a}^*)$, and so in a sense $\Lambda(\boldsymbol{a}^*)$ is an $(n-1)$-dimensional lattice. The following corollary makes those remarks more precise.

COROLLARY. *Let $\boldsymbol{b}^* = (b_1^*, \ldots, b_n^*)$ be a primitive point of $\Lambda^*$ and suppose that $b_n^* \neq 0$. Then the set of $(n-1)$-dimensional vectors $\boldsymbol{a}' = (a_1, \ldots, a_{n-1})$ such that for some $a_n$ the vector $\boldsymbol{a} = (a_1, \ldots, a_n)$ is in $\Lambda$ and satisfies $\boldsymbol{b}^*\boldsymbol{a} = 0$ is an $(n-1)$-dimensional lattice $\mathsf{M}$ of determinant $d(\mathsf{M}) = |b_n^*|\, d(\Lambda)$.*

We note that $\mathsf{M}$ is the projection on $x_n = 0$ of the set $\Lambda(\boldsymbol{b}^*)$ just defined. Since $b_n^* \neq 0$, if $a_n$ exists it is uniquely determined by $a_1, \ldots, a_{n-1}$, and the condition $\boldsymbol{b}^*\boldsymbol{a} = 0$.

We may suppose that $\boldsymbol{b}^* = \boldsymbol{b}_n^*$, where $\boldsymbol{b}_1^*, \ldots, \boldsymbol{b}_n^*$ is a basis for $\Lambda^*$ and $\boldsymbol{b}_j$ is the polar basis. After what was said before the enunciation of the corollary, it is clear that the $(n-1)$-dimensional vectors $\boldsymbol{b}_j'$ formed by taking the first $n-1$ coordinates of $\boldsymbol{b}_j$ are a basis for $\mathsf{M}$. We now show that

$$b_n^* \det(\boldsymbol{b}_1, \ldots, \boldsymbol{b}_n) = \det(\boldsymbol{b}_1', \ldots, \boldsymbol{b}_{n-1}'). \tag{1}$$

If in the determinant in the left the $n$-th coordinate $x_n$ is replaced by $\boldsymbol{b}_n^*\boldsymbol{x}$ for $\boldsymbol{x} = \boldsymbol{b}_1, \ldots, \boldsymbol{b}_n$, the value of the determinant is multiplied by $b_{nn}^* = b_n^*$. Since $\boldsymbol{b}_n^*\boldsymbol{b}_j = 0$ for $1 \leqq j \leqq n-1$ and $\boldsymbol{b}_n^*\boldsymbol{b}_n = 1$, the equation (1) follows at once. In particular $|b_n^*|\, d(\Lambda) = d(\mathsf{M})$, as required.

**I.5.3.** Finally we must note the effect of homogeneous linear transformations on the relationship between polar pairs of lattices. Let

$$\boldsymbol{X} = \boldsymbol{\tau}\boldsymbol{x} \tag{1}$$

be a non-singular homogeneous linear transformation given by

$$X_i = \sum_j \tau_{ij} x_j$$

where

$$\det(\boldsymbol{\tau}) = \det(\tau_{ij}) \neq 0. \tag{2}$$

If $\boldsymbol{Y}$ is any vector, we have

$$\boldsymbol{Y}\boldsymbol{X} = \sum_i Y_i X_i = \sum_{i,j} Y_i \tau_{ij} x_j.$$

Hence

$$\boldsymbol{Y}\boldsymbol{X} = \boldsymbol{y}\boldsymbol{x}, \tag{3}$$

where

$$y_j = \sum_i Y_i \tau_{ij} \qquad (1 \leqq j \leqq n). \tag{4}$$

Since $\det(\boldsymbol{\tau}) \neq 0$, by hypothesis, the equations (4) define $\boldsymbol{Y}$ as a function of $\boldsymbol{y}$. We write

$$\boldsymbol{Y} = \boldsymbol{\tau}^* \boldsymbol{y},$$

where $\boldsymbol{\tau}^*$ is called the transformation polar to $\boldsymbol{\tau}$.

LEMMA 7. *Let $\boldsymbol{\tau}$ be a non-singular homogeneous linear transformation, $\Lambda$ a lattice, and $\boldsymbol{\tau}\Lambda$ the lattice of points $\boldsymbol{\tau x}$, $\boldsymbol{x} \in \Lambda$. Then the polar lattice of $\boldsymbol{\tau}\Lambda$ is $\boldsymbol{\tau}^*\Lambda^*$, where $\boldsymbol{\tau}^*$ and $\Lambda^*$ are respectively polar to $\boldsymbol{\tau}$ and $\Lambda$.*

This follows at once from Lemma 5 and equation (3) above, where

$$\boldsymbol{X} = \boldsymbol{\tau x}, \quad \boldsymbol{Y} = \boldsymbol{\tau}^* \boldsymbol{y}.$$

Chapter II

# Reduction

**II.1. Introduction.** In investigating the values taken by an algebraic form $f(\boldsymbol{x})$ for integer values of the variables it is often useful to substitute for $f$ a form equivalent to it (in the sense of Chapter I, § 4) which bears a special relation to the problem under consideration. This process is independent of the geometrical notions introduced by MINKOWSKI and depends only on the properties of bases of lattices developed in Chapter I. Indeed only the lattice $\Lambda_0$ of integer vectors comes into consideration.

It is convenient to collect together in one chapter the various applications of reduction. The later parts of the chapter involve some moderately heavy computation. The beginner might well omit all after the enunciation of the results in § 4.2. Indeed the next few chapters are practically independent of Chapter II, which might well have been deferred until later.

In § 2 we discuss the general method. In the rest of the chapter we shall be mainly occupied in investigating

$$M(f) = \inf_{\substack{\boldsymbol{u} \neq \boldsymbol{o} \\ \boldsymbol{u} \text{ integral}}} |f(\boldsymbol{u})|$$

where $f(\boldsymbol{x})$ is a form of a special type. Definite and indefinite quadratic forms are treated in §§ 3.4 respectively and binary cubic forms in § 5.

The methods of this chapter have been successfully applied to related problems: for example, when $f(\boldsymbol{x})$ is indefinite, to the estimation of

$$\inf f(\boldsymbol{u})$$

over integer vectors $\boldsymbol{u} \neq \boldsymbol{o}$ for which $f(\boldsymbol{u})$ is positive [either in the strict sense $f(\boldsymbol{u}) > 0$ or the weak sense $f(\boldsymbol{u}) \geqq 0$: two distinct problems in general] but we shall do this only for binary forms.

A table listing the known results about quadratic forms is given in an appendix. We shall be considering quadratic forms later from other points of view.

DAVENPORT and ROGERS (1950a) have shown that in many cases not merely one but infinitely many integer points $\boldsymbol{u}$ exist such that $f(\boldsymbol{u})$ satisfies the inequalities stated. This requires deeper methods than those used here and will be discussed in Chapter X.

It should be remarked that there is a classical theory of reduction for indefinite binary quadratic forms which we do not discuss here. Although it comes into the general scope of reduction as defined here, that is the choosing of bases with special properties, it is best understood after the discussion of Chapter III. It is closely related to continued fraction theory. See Chapter X, § 8.

**II.2. The basic process.** We first discuss the standard procedure for positive definite forms $f(\boldsymbol{x})$; that is for forms such that $f(\boldsymbol{x}) > 0$ for all real vectors $\boldsymbol{x} \neq \boldsymbol{o}$.

We note first that if $f(\boldsymbol{x})$ is positive definite of degree $r$, say, then there is a constant $\varkappa > 0$ such that

$$f(\boldsymbol{x}) \geqq \varkappa |\boldsymbol{x}|^r \tag{1}$$

for all real $\boldsymbol{x}$, where we have written

$$|\boldsymbol{x}| = (x_1^2 + \cdots + x_n^2)^{\frac{1}{2}}.$$

For on the surface of the sphere $|\boldsymbol{x}| = 1$ the continuous function $f(\boldsymbol{x})$ must attain its lower bound $\varkappa$, so $\varkappa > 0$; and then (1) follows by homogeneity. In particular, there are only a finite number of integral vectors* $\boldsymbol{u}$ such that $f(\boldsymbol{u})$ is less than any given number.

We now choose a basis for the lattice $\Lambda_0$ of integral vectors with respect to the positive definite form $f(\boldsymbol{x})$ as follows. Let $\boldsymbol{e}_1' \neq 0$ be one of those integral vectors $\boldsymbol{u}$ for which $f(\boldsymbol{u})$ is as small as possible. By the argument of the last paragraph such $\boldsymbol{u}$ exist, and there are only a finite number of them. If $\boldsymbol{e}_1'$ were of the shape $\boldsymbol{e}_1' = k\boldsymbol{a}$, $\boldsymbol{a} \in \Lambda_0$, where $k > 1$ is an integer we should have

$$0 < f(\boldsymbol{a}) = k^{-r} f(\boldsymbol{e}_1') < f(\boldsymbol{e}_1'),$$

* i.e. vectors whose co-ordinates are rational integers.

contrary to the definition of $e_1'$. Hence by Corollary 3 to Theorem I of Chapter I, we may extend $e_1'$ to a basis $e_1', b_2, \ldots, b_n$ of the lattice $\Lambda_0$ of integer vectors. We now choose $e_j'$ $(2 \leqq j \leqq n)$ in succession. Suppose that $e_1', \ldots, e_{j-1}'$ have already been chosen and are extensible to a base $e_1', \ldots, e_{j-1}', b_j, \ldots, b_n$ of $\Lambda_0$. Then $e_j'$ is one of the finite number of vectors with the property that $e_1', \ldots, e_j'$ is extensible to a base of $\Lambda_0$ and for which $f(e_j')$ is as small as possible. Such $e_j'$ exist but are finite in number, by argument used for $e_1'$. In this way we obtain a base $e_1', \ldots, e_n'$: and for any given $f(x)$ there are only a finite number of such bases.

If the function $f(x)$ is such that we may indeed choose

$$e_j' = e_j = (\overbrace{0, \ldots, 0}^{j-1}, 1, \overbrace{0, \ldots, 0}^{n-j}) \qquad (1 \leqq j \geqq n)$$

for the above basis, then $f(x)$ is said to be reduced (in the sense of MINKOWSKI). The above proof shows that every positive definite form is equivalent (in the sense introduced in Chapter I, § 4) to at least one and to at most a finite number of reduced forms.

We may make the definition of a reduced form more explicit. By Corollary 4 to Theorem I of Chapter I (or by Lemma 2 of Chapter I), a necessary and sufficient condition that $e_1, \ldots, e_{j-1}$ and the integral vector $u = (u_1, \ldots, u_n)$ be extensible to a basis for $\Lambda_0$ is that

$$\text{g.c.d.}\,(u_j, \ldots, u_n) = 1. \tag{2}$$

Hence the form $f(x)$ is reduced if and only if

$$f(u_1, \ldots, u_n) \geqq f(e_j)$$

for all $j$ and for all integers $u_1, \ldots, u_n$ satisfying (2).

**II.2.2.** When the form $f(x)$ is not definite, then there is no generally valid procedure to replace the reduction procedure for definite forms.

If we know (or may assume) that $f(u)$ does not assume arbitrarily small values for integral $u \neq o$ then it is possible to salvage something of the reduction procedure. Let $\varepsilon > 0$ be chosen arbitrarily small. By hypothesis,

$$M_1 = \inf_{\substack{u \neq o \\ \text{integral}}} |f(u)| > 0.$$

Hence we may find an integral $e_1' \neq 0$ such that

$$|f(e_1')| \leqq M_1/(1-\varepsilon).$$

Without loss of generality $e_1'$ is not of the form $ka$ where $a$ is integral and $k > 1$ is a rational integer. If $e_1', \ldots, e_{j-1}'$ have already been found, write

$$M_j = \inf |f(\boldsymbol{u})|$$

where the infimum is over all integral vectors $\boldsymbol{u}$ such that $\boldsymbol{e}_1', \ldots, \boldsymbol{e}_{j-1}', \boldsymbol{u}$ is extensible to a basis for $\Lambda_0$. Then

$$M_j \geqq M_1 > 0,$$

and so we may choose $\boldsymbol{e}_j'$ so that $\boldsymbol{e}_1', \ldots, \boldsymbol{e}_j'$ is extensible to a basis and

$$|f(\boldsymbol{e}_j')| \leqq M_j/(1-\varepsilon).$$

Let $f'(\boldsymbol{x})$ be the equivalent form for which

$$f(\boldsymbol{e}_j') = f'(\boldsymbol{e}_j).$$

Then we have

$$|f'(u_1, \ldots, u_n)| \geqq (1-\varepsilon)\,|f'(\boldsymbol{e}_j)|$$

for all sets of integers $u_1, \ldots, u_n$ such that g.c.d. $(u_j, \ldots, u_n) = 1$. But, of course, there is no reason to suppose that there are only a finite number of $f'$ with this property and equivalent to a given $f$.

An alternative procedure which is sometimes possible is to find some other form $g(\boldsymbol{x})$, related to our given $f$, which is definite and to reduce $g(\boldsymbol{x})$. We shall do this for binary cubic forms in § 6. This method goes back to HERMITE, who applied it to indefinite quadratic forms as follows.

Let $f(\boldsymbol{x})$ be an indefinite quadratic form of signature $(r, n-r)$, so that, as before,

$$f(\boldsymbol{x}) = X_1^2 + \cdots + X_r^2 - X_{r+1}^2 - \cdots - X_n^2, \tag{1}$$

where the $X_j$ are linear forms in $x_1, \ldots, x_n$. Then

$$g(\boldsymbol{x}) = X_1^2 + \cdots + X_r^2 + X_{r+1}^2 + \cdots + X_n^2 \tag{2}$$

is a definite quadratic form with the same determinant, except, perhaps, for sign. The forms $X_1, \ldots, X_n$ are not uniquely determined by $f(\boldsymbol{x})$ but we say that $f(\boldsymbol{x})$ is reduced (in the sense of HERMITE) if the form $g(\boldsymbol{x})$ is reduced in the sense of MINKOWSKI for some choice of $X_1, \ldots, X_n$. Clearly $f(\boldsymbol{x})$ is always equivalent to a reduced form, since we may choose any representation (1) and then apply the transformation which reduces $g(\boldsymbol{x})$. Reduction more or less of this kind was first introduced by HERMITE, and has been further discussed, amongst others, by SIEGEL (1940a), as a tool for investigating the arithmetical properties of quadratic forms. In general a form $f(\boldsymbol{x})$ is equivalent to infinitely many HERMITE-reduced forms, but SIEGEL shows that it is equivalent to only finitely many if the coefficients of $f(\boldsymbol{x})$ are all rational.

We note here that the relationship between (1) and (2) allows estimates for the minimum of a definite form to be extended to an indefinite one, since clearly $|f(\boldsymbol{x})| \leq g(\boldsymbol{x})$ for all real vectors $\boldsymbol{x}$. But in general, the information so obtained is quite weak.

**II.3. Definite quadratic forms.** We shall be considering definite quadratic forms from many different points of view in the course of this book. Here we see what can be done by reduction methods alone. The study of reduction is of great importance in the arithmetical theory of quadratic forms, see WEYL (1940a) or VAN DER WAERDEN (1956a), who give references to earlier literature. Here we are concerned only with the minima of forms.

Let

$$f(x_1, x_2) = f_{11} x_1^2 + 2 f_{12} x_1 x_2 + f_{22} x_2^2$$

be a positive definite quadratic form. We wish to prove that there are integers $(u_1, u_2) \neq (0, 0)$ such that

$$f(u_1, u_2) \leq (4D/3)^{\frac{1}{2}},$$

where

$$D = f_{11} f_{22} - f_{12}^2.$$

By taking an equivalent form, if need be, we may suppose that

$$M(f) = \inf_{\substack{u_1, u_2 \\ \text{integers not both } 0}} f(u_1, u_2) = f_{11}.$$

We have

$$f(x_1, x_2) = f_{11}\left(x_1 + \frac{f_{12}}{f_{11}} x_2\right)^2 + \frac{D}{f_{11}} x_2^2.$$

Put $u_2 = 1$ and choose for $u_1$ an integer such that

$$\left| u_1 + \frac{f_{12}}{f_{11}} \right| \leq \frac{1}{2}. \tag{1}$$

Then, on the one hand,

$$f(u_1, 1) \geq f_{11},$$

and, on the other,

$$f(u_1, 1) \leq \frac{1}{4} f_{11} + \frac{D}{f_{11}}. \tag{2}$$

Hence

$$\frac{D}{f_{11}} \geq \frac{3}{4} f_{11},$$

that is

$$f_{11}^2 \leq 4D/3,$$

as required. That $\leq$ here cannot be replaced by $<$ is shown by the form

$$f_0(x_1, x_2) = x_1^2 + x_1 x_2 + x_2^2$$

for which $D = \frac{3}{4}$ and $f(u_1, u_2) \geq 1$ for integers $(u_1, u_2) \neq (0, 0)$. It is not difficult to show by examining when equality can occur in the above argument that $\leqq$ can be replaced by $<$ unless $f$ is equivalent to a multiple of $f_0$. We do not go into details, since we shall prove this later more simply.

**II.3.2.** As HERMITE noted, this argument can be extended to prove the following theorem.

THEOREM I. *A non-singular quadratic form*

$$f(\boldsymbol{x}) = \sum f_{ij} x_i x_j$$

*represents a value* $f(\boldsymbol{u})$ *with*

$$|f(\boldsymbol{u})| \leqq (\tfrac{4}{3})^{(n-1)/2} |D|^{1/n}, \tag{1}$$

*where* $\boldsymbol{u} \neq \boldsymbol{o}$ *is integral and*

$$D = \det(f_{ij}).$$

By the remarks at the end of § 2.2 we may suppose, without loss of generality, that $f(\boldsymbol{x})$ is positive definite. We may now suppose, as before, that

$$f_{11} \leqq f(\boldsymbol{u})$$

for all integral $\boldsymbol{u} \neq \boldsymbol{o}$. Then

$$f(\boldsymbol{x}) = f_{11}\left(x_1 + \frac{f_{12}}{f_{11}} x_2 + \cdots + \frac{f_{1n}}{f_{11}} x_n\right)^2 + g(x_2, \ldots, x_n),$$

where $g(x_2, \ldots, x_n)$ is a definite quadratic form of determinant $D/f_{11}$. Since we may suppose the result already proved for forms in $n-1$ variables, there are integers $u_2, \ldots, u_n$ not all 0 such that

$$g(u_2, \ldots, u_n) \leqq \left(\frac{4}{3}\right)^{\frac{1}{2}(n-2)} \left(\frac{D}{f_{11}}\right)^{1/(n-1)}.$$

Choose the integer $u_1$ so that

$$\left|u_1 + \frac{f_{12}}{f_{11}} u_2 + \cdots + \frac{f_{1n}}{f_{11}} u_n\right| \leqq \frac{1}{2}.$$

Then

$$f_{11} \leqq f(\boldsymbol{u}) \leqq \frac{f_{11}}{4} + \left(\frac{4}{3}\right)^{\frac{1}{2}(n-2)} \left(\frac{D}{f_{11}}\right)^{1/(n-1)},$$

and so

$$f_{11} \leqq \left(\frac{4}{3}\right)^{\frac{1}{2}(n-1)} D^{1/n}.$$

This proves the assertion. Unfortunately, the constant $(\frac{4}{3})^{\frac{1}{2}(n-1)}$ is the best possible only for $n=2$. We shall show below that it is not the best possible for $n=3$, and since the above proof is by induction it cannot be best possible for $n \geqq 3$. It is possible to modify the argument to give the best possible result for $n=3$ [for a neat version of this see MORDELL (1948a)], but we shall not do this. Instead we give a more elegant, if more artificial, treatment depending on a more detailed examination of reduced forms which goes back essentially at least as far as GAUSS.

**II.3.3.** We start with the consideration of a positive definite binary form which is reduced in the sense of MINKOWSKI:

$$f(x_1, x_2) = f_{11} x_1^2 + 2f_{12} x_1 x_2 + f_{22} x_2^2.$$

After the substitution $x_1 \to x_1$, $x_2 \to -x_2$ if need be, we may suppose without loss of generality that

$$f_{12} \geqq 0. \tag{1}$$

By the definition of reduction,

$$f_{22} = f(0, 1) \geqq f(1, 0) = f_{11} \tag{2}$$

and

$$f(-1, 1) \geqq f(0, 1),$$

that is

$$2f_{12} \leqq f_{11}. \tag{3}$$

By (1), (2), (3) we have

$$4D - 3f_{11}f_{22} = f_{11}f_{22} - 4f_{12}^2 \geqq f_{11}^2 - 4f_{12}^2 \geqq 0;$$

and so

$$f_{11}^2 \leqq f_{11}f_{22} \leqq \tfrac{4}{3} D.$$

The sign of equality is required only when $f_{11} = f_{22} = 2f_{12}$; i.e. when

$$f(\boldsymbol{x}) = f_{11}(x_1^2 + x_1 x_2 + x_2^2).$$

Before going on to ternary forms, we note that any form satisfying (1), (2), (3) is reduced. This is a special case of the general theorem that MINKOWSKI-reduced forms can be characterised by a finite set of inequalities, but here it is easy to verify directly.

Let $u_1, u_2$ be integers neither of which is 0. If $|u_1| \geqq |u_2|$ we have

$$\begin{aligned} f(u_1, u_2) &= |u_1| \{f_{11}|u_1| \pm 2f_{12}|u_2|\} + f_{22}u_2^2 \\ &\geqq |u_1| \{f_{11}|u_1| - 2f_{12}|u_1|\} + f_{22}u_2^2 \\ &= u_1^2(f_{11} - 2f_{12}) + f_{22}u_2^2 \\ &\geqq f_{11} - 2f_{12} + f_{22} = f(-1, 1); \end{aligned}$$

and if $0 < |u_1| \leqq |u_2|$ the same inequality follows on reversing the rôles of $u_1$ and $u_2$. Since $f_{11} - 2f_{12} + f_{22} \geqq f_{22}$, by (3), we have shown that $f(\boldsymbol{x})$ is reduced.

In particular, if $t$ is any number $\geqq \frac{3}{4}$ then the form

$$f_t = x_1^2 + x_1 x_2 + (t + \tfrac{1}{4}) x_2^2$$

is reduced. Since

$$M(f_t) = f(1, 0) = 1$$

and the determinant $D$ of $f_t$ is $t$, we see that

$$M(f)/D^{\frac{1}{2}}$$

may take any value $t^{-\frac{1}{2}} \leq (\frac{4}{3})^{\frac{1}{2}}$. This is in striking contrast with the behaviour of indefinite quadratic forms (see § 4).

For later convenience we collect what has been proved so far and express it as a theorem.

THEOREM II. *A positive definite binary quadratic form*

$$f_{11} x_1^2 + 2 f_{12} x_1 x_2 + f_{22} x_2^2$$

*is reduced if and only if*

$$|2 f_{12}| \leq f_{11} \leq f_{22}.$$

*The three smallest values taken by $f(\boldsymbol{u})$ for a reduced form and integral $\boldsymbol{u} \neq \boldsymbol{o}$ are $f_{11}, f_{22}$ and $f_{11} - 2|f_{12}| + f_{22}$, where*

$$f_{11} \leq f_{22} \leq f_{11} - 2|f_{12}| + f_{22}.$$

*For a reduced form*

$$f_{11} f_{22} \leq 4D/3,$$

*where*

$$D = f_{11} f_{22} - f_{12}^2.$$

*The ratio $\varrho = f_{11}/D^{\frac{1}{2}}$ may take any value in the interval*

$$0 < \varrho \leq (\tfrac{4}{3})^{\frac{1}{2}}.$$

**II.3.4.** We now consider ternary quadratic forms. As we shall later be considering definite quadratic forms in a wider context (Chapter V, § 9, see also Chapter IX, § 3.3) we content ourselves with the following.

THEOREM III. A. *Let*

$$f(\boldsymbol{x}) = \sum f_{ij} x_i x_j \qquad (f_{ij} = f_{ji})$$

*be a positive definite ternary quadratic form. Then there is an integral vector $\boldsymbol{u} \neq \boldsymbol{o}$ such that*

$$f(\boldsymbol{u}) \leq (2D)^{\frac{1}{3}},$$

*where*

$$D = D(f) = \det(f_{ij}).$$

B. *If $f(\boldsymbol{x})$ is reduced, then*

$$f_{11} f_{22} f_{33} \leq 2D.$$

C. *The signs of equality are required when and only when $f(\boldsymbol{x})$ is equivalent to a multiple of*

$$f_0(\boldsymbol{x}) = x_1^2 + x_2^2 + x_3^2 + x_2 x_3 + x_3 x_1 + x_1 x_2.$$

We note again that we get as good an estimate for $f_{11} f_{22} f_{33}$ as we do for $f_{11}^3$. This will be put in a wider setting in Chapter VIII, § 2.

Since $f_0(\boldsymbol{u})$ is an integer we have $f_0(\boldsymbol{u}) \geqq 1$. Since $D(f_0) = \frac{1}{2}$, this shows that the equality signs are required for $f_0$. Part A of the theorem follows from the rest. Hence we need only prove Part B and that equality in B occurs only for multiples of $f_0$.

Following GAUSS (1831a) we distinguish two cases. Suppose first that

$$f_{12} f_{23} f_{31} \geqq 0.$$

Then after a substitution

$$x_i \to \pm x_i$$

we may suppose without loss of generality that

$$f_{12} \geqq 0, \quad f_{23} \geqq 0, \quad f_{31} \geqq 0.$$

Write

$$\vartheta_{ij} = f_{ii} - 2 f_{ij} \qquad (f_{ij} = f_{ji}). \tag{1}$$

Then

$$\vartheta_{ij} \geqq 0$$

for all $i \neq j$ since $f$ is reduced. For example

$$f(1, -1, 0) \geqq f(1, 0, 0)$$

gives $\vartheta_{21} \geqq 0$. We have identically

$$2D - f_{11} f_{22} f_{33} = \vartheta_{32} \vartheta_{21} \vartheta_{13} + \sum \{f_{11} f_{23} \vartheta_{23} + f_{23} \vartheta_{13} \vartheta_{21}\}, \tag{2}$$

where the sum is over cyclic permutations of 1, 2, 3; as is readily verified on expressing both sides in terms of the $f_{ij}$ alone[1]. Since all the terms on the right-hand side of (2) are non-negative, we have

$$f_{11}^3 \leqq f_{11} f_{22} f_{33} \leqq 2D, \tag{3}$$

as required.

The other case is when $f_{12} f_{23} f_{31} \leqq 0$, and then we may suppose that

$$f_{12} \leqq 0, \quad f_{23} \leqq 0, \quad f_{31} \leqq 0.$$

We write now

$$\psi_{ij} = f_{ii} + 2 f_{ij}$$

and

$$\omega_i = f(1, 1, 1) - f_{ii}.$$

[1] This is an application of LITTLEWOOD's Principle: all identities are trivial (once they have been written down by someone else).

Then $\psi_{ij} \geqq 0$ and $\omega_i \geqq 0$, since $f$ is reduced. Then identically

$$\left.\begin{aligned} &6D - 3f_{11}f_{22}f_{33} = \psi_{23}\psi_{31}\psi_{12} + \\ &+ 2\psi_{32}\psi_{13}\psi_{21} + \sum \{f_{11}(-f_{23})(\psi_{23} + 2\omega_1) + (-f_{23})\psi_{13}\psi_{21}\}. \end{aligned}\right\} \quad (4)$$

Again all the terms on the right-hand side are non-negative, so (3) holds.

We leave to the reader an examination of when equality can occur. A rather tedious investigation of cases shows that it can occur only when

$$f_{11} = f_{22} = f_{33}$$

and either $2f_{23} = 2f_{31} = 2f_{12} = \pm 1$, or one of $2f_{23}, 2f_{31}, 2f_{12}$ vanishes and the remaining two are equal to $\pm 1$. But all these forms are equivalent to $f_{11}f_0(\boldsymbol{x})$, as is readily verified. For example,

$$x_1^2 + x_2^2 + x_3^2 + x_1x_2 + x_2x_3 = f_0(x_1, x_2 + x_3, -x_3).$$

Gauss lists several other identities which could be used instead of those here.

**II.4. Indefinite quadratic forms.** These will also be considered again and again throughout the book from different points of view. A table listing known results is given in Appendix A. We do not here carry the reduction argument as far as it will go, but only far enough to illustrate the different nature of the results from those obtained in the definite case.

We shall continue to use the notation

$$M(f) = \inf_{\substack{\boldsymbol{u} \neq \boldsymbol{o} \\ \text{integral}}} |f(\boldsymbol{u})|,$$

where $f(\boldsymbol{x})$ is a form in any number of variables, and write

$$D = D(f) = \det(f_{ij})$$

for a quadratic form $\sum f_{ij}x_ix_j = f(\boldsymbol{x})$.

There are two characteristic differences between the behaviour of $M(f)$ for definite and indefinite forms. For definite binary forms we saw that $M(f)/|D(f)|^{\frac{1}{2}}$ could take any value $\varrho$ in an interval

$$0 < \varrho \leqq (\tfrac{4}{3})^{\frac{1}{2}},$$

where $(\frac{4}{3})^{\frac{1}{2}}$ was the maximum possible value. It is not difficult to verify that definite quadratic forms in any number of variables behave similarly, cf. Chapter V, Lemma 6. The first difference in the behaviour of indefinite quadratic forms is rather trivial: it is quite possible that $M(f) = 0$, and this may occur either because there is an integral $\boldsymbol{u} \neq \boldsymbol{o}$ such that $f(\boldsymbol{u}) = 0$, or because there are integral $\boldsymbol{u} \neq \boldsymbol{o}$ such that $f(\boldsymbol{u})$

is arbitrarily small but not 0. The second difference is deeper: the values of $M(f)/|D(f)|^{\frac{1}{n}}$ do not fill the complete interval up to the maximum possible value.

The position for indefinite binary quadratic forms has been the most investigated. Here a very great deal is known about the possible values of $M(f)/|D(f)|^{\frac{1}{2}}$. The greatest value is $(\frac{4}{5})^{\frac{1}{2}}$, given by the multiples of $x_1^2+x_1x_2-x_2^2$. Otherwise $M(f)\leqq(\frac{1}{2})^{\frac{1}{2}}|D(f)|^{\frac{1}{2}}$. A well-known theorem of MARKOFF ("the MARKOFF chain") states that there are only denumerably many possible values of $M(f)/|D(f)|^{\frac{1}{2}}$ greater than $\frac{2}{3}$. There are certainly intervals to the left of $\frac{2}{3}$ which contain no values of $M(f)/|D(f)|^{\frac{1}{2}}$. The author has given a proof of the Markoff chain theorem in his Cambridge Tract [CASSELS (1956a)], to which the reader is referred for references for the various statements made in this paragraph. Here we shall be content with finding the two largest possible values of $M(f)/|D(f)|^{\frac{1}{2}}$.

There is a similar state of affairs for ternary quadratics but much less is known. The most complete information is due to VENKOV (1945a) who has found the eleven largest values of $M(f)/|D(f)|^{\frac{1}{3}}$, but they do not seem to follow any general pattern, except that they are all given by forms with integral coefficients. There are two unsolved problems about indefinite ternaries which appear completely intractable. It is not known whether there are forms $f$ with $M(f)>0$ which are not multiples of integral forms; and it is not known whether the set of values of $M(f)/|D(f)|^{\frac{1}{3}}$ has any limit point other than 0. These two problems are closely related [CASSELS and SWINNERTON-DYER (1955a); see also Chapter X, Theorem XII].

This phenomenon of "successive minima" (not to be confused with the "successive minima" of a lattice with respect to a point set which is discussed in Chapter VIII) occurs very widely with indefinite forms. It takes a great many different shapes and a general theory hardly exists*. It is not possible to predict when it occurs: for example it does not occur in the problems discussed in § 4.5 or § 5.

It is not difficult to see how "successive minima" can occur. An inequality of the type $|f(\boldsymbol{u})|\geqq 1$, where $f(\boldsymbol{x})$ is an indefinite form and $\boldsymbol{u}$ is an integer vector, is really a pair of alternatives

either $$f(\boldsymbol{u})\geqq 1$$

or $$f(\boldsymbol{u})\leqq -1.$$

Each of these inequalities may be regarded as a linear inequality in the coefficients of $f$. If we consider a large number of different $\boldsymbol{u}$ then

* MAHLER has shown that the minima form a closed set. In fact this follows at once from his compactness theorem of Chapter V.

the various pairs of alternatives are a priori independent. It may turn out, on combining the various alternatives, that some combinations of alternatives are altogether impossible while other combinations of alternatives define a form $f$ uniquely. An example may make this clearer. Suppose that we are interested in binary quadratic forms for which $M(f)=1$ and $f(1,0)=1$. Such a form has the shape

$$\left.\begin{aligned} &f(x)= \\ &x_1^2+\alpha x_1 x_2+\beta x_2^2, \end{aligned}\right\} \quad (1)$$

where the coefficients $\alpha$ and $\beta$ are to be investigated. The only such form which satisfies the inequalities

$$\begin{aligned} f(0,1) &\leqq -1, \\ f(1,1) &\geqq +1, \\ f(2,-1) &\geqq +1, \end{aligned}$$

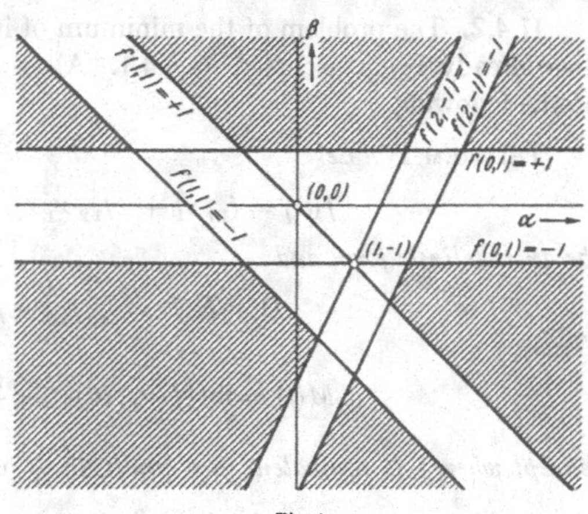

Fig. 4

is $x_1^2+x_1x_2-x_2^2$, as the reader will easily verify. Hence any other form with $f(1,0)=1$ and $M(f)=1$ must satisfy at least one of the inequalities $f(0,1)\geqq +1$, $f(1,1)\leqq -1$, $f(2,-1)\leqq -1$. The form $x_1^2+x_1x_2-x_2^2$ is thus in a strong sense isolated from all other forms (1) with $M(f)=1$. For example if $\alpha$ and $\beta$ are plotted as cartesian coordinates for the form $f$, a condition $|f(u_1,u_2)|\geqq 1$ excludes a strip of the plane between two parallel lines. The three conditions

$$|f(0,1)|\geqq 1, \quad |f(1,1)|\geqq 1, \quad |f(2,-1)|\geqq 1$$

exclude three strips. What is left consists of the point $(1,-1)$ and a number of infinite regions which are separated from the point by one of the strips (see Fig. 4).

In the actual proofs, this general principle tends to be obscured. If $f$ is an indefinite form and $M(f)=1$ there is not necessarily an integral vector $\boldsymbol{u}$ with $|f(\boldsymbol{u})|=1$, though there are integral vectors with $1\leqq|f(\boldsymbol{u})|<1+\varepsilon$ for any given $\varepsilon>0$, and further devices must be used to deal with this. The difficulty is that if $t>1$, then the form $tf(\boldsymbol{x})=f'(\boldsymbol{x})$ satisfies the same choice of inequalities "$f(\boldsymbol{u})\geqq 1$ or $f(\boldsymbol{u})\leqq -1$" as the original $f(\boldsymbol{x})$. Here $t$ might be arbitrarily close to 1, that is, the coefficients of $f'(\boldsymbol{x})$ might be arbitrarily close those of $f(\boldsymbol{x})$. Hence to pin down $f(\boldsymbol{x})$ uniquely we must some-how make use of the normalization $M(f)=1$. We do this by first finding the determinant of the form in question and then using this as part of our information. The actual proofs will make the details clearer.

We shall later deal with isolation of this type from a more sophisticated point of view (Chapter X). The treatment there will also help to show why the additional devices just mentioned are effective.

**II.4.2.** The problem of the minimum of indefinite binary quadratics has already been discussed in § 4.1. All we shall actually prove here is the following.

THEOREM IV. *Let*

$$f(\boldsymbol{x}) = f_{11} x_1^2 + 2 f_{12} x_1 x_2 + f_{22} x_2^2 \tag{1}$$

*be an indefinite form and*

$$D = D(f) = f_{11} f_{22} - f_{12}^2.$$

*Then*

$$M(f) = \inf |f(u_1, u_2)| \leqq \left| \frac{100 D}{221} \right|^{\frac{1}{2}}, \tag{2}$$

*except when $f$ is equivalent to a multiple of one of the two forms*

$$f_0(\boldsymbol{x}) = x_1^2 + x_1 x_2 - x_2^2, \tag{3}$$

$$f_1(\boldsymbol{x}) = x_1^2 - 2x_2^2 \tag{4}$$

*for which $M(f) = 1$ and $|D| = \frac{5}{4}$, 2 respectively.*

That $f_0$ and $f_1$ are exceptional is clear, since they both represent only non-zero integers for integral $\boldsymbol{u} \neq \boldsymbol{o}$. The constant $\frac{100}{221}$ in (2) cannot, in fact, be improved since the next form of the MARKOFF chain is

$$f_2 = 5 x_1^2 + 11 x_1 x_2 - 5 x_2^2$$

which has $D(f_2) = -221/4$ and can be shown to have $M(f_2) = 5$.

We now prove Theorem IV. If $M(f) = 0$ there is nothing to prove. Otherwise, we may suppose, without loss of generality, that

$$M(f) = 1,$$

by considering $tf$ instead of $f$, where $t$ is a suitable number. By the general argument of § 2.2, there is a form $g(\boldsymbol{x}) = g_\varepsilon(\boldsymbol{x})$ equivalent to $f(\boldsymbol{x})$ for which

$$1 \leqq |g(1,0)| < (1-\varepsilon)^{-1},$$

where $\varepsilon$ is any given positive number in the range $0 < \varepsilon < 1$. Put

$$\pm g(1,0) = (1-\eta)^{-1},$$

where

$$0 \leqq \eta = \eta_\varepsilon < \varepsilon < 1.$$

Since the equivalent forms $f$ and $g$ have the same determinant $D$, we may write $g(\boldsymbol{x})$ in the shape

$$\pm g(\boldsymbol{x}) = \frac{(x_1 + \alpha x_2)^2}{1-\eta} - |D|(1-\eta)x_2^2, \tag{5}$$

where $\alpha = \alpha_\varepsilon$ is a real number, which may be supposed to satisfy

$$0 \leqq \alpha \leqq \tfrac{1}{2} \tag{6}$$

on replacing $x_1$ by $\pm x_1 + v x_2$ with a suitable integer $v$. Since $M(f) = 1$, we have either

$$\frac{(u_1 + \alpha_\varepsilon u_2)^2}{1-\eta_\varepsilon} - |D|(1-\eta_\varepsilon)u_2^2 \geqq 1 \tag{7}$$

or

$$\frac{(u_1 + \alpha_\varepsilon u_2)^2}{1-\eta_\varepsilon} - |D|(1-\eta_\varepsilon)u_2^2 \leqq -1 \tag{8}$$

for each pair of integers $u_1, u_2$ not both 0. Of course as $\varepsilon$ changes there is no reason to suppose that for fixed $\boldsymbol{u}$ the same alternative (7) or (8) always holds.

We consider various suitable pairs of integers $u_1, u_2$ and must consider various cases according as (7) or (8) holds for the integers in question. Since we wish to single out the forms (3) and (4), we naturally choose values of $\boldsymbol{u}$ such that $f_0(\boldsymbol{u}) = \pm 1$ or $f_1(\boldsymbol{u}) = \pm 1$.

In the first place, (7) cannot hold with $(u_1, u_2) = (0, 1)$ since by (6) it would imply $|D| < 0$, at least when $\eta$ is small enough. Hence on putting $(u_1, u_2) = (0, 1)$ in (8), we have

$$(1-\eta)^2|D| \geqq (1-\eta) + \alpha^2 \tag{9}$$

for all $\varepsilon$ less than some $\varepsilon_0 > 0$.

We now consider the two possibilities when $(u_1, u_2) = (1, 1)$. Suppose, first, that there are arbitrarily small values of $\varepsilon$ such that (7) holds. For these $\varepsilon$ we have, suppressing the suffix $\varepsilon$, that

$$(1-\eta)^2|D| \leqq -(1-\eta) + (1+\alpha)^2. \tag{10}$$

On eliminating $|D|$ between (9) and (10) we have

$$2\alpha \geqq 1 - 2\eta$$

and so

$$\tfrac{1}{2} - \eta \leqq \alpha \leqq \tfrac{1}{2} \tag{11}$$

by (6). On substituting this in (9) and (10), it follows that $|D|$ can differ from $\frac{5}{4}$ at most by terms of the order of $\eta$. But now $|D|$ is independent of $\eta$ and either $\eta = 0$ or $\eta > 0$ can be made arbitrarily small. Hence $|D| = \frac{5}{4}$. We now revert to one particular $g(\boldsymbol{x}) = g_\varepsilon(\boldsymbol{x})$ for which (10) is true, where now we have the additional information

$|D| = \frac{5}{4}$. On substituting $D = \frac{5}{4}$, $\alpha \geqq \frac{1}{2} - \eta$ in (9), we have

$$\eta^2 - 2\eta \geqq 0.$$

Since $\eta < 2$, this implies $\eta = 0$. Hence $\alpha = \frac{1}{2}$ and

$$\pm g(x_1, x_2) = (x_1 + \tfrac{1}{2} x_2)^2 - \tfrac{5}{4} x_2^2 = f_0(x_1, x_2).$$

Otherwise (10) cannot hold, when $\varepsilon$ is small enough; and so for all $\varepsilon$ less than some $\varepsilon_1 > 0$ we have (8) with $\boldsymbol{u} = (1, 1)$, that is

$$(1-\eta)^2 |D| \geqq (1-\eta) + (1+\alpha)^2. \tag{12}$$

We now consider the possibilities for $\boldsymbol{u} = (-3, 2)$. Note that $f_1(-3, 2) = 1$, where $f_1$ is given by (4). If there are arbitrarily small values of $\varepsilon$ such that (7) holds with $\boldsymbol{u} = (-3, 2)$, then for these $\varepsilon$

$$4(1-\eta)^2 |D| \leqq -(1-\eta) + (-3 + 2\alpha)^2. \tag{13}$$

On eliminating $|D|$ between (12) and (13) we have $4\alpha \leqq \eta$, so $0 \leqq 4\alpha \leqq \eta$ by (6). On substituting in (12) and (13) and using the fact that $\eta = 0$ or $\eta$ can be made arbitrarily small and positive, we find that $|D| = 2$. Finally on putting $|D| = 2$, $\alpha \geqq 0$ in (12) we get $\eta = 0$, so $\alpha = 0$ and

$$\pm g(\boldsymbol{x}) = x_1^2 - 2x_2^2 = f_1(\boldsymbol{x}).$$

Otherwise for all $\varepsilon$ less than some $\varepsilon_2 > 0$ we must have (8) with $\boldsymbol{u} = (-3, 2)$, that is

$$4(1-\eta)^2 |D| \geqq (1-\eta) + (-3 + 2\alpha)^2. \tag{14}$$

But now the right-hand sides of (12) and (14) increase and decrease respectively in $0 \leqq \alpha \leqq \frac{1}{2}$. If $\alpha \leqq \frac{1}{10}$ we use (14) and if $\alpha \geqq \frac{1}{10}$ we use (12). In either case we obtain $|D| \geqq 2.21 + O(\eta)$, so $|D| \geqq 2.21$ since $|D|$ is independent of $\eta$.

It is at first sight remarkable in these proofs that the inequalities obtained show that $\eta = 0$. As already mentioned, this is tied up with the phenomenon of "isolation" which we shall discuss more fully later.

**II.4.3.** We consider now the "one-sided" problem for indefinite binary quadratic forms. In contrast with § 4.2 there is here no set of successive minima. Theorem V A, which we now enunciate, is a special case of Theorem IX of Chapter XI and is due to MAHLER.

THEOREM V. A. *Let*

$$f(\boldsymbol{x}) = f_{11} x_1^2 + 2f_{12} x_1 x_2 + f_{22} x_2^2$$

*be an indefinite quadratic form and*

$$D = f_{11} f_{22} - f_{12}^2.$$

*Then there is an integral vector $\boldsymbol{u} \neq \boldsymbol{o}$ such that*

$$0 < f(\boldsymbol{u}) \leq 2|D|^{\frac{1}{2}}. \tag{1}$$

*The sign of equality is required when and only when $f$ is equivalent to a multiple of*

$$f_0(\boldsymbol{x}) = x_1 x_2.$$

B. *For any $\varepsilon > 0$ there are infinitely many forms, not equivalent to multiples of each other, such that*

$$M_+(f) = \inf_{\substack{f(\boldsymbol{u}) > o \\ \boldsymbol{u} \text{ integral}}} f(\boldsymbol{u}) > (2-\varepsilon)|D|^{\frac{1}{2}}. \tag{2}$$

We first prove A. That $f_0 = x_1 x_2$ is exceptional is obvious, so we need only prove (1) and that equality can occur only when stated. As in § 4.2, we may suppose that

$$M_+(f) = 1,$$

where $M_+(f)$ is defined by (2). Hence, as in § 4.2, there is a form

$$g(\boldsymbol{x}) = \frac{(x_1 + \alpha x_2)^2}{1-\eta} - (1-\eta)|D| x_2^2$$

equivalent to $f$, where

$$0 \leq \alpha \leq \tfrac{1}{2}$$

and $\eta \geq 0$ can be made arbitrarily small *. Suppose, first, that $g(-1, 1) \geq 1$. Then

$$(1-\eta)^2 |D| \leq (1-\alpha)^2 - (1-\eta) \leq \eta,$$

which is impossible if $\eta$ is small enough, since $|D|$ is independent of $\eta$. Hence $g(-1, 1) \leq 0$, that is

$$|D| \geq \frac{(1-\alpha)^2}{(1-\eta)^2} \geq \frac{1}{4},$$

the sign of equality being required only when $\alpha = \frac{1}{2}$, $\eta = 0$; that is when

$$g(\boldsymbol{x}) = (x_1 + \tfrac{1}{2} x_2)^2 - \tfrac{1}{4} x_2^2 = x_1(x_1 + x_2) = f_0(x_1, x_1 + x_2).$$

It remains to prove B. It will be shown in § 4.4 that the forms

$$f_k(x) = k(x_1^2 + x_1 x_2) - x_2^2$$

have

$$M_+(f_k) = k$$

when $k$ is a positive integer. Since

$$|D(f_k)| = \tfrac{1}{4} k^2 + k,$$

* More precisely, we should work with a family of forms $g_\varepsilon(\boldsymbol{x})$ as in § II 4.2. Having once carried out this type of proof in full rigour, in the rest of this chapter we shall be more informal.

the ratio

$$M_+(f_k)/|D(f_k)|^{\frac{1}{2}}$$

may be arbitrarily close to 2.

Another simple proof of B would be by means of continued fractions.

**II.4.4.** As an interpolation between the problems of § 4.2 and 4.3 one may consider the forms $f(\boldsymbol{x})$ such that there is no integral point $\boldsymbol{u} \neq \boldsymbol{o}$ in

$$-a < f(\boldsymbol{u}) < b,$$

where $a$ and $b$ are given positive numbers.

For some values of $a$ and $b$ one may deduce the least possible value of $D(f)$ from the results of § 4.2. For example[1] if

$$a = 1, \quad b = \frac{11}{10}$$

we certainly have

$$M(f) \geqq 1,$$

and so by Theorem IV either

$$|D(f)| \geqq 2$$

or $f$ is equivalent to

$$t(x_1^2 + x_1 x_2 - x_2^2)$$

for some $t$. In the second case it is clearly enough that $t \geqq \frac{11}{10}$. The corresponding determinant is $\left(\frac{11}{10}\right)^2 \cdot \frac{5}{4} < 2$. Hence we have an isolated first minimum. Note that the form with the least $|D|$ does not take any values in the neighbourhood of $-a$.

For any given values of $a$ and $b$ the techniques of §§ 4.2, 4.3 sometimes apply. For example, the minimum determinant when $a = 5$, $b = 3$ is $|D| = 24$ given by $3x_1^2 - 8x_2^2$; this being isolated. The verification of this statement is left to the reader. Here we shall prove only the following theorem due essentially to SEGRE (1945a).

THEOREM VI. *Let*

$$f(\boldsymbol{x}) = f_{11}x_1^2 + 2f_{12}x_1x_2 + f_{22}x_2^2 \tag{1}$$

*have determinant*

$$D(f) = f_{11}f_{22} - f_{12}^2 < 0. \tag{2}$$

*Suppose that there is no integral* $\boldsymbol{u} \neq \boldsymbol{o}$ *such that*

$$-a < f(\boldsymbol{u}) < b, \tag{3}$$

*where* $a > 0$, $b > 0$. *Then*

$$|D| \geqq ab + \tfrac{1}{4}\max(a^2, b^2). \tag{4}$$

[1] This remark was made to the author by Professor C. A. ROGERS.

*If $b>a$, the sign of equality is required when and only when*

$$k=b/a \tag{5}$$

*is an integer and*

$$f(\boldsymbol{x})=a f_k(\boldsymbol{x}), \tag{6}$$

*where*

$$f_k(\boldsymbol{x})=k(x_1^2+x_1x_2)-x_2^2. \tag{7}$$

For $k=1$, Theorem VI is contained in Theorem IV. When $k$ is not an integer, an explicit improvement of (4) can be given. When $k$ is an integer, there is isolation and much more is in fact known [SAWYER (1953a), TORNHEIM (1955a)]. When $b\leqq a$ the cases of equality may, of course, be deduced from the theorem by interchanging $a$ and $b$.

We may suppose without loss of generality that

$$a=1, \quad b=k,$$

where at first $k$ is not necessarily an integer. Let

$$c=M_+(f)=\inf_{f(\boldsymbol{u})>0} f(\boldsymbol{u}),$$

so that

$$c\geqq k.$$

As in § 4.2 there is a form $g(\boldsymbol{x})$ equivalent to $f(\boldsymbol{x})$ of the shape

$$g(\boldsymbol{x})=\frac{c}{1-\eta}(x_1+\alpha x_2)^2-\frac{|D|}{c}(1-\eta)x_2^2,$$

where

$$0\leqq\alpha\leqq\tfrac{1}{2}$$

and $\eta\geqq 0$ may be chosen arbitrarily small.

Clearly $g(0,1)<c$, so $g(0,1)\leqq -1$. Hence $g(1,-1)<c$, and so $g(1,-1)\leqq -1$, that is

$$|D|\geqq\frac{c}{1-\eta}+\frac{c^2}{(1-\eta)^2}(1-\alpha)^2.$$

Hence

$$|D|\geqq c+\tfrac{1}{4}c^2\geqq k+\tfrac{1}{4}k^2$$

with equality only when

$$\eta=0, \quad \alpha=\tfrac{1}{2}, \quad c=k,$$

so

$$g(\boldsymbol{x})=f_k(\boldsymbol{x}).$$

It remains to see whether $f_k(\boldsymbol{x})$ has any integral solutions $\boldsymbol{u}\neq\boldsymbol{o}$ of

$$-1<f_k(\boldsymbol{u})<k. \tag{8}$$

Since $f_k(1, 0) > 0$ but $f_k(1, x_2) \to -\infty$ as $x_2 \to +\infty$, there must be some integer $v \geqq 0$ such that

$$f_k(1, v) \geqq 0 > f_k(1, v+1).$$

If (8) were insoluble, we should have

$$f_k(1, v) \geqq k, \quad f_k(1, v+1) \leqq -1:$$

that is

$$v(k-v) \geqq 0, \quad (v+2)(k-v) \leqq 0.$$

This is possible only when $v = k$, i.e. when $k$ is an integer.

It remains only to show that when $k$ is an integer there is no integral $\boldsymbol{u} \neq \boldsymbol{o}$ such that $-1 < f_k(\boldsymbol{u}) < k$. Since the roots $\vartheta$ of $f_k(\vartheta, 1) = 0$ are irrational, it is impossible that $f_k(\boldsymbol{u}) = 0$. Hence we must deduce a contradiction from

$$0 < f_k(\boldsymbol{u}) < k. \tag{9}$$

If there are several solutions of (9) we choose one for which the integer $|u_1|$ is as small as possible. Clearly

$$u_1 \neq 0.$$

We require the identities

$$\begin{aligned} f_k(\boldsymbol{x}) &= f_k\{(k+1)x_1 - x_2, \ -kx_1 + x_2\} \\ &= f_k\{x_1 + x_2, \ kx_1 + (k+1)x_2\} \\ &= (k+2)x_1\{(k+1)x_1 - x_2\} - \{(k+1)x_1 - x_2\}^2 - x_1^2 \\ &= (k+2)x_1(x_1 + x_2) - (x_1 + x_2)^2 - x_1^2. \end{aligned}$$

Since $f_k(\boldsymbol{u}) > 0$, the last of these identities shows that

$$\begin{aligned} u_1\{(k+1)u_1 - u_2\} &> 0 \\ u_1(u_1 + u_2) \qquad &> 0. \end{aligned}$$

On writing $-\boldsymbol{u}$ for $\boldsymbol{u}$ if necessary, we thus have

$$u_1 > 0, \quad (k+1)u_1 > u_2 > -u_1. \tag{10}$$

From the first two identities and the minimal property of $|u_1|$, we have

$$\begin{aligned} |u_1 + u_2| &\geqq u_1, \\ |(k+1)u_1 - u_2| &\geqq u_1: \end{aligned}$$

and so, by (10),

$$0 \leqq u_2 \leqq k u_1, \quad 0 < u_1.$$

But then

$$f_k(\boldsymbol{u}) = k u_1^2 + u_2(k u_1 - u_2) \geqq k u_1^2 \geqq k.$$

Hence our assumption (9) was false.

By considering $f(1, v)$ for all integers $v$, the estimate (4) may be improved when $k$ is not an integer. Since $f_k$ is the only form $f$ satisfying $f(1,0)=k$ and

$$f(0,1)\leqq -1, \quad f(1,k+1)\leqq -1,$$

$$f(-1,1)\leqq -1, \quad f(1,k)\geqq k,$$

the form $f_k$ gives an isolated first minimum when $k$ is an integer. The proof of these statements is left to the reader (cf. papers quoted at the beginning of § 4.4).

**II.4.5.** We now consider indefinite ternary forms. As already noted (§ 4.1) there is a set of successive minima, the first eleven having been found by VENKOV (1945a). There is a derivation of the first four minima due to OPPENHEIM in DICKSON (1930a) and a neat proof of the first minimum only by DAVENPORT (1947a). Here we shall prove only the following result.

THEOREM VII. *Let*

$$f(\boldsymbol{x}) = \sum f_{ij} x_i x_j \tag{1}$$

*be an indefinite ternary quadratic form with determinant*

$$D(f) = \det(f_{ij}) \neq 0. \tag{2}$$

*Then*

$$M(f) = \inf_{\substack{\boldsymbol{u}\neq\boldsymbol{o}\\ \text{integral}}} |f(\boldsymbol{u})| \leqq |\tfrac{2}{3}D|^{\frac{1}{3}}, \tag{3}$$

*except when $f$ is equivalent to a multiple of*

$$f_0 = x_1^2 + x_1x_2 - x_2^2 - x_2x_3 + x_3^2. \tag{4}$$

*Further,*

$$M(f_0) = 1, \quad D(f_0) = \tfrac{3}{2}. \tag{5}$$

We first prove (5). Since $f_0(\boldsymbol{u})$ is an integer when $\boldsymbol{u}\neq\boldsymbol{o}$ is integral, it is enough to show that $f_0(\boldsymbol{u})\neq 0$. Now

$$4f_0(\boldsymbol{u}) = (2u_1+u_2)^2 + (2u_3-u_2)^2 - 6u_2^2.$$

Hence it is enough to show that there are no integral solutions of

$$v_1^2 + v_3^2 = 6v_2^2$$

other than $v_1=v_2=v_3=0$. We may suppose that $v_1, v_2, v_3$ have no common factor. Then clearly $v_1$ and $v_3$ must be divisible by 3. Then $v_1^2+v_3^2$ must be divisible by 9, so $v_2$ is divisible by 3; a contradiction.

That the constant $\frac{2}{3}$ in (3) cannot be further improved is shown by

$$f_1(\boldsymbol{x}) = x_1^2 + x_1x_2 - x_2^2 - 2x_3^2.$$

The reader should have no difficulty in modifying the proof to show that this is the only case when there is equality in (3) and that it is isolated.

We may suppose as before that

$$M(f) = 1 \tag{6}$$

and, by taking $-f$ for $f$ if necessary, that

$$D < 0. \tag{7}$$

We have to show that $f$ is equivalent to $f_0$ or $D \leqq -\frac{5}{2}$. It is convenient to enunciate steps of the proof as propositions.

PROPOSITION 1. *Either*

$$M_+(f) = \inf_{f(\boldsymbol{u})>0} f(\boldsymbol{u}) = 1 \tag{8}$$

*or*

$$D \leqq -\tfrac{7}{2}. \tag{9}$$

If (6) is true but (8) is false, there must be integral $\boldsymbol{u}$ such that $f(\boldsymbol{u}) = -(1-\eta)^{-1}$, where $\eta \geqq 0$ may be chosen arbitrarily small. Hence $f(\boldsymbol{x})$ is equivalent to a form $g(\boldsymbol{x})$ of the shape

$$(1-\eta)\, g(\boldsymbol{x}) = -(x_1 + \alpha x_2 + \beta x_3)^2 + h(x_2, x_3),$$

where $\alpha, \beta$ are real numbers and the form

$$h(\boldsymbol{x}) = h_{22} x_2^2 + 2h_{23} x_2 x_3 + h_{33} x_3^2$$

must be positive definite. The determinant of $h(\boldsymbol{x})$ is

$$h_{22} h_{33} - h_{23}^2 = -(1-\eta)^3 D = (1-\eta)^3 |D|.$$

After a transformation on the variables $x_2$, $x_3$, we may suppose that $h(\boldsymbol{x})$ is reduced; and so

$$h_{22}^2 \leqq \tfrac{4}{3}(1-\eta)^3 |D| \tag{10}$$

by Theorem II.

We now consider the indefinite binary form

$$G(x_1, x_2) = (1-\eta)\, g(x_1, x_2, 0) = -(x_1 + \alpha x_2)^2 + h_{22} x_2^2,$$

of determinant $-h_{22}$. Clearly

$$M(G) \geqq (1-\eta)\, M(g) = 1 - \eta.$$

Hence, by Theorem IV, either

$$h_{22} \geqq \frac{221}{100}(1-\eta)^2 \tag{11}$$

or $G(x_1, x_2)$ is equivalent to one of $t(x_1^2 + x_1 x_2 - x_2^2)$ or $t(x_1 - 2x_2^2)$ for some number $t$ with $|t| \geqq (1-\eta)$. If the second alternative holds, we must have $t = -1$, since $G(1, 0) = -1$. Then there are integral $u_1, u_2$ such that $G(u) = +1$, i.e. $g(u_1, u_2, 0) = (1-\eta)^{-1}$, so

$$M_+(f) = M_+(g) = 1$$

since $\eta \geqq 0$ may be chosen arbitrarily small. Otherwise the first alternative, namely (11), holds, and so, by (10),

$$|D| \geqq (1-\eta) \cdot \frac{3}{4} \cdot \left(\frac{221}{100}\right)^2 > \frac{7}{2}.$$

This proves the proposition.

We may now suppose that

$$M_+(f) = 1. \tag{12}$$

As before, there is a form $g$ equivalent to $f$ such that

$$(1-\eta)\, g(\boldsymbol{x}) = (x_1 + \alpha x_2 + \beta x_3)^2 + h(x_2, x_3),$$

where $\eta \geqq 0$ may be chosen arbitrarily small, and the form

$$h(x_2, x_3) = h_{22} x_2^2 + 2h_{23} x_2 x_3 + h_{33} x_3^2 \tag{13}$$

is now indefinite and has determinant

$$h_{22} h_{33} - h_{23}^2 = (1-\eta)^3 D < 0. \tag{14}$$

PROPOSITION 2. *If $u_2, u_3$ are integers not both 0, then either*

$$h(u_2, u_3) \geqq \tfrac{3}{4} - \eta, \tag{15}$$

*or*

$$h(u_2, u_3) \leqq -2 + \eta, \tag{16}$$

*or*

$$-\tfrac{5}{4} - \eta \leqq h(u_2, u_3) \leqq -\tfrac{5}{4} + \eta. \tag{17}$$

*Further, if (17) holds there is an integer $v$ such that*

$$|v + \tfrac{1}{2} - (\alpha u_2 + \beta u_3)| \leqq \tfrac{1}{2}\eta. \tag{18}$$

We must first show that there are no integral solutions $\boldsymbol{u} \neq 0$ of

$$-2 + \eta < h(u_2, u_3) < -\tfrac{5}{4} - \eta,$$
$$-\tfrac{5}{4} + \eta < h(u_2, u_3) \leqq -1 + \eta,$$
$$-1 + \eta < h(u_2, u_3) < \tfrac{3}{4} - \eta.$$

We may clearly choose the integer $u_1$ so that respectively

$$1 \leqq |u_1 + \alpha u_2 + \beta u_3| \leqq \tfrac{3}{2},$$
$$\tfrac{1}{2} \leqq |u_1 + \alpha u_2 + \beta u_3| \leqq 1,$$

and

$$0 \leqq |u_1 + \alpha u_2 + \beta u_3| \leqq \tfrac{1}{2}.$$

Then in each case we have

$$(1-\eta)\,|g(u)| < 1 - \eta,$$

contrary to hypothesis.

Suppose that (17) holds. There is an integer $t$ and a real number $\tau$ such that by choice of sign

$$\alpha u_2 + \beta u_3 = t \pm \tau, \quad 0 \leqq \tau \leqq \tfrac{1}{2}.$$

We may clearly choose integers $u_1', u_1''$ so that

$$|u_1' + \alpha u_2 + \beta u_3| = 1 - \tau$$

$$|u_1'' + \alpha u_2 + \beta u_3| = 1 + \tau.$$

Then

$$-\eta + g(u_1', u_2, u_3) \leqq 0 \leqq g(u_1'', u_2, u_3) + \eta,$$

and so

$$h(u_2, u_3) + (1 - \tau)^2 = g(u_1', u_2, u_3) \leqq -1 + \eta, \tag{19}$$

$$h(u_2, u_3) + (1 + \tau)^2 = g(u_1'', u_2, u_3) \geqq 1 - \eta. \tag{20}$$

By subtracting (19) from (20) we have

$$\tfrac{1}{2}(1 - \eta) \leqq \tau \leqq \tfrac{1}{2}.$$

This is equivalent to (18) and so proves the proposition.

COROLLARY. *If* (17) *holds, then* $u_2$ *and* $u_3$ *cannot have a common factor except* $\pm 1$.

For if $u_2 = v u_2'$, $u_3 = v u_3'$, where $v > 1$, none of (15), (16) or (17) would be satisfied by $h(u_2', u_3')$.

PROPOSITION 3. *Either*

$$|D| \geqq \tfrac{5}{2} \tag{21}$$

*or, after an equivalence transformation, we may suppose that*

$$-\tfrac{5}{4} - \eta \leqq h(1, 0) \leqq -\tfrac{5}{4} + \eta, \tag{22}$$

$$h(1, -1) \geqq \tfrac{3}{4} - \eta, \tag{23}$$

$$-\tfrac{5}{4} - \eta \leqq h(1, 1) \leqq -\tfrac{5}{4} + \eta, \tag{24}$$

$$h(2, -1) \leqq -2 + \eta, \tag{25}$$

$$|\alpha - \tfrac{1}{2}| \leqq \tfrac{1}{2}\eta, \tag{26}$$

$$|\beta| \leqq \eta \tag{27}$$

*provided that* $\eta$ *is less than some absolute constant* $\eta_0 > 0$.

Suppose, first, that there are no solutions of

$$-2+\eta<h(u_2,u_3)<\tfrac{3}{4}-\eta.$$

Then, by SEGRE's Theorem VI, we must have

$$|h_{22}h_{33}-h_{23}^2|\geqq\tfrac{1}{4}(2-\eta)^2+(2-\eta)(\tfrac{3}{4}-\eta)=\tfrac{5}{4}(2-\eta)(1-\eta).$$

Hence, by (14),

$$|D|\geqq\frac{5(2-\eta)}{4(1-\eta)^2}\geqq\frac{5}{2}.$$

Otherwise by Proposition 2 there is a solution of $|h(u_2,u_3)+\tfrac{5}{4}|\leqq\eta$ and by Proposition 2, Corollary we may suppose, after a suitable transformation on $x_2, x_3$, that

$$-\tfrac{5}{4}-\eta\leqq h(1,0)=h_{22}\leqq-\tfrac{5}{4}+\eta. \tag{28}$$

After a further substitution of the type $x_2\rightarrow\pm x_2+vx_3$, where $v$ is an integer, we may suppose further that

$$0\geqq 2h_{23}\geqq h_{22}\geqq-\tfrac{5}{4}-\eta. \tag{29}$$

We now consider $h(u_2,u_3)$ for various choices of $u_2, u_3$. If $h(0,1)\leqq-\tfrac{5}{4}+\eta$; that is $h_{33}\leqq-\tfrac{5}{4}+\eta$, we should have

$$h_{22}h_{33}-h_{23}^2>0,$$

contrary to the assumption that $h$ is an indefinite form. Hence $h_{33}>-\tfrac{5}{4}+\eta$, and so, by Proposition 2,

$$h_{33}=h(0,1)\geqq\tfrac{3}{4}-\eta.$$

But now, by (29),

$$h(1,-1)\geqq h_{22}+h_{33}>-\tfrac{5}{4}+\eta,$$

and so, by Proposition 2 again,

$$h_{22}-2h_{23}+h_{33}=h(1,-1)\geqq\tfrac{3}{4}-\eta. \tag{30}$$

Hence

$$h(1,1)=h(1,-1)+4h_{23}>-2+\eta$$

by (29).

We now consider the two remaining possibilities for $h(1,1)$ allowed by Proposition 2. Suppose, first, that

$$h(1,1)=h_{22}+2h_{23}+h_{33}\geqq\tfrac{3}{4}-\eta,$$

so

$$h_{33}\geqq\tfrac{3}{4}-\eta-h_{22}\geqq 2-2\eta.$$

Then, by (14),

$$(1-\eta)^3|D|=h_{23}^2-h_{22}h_{33}\geqq-h_{22}h_{33},$$

so

$$|D| \geqq \frac{(1-\frac{4}{8}\eta)}{(1-\eta)^2} \cdot \frac{5}{2} \geqq \frac{5}{2},$$

which is all we require. We may therefore suppose that

$$-\tfrac{5}{4} - \eta \leqq h(1,1) = h_{22} + 2h_{23} + h_{33} \leqq -\tfrac{5}{4} + \eta.$$

We now invoke the part of Proposition 2 referring to $\alpha$ and $\beta$ with $(u_2, u_3) = (1, 0)$ and $(1, 1)$. Hence there are integers $v'$ and $v''$ such that

$$|v' + \tfrac{1}{2} - \alpha| \leqq \tfrac{1}{2}\eta,$$

$$|v'' + \tfrac{1}{2} - (\alpha + \beta)| \leqq \tfrac{1}{2}\eta.$$

After a substitution of $x_1 + v' x_2 + (v'' - v') x_3$ for $x_1$ we may suppose indeed that

$$|\tfrac{1}{2} - \alpha| \leqq \tfrac{1}{2}\eta,$$

$$|\tfrac{1}{2} - (\alpha + \beta)| \leqq \tfrac{1}{2}\eta.$$

Then

$$|\beta| \leqq \eta.$$

We now consider

$$h(2, -1) = h(1,1) + 3h_{22} - 6h_{23} \leqq h(1,1) \leqq -\tfrac{5}{4} + \eta.$$

We cannot have $h(2, -1) \geqq -\frac{5}{4} - \eta$, since then by Proposition 2 the fractional part of $2\alpha - \beta$ would be about $\frac{1}{2}$, while we know that $2\alpha - \beta$ is $1 + O(\eta)$. Hence

$$4h_{22} - 4h_{23} + h_{33} = h(2, -1) \leqq -2 + \eta.$$

This completes the proof of the assertions of Proposition 3.

We now conclude the proof of the theorem. The inequalities (22) to (25) of Proposition 3 are linear inequalities in $h_{22}, h_{23}, h_{33}$. Put

$$h_{22} = -\tfrac{5}{4} + \lambda\eta, \tag{31}$$

$$h_{23} = -\tfrac{1}{2} + \mu\eta, \tag{32}$$

$$h_{33} = 1 + \nu\eta. \tag{33}$$

Then (22) to (25) become

$$|\lambda| \leqq 1, \tag{34}$$

$$\lambda - 2\mu + \nu \geqq -1, \tag{35}$$

$$|\lambda + 2\mu + \nu| \leqq 1, \tag{36}$$

$$4\lambda - 4\mu + \nu \leqq 1. \tag{37}$$

Hence

$$2\nu = (\lambda - 2\mu + \nu) + (\lambda + 2\mu + \nu) - 2\lambda \geqq -4,$$

$$3\nu = 4\lambda - 4\mu + \nu + 2(\lambda + 2\mu + \nu) - 6\lambda \leqq 9,$$

so

$$|\nu| \leqq 3.$$

Hence

$$|\mu| \leqq 3,$$

by (36). Hence and by (14),

$$(1-\eta)^3 |D| = h_{23}^2 - h_{22} h_{33} = \tfrac{3}{2} + (-\lambda - \mu + \tfrac{5}{4}\nu)\eta + O(\eta^2). \quad (38)$$
$$= \tfrac{3}{2} + O(\eta)$$

But $D$ is independent of $\eta$, so*

$$|D| = \tfrac{3}{2}.$$

Suppose, if possible, that $\eta \neq 0$. On putting $|D| = \frac{3}{2}$ in (38) we have

$$-\lambda - \mu + \tfrac{5}{4}\nu = -\tfrac{9}{2} + O(\eta).$$

For small enough $\eta$ this contradicts (34), (35) and (36), since they give

$$-\lambda - \mu + \tfrac{5}{4}\nu = -\tfrac{9}{4}\lambda + \tfrac{7}{8}(\lambda - 2\mu + \nu) + \tfrac{3}{8}(\lambda + 2\mu + \nu)$$
$$\geqq -\tfrac{9}{4} - \tfrac{7}{8} - \tfrac{3}{8} = -\tfrac{7}{2}.$$

Hence $\eta = 0$, so by (13), (26), (27), (31), (32), (33), we have

$$g(\boldsymbol{x}) = (x_1 + \tfrac{1}{2}x_2)^2 - \tfrac{5}{4}x_2^2 - x_2 x_3 + x_3^2 = f_0(\boldsymbol{x}).$$

Since $g(\boldsymbol{x})$ is equivalent to $f(\boldsymbol{x})$, this concludes the proof of Theorem VII.

**II.5. Binary cubic forms.** We must first consider briefly the algebra associated with a binary cubic form

$$f(x_1, x_2) = a x_1^3 + b x_1^2 x_2 + c x_1 x_2^2 + d x_2^3. \quad (1)$$

Such a form may always be split up into linear factors with real or complex coefficients:

$$f(x_1, x_2) = \prod_{1 \leqq j \leqq 3} (\vartheta_j x_1 + \psi_j x_2). \quad (2)$$

With the form is associated the discriminant

$$D(f) = \prod_{1 \leqq j < k \leqq 3} \{\vartheta_j \psi_k - \vartheta_k \psi_j\}^2. \quad (3)$$

It is easily verified that

$$D(f) = 18abcd + b^2c^2 - 4ac^3 - 4db^3 - 27a^2d^2 \quad (4)$$

(see § 5.2). From (3) it follows that $D(f) = 0$ if and only if $f(x_1, x_2)$ has a repeated linear factor. Forms $f$ with $D(f) = 0$ are called singular.

The discriminant $D(f)$ is an invariant of the cubic, in the sense that if

$$f'(x_1, x_2) = f(\alpha x_1 + \beta x_2, \gamma x_1 + \delta x_2) \quad (5)$$

---

* More precisely, we should have worked with a family of forms $g_\varepsilon(\boldsymbol{x})$ as in § II 4.2, each form with its own $\eta = \eta_\varepsilon$ and $0 \leqq \eta < \varepsilon$. Then $\lambda, \mu, \nu$ depend on $\varepsilon$, but (38) is true for all sufficiently small $\varepsilon$.

identically for some numbers $\alpha, \beta, \gamma, \delta$, then

$$D(f) = (\alpha\delta - \beta\gamma)^6 D(f), \tag{6}$$

as follows at once from (3) and the fact that

$$f'(x_1, x_2) = \prod_j (\vartheta_j' x_1 + \psi_j' x_2), \tag{7}$$

where

$$\vartheta_j' = \alpha\vartheta_j + \gamma\psi_j, \quad \psi_j' = \beta\vartheta_j + \delta\psi_j. \tag{8}$$

In particular, $D(f') = D(f)$ if $f$ and $f'$ are equivalent, since then (5) holds for some integers $\alpha, \beta, \gamma, \delta$ with $\alpha\delta - \beta\gamma = \pm 1$.

If $a, b, c, d$ are real, then either all the ratios $\psi_j/\vartheta_j$ are real or two of them are conjugate complex and the third is real, since roots $\xi$ of an equation $f(\xi, 1) = 0$ with real coefficients occur in complex conjugate pairs. This subdivides the real non-singular binary cubic forms into two essentially distinct types. We show now that two forms in the same type may be transformed into each other by a transformation of the type (7) with real $\alpha, \beta, \gamma, \delta$. It is enough to show that all forms $f$ of a given type may be transformed into an $f'$ which is fixed for the type. We may suppose without loss of generality that either

$$\vartheta_j, \psi_j \quad \text{are all real} \tag{$9_1$}$$

or

$$\vartheta_3, \psi_3 \quad \text{are real, and} \quad \vartheta_2 = \overline{\vartheta}_1, \ \psi_2 = \overline{\psi}_1 \tag{$9_2$}$$

in our two respective cases, where the bar denotes the complex conjugate. Clearly these two cases are characterised by $D > 0$ and $D < 0$ respectively. There exist numbers $\lambda_1, \lambda_2, \lambda_3$ not all 0 such that

$$\left.\begin{aligned} \lambda_1\psi_1 + \lambda_2\psi_2 + \lambda_3\psi_3 &= 0 \\ \lambda_1\vartheta_1 + \lambda_2\vartheta_2 + \lambda_3\vartheta_3 &= 0. \end{aligned}\right\} \tag{10}$$

If, say, $\lambda_3 = 0$, we should have

$$\vartheta_1\psi_2 - \vartheta_2\psi_1 = 0,$$

and so $D(f) = 0$ by (3), contrary to the hypothesis that $f$ is non-singular. Hence $\lambda_1\lambda_2\lambda_3 \neq 0$ and we may suppose, without loss of generality, by multiplying $\lambda_1, \lambda_2, \lambda_3$ by a common factor, that

$$\lambda_1\lambda_2\lambda_3 = 1. \tag{11}$$

We now distinguish the two cases according as ($9_1$) or ($9_2$) holds. If ($9_1$) holds, we may suppose that $\lambda_1, \lambda_2, \lambda_3$ are real and put

$$X_j = -\lambda_j(\vartheta_j x_1 + \psi_j x_2) \qquad (j = 1, 2).$$

Then

$$\lambda_3(\vartheta_3 x_1 + \psi_3 x_2) = X_1 + X_2,$$

and so, by (11),

$$f(x_1, x_2) = X_1 X_2 (X_1 + X_2). \tag{12}$$

If $(9_2)$ holds, we may suppose that $\lambda_2 = \bar{\lambda}_1$, $\lambda_3 = \bar{\lambda}_3$ and put

$$\left.\begin{aligned} \varrho X_1 + \varrho^2 X_2 &= \lambda_1(\vartheta_1 x_1 + \psi_1 x_2) \\ \varrho^2 X_1 + \varrho X_2 &= \lambda_2(\vartheta_2 x_1 + \psi_2 x_2), \end{aligned}\right\} \tag{13}$$

where $\varrho$ is a complex cube root of 1. Then, by (10),

$$X_1 + X_2 = \lambda_3(\vartheta_3 x_1 + \psi_3 x_2) \tag{14}$$

and

$$f(x_1, x_2) = X_1^3 + X_2^3. \tag{15}$$

The coefficients $\alpha, \beta, \gamma, \delta$ in

$$X_1 = \alpha x_1 + \beta x_2, \quad X_2 = \gamma x_1 + \delta x_2$$

are real, since the two equations (13) here are complex conjugates one of the other.

In the sense of § 4 of Chapter I the values taken by non-singular binary cubic forms are the values taken by the function

$$\varphi(\boldsymbol{X}) = X_1 X_2 (X_1 + X_2)$$

or

$$\varphi(\boldsymbol{X}) = X_1^3 + X_2^3$$

at the points of a lattice. The reader will have no difficulty in verifying that there is a corresponding result for singular cubic forms, with

$$\varphi(\boldsymbol{X}) = X_1^2 X_2,$$
$$\varphi(\boldsymbol{X}) = X_1^3,$$

according as only two or all three of the linear forms $\vartheta_j x_1 + \psi_j x_2$ are multiples of each other.

It was first shown by MORDELL (1943b) that if $f$ is a real cubic form, then there is an integer vector $\boldsymbol{u} \neq \boldsymbol{o}$ such that

$$|f(\boldsymbol{u})| \leq \left\{\begin{matrix} \left|\dfrac{D}{49}\right|^{\frac{1}{4}}, \\ \left|\dfrac{D}{23}\right|^{\frac{1}{4}}, \\ \varepsilon, \end{matrix}\right\} \tag{16}$$

according as $D > 0$, $D < 0$ or $D = 0$, where $\varepsilon$ is an arbitrarily small positive number. The third case, when $f(\boldsymbol{x})$ is singular, may be dealt with trivially by MINKOWSKI's theorem of the next chapter, so we do not discuss it here. That the coefficients 49, 23 are best possible in their respective cases is shown by the binary cubic forms

$$x_1^3 + x_1^2 x_2 - 2x_1 x_2^2 - x_2^3 \tag{17}$$

and

$$x_1^3 - x_1 x_2^2 - x_2^3. \tag{18}$$

These have discriminants 49 and 23 respectively. Since they do not represent 0 and represent integer values for integer vectors $\boldsymbol{u}$, the $\leqq$ in (16) cannot be replaced by $<$. It will be shown that $<$ may be taken in (16) for all forms not equivalent to (17) and (18).

The results (16) were not first obtained by reduction arguments. DAVENPORT (1945a, b) has however given simple proofs by such arguments.

This treatment consists in defining a binary cubic form as being reduced if a certain definite quadratic form associated with it is reduced: it is necessary to choose different quadratic forms according as $D>0$ or $D<0$. DAVENPORT then shows for a reduced form that either (16) is true with strict inequality for one of a prescribed set of $\boldsymbol{u}$, or $f(\boldsymbol{x})$ is one of the forms (17), (18). We give the proof for $D>0$ in full but only sketch that for $D<0$ since we shall later be using the case $D<0$ to illustrate another technique.

It was shown by DAVENPORT (1941b) that neither the 49 nor the 23 is isolated. We do not give the proof, which depends essentially on the fact that although a cubic form $f(\boldsymbol{x})$ is always indefinite the area of the region

$$|f(\boldsymbol{x})| < 1$$

is finite, and the forms (17), (18) take the values $\pm 1$ only a finite number of times: in contrast to the situation with indefinite quadratic forms.

**II.5.2.** In order to enunciate DAVENPORT's result we must first introduce a quadratic form associated with a cubic form

$$f(x_1, x_2) = a x_1^3 + b x_1^2 x_2 + c x_1 x_2^2 + d x_2^3 \tag{1}$$

$$= \prod_{1 \leqq j \leqq 3} (\vartheta_j x_1 + \psi_j x_2), \tag{2}$$

namely the hessian

$$h(x_1, x_2) = \frac{1}{4}\left\{\left(\frac{\partial^2 f}{\partial x_1 \partial x_2}\right)^2 - \frac{\partial^2 f}{\partial x_1^2}\,\frac{\partial^2 f}{\partial x_2^2}\right\} \tag{3}$$

$$= A x_1^2 + B x_1 x_2 + C x_2^2, \tag{4}$$

where

$$A = b^2 - 3ac, \quad B = bc - 9ad, \quad C = c^2 - 3bd. \tag{5}$$

On evaluating the partial differentials by (2), a brief calculation shows that

$$h(x_1, x_2) = \sum (\vartheta_2\psi_3 - \vartheta_3\psi_2)^2 (\vartheta_1 x_1 + \psi_1 x_2)^2, \tag{6}$$

the sum being taken over all cyclic permutations of 1, 2, 3.

We now show that the hessian is a covariant of the form $f(x_1, x_2)$; that is if $\alpha, \beta, \gamma, \delta$ are real numbers with

$$\alpha\delta - \beta\gamma = \pm 1, \tag{7}$$

then the hessian of the form $f'(x_1, x_2)$ defined by

$$f'(x_1, x_2) = f(\alpha x_1 + \beta x_2, \gamma x_1 + \delta x_2)$$

is

$$h'(x_1, x_2) = h(\alpha x_1 + \beta x_2, \gamma x_1 + \delta x_2).$$

Indeed this follows at once from (6) and the expressions (7), (8) of § 5.1, on noting that

$$\vartheta'_j\psi'_k - \vartheta'_k\psi'_j = (\alpha\delta - \beta\gamma)(\vartheta_j\psi_k - \vartheta_k\psi_j) = \pm(\vartheta_j\psi_k - \vartheta_k\psi_j),$$

on using (7).

From either (5) or (6) we see that the determinant of $h(x_1, x_2)$ is

$$AC - \tfrac{1}{4}B^2 = \tfrac{3}{4}D(f). \tag{8}$$

In particular, $h(x_1, x_2)$ is definite when and only when $D > 0$, i.e. when $f$ is a product of three real linear forms [when the $\vartheta_j, \psi_j$ are real the form (6) is clearly positive definite, but the converse is not so clear without using (8)].

When the $\vartheta_j, \psi_j$ are real, the form $f$ was said by HERMITE to be reduced when the definite quadratic form $h$ is reduced in the sense of MINKOWSKI[1].

Every form with real $\vartheta_j, \psi_j$ is equivalent to a reduced form. For the transformation which reduces the $h(\boldsymbol{x})$ in MINKOWSKI's sense also reduces $f(\boldsymbol{x})$; since $h(\boldsymbol{x})$ is a covariant of $f(\boldsymbol{x})$, as we have seen. Further, this reduction can be carried out in only a finite number of ways since we saw that a definite quadratic form can be reduced by only a finite number of transformations.

**II.5.3.** We may now enunciate and prove DAVENPORT's theorem:

THEOREM VIII. *Let $f(\boldsymbol{x})$ be a binary cubic form with discriminant $D > 0$ which is reduced in the sense of* HERMITE *(§ 5.2). Then*

$$\min\{|f(1,0)|, |f(0,1)|, |f(1,1)|, |f(1,-1)|\} \leqq \left(\frac{D}{49}\right)^{\frac{1}{4}}. \tag{1}$$

[1] He could not put it this way, of course!

*The sign of equality is needed only when*

$$\pm f(x_1, \pm x_2) = x_1^3 + x_1^2 x_2 - 2 x_1 x_2^2 - x_2^3 \tag{2}$$

*or*

$$\pm f(x_1, \pm x_2) = x_1^3 + 2 x_1^2 x_2 - x_1 x_2^2 - x_2^3, \tag{3}$$

*the $\pm$ signs being independent.*

DAVENPORT actually proved that if $f(x_1, x_2)$ is reduced, then at least one of the five products

$$|f(1,0)\,f(0,1)|, \quad |f(1,0)\,f(1,1)|, \quad |f(0,1)\,f(1,1)|,$$
$$|f(1,0)\,f(1,-1)|, \quad |f(0,1)\,f(1,-1)|$$

is $\leqq (D/49)^{\frac{1}{2}}$, with equality only for the forms (2) and (3), as before. We shall follow CHALK (1949) and prove another generalisation. Let

$$h(x_1, x_2) = A\,x_1^2 + B\,x_1 x_2 + C\,x_2^2$$

be the hessian of $f(\boldsymbol{x})$, so that

$$0 \leqq B \leqq A \leqq C, \quad A > 0 \tag{4}$$

CHALK'S result is that

$$\min\{|f(1,0)|,\ |f(0,1)|,\ |f(1,1)|,\ |f(1,-1)|\} \leqq \left(\frac{A}{7}\right)^{\frac{1}{2}},$$

with equality only for the forms (2) and (3). Since $4AC - B^2 \geqq 3A^2$, and $4AC - B^2 = 3D(f)$ by (8) of § 5.2, this will be a stronger result than Theorem VIII.

We may suppose by homogeneity that

$$A = 7. \tag{5}$$

We must then deduce a contradiction from

$$|f(1,0)| \geqq 1, \quad |f(0,1)| \geqq 1, \quad |f(1,1)| \geqq 1, \quad |f(1,-1)| \geqq 1,$$

except for the forms (2) and (3). On writing

$$f(\boldsymbol{x}) = a\,x_1^3 + b\,x_1^2 x_2 + c\,x_1 x_2^2 + d\,x_2^3$$

these inequalities are

$$|a| \geqq 1, \quad |d| \geqq 1, \tag{6}$$

$$|a+b+c+d| \geqq 1, \quad |a-b+c-d| \geqq 1. \tag{7}$$

By taking $-f$ for $f$ we may suppose that

$$a \geqq 1. \tag{8}$$

We shall require the identities

$$A = b^2 - 3ac, \quad B = bc - 9ad, \quad C = c^2 - 3bd, \tag{9}$$

from which follow

$$Bc - Cb = 3Ad, \quad Bb - Ac = 3Ca. \tag{10}$$

From (4), (5) and (9) we have

$$0 \leqq bc - 9ad \leqq 7. \tag{11}$$

Suppose, if possible, that $d > 0$. Then (11) gives

$$bc \geqq 9ad \geqq 9. \tag{12}$$

If $b \geqq c > 0$ there is a contradiction with $(10_1)$ and if $c \geqq b > 0$ there is a contradiction with $(10_2)$, so

$$b < 0, \quad c < 0.$$

Then we should have

$$\begin{aligned} A &= b^2 - 3ac \\ &= |b|^2 + \tfrac{3}{2}|ac| + \tfrac{3}{2}|ac| \\ &\geqq 3\left(\tfrac{9}{4}a^2b^2c^2\right)^{\frac{1}{3}} \end{aligned}$$

by the inequality of the arithmetic and geometric means; and so, by (12),

$$A \geqq 3\left(\frac{729}{4}\right)^{\frac{1}{3}} > 7$$

in contradiction with the normalization $A = 7$.

Hence we may suppose that

$$d < 0,$$

and so, by (11),

$$bc \leqq 7 - 9a|d| \leqq -2. \tag{13}$$

If $b < 0 < c$ we have a contradiction with $(10_2)$, so

$$c < 0 < b,$$

and (13) becomes

$$b|c| \geqq 9a|d| - 7 \geqq 2. \tag{14}$$

Further, (5) becomes

$$7 = A = b^2 + 3a|c| \geqq b^2 + 3|c|. \tag{15}$$

On substituting (14) in (15), we have

$$7 \geqq b^2 + 3|c| \geqq b^2 + 6/b,$$

and so

$$1 \leqq b \leqq 2. \tag{16}$$

Similarly we have

$$7 \geqq \frac{4}{c^2} + 3|c|,$$

and so

$$1 \leqq -c \leqq 2. \tag{17}$$

Clearly a sign of equality can hold in (16) or (17) only if

$$a = -d = 1, \quad bc = -2. \tag{18}$$

From (14), (16) and (17) we now have

$$9a|d| \leqq 7 + |bc| \leqq 11$$

and so

$$a \leqq \frac{11}{9}, \quad |d| \leqq \frac{11}{9}.$$

But now

$$a - b + c - d \leqq \frac{11}{9} - 1 - 1 + \frac{11}{9} < 1,$$

and so

$$a - b + c - d \leqq -1. \tag{19}$$

We now consider the two possibilities for $f(1, 1)$. If

$$a + b + c + d \leqq -1, \tag{20}$$

then on adding (19) and (20) we have

$$a - |c| \leqq -1, \quad \text{so} \quad |c| \geqq 1 + a \geqq 2.$$

Comparison with (17) shows that $|c| = 2$, and, since there is equality in (17), we must have (18); that is

$$a = -d = 1, \quad b = 1, \quad c = -2.$$

Similarly, if

$$a + b + c + d \geqq +1,$$

then

$$b + d \geqq +1, \quad \text{so} \quad b \geqq 2;$$

and we have

$$a = -d = 1, \quad b = 2, \quad c = -1.$$

This concludes the proof of the theorem.

**II.5.4.** When the binary cubic form $f$ has discriminant $D(f) < 0$ the hessian form is indefinite, and so a reduction of the hessian does not single out a finite number of reduced forms from amongst the forms equivalent to $f$. However, if $D < 0$ then only one of the linear factors of $f$ is real, and $f$ may be put in the shape

$$f(x_1, x_2) = (\vartheta_3 x_1 + \psi_3 x_2)(P x_1^2 + Q x_1 x_2 + R x_2^2), \tag{1}$$

where the form $P x_1^2 + Q x_1 x_2 + R x_2^2$ is positive definite, since it is the product of two conjugate forms with complex coefficients. DAVENPORT following earlier workers calls such a form reduced if the quadratic form

$$P x_1^2 + Q x_1 x_2 + R x_2^2$$

is MINKOWSKI-reduced, that is

$$|Q| \leqq P \leqq R, \tag{2}$$

and, further,

$$\vartheta_3 \psi_3 \geqq 0. \tag{3}$$

The last condition may be achieved by changing the sign of $x_2$ if need be, which does not affect (2). DAVENPORT (1945b) proves

THEOREM IX. *If $f(\boldsymbol{x})$ is binary cubic form with discriminant $D(f) < 0$, then there are integers $\boldsymbol{u} \neq \boldsymbol{o}$ such that*

$$|f(\boldsymbol{u})| \leqq \left|\frac{D}{23}\right|^{\frac{1}{4}}.$$

*If, further, $f(\boldsymbol{x})$ is reduced, then*

$$\min[|f(1,0)|, |f(0,1)|, |f(1,-1)|, |f(1,-2)|] \leqq \left|\frac{D}{23}\right|^{\frac{1}{4}},$$

*with equality only when*

$$f(x_1, x_2) = a(x_1^3 + x_1^2 x_2 + 2 x_1 x_2^2 + x_2^3).$$

We only sketch the proof and refer to the original memoire for the details. We later give another proof of the first paragraph of the theorem (Chapter III, Theorem VII).

We have to show that $D(f) \leqq -23$ when

$$|f(1,0)| \geqq 1, \qquad |f(0,1)| \geqq 1,$$
$$|f(1,-1)| \geqq 1, \qquad |f(1,-2)| \geqq 1,$$

i.e. when

$$P|\vartheta_3| \geqq 1, \qquad R|\psi_3| \geqq 1, \tag{$4_1$}$$

$$|\vartheta_3 - \psi_3|\,(P - Q + R) \geqq 1, \tag{$4_2$}$$

$$|\vartheta_3 - 2\psi_3|\,(P - 2Q + 4R) \geqq 1, \tag{$4_3$}$$

since $P - Q + R$, $P - 2Q + 4R$ are positive by the positive definiteness of the quadratic form. For fixed $\vartheta_3$ and $\psi_3$, the inequalities (2) and (4) restrict the point $P, Q, R$ in 3-dimensional euclidean space to lie in a certain infinite region $\mathscr{S}$ bounded by planes. DAVENPORT shows, further, that

$$-D(f) = \{P\psi_3^2 - Q\vartheta_3\psi_3 + R\psi_3^2\}^2 (4PR - Q^2),$$

and that $|D(f)|^{\frac{1}{2}}$ is a convex function of $(P, Q, R)$ for fixed $\vartheta_3, \psi_3$. Hence the maximum of $D(f)$ is attained at the vertices of $\mathcal{S}$, where three of the plane faces meet [since it is easily seen that $|D| \to \infty$ as $\max(|P|, |Q|, |R|) \to \infty$]. The proof then follows from a rather tricky estimation of $D(f)$ at the vertices of $\mathcal{S}$.

**II.6. Other forms.** We briefly survey here results on the reduction of forms other than those already discussed.

**II.6.2.** For binary forms of degree $n \geqq 4$ there is more than one invariant. For example, a binary quartic form $f(x_1, x_2)$ which is the product of two pairs of complex conjugate linear forms may be reduced to the shape

$$\varphi(\mathbf{X}) = \varphi(X_1, X_2) = X_1^4 + 6\mu X_1^2 X_2^2 + X_2^4,$$

where

$$X_1 = \alpha x_1 + \beta x_2, \quad X_2 = \gamma x_1 + \delta x_2,$$

for some real $\alpha, \beta, \gamma, \delta$ and $\mu = \mu(f)$ is a real number lying in

$$|\mu| < \tfrac{1}{3}.$$

Two forms with different $\mu$ cannot be transformed into each other by a homogeneous linear transformation of the variables. Further, $\mu(f)$ is an absolute invariant in the sense that $\mu(tf) = \mu(f)$, where $t$ is any number. Of course we still also have the discriminant

$$D(f) = \prod_{1 \leqq j < k < 4} (\vartheta_j \psi_k - \vartheta_k \psi_j)^2,$$

where

$$f(x_1, x_2) = \prod_j (\vartheta_j x_1 + \psi_j x_2).$$

The problem for definite binary quartics was solved independently by Davis (1951a) and Černý (1952a) in the sense that they found the best possible function $\gamma(\mu)$ of $\mu$ such that every form $f$ with invariant $\mu$ has

$$\inf_{\substack{\mathbf{u} \neq \mathbf{o} \\ \text{integral}}} f(\mathbf{u}) \leqq \gamma(\mu) \{D(f)\}^{\frac{1}{6}}.$$

Davis (1951a) also gives some results for indefinite binary quartic and full references to earlier work. It is no longer true, as it was for quadratic and cubic forms, that forms $f$ with $D(f) = 0$ assume arbitrarily small values. This case was completely elucidated by Davenport (1950a).

The methods of these authors combines reduction techniques with other tools drawn from the geometry of numbers.

There does not seem to be any systematic work on binary forms of degree greater than 4.

**II.6.3.** The only other types of forms $f(x_1, \ldots, x_m)$ of degree $n$ with $m>2$, $n>2$ for which the best estimate of

$$M(f) = \inf_{\substack{u \neq o \\ \text{integral}}} |f(u)|$$

is known appear to be the ternary cubic forms with real coefficients which are expressible as the product of three real linear forms:

$$f(x_1, x_2, x_3) = \prod_{1 \leq j \leq 3} (\vartheta_{j1} x_1 + \vartheta_{j2} x_2 + \vartheta_{j3} x_3),$$

where either all the $\vartheta_{jk}$ are real (first type) or $\vartheta_{31}, \vartheta_{32}, \vartheta_{33}$ are real and $\vartheta_{2k} = \bar{\vartheta}_{1k}$ $(1 \leq k \leq 3)$. There is an invariant

$$D(f) = \left\{\det_{j,k} (\vartheta_{jk})\right\}^2.$$

This is the only invariant in each type, since there are obvious real transformations taking $f$ into

$$X_1 X_2 X_3$$

and

$$X_1(X_2^2 + X_3^2),$$

respectively. The two types are distinguished by $D>0$ and $D<0$ respectively. The following two results are known:

THEOREM X. *Let $f(x_1, x_2, x_3)$ be a factorisable ternary cubic form with $D(f)>0$. Then there exist integers $u \neq o$ such that*

$$|f(u)| < \frac{D^{\frac{1}{2}}}{9.1},$$

*except when $f$ is equivalent to a multiple of one of the forms*

$$f_{49} = x_1^3 + x_2^3 + x_3^3 - x_1^2 x_2 + 5 x_1^2 x_3 - 2 x_1 x_2^2 + 6 x_1 x_3^2 - 2 x_2 x_3^2 - x_2^2 x_3 - x_1 x_2 x_3,$$

$$f_{81} = x_1^3 + x_2^3 + x_3^3 + 6 x_1^2 x_3 - 3 x_1 x_2^2 + 9 x_1 x_2^2 - 3 x_2 x_3^2 - 3 x_1 x_2 x_3,$$

*for which $M(f) = 1$ and $D(f) = 49, 81$ respectively.*

THEOREM XI. *Let $f(x)$ be a factorisable ternary cubic form with $D(f)<0$. Then there exist integers $u \neq o$ such that*

$$|f(u)| \leq \left|\frac{D}{23}\right|^{\frac{1}{2}}.$$

*The sign of equality is needed when and only when $f(x)$ is equivalent to a multiple of the form*

$$f_{23} = x_1^3 + x_2^3 + x_3^3 + 2 x_1^2 x_3 - x_1 x_2^2 + x_1 x_3^2 - x_2 x_3^2 - 3 x_1 x_2 x_3.$$

We note that $f_{49}, f_{81}$ and $f_{23}$ are all of the shape

$$\mathrm{Norm}(x_1 + \varphi x_2 + \psi x_3),$$

where $1, \varphi, \psi$ are a basis for the integers of a cubic field. We shall discuss later the reasons why this might have been expected (Chapter X). For $f_{49}, f_{81}$ and $f_{23}$ we have $\psi = \varphi^2$, and $\varphi$ satisfies the respective equations:

$$\varphi^3 + \varphi^2 - 2\varphi - 1 = 0,$$

$$\varphi^3 - 3\varphi - 1 = 0,$$

and

$$\varphi^3 - \varphi - 1 = 0.$$

[By Norm is meant the product of the three forms obtained from the given one by inserting the three pairs of conjugate values for $\varphi$ and $\psi$.] The first equation here corresponds in an obvious way to the form in Theorem VIII. The third equation here corresponds to the binary form

$$x_1^3 - x_1 x_2^2 - x_2^3$$

which is equivalent to that in Theorem IX on making the substitution $x_1 \to x_1,\ x_2 \to -x_1 - x_2$.

For $D > 0$ Theorem X gives the first two successive minima and shows that the second minimum is isolated. The first minimum in Theorem XI is not isolated; but there is a weaker sense in which it is isolated [DAVENPORT and ROGERS (1950a, especially Theorem 14): see also Chapter X]. Theorem X was obtained by DAVENPORT (1943a). He had already obtained the first minimum [DAVENPORT (1938a) and a simpler proof in DAVENPORT (1941a)]. A slightly weaker form of Theorem XI in which $|D/23|^{\frac{1}{3}} + \varepsilon$ with arbitrarily small $\varepsilon > 0$ appears instead of $|D/23|^{\frac{1}{3}}$ was given by DAVENPORT (1943a); the full form is in DAVENPORT and ROGERS (1950a). CHALK and ROGERS (1951a) showed that every factorisable ternary cubic form with $D > 0$ is either equivalent to a multiple of $f$ or to a form $g(\boldsymbol{x})$ with

$$|g(1,0,0)\, g(0,1,0)\, g(0,0,1)| \leqq \left(\frac{D^{\frac{1}{2}}}{7.1}\right)^3.$$

This is analogue of the results about the products of the diagonal terms of definite quadratic forms obtained in § 3.

We do not prove Theorems X and XI here, since in Chapter X, following MORDELL, we deduce Theorems X, XI from the corresponding results for binary cubics (in which, as the reader will have noticed, the integers 49 and 23 also occur). It is however worth sketching the reduction which DAVENPORT use to prove Theorem X:

Let $f(\boldsymbol{x})$ be a factorisable ternary cubic with $D > 0$, where we may suppose, without loss of generality for our purpose that $M(f) = 1$. Hence $f$ is equivalent to a form $g$ such that

$$g(1,0,0) = (1-\eta)^{-1},$$

where $\eta \geqq 0$ is arbitrarily small. Hence we may write

$$(1-\eta)\, g(\boldsymbol{x}) = \prod_{j=1}^{3} (x_1 + \alpha_j x_2 + \beta_j x_3).$$

We consider also the quadratic form

$$h(\boldsymbol{x}) = \sum_j (x_1 + \alpha_j x_2 + \beta_j x_3)^2.$$

From the inequality of the arithmetic and geometric means $h(\boldsymbol{u}) \geqq 3(1-\eta)^{\frac{2}{3}}$ for all integers $\boldsymbol{u} \neq \boldsymbol{o}$, and it is easy to verify that in fact $h(\boldsymbol{u}) \geqq 3$, with equality only when $\boldsymbol{u} = (1, 0, 0)$. Hence $h(\boldsymbol{x})$ may be reduced in the sense of MINKOWSKI by a transformation of the type

$$\begin{aligned} x_1 &\to x_1 + v_{12} x_2 + v_{13} x_3 \\ x_2 &\to \phantom{x_1 +{}} v_{22} x_2 + v_{23} x_3 \\ x_3 &\to \phantom{x_1 +{}} v_{32} x_2 + v_{33} x_3 \end{aligned}$$

where the $v_{ij}$ are integers and $v_{22}v_{33} - v_{23}v_{32} = \pm 1$. Since $h(\boldsymbol{x})$ has determinant $(1-\eta)^6 D(f)$ and is reduced, we have bounds for the coefficients. The proof now continues by an intricate and delicate chain of computations using these bounds and the fact that $|g(\boldsymbol{u})| \geqq 1$ for all integers $\boldsymbol{u} \neq \boldsymbol{o}$.

DAVENPORT'S treatment of Theorem XI starts off with a similar reduction but the completion of the proof requires different ideas and the detailed consideration of an intractable 2-dimensional figure.

**II.6.4.** The corresponding problem for the product of $n > 3$ homogeneous forms in $n$ variables has been much worked on. Estimates but no precise results are known, and these estimates were obtained by other methods. We shall consider the case of large $n$ in Chapter IX, § 8. The best estimates for $n = 4, 5$ in print appear to be those of ŽILINSKAS (1941a) and GODWIN (1950a) respectively; but GODWIN refers to a better estimate for $n = 4$, presumably the Vienna dissertation of G. BÖHM (1942) also mentioned in KELLER'S encyclopedia article [KELLER (1954a)] but unavailable to me.

There is however a striking result of CHALK on the product of the values taken by $n$ linear forms when these values are positive. He shows that if $L_1, \ldots, L_n$ are $n$ linear forms in $n$ variables $\boldsymbol{x} = (x_1, \ldots, x_n)$ with determinant $\Delta \neq 0$, then there exist integers $\boldsymbol{u} \neq \boldsymbol{o}$ such that

$$L_j(\boldsymbol{u}) > 0 \qquad (1 \leqq j \leqq n), \tag{1}$$

$$\prod_j L_j(\boldsymbol{u}) \leqq |\Delta|. \tag{2}$$

That the implied constant 1 on the right-hand side of (2) is the best possible is shown by the simple example $L_j = x_j$. CHALK'S theorem is indeed more general than the form given here since it refers to the product of inhomogeneous linear forms. Consequently we do not prove it here, but later in Chapter XI, § 4.

## Chapter III

# Theorems of BLICHFELDT and MINKOWSKI

**III.1. Introduction.** The whole of the geometry of numbers may be said to have sprung from MINKOWSKI'S convex body theorem. In its crudest sense this says that if a point set $\mathscr{S}$ in $n$-dimensional euclidean space is symmetric about the origin (i.e. contains $-\boldsymbol{x}$ when it contains $\boldsymbol{x}$) and convex [i.e. contains the whole line-segment

$$\lambda \boldsymbol{x} + (1-\lambda)\boldsymbol{y} \qquad (0 \leqq \lambda \leqq 1)$$

when it contains $\boldsymbol{x}$ and $\boldsymbol{y}$] and has volume $V > 2^n$, then it contains an integral point $\boldsymbol{u}$ other than the origin. In this way we have a link between the "geometrical" properties of a set — convexity, symmetry and volume — and an "arithmetical" property, namely the existence of an integral point in $\mathscr{S}$. Another form of the same theorem, which is more general only in appearance, states that if $\Lambda$ is a lattice of determinant $d(\Lambda)$ and $\mathscr{S}$ is convex and symmetric about the origin, as before, then $\mathscr{S}$ contains a point of $\Lambda$ other than the origin, provided that the volume $V$ of $\mathscr{S}$ is greater than $2^n d(\Lambda)$. In § 2 we shall prove MINKOWSKI'S theorem and some refinements. We shall not follow MINKOWSKI'S own proof but deduce his theorem from one of BLICHFELDT, which has important applications of its own and which is intuitively practically obvious: if a point set $\mathscr{R}$ has volume strictly greater than $d(\Lambda)$ then it contains two distinct points $\boldsymbol{x}_1$ and $\boldsymbol{x}_2$ whose difference $\boldsymbol{x}_1 - \boldsymbol{x}_2$ belongs to $\Lambda$.

The theorems of BLICHFELDT and MINKOWSKI may be regarded as statements about the characteristic functions of a set $\mathscr{S}$, that is the function $\chi(\boldsymbol{x})$ which is 1 if $\boldsymbol{x} \in \mathscr{S}$ but otherwise 0. There are generalisations of the theorems of BLICHFELDT and MINKOWSKI to non-negative functions $\psi(\boldsymbol{x})$ due to SIEGEL and RADO. These we present in § 3. We do not in fact use these theorems later.

In § 4 we use MINKOWSKI'S theorem to obtain a characterisation of a lattice which is independent of the notion of a basis: a lattice is any set of points $\Lambda$ in $n$-dimensional space which (i) contains $n$ linearly independent vectors, (ii) is a group under addition, i.e. if $\boldsymbol{x}$ and $\boldsymbol{y}$ are in $\Lambda$ so are $\boldsymbol{x} \pm \boldsymbol{y}$, and (iii) has only the origin in some sphere $x_1^2 + \cdots + x_n^2 < \eta^2$, where $\eta > 0$.

In § 5 we introduce the notion of the lattice constant $\Delta(\mathscr{S})$ of a set $\mathscr{S}$. This is a number with the property that every lattice $\Lambda$ with $d(\Lambda) < \Delta(\mathscr{S})$ has a point other than $\boldsymbol{o}$ in $\mathscr{S}$, while there are lattices whose determinant $d(\Lambda)$ is arbitrarily near to $\Delta(\mathscr{S})$ with no other point than $\boldsymbol{o}$ in $\mathscr{S}$. In § 6 we discuss at length a method due to MORDELL which uses MINKOWSKI's convex body theorem to evaluate or estimate $\Delta(\mathscr{S})$ for sets which may or may not be convex. The idea is, roughly speaking, to show that if a lattice $\Lambda$ of given determinant $d(\Lambda)=\Delta_0$ has no points except $\boldsymbol{o}$ in $\mathscr{S}$, then at least $\Lambda$ must have points in various sets abutting on $\mathscr{S}$. Since these points belong to $\Lambda$, so do linear combinations of them. These combinations must be either $\boldsymbol{o}$ or lie outside $\mathscr{S}$. In this way more and more information about these points of $\Lambda$ near $\mathscr{S}$ is obtained, until there is a contradiction; the contradiction showing that every lattice $\Lambda$ with determinant $d(\Lambda)=\Delta_0$ has a point in $\mathscr{S}$. This method is particularly effective in 2 dimensions, since the relationship of the various points to each other then springs to the eye. Consequently in § 6.2 we give a series of simple lemmas about 2-dimensional lattices which are non-the-less useful tools. MORDELL's method is applied, amongst other things, to finding $\Delta(\mathscr{S})$ when $\mathscr{S}$ is the region

$$|X_1^3+X_2^3|<1. \tag{1}$$

This is equivalent to finding the lower bound of the values taken by a binary cubic form with negative discriminant. This question was discussed but not answered in Chapter II. The proof given here is a conflation of several given by MORDELL. It uses essentially the algebraic background. We remark in passing that MORDELL (1946a) has shown that the result obtained generalizes to all regions which look sufficiently like (1). Similarly, BAMBAH (1951a) has proved a result to show that all sets which look sufficiently like

$$|X_1X_2(X_1+X_2)|<1 \tag{2}$$

do, in fact behave like (2). The set (2) corresponds to binary cubic forms with positive discriminant in the same way as (1) does to those with negative discriminant. For example BAMBAH's result applies to regions $\mathscr{S}$ with hexagonal symmetry and six asymptotes at angles $\pi/3$, the set of points between two asymptotes which do not belong to $\mathscr{S}$ being convex. Compare Chapter X, § 3.3.

Finally, in § 7 we use MINKOWSKI's theorem to obtain some results about the representations of numbers by quadratic forms; for example that every prime $p=4m+1$ can be expressed as the sum of the squares of two integers; $p=u_1^2+u_2^2$. This is all rather aside from the main theme of the book but the proofs are so elementary and so striking that they deserve to be better known.

**III.1.2.** It is convenient to introduce here some important definitions and notions.

The length of a vector $\boldsymbol{x}=(x_1,\ldots,x_n)$, namely

$$(x_1^2+\cdots+x_n^2)^{\frac{1}{2}}$$

will, as usual, be denoted by

$$|\boldsymbol{x}|.$$

It satisfies the "triangle inequality"

$$|\boldsymbol{x}+\boldsymbol{y}| \leqq |\boldsymbol{x}| + |\boldsymbol{y}|$$

for all vectors $\boldsymbol{x}$ and $\boldsymbol{y}$. The length of a vector is not an invariant under all unimodular transformations, unlike most of the concepts we work with, but we shall be concerned only with the topology induced by the metric $|\boldsymbol{x}|$ and not the metric itself. Let

$$y_i = \sum \alpha_{ij} x_j \qquad (1 \leqq i,\ j \leqq n) \tag{1}$$

be a real transformation of determinant

$$\det(\alpha_{ij}) \neq 0. \tag{2}$$

Clearly

$$|\boldsymbol{y}|^2 = \sum_i \Big(\sum_j \alpha_{ij} x_j\Big)^2 \leqq n^3 A^2 \sum x_j^2 = n^3 A^2 |\boldsymbol{x}|^2,$$

where

$$A = \max |\alpha_{ij}|.$$

Since $\det(\alpha_{ij}) \neq 0$, we may solve (1) for the $x_j$ and obtain, say,

$$x_i = \sum_j \beta_{ij} y_j. \tag{3}$$

Then similarly

$$|\boldsymbol{x}|^2 \leqq n^3 B^2 |\boldsymbol{y}|^2,$$

where

$$B = \max |\beta_{ij}|.$$

Hence there exist constants $c_1, c_2$ independent of $\boldsymbol{x}$ and $\boldsymbol{y}$ such that[1]

$$0 < c_1 \leqq \frac{|\boldsymbol{x}|}{|\boldsymbol{y}|} \leqq c_2 < \infty. \tag{4}$$

We shall often make use of the following consequences without explicit reference.

LEMMA 1. *Let $\Lambda$ be a lattice in n-dimensional space. Then there exist constants $\eta_1, \eta_2$ depending only on $\Lambda$ with the following properties*

*(i) If $\boldsymbol{u} \in \Lambda$, $\boldsymbol{v} \in \Lambda$ and $|\boldsymbol{u}-\boldsymbol{v}| < \eta_1$, then $\boldsymbol{u}$ and $\boldsymbol{v}$ are identical:*

*(ii) The number $N(R)$ of points of $\Lambda$ in a sphere $|\boldsymbol{x}| < R$ is at most $\eta_2(R^n+1)$.*

Both of these statements are trivially true for the lattice $\Lambda_0$ of points with integer coordinates. But now (cf. § 3 of Chapter I) if $\Lambda$

[1] This is a particular case of a result to be proved later (Chapter IV, Lemma 2 Corollary).

is any lattice with basis

$$\boldsymbol{b}_j = (\beta_{1j}, \ldots, \beta_{nj}) \qquad (1 \leq j \leq n),$$

then the points of $\Lambda$ are just the points (3) with $\boldsymbol{y} \in \Lambda_0$. The truth of (i), (ii) in general now follows at once from (4) and the truth of (i), (ii) for $\Lambda_0$.

**III.1.3.** We say that a sequence of vectors $\boldsymbol{x}_r$ $(r = 1, 2, \ldots)$ converges to the vector $\boldsymbol{x}'$ as limit if

$$\lim |\boldsymbol{x}_r - \boldsymbol{x}'| = 0$$

in the usual sense. Clearly a necessary and sufficient condition for this is that the co-ordinates of $\boldsymbol{x}_r$ should converge to the corresponding co-ordinates of $\boldsymbol{x}'$, since clearly

$$\max |x_j| \leq |\boldsymbol{x}| \leq n^{\frac{1}{2}} \max |x_j|$$

for any vector $\boldsymbol{x} = (x_1, \ldots, x_n)$. An immediate consequence of Lemma 1 (i) is that a sequence of vectors $\boldsymbol{u}_r$ of a lattice $\Lambda$ can converge only if $\boldsymbol{u}_r$ is the same for all sufficiently large $r$, say

$$\boldsymbol{u}_r = \boldsymbol{u}' \qquad (\text{all } r \geq r_0).$$

A set $\mathscr{S}$ of points is said to be compact if every sequence of points $\boldsymbol{x}_r \in \mathscr{S}$ contains a subsequence $\boldsymbol{y}_s = \boldsymbol{x}_{r_s}$ $(r_1 < r_2 < \cdots)$ which converges to a limit in $\mathscr{S}$:

$$\lim_{s \to \infty} \boldsymbol{y}_s = \boldsymbol{y}' \in \mathscr{S}.$$

A classical theorem of WEIERSTRASS states that a set $\mathscr{S}$ in $n$-dimensional euclidean space is compact if and only if it is both bounded (i.e. contained in a sphere $|\boldsymbol{x}| < R$ for some sufficiently large $R$) and closed (i.e. if $\boldsymbol{x}_r \in \mathscr{S}$ $(1 \leq r < \infty)$ and $\boldsymbol{x}' = \lim \boldsymbol{x}_r$ exists, then $\boldsymbol{x}' \in \mathscr{S}$).

For the sake of completeness we give a proof of WEIERSTRASS's theorem. Suppose first that $\mathscr{S}$ is a compact set. If $\mathscr{S}$ were unbounded, we could find a sequence of points $\boldsymbol{x}_r \in \mathscr{S}$ such that $|\boldsymbol{x}_r| \to \infty$, and then it clearly cannot contain a convergent subsequence. Hence a compact set $\mathscr{S}$ is bounded. If $\mathscr{S}$ were not closed, we could find a sequence of points $\boldsymbol{x}_r \in \mathscr{S}$ such that $\lim \boldsymbol{x}_r = \boldsymbol{x}'$ is not in $\mathscr{S}$. Clearly every subsequence of the original sequence tends to $\boldsymbol{x}'$. Hence a compact set $\mathscr{S}$ is closed. Now let $\mathscr{S}$ be a set which is both bounded and closed. We shall show that $\mathscr{S}$ is compact. Let $\boldsymbol{x}_r$ $(1 \leq r < \infty)$ be a sequence of points of $\mathscr{S}$. We may suppose that originally all the $\boldsymbol{x}_r$ are contained in a $n$-dimensional cube $\mathscr{C}_0$ of side $2R$ for some $R$. This cube may be dissected into $2^n$ cubes of side $R$ by taking planes through the centre of $\mathscr{C}_0$ parallel to the faces. For definiteness we take the cubes of side $\mathscr{C}_0$ to be closed, that is to include their boundary points. At least one of the cubes of side $R$ must contain $\boldsymbol{x}_r$ for infinitely many $r$. Let $\mathscr{C}_1$ be one of these. On repeating the original process with $\mathscr{C}_1$ instead of $\mathscr{C}_0$ we obtain a cube $\mathscr{C}_2$ of side $\frac{1}{2}R$ contained in $\mathscr{C}_1$ which contains $\boldsymbol{x}_r$ for infinitely many $r$. And so on. In this way we obtain a sequence of cubes $\mathscr{C}_s$ $(0 \leq s < \infty)$ of side

$2^{1-s}R$, such that $\mathscr{C}_{s+1}$ is contained in $\mathscr{C}_s$. Each $\mathscr{C}_s$ contains $\boldsymbol{x}_r$ for infinitely many $r$. The cubes $\mathscr{C}_s$ define a point $\boldsymbol{x}'$ which is contained in all of them. We may now find a subsequence $\boldsymbol{x}_{r_s}$ tending to $\boldsymbol{x}'$ as follows: $\boldsymbol{x}_{r_1}$ is any point of the original sequence in $\mathscr{C}_0$: if $r_1, \ldots, r_s$ have already been fixed with

$$r_1 < r_2 < \cdots < r_s,$$

then $r_{s+1}$ is any one of the infinitely many indices $r > r_s$ such that $\boldsymbol{x}_r$ is in $\mathscr{C}_s$. Finally, since

$$\boldsymbol{x}' = \lim_{s\to\infty} \boldsymbol{x}_{r_s},$$

the point $\boldsymbol{x}'$ is in $\mathscr{S}$, since $\mathscr{S}$ is assumed closed.

There is a form of WEIERSTRASS' Theorem which is apparently more general. Let

$$\boldsymbol{x}_{kr} \quad (1 \leqq k \leqq m,\ 1 \leqq r < \infty)$$

be a sequence of sets $\mathsf{A}_r$ of $m$ points $\boldsymbol{x}_{kr}$ in a compact set $\mathscr{S}$. Then there is a increasing sequence $r_1 < r_2 < \cdots$ of integers such that all the limits

$$\lim_{s\to\infty} \boldsymbol{x}_{k r_s}$$

exist and are in $\mathscr{S}$. For if

$$\boldsymbol{x}_{kr} = (x_{1kr}, \ldots, x_{nkr}),$$

the sets $\mathsf{A}_r$ may be represent by points $\boldsymbol{X}_r$ with coordinates $x_{jkr}$ $(1 \leqq j \leqq n,\ 1 \leqq k \leqq m)$ in $nm$-dimensional space. Clearly the set $\mathscr{S}_m$ of points $\boldsymbol{X} = (x_{jk})$ with

$$(x_{1k}, \ldots, x_{nk}) \in \mathscr{S} \qquad (1 \leqq k \leqq n)$$

is bounded and closed if $\mathscr{S}$ is. Hence the points $\boldsymbol{X}_r$ have a convergent subsequence $\boldsymbol{X}_{r_s}$. Then the $r_s$ clearly do what is required.

[Alternatively one could make use of the so-called diagonal process. First pick out a subsequence

$$\mathsf{A}_{r_s} = \mathsf{B}_s = (\boldsymbol{y}_{1s}, \ldots, \boldsymbol{y}_{ms})$$

of the $\mathsf{A}_r$ such that $\boldsymbol{y}_{1s}$ is convergent. Then pick out a subsequence $\mathsf{C}_t = (\boldsymbol{z}_{1t}, \ldots, \boldsymbol{z}_{mt})$ of the $\mathsf{B}_s$ such that $\boldsymbol{z}_{2t}$ is convergent. The sequence $\boldsymbol{z}_{1t}$ is also convergent, being a subsequence of the convergent sequence $\boldsymbol{y}_{1s}$. And so on. After $m$ repetitions of the process one obtains the required subsequence.]

**III.1.4.** By volume we shall mean in this book LEBESGUE measure unless the contrary is stated. We shall however have no need of any of the more recondite properties of measure; the sets we shall be mainly concerned with have a volume by any definition, for example the interiors of cubes or ellipsoids.

**III.2. BLICHFELDT's and MINKOWSKI's theorems.** We use the notation and results of Chapter I. To BLICHFELDT is due the realization that the following almost intuitive result forms a basis for a great portion of the geometry of numbers [BLICHFELDT (1914a)].

THEOREM I. *Let $m$ be a positive integer, $\Lambda$ a lattice with determinant $d(\Lambda)$, and $\mathscr{S}$ a point-set of volume $V(\mathscr{S})$, possibly $V(\mathscr{S})=\infty$. Suppose that* **either**

$$V(\mathscr{S}) > m\,d(\Lambda), \tag{1}$$

**or**

$$V(\mathscr{S}) = m\,d(\Lambda) \tag{2}$$

*and $\mathscr{S}$ is compact. Then there exist $m+1$ distinct points $\boldsymbol{x}_1, \ldots, \boldsymbol{x}_{m+1}$ of $\mathscr{S}$ such that the differences $\boldsymbol{x}_i - \boldsymbol{x}_j$ are all in $\Lambda$.*

Let $\boldsymbol{b}_1, \ldots, \boldsymbol{b}_n$ be any basis of $\Lambda$ and let $\mathscr{P}$ be the generalized parallelopiped of points

$$y_1\boldsymbol{b}_1 + \cdots + y_n\boldsymbol{b}_n \qquad (0 \leqq y_j < 1,\ 1 \leqq j \leqq n).$$

Then $\mathscr{P}$ has volume

$$V(\mathscr{P}) = |\det(\boldsymbol{b}_1, \ldots, \boldsymbol{b}_n)| = d(\Lambda). \tag{3}$$

Every point $\boldsymbol{x}$ in space may be put in the shape

$$\boldsymbol{x} = \boldsymbol{u} + \boldsymbol{v}, \quad \boldsymbol{u} \in \Lambda, \quad \boldsymbol{v} \in \mathscr{P},$$

and this expression is unique, since the points of $\Lambda$ are just the $y_1\boldsymbol{b}_1 + \cdots + y_n\boldsymbol{b}_n$, where $y_1, \ldots, y_n$ are integers.

This parallelopiped $\mathscr{P}$ will play an important part later (Chapter VII), where it will be called a fundamental parallelopiped for $\Lambda$.

For each $\boldsymbol{u} \in \Lambda$ let $\mathscr{R}(\boldsymbol{u})$ be the set of points $\boldsymbol{v}$ such that

$$\boldsymbol{v} \in \mathscr{P}, \quad \boldsymbol{v} + \boldsymbol{u} \in \mathscr{S}.$$

Clearly the corresponding volumes $V\{\mathscr{R}(\boldsymbol{u})\}$ satisfy

$$\sum_{\boldsymbol{u}} V\{\mathscr{R}(\boldsymbol{u})\} = V(\mathscr{S}). \tag{4}$$

Suppose now that the first alternative holds, namely $V(\mathscr{S}) > m\,d(\Lambda)$, so that (4) implies

$$\sum_{\boldsymbol{u}} V\{\mathscr{R}(\boldsymbol{u})\} > m\,d(\Lambda) = m\,V(\mathscr{P}).$$

Since the $\mathscr{R}(\boldsymbol{u})$ are all contained in $\mathscr{P}$, there must be at least one point $\boldsymbol{v}_0 \in \mathscr{P}$ which belongs to at least $m+1$ of the $\mathscr{R}(\boldsymbol{u})$, say

$$\boldsymbol{v}_0 \in \mathscr{R}(\boldsymbol{u}_j) \quad (1 \leqq j \leqq m+1),$$

where the $\boldsymbol{u}_j$ are distinct. Then the points

$$\boldsymbol{x}_j = \boldsymbol{v}_0 + \boldsymbol{u}_j$$

are in $\mathscr{S}$ by the definition of $\mathscr{R}(\boldsymbol{u})$, and

$$\boldsymbol{x}_i - \boldsymbol{x}_j = \boldsymbol{u}_i - \boldsymbol{u}_j \begin{cases} \in \Lambda \\ \neq \boldsymbol{o} \quad (i \neq j). \end{cases}$$

This proves the theorem for the first alternative.

Suppose now that the second alternative holds. Let $\varepsilon_r$ $(1 \leqq r < \infty)$ be a sequence of positive numbers and

$$\lim \varepsilon_r = 0.$$

For each $r$, the set $(1+\varepsilon_r)\mathscr{S}$ of points $(1+\varepsilon_r)\boldsymbol{x}$, $\boldsymbol{x} \in \mathscr{S}$ clearly has volume

$$(1+\varepsilon_r)^n V(\mathscr{S}) > V(\mathscr{S}) = m\, d(\Lambda).$$

Hence, by what we have already proved, there exist points

$$\boldsymbol{x}_{jr} \in (1+\varepsilon_r)\mathscr{S} \qquad (1 \leqq j \leqq m+1)$$

such that

$$\left.\boldsymbol{u}_r(i,j) \quad (\text{say}) = \boldsymbol{x}_{ir} - \boldsymbol{x}_{jr} \begin{cases} \in \Lambda \\ \neq \boldsymbol{o} \quad (i \neq j). \end{cases}\right\} \tag{5}$$

By extracting suitable subsequences of the original sequences, and then calling them $\varepsilon_r$, $\boldsymbol{x}_{jr}$ again to avoid introducing new notation, we may suppose, without loss of generality, that

$$\lim_{r\to\infty} \boldsymbol{x}_{jr} = \boldsymbol{x}'_j \qquad (1 \leqq j \leqq m+1)$$

all exist. Since $\mathscr{S}$ is now assumed to be compact, the $\boldsymbol{x}'_j$ are in $\mathscr{S}$. Then, by (5),

$$\boldsymbol{x}'_i - \boldsymbol{x}'_j = \lim_{r\to\infty} \boldsymbol{u}_r(i,j).$$

But now the $\boldsymbol{u}_r(i,j)$ are in $\Lambda$. Hence (cf. § 1.3) $\boldsymbol{u}_r(i,j)$ is independent of $r$ from some stage onwards:

$$\boldsymbol{u}_r(i,j) = \boldsymbol{u}'(i,j) \qquad (r \geqq r_0).$$

Hence

$$\left.\boldsymbol{x}'_i - \boldsymbol{x}'_j = \boldsymbol{u}'(i,j) \begin{cases} \in \Lambda \\ \neq \boldsymbol{o} \quad (i \neq j), \end{cases}\right\},$$

as required.

For later reference (Chapter VII) we note that in the proof for the first alternative we have implicitly proved the following:

COROLLARY. *Let $\mathscr{S}$ be any set of points and let $\mathscr{S}_1$ be the set of points $\boldsymbol{v}$ of the fundamental parallelopiped which can be put in the shape*

$$\boldsymbol{v} = \boldsymbol{x} - \boldsymbol{u}, \qquad \boldsymbol{x} \in \mathscr{S}, \ \boldsymbol{u} \in \Lambda.$$

*Then*

$$V(\mathscr{S}_1) \leqq V(\mathscr{S}).$$

*If no difference* $\boldsymbol{x}_1 - \boldsymbol{x}_2$ *between distinct points of* $\mathscr{S}$ *belongs to* $\Lambda$ *then*

$$V(\mathscr{S}_1) = V(\mathscr{S}).$$

The first paragraph is clear. The second follows since then no two $\mathscr{R}(\boldsymbol{u})$ overlap.

**III.2.2.** From Theorem I we deduce almost at once the following theorem which is due, at least[1] for $m=1$ to MINKOWSKI ("MINKOWSKI'S convex body theorem").

THEOREM II. *Let* $\mathscr{S}$ *be a point set of volume* $V(\mathscr{S})$ *(possibly infinite) which is symmetric*[2] *about the origin and convex*[2]. *Let* $m$ *be an integer and let* $\Lambda$ *be a lattice of determinant* $d(\Lambda)$. *Suppose that* **either**

$$V(\mathscr{S}) > m\, 2^n d(\Lambda),$$

**or**

$$V(\mathscr{S}) = m\, 2^n d(\Lambda)$$

*and* $\mathscr{S}$ *is compact. Then* $\mathscr{S}$ *contains at least* $m$ *pairs of points* $\pm \boldsymbol{u}_j$ $(1 \leqq j \leqq m)$ *which are distinct from each other and from* $\boldsymbol{o}$.

Again we note that the possibility of infinite volume is not excluded.

Theorem I applies to the set $\frac{1}{2}\mathscr{S}$ of points $\frac{1}{2}\boldsymbol{x}$, $\boldsymbol{x} \in \mathscr{S}$ which has volume $2^{-n}V(\mathscr{S})$. Hence there exist $m+1$ distinct points

$$\tfrac{1}{2}\boldsymbol{x}_j \in \tfrac{1}{2}\mathscr{S} \qquad (1 \leqq j \leqq m+1),$$

such that

$$\tfrac{1}{2}\boldsymbol{x}_i - \tfrac{1}{2}\boldsymbol{x}_j \begin{Bmatrix} \in \Lambda \\ \neq \boldsymbol{o} \quad (i \neq j) \end{Bmatrix}.$$

We introduce an ordering of the real vectors and write

$$\boldsymbol{x}_1 > \boldsymbol{x}_2$$

if the first non-zero component of $\boldsymbol{x}_1 - \boldsymbol{x}_2$ is positive. We may suppose without loss of generality that

$$\boldsymbol{x}_1 > \boldsymbol{x}_2 > \cdots > \boldsymbol{x}_{m+1}.$$

Put

$$\boldsymbol{u}_j = \tfrac{1}{2}\boldsymbol{x}_j - \tfrac{1}{2}\boldsymbol{x}_{m+1}.$$

Then clearly

$$\boldsymbol{o}, \pm \boldsymbol{u}_1, \ldots, \pm \boldsymbol{u}_m$$

---

[1] The general case is apparently due to VAN DER CORPUT (1936a).

[2] For the definition of these terms see § 1.1.

are all distinct. But $-\boldsymbol{x}_{m+1} \in \mathscr{S}$ since $\boldsymbol{x}_{m+1} \in \mathscr{S}$ and $\mathscr{S}$ is symmetric. Hence

$$\boldsymbol{u}_j = \tfrac{1}{2}\boldsymbol{x}_j + \tfrac{1}{2}(-\boldsymbol{x}_{m+1}) \in \mathscr{S}$$

by the convexity of $\mathscr{S}$. This proves the theorem.

For later use we note the

COROLLARY. *Let $\mathscr{S}$ be symmetric about the origin and convex. A necessary and sufficient condition that $\mathscr{S}$ contain a point of $\Lambda$ other than $\boldsymbol{o}$ is that there exist two distinct points $\frac{1}{2}\boldsymbol{x}_1$, $\frac{1}{2}\boldsymbol{x}_2 \in \frac{1}{2}\mathscr{S}$ whose difference $\frac{1}{2}\boldsymbol{x}_1 - \frac{1}{2}\boldsymbol{x}_2$ is in $\mathscr{S}$.*

If $\mathscr{S}$ contains the point $\boldsymbol{a} \in \Lambda$ then $\frac{1}{2}\mathscr{S}$ contains the two points $\frac{1}{2}\boldsymbol{a}$ and $-\frac{1}{2}\boldsymbol{a}$ whose difference is $\boldsymbol{a}$; which proves part of the corollary. Conversely, as in the proof of the theorem, if $\frac{1}{2}\boldsymbol{x}_1$, $\frac{1}{2}\boldsymbol{x}_2$ are given, then $\frac{1}{2}\boldsymbol{x}_1 - \frac{1}{2}\boldsymbol{x}_2$ is in $\mathscr{S}$.

Theorem II is the best possible of its kind for any $m$. For example the convex symmetric set

$$|x_1| < m, \quad |x_j| < 1 \qquad (2 \leqq j \leqq n),$$

has volume $m 2^n$ but contains only $m-1$ pairs of points of the lattice $\Lambda_0$ of integral points other than $\boldsymbol{o}$ namely

$$\pm(u, 0, \ldots, 0) \qquad (1 \leqq u \leqq m-1).$$

We shall return in Chapter IX to the general problem of finding convex symmetric sets of volume $2^n d(\Lambda)$ which do not contain any lattice points other than the origin.

**III.2.3.** Important examples of a convex symmetric point set are those sets $\mathscr{S}$ defined by a set of inequalities of the type

$$|a_{l1} x_1 + \cdots + a_{ln} x_n| < c_l \quad \text{or} \quad \leqq c_l \qquad (1 \leqq l \leqq L),$$

where the $a_{lj}$ are real or complex numbers. Such a set is clearly symmetric. It is also convex, since if $\boldsymbol{x}, \boldsymbol{y}$ are in $\mathscr{S}$ and

$$\boldsymbol{z} = \lambda \boldsymbol{x} + (1-\lambda)\boldsymbol{y} \qquad (0 \leqq \lambda \leqq 1),$$

then clearly

$$\left|\sum a_{lj} z_j\right| \leqq \lambda \left|\sum_j a_{lj} x_j\right| + (1-\lambda)\left|\sum_j a_{lj} y_j\right| \leqq \max\left\{\left|\sum_j a_{lj} x_j\right|, \left|\sum_j a_{lj} y_j\right|\right\}.$$

For sets $\mathscr{S}$ of this kind one can relax the condition of compactness in Theorem II somewhat. We enunciate the theorem for the most important case when the $a_{lj}$ are all real. It will be observed that the argument might be used for a wide class of convex sets $\mathscr{S}$.

THEOREM III. *Let $\Lambda$ be an n-dimensional lattice of determinant $d(\Lambda)$ and let $a_{ij}$ $(1 \leqq i, j \leqq n)$ be real numbers. Suppose that*[1] $c_j > 0$ $(1 \leqq j \leqq n)$ *are numbers such that*

$$c_1 \ldots c_n \geqq |\det(a_{ij})| \, d(\Lambda). \tag{1}$$

*Then there is a point $\boldsymbol{u} \in \Lambda$ other than $\boldsymbol{o}$ satisfying*

$$\left.\begin{array}{l} |\sum a_{1j} u_j| \leqq c_1 \\ |\sum a_{ij} u_j| < c_i \quad (2 \leqq i \leqq n). \end{array}\right\} \tag{2}$$

Suppose, first, that

$$\det(a_{ij}) \neq 0.$$

Then (cf. Chapter I, § 3) the points $\boldsymbol{X} = (X_1, \ldots, X_n)$ defined by

$$X_i = \sum_j a_{ij} x_j \qquad \boldsymbol{x} \in \Lambda$$

form a lattice M of determinant

$$d(\mathsf{M}) = |\det(a_{ij})| \, d(\Lambda). \tag{3}$$

The inequalities (2) become

$$\left.\begin{array}{l} |X_1| \leqq c_1 \\ |X_i| < c_i \quad (2 \leqq i \leqq n). \end{array}\right\} \tag{4}$$

These define a set $\mathscr{S}$ in the space of $\boldsymbol{X}$ of volume $2^n c_1 \ldots c_n$. Hence if there is strict inequality in (1) the theorem follows from the first alternative in Theorem II. Let now $\varepsilon$ be any number in

$$0 < \varepsilon < 1.$$

Even if there is equality in (1), there is certainly a point $\boldsymbol{X}_\varepsilon \in \mathsf{M}$ other than $\boldsymbol{o}$, with co-ordinates $(X_{1\varepsilon}, \ldots, X_{n\varepsilon})$, such that

$$\begin{array}{l} |X_{1\varepsilon}| \leqq c_1 + \varepsilon < c_1 + 1 \\ |X_{i\varepsilon}| < c_i \quad (2 \leqq i \leqq n). \end{array}$$

But now there are only a finite number of possibilities for $\boldsymbol{X}_\varepsilon$, by Lemma 1 (ii). Since $\varepsilon$ is arbitrarily small, one of those possibilities must therefore satisfy (4). This proves the theorem unless $\det(a_{ij}) = 0$. But then it is readily verified that (2) defines a region of infinite volume, and so Theorem II certainly applies.

**III.3. Generalisations to non-negative functions**[2]. The results of § 2 may to some extent be generalised to non-negative functions $\psi(\boldsymbol{x})$

[1] $c_j > 0$ follows from (1) except when $\det(a_{ij}) = 0$. But we do not exclude this.
[2] The results of § 3 will not be used later.

of a vector variable $\boldsymbol{x}$. We suppose that $\psi(\boldsymbol{x})$ is integrable and write

$$V(\psi) = \int_{-\infty < x_i < \infty} \psi(\boldsymbol{x})\, d\boldsymbol{x}, \tag{1}$$

where

$$d\boldsymbol{x} = d x_1 \ldots d x_n.$$

This notation is justified, since if $\psi$ is the characteristic function of a set $\mathscr{S}$, that is,

$$\psi(\boldsymbol{x}) = \begin{cases} 1 & \text{if } \boldsymbol{x} \in \mathscr{S} \\ 0 & \text{otherwise,} \end{cases} \tag{2}$$

then $V(\psi)$ is just the volume $V(\mathscr{S})$ of $\mathscr{S}$.

We now have the following simple analogue of BLICHFELDT'S Theorem I:

THEOREM IV. *Let $\psi(\boldsymbol{x})$ be a non-negative integrable function and let $\Lambda$ be a lattice of determinant $d(\Lambda)$. Then there is certainly a point $\boldsymbol{v}_0$ such that*

$$d(\Lambda) \sum_{\boldsymbol{u} \in \Lambda} \psi(\boldsymbol{v}_0 + \boldsymbol{u}) \geqq V(\psi). \tag{3}$$

Before proving Theorem IV we note that it certainly implies the first alternative form of Theorem I. For if $\psi$ is the characteristic function of a set $\mathscr{S}$ and $V(\psi) = V(\mathscr{S}) > m\, d(\Lambda)$ for some integer $m$, then (3) gives

$$\sum_{\boldsymbol{u}} \psi(\boldsymbol{v}_0 + \boldsymbol{u}) > m,$$

and so

$$\sum_{\boldsymbol{u}} \psi(\boldsymbol{v}_0 + \boldsymbol{u}) \geqq m + 1,$$

since now $\psi(\boldsymbol{x})$ is given by (2). But this means that there are $m+1$ distinct vectors $\boldsymbol{u}_j$ such that $\boldsymbol{v}_0 + \boldsymbol{u}_j \in \mathscr{S}$, and this is just the conclusion of Theorem I.

The proof of Theorem IV follows that of Theorem I. Let $\boldsymbol{b}_1, \ldots, \boldsymbol{b}_n$ be a base of $\Lambda$, and $\mathscr{P}$, as before, the set of

$$y_1 \boldsymbol{b}_1 + \cdots + y_n \boldsymbol{b}_n \qquad (0 \leqq y_j < 1);$$

so that every $\boldsymbol{x}$ is uniquely of the shape

$$\boldsymbol{x} = \boldsymbol{v} + \boldsymbol{u}, \qquad \boldsymbol{v} \in \mathscr{P},\ \boldsymbol{u} \in \Lambda.$$

Then

$$\begin{aligned} V(\psi) &= \int \psi(\boldsymbol{x})\, d\boldsymbol{x} \\ &= \sum_{\boldsymbol{u} \in \Lambda} \int_{\boldsymbol{v} \in \mathscr{P}} \psi(\boldsymbol{u} + \boldsymbol{v})\, d\boldsymbol{v} \\ &= \int_{\boldsymbol{v} \in \mathscr{P}} \Big\{ \sum_{\boldsymbol{u} \in \Lambda} \psi(\boldsymbol{u} + \boldsymbol{v}) \Big\}\, d\boldsymbol{v}. \end{aligned}$$

Since $\mathscr{P}$ has volume $V(\mathscr{P}) = d(\Lambda)$, the theorem now follows at once.

II.3.2. SIEGEL (1935a) has given a stronger form of Theorem IV which has, however, remained rather sterile of applications. For notational simplicity we enunciate it only for the lattice $\Lambda_0$ of integral vectors. The function

$$\varphi(\boldsymbol{v}) = \sum_{\boldsymbol{u}\in\Lambda_0} \psi(\boldsymbol{v}+\boldsymbol{u}) \tag{1}$$

is periodic by definition. Its Fourier coefficients $c(\boldsymbol{p}) = c(p_1, \ldots, p_n)$, where $\boldsymbol{p}\in\Lambda_0$, are given by

$$c(\boldsymbol{p}) = \int_{\mathscr{P}} \varphi(\boldsymbol{v})\, e^{-2\pi i(\boldsymbol{p}\boldsymbol{v})}\, d\boldsymbol{v}, \tag{2}$$

where $(\boldsymbol{p}\boldsymbol{v})$ denotes the scalar product

$$p_1 v_1 + \cdots + p_n v_n.$$

On substituting (1) in (2), we have

$$c(\boldsymbol{p}) = \int_{\substack{-\infty<x_j<\infty \\ (1\leqq j\leqq n)}} \psi(\boldsymbol{x})\, e^{-2\pi i(\boldsymbol{p}\boldsymbol{x})}\, d\boldsymbol{x}, \tag{3}$$

since $\boldsymbol{p}\boldsymbol{u}$ is an integer when $\boldsymbol{p}\in\Lambda_0$, $\boldsymbol{u}\in\Lambda_0$. In particular,

$$\int_{\mathscr{P}} \varphi(\boldsymbol{v})\, d\boldsymbol{v} = c(\boldsymbol{o}) = V(\psi). \tag{4}$$

But now, by a fundamental theorem in the theory of Fourier series,

$$\int_{\mathscr{P}} \varphi^2(\boldsymbol{v})\, d\boldsymbol{v} = \sum_{\boldsymbol{p}\in\Lambda_0} |c(\boldsymbol{p})|^2. \tag{5}$$

Since $\varphi(\boldsymbol{u}) \geqq 0$ for all $\boldsymbol{v}$, there must be some $\boldsymbol{v}_0$ such that

$$\int_{\mathscr{P}} \varphi^2(\boldsymbol{v})\, d\boldsymbol{v} \geqq \varphi(\boldsymbol{v}_0) \int_{\mathscr{P}} \varphi(\boldsymbol{v})\, d\boldsymbol{v} = \varphi(\boldsymbol{v}_0)\, V(\psi). \tag{6}$$

On substituting the definition of $\varphi(\boldsymbol{v}_0)$ and the values (3), (4), (5) in (6) we have

$$\sum_{\boldsymbol{u}\in\Lambda_0} \psi(\boldsymbol{v}_0+\boldsymbol{u}) = \varphi(\boldsymbol{v}_0) \geqq V(\psi) + \{V(\psi)\}^{-1} \sum_{\substack{\boldsymbol{p}\in\Lambda_0 \\ \boldsymbol{p}\neq\boldsymbol{o}}} \left| \int \psi(\boldsymbol{x})\, e^{-2\pi i(\boldsymbol{p}\boldsymbol{x})}\, d\boldsymbol{x} \right|^2. \tag{7}$$

This is SIEGEL's inequality.

When a general lattice $\Lambda$ is substituted for $\Lambda_0$ on the left-hand side of (7) then $\Lambda^*$ must be read for $\Lambda_0$ on the right-hand side, where $\Lambda^*$ is the polar lattice of $\Lambda$ defined in Chapter I, § 5.

III.3.3. We now give RADO's generalisation of MINKOWSKI's convex body theorem II. [RADO (1946a), see also CASSELS (1947a).] RADO considered very generally a homogeneous linear mapping $\boldsymbol{\lambda}$ of $n$-dimensional vector space into itself given by

$$X_i = \sum \lambda_{ij} x_j \tag{1}$$

when $\boldsymbol{X} = \boldsymbol{\lambda}\boldsymbol{x}$. We write $\det(\boldsymbol{\lambda}) = \det(\lambda_{ij})$.

THEOREM V. *Let $\psi(\boldsymbol{x})$ be a non-negative function of the vector $\boldsymbol{x}$ in $n$-dimensional space which vanishes outside a bounded set, and suppose that*

$$\psi(\boldsymbol{\lambda}\boldsymbol{x} - \boldsymbol{\lambda}\boldsymbol{y}) \geqq \min\{\psi(\boldsymbol{x}), \psi(\boldsymbol{y})\} \tag{2}$$

*for all real vectors* $\boldsymbol{x}$ *and* $\boldsymbol{y}$. *Then*

$$\psi(\boldsymbol{o}) + \frac{1}{2} \sum_{\substack{\boldsymbol{u} \in \Lambda \\ \neq \boldsymbol{o}}} \psi(\boldsymbol{u}) \geq \frac{|\det(\boldsymbol{\lambda})|}{d(\Lambda)} V(\psi) \tag{3}$$

*for any lattice* $\Lambda$, *where*

$$V(\psi) = \int_{\substack{-\infty < x_j < \infty \\ (1 \leq j \leq n)}} \psi(\boldsymbol{x}) \, d\boldsymbol{x}.$$

Before proving Theorem V we note that it does in fact imply the first alternative part of Theorem II. Let $\psi(\boldsymbol{x})$ be the characteristic function of a convex symmetric set $\mathscr{S}$, so that $V(\psi) = V(\mathscr{S})$. For $\boldsymbol{\lambda}$ we merely take $\boldsymbol{\lambda x} = \frac{1}{2}\boldsymbol{x}$, so that $\det(\boldsymbol{\lambda}) = (\frac{1}{2})^n$. The condition (2) is certainly satisfied, since the right-hand side of (2) is 0 unless both $\boldsymbol{x}$ and $\boldsymbol{y}$ are in $\mathscr{S}$; and then

$$\boldsymbol{\lambda x} - \boldsymbol{\lambda y} = \tfrac{1}{2}\boldsymbol{x} + \tfrac{1}{2}(-\boldsymbol{y})$$

is also in $\mathscr{S}$ by the convexity and symmetry. On the other hand the left-hand side of (3) is $p+1$, where $p$ is the number of distinct pairs $\pm \boldsymbol{u} \in \Lambda$ in $\mathscr{S}$ other than $\boldsymbol{o}$. Hence if $V(\psi) > m 2^n d(\Lambda)$, we have $1+p>m$, that is $p \geq m$; which is the conclusion of Theorem II.

To prove Theorem V we need an elementary combinatorial lemma.

LEMMA 2. *Given any sequence of distinct vectors*

$$\{\boldsymbol{z}\} : \boldsymbol{z}_0, \boldsymbol{z}_1, \ldots, \boldsymbol{z}_r, \ldots,$$

*we can construct another sequence*

$$\{\boldsymbol{w}\} : \boldsymbol{w}_0, \boldsymbol{w}_1, \ldots, \boldsymbol{w}_r, \ldots$$

*satisfying the following three conditions:*

*(i)* $\boldsymbol{w}_0 = \boldsymbol{o}$,

*(ii)* $\boldsymbol{w}_r \neq \pm \boldsymbol{w}_s$ *if* $r \neq s$,

*(iii) every* $\boldsymbol{w}_r$ *is the difference between two of the first* $r+1$ *elements of* $\{\boldsymbol{z}\}$, *say*

$$\boldsymbol{w}_r = \boldsymbol{z}_{l_r} - \boldsymbol{z}_{m_r} \qquad (l_r \leq r,\ m_r \leq r). \tag{4}$$

We introduce an ordering of real vectors and write

$$\boldsymbol{x}_1 > \boldsymbol{x}_2$$

if the first non-zero coordinate of $\boldsymbol{x}_1 - \boldsymbol{x}_2$ is positive. If $\boldsymbol{x}_1 \neq \boldsymbol{x}_2$ then either $\boldsymbol{x}_1 > \boldsymbol{x}_2$ or $\boldsymbol{x}_2 > \boldsymbol{x}_1$. We construct $\boldsymbol{w}_0, \ldots, \boldsymbol{w}_r, \ldots$ in turn, so that

$$\boldsymbol{w}_r > \boldsymbol{o} \qquad (r > 0).$$

The vector $\boldsymbol{w}_0$ is given. Suppose that $\boldsymbol{w}_0, \ldots, \boldsymbol{w}_{r-1}$ have already been constructed, where $r \geq 1$. There is a unique permutation $\boldsymbol{z}_{k_j}$ $(0 \leq j \leq r)$

of the vectors $z_j$ $(0 \leqq j \leqq r)$ so that

$$z_{k_0} < z_{k_1} < \cdots < z_{k_r}.$$

The $r$ vectors

$$z_{k_j} - z_{k_0} \qquad (j = 1, 2, \ldots, r)$$

are distinct from each other and from $\boldsymbol{o}$. Hence we may choose as $\boldsymbol{w}_r$ one of them which is distinct also from the $r-1$ vectors $\boldsymbol{w}_1, \ldots, \boldsymbol{w}_{r-1}$. Since $\boldsymbol{w}_j > \boldsymbol{o}$ $(1 \leqq j \leqq r)$ we cannot have $\boldsymbol{w}_r = -\boldsymbol{w}_j$. Hence the $\boldsymbol{w}_r$ do what is required.

Theorem V, will be an almost immediate consequence of the following Lemma.

LEMMA 3. *Suppose that* (2) *holds and that* $\det(\boldsymbol{\lambda}) \neq 0$, *so that a transformation* $\boldsymbol{\lambda}^{-1}$ *reciprocal to* $\boldsymbol{\lambda}$ *exists. Then*

$$\sum_{\boldsymbol{u} \in \Lambda} \psi(\boldsymbol{\lambda}^{-1}\boldsymbol{u} + \boldsymbol{\lambda}^{-1}\boldsymbol{t}) \leqq \psi(\boldsymbol{o}) + \tfrac{1}{2} \sum_{\substack{\boldsymbol{u} \in \Lambda \\ \neq \boldsymbol{o}}} \psi(\boldsymbol{u}) \tag{5}$$

*for every real vector* $\boldsymbol{t}$.

For fixed $\boldsymbol{t}$ let $\boldsymbol{z}_r$ be the sequence of vectors $\boldsymbol{z}$ of $\Lambda$ such that $\psi(\boldsymbol{\lambda}^{-1}\boldsymbol{z} + \boldsymbol{\lambda}^{-1}\boldsymbol{t}) > 0$ arranged so that

$$\psi(\boldsymbol{\lambda}^{-1}\boldsymbol{z}_r + \boldsymbol{\lambda}^{-1}\boldsymbol{t}) \geqq \psi(\boldsymbol{\lambda}^{-1}\boldsymbol{z}_s + \boldsymbol{\lambda}^{-1}\boldsymbol{t}) \qquad (r \leqq s). \tag{6}$$

Let $\boldsymbol{w}_r$ be the corresponding sequence defined by Lemma 2. We apply (2) with

$$\boldsymbol{x} = \boldsymbol{x}_r = \boldsymbol{\lambda}^{-1}\boldsymbol{z}_{l_r} + \boldsymbol{\lambda}^{-1}\boldsymbol{t}$$
$$\boldsymbol{y} = \boldsymbol{y}_r = \boldsymbol{\lambda}^{-1}\boldsymbol{z}_{m_r} + \boldsymbol{\lambda}^{-1}\boldsymbol{t},$$

where $l_r$ and $m_r$ are defined by (4). Then

$$\min\{\psi(\boldsymbol{x}_r), \psi(\boldsymbol{y}_r)\} \geqq \psi(\boldsymbol{\lambda}^{-1}\boldsymbol{z}_r + \boldsymbol{\lambda}^{-1}\boldsymbol{t}) \tag{7}$$

by (6), and since $l_r \leqq r$, $m_r \leqq r$. But now, by (4) again,

$$\boldsymbol{\lambda}(\boldsymbol{x}_r - \boldsymbol{y}_r) = \boldsymbol{w}_r$$

and so, by (2) and (7)

$$\psi(\boldsymbol{w}_r) \geqq \psi(\boldsymbol{\lambda}^{-1}\boldsymbol{z}_r + \boldsymbol{\lambda}^{-1}\boldsymbol{t}).$$

Similarly, on interchanging $\boldsymbol{x}_r$ and $\boldsymbol{y}_r$, we obtain

$$\psi(-\boldsymbol{w}_r) \geqq \psi(\boldsymbol{\lambda}^{-1}\boldsymbol{z}_r + \boldsymbol{\lambda}^{-1}\boldsymbol{t}).$$

Hence, since $\psi \geqq 0$, we have

$$\left.\begin{aligned} \sum_{\boldsymbol{u} \in \Lambda} \psi(\boldsymbol{u}) &\geqq \psi(\boldsymbol{w}_0) + \sum_{r>0} \{\psi(\boldsymbol{w}_r) + \psi(-\boldsymbol{w}_r)\} \\ &\geqq \psi(\boldsymbol{\lambda}^{-1}\boldsymbol{z}_0 + \boldsymbol{\lambda}^{-1}\boldsymbol{t}) + 2\sum_{r>0} \psi(\boldsymbol{\lambda}^{-1}\boldsymbol{z}_r + \boldsymbol{\lambda}^{-1}\boldsymbol{t}) \\ &= -\psi(\boldsymbol{\lambda}^{-1}\boldsymbol{z}_0 + \boldsymbol{\lambda}^{-1}\boldsymbol{t}) + 2\sum_{\boldsymbol{u} \in \Lambda} \psi(\boldsymbol{\lambda}^{-1}\boldsymbol{u} + \boldsymbol{\lambda}^{-1}\boldsymbol{t}), \end{aligned}\right\} \tag{8}$$

since every vector $\boldsymbol{u} \in \Lambda$ with $\psi(\boldsymbol{\lambda}^{-1}\boldsymbol{u}+\boldsymbol{\lambda}^{-1}\boldsymbol{t}) > 0$ occurs as a $\boldsymbol{z}_r$. But now (2) with $\boldsymbol{y}=\boldsymbol{x}$ implies that $\psi(\boldsymbol{o}) \geqq \psi(\boldsymbol{x})$ for any $\boldsymbol{x}$, and in particular

$$\psi(\boldsymbol{\lambda}^{-1}\boldsymbol{z}_0 + \boldsymbol{\lambda}^{-1}\boldsymbol{t}) \leqq \psi(\boldsymbol{o}). \tag{9}$$

The truth of the lemma follows now at once from (8) and (9).

Finally Theorem V follows from (5) on integrating with respect to $\boldsymbol{t}$ over a fundamental parallelopiped $\mathscr{P}$ of $\Lambda$ defined as in § 2.1. The left-hand side becomes

$$\int_{\mathscr{P}} \Big\{ \sum_{\boldsymbol{u} \in \Lambda} \psi(\boldsymbol{\lambda}^{-1}\boldsymbol{u} + \boldsymbol{\lambda}^{-1}\boldsymbol{t}) \Big\} d\boldsymbol{t} = \int_{\substack{-\infty < t_j < \infty \\ (1 \leqq j \leqq n)}} \psi(\boldsymbol{\lambda}^{-1}\boldsymbol{t})\, d\boldsymbol{t} = |\det(\boldsymbol{\lambda})|\, V(\psi).$$

The right-hand side of (5) is independent of $\boldsymbol{t}$ and so, on integrating with respect to $\boldsymbol{t}$, is merely multiplied by $V(\mathscr{P}) = d(\Lambda)$. This proves the theorem.

RADO (1946a) discusses the homogeneous linear transformations $\boldsymbol{\lambda}$ for which there is a function $\psi(\boldsymbol{x})$ which is not identically 0 satisfying (2). It turns out that $\boldsymbol{\lambda}$ must satisfy pretty stringent conditions, and that taking multiplication by $\frac{1}{2}$ for $\boldsymbol{\lambda}$ is in a sense on the borderline of what is possible.

**III.4. Characterisation of lattices.** We are now in a position to give a characterisation of lattices in which the notion of a basis does not appear.

THEOREM VI. *A necessary and sufficient condition that a set of points $\Lambda$ in n-dimensional euclidean space be a lattice is that it should have the following three properties:*

*(i) If $\boldsymbol{a}$ and $\boldsymbol{b}$ are in $\Lambda$ then $\boldsymbol{a} \pm \boldsymbol{b}$ is in $\Lambda$.*

*(ii) $\Lambda$ contains n linearly independent points $\boldsymbol{a}_1, \ldots, \boldsymbol{a}_n$.*

*(iii) There exists a constant $\eta > 0$ such that $\boldsymbol{o}$ is the only point of $\Lambda$ in the sphere*

$$|\boldsymbol{x}| < \eta,$$

*where, as usual,*

$$|\boldsymbol{x}| = (x_1^2 + \cdots + x_n^2)^{\frac{1}{2}}.$$

By the definition and Lemma 1 every lattice satisfies (i), (ii), (iii). It remains to show that any set $\Lambda$ satisfying (i), (ii) and (iii) is a lattice.

We note first that it follows by induction from (i) that if $\boldsymbol{c}_1, \ldots, \boldsymbol{c}_m$ are any points of $\Lambda$ and $u_1, \ldots, u_m$ are integers, then

$$u_1 \boldsymbol{c}_1 + \cdots + u_m \boldsymbol{c}_m \in \Lambda.$$

Secondly, we show that if

$$\boldsymbol{c}_j = (c_{1j}, \ldots, c_{nj}) \qquad (1 \leqq j \leqq n+1)$$

are $n+1$ points of $\Lambda$, then there are integers $u_j$ $(1 \leqq j \leqq n+1)$ not all 0 such that

$$\sum u_j \boldsymbol{c}_j = 0.$$

For by Theorem II there certainly exist points $(u_1, \ldots, u_{n+1}) \neq \boldsymbol{o}$ of the $(n+1)$-dimensional lattice $\Lambda_0$ of integral vectors in the convex symmetric $(n+1)$-dimensional set $\mathscr{S}$ of infinite volume defined by the $n$ inequalities

$$\left| \sum_{1 \leqq j \leqq n+1} c_{ij} u_j \right| < \eta/n \qquad (1 \leqq i \leqq n).$$

Put

$$\boldsymbol{d} = \sum u_j \boldsymbol{c}_j,$$

so that trivially

$$|\boldsymbol{d}| < \eta.$$

Then $\boldsymbol{d} = \boldsymbol{o}$ by property (iii), as was required.

Now let $\mathsf{M}_1$ be the lattice with the basis $\boldsymbol{a}_1, \ldots, \boldsymbol{a}_n$ given by (ii). Then $\mathsf{M}_1$ is a subset of $\Lambda$. If $\Lambda$ coincides with $\mathsf{M}_1$ there is nothing to prove. If not, there is some vector $\boldsymbol{b}$ in $\Lambda$ but not in $\mathsf{M}_1$. But now, on applying the result of the previous paragraph to the $n+1$ vectors $\boldsymbol{a}_1, \ldots, \boldsymbol{a}_n$ and $\boldsymbol{b}$, there must be integers $u_1, \ldots, u_n$ and $v$ not all 0 such that

$$v\boldsymbol{b} = u_1 \boldsymbol{a}_1 + \cdots + u_n \boldsymbol{a}_n. \tag{1}$$

Here $v \neq 0$, since $\boldsymbol{a}_1, \ldots, \boldsymbol{a}_n$ are linearly independent. Further, $v \neq \pm 1$ since $\boldsymbol{b}$ is not in $\mathsf{M}_1$ by hypothesis. We may suppose that $\boldsymbol{b}$ is chosen so that $|v|$ in (1) is as small as possible. Let $p$ be a prime divisor of $v$ and write

$$v = p v_1 \qquad \boldsymbol{b}_1 = v_1 \boldsymbol{b}.$$

Then

$$p \boldsymbol{b}_1 = u_1 \boldsymbol{a}_1 + \cdots + u_n \boldsymbol{a}_n,$$

where not all of $u_1, \ldots, u_n$ are divisible by $p$ since $\boldsymbol{b}_1$ is not in $\mathsf{M}_1$ (because $v$ was chosen minimal). Without loss of generality, $p$ does not divide $u_1$, and so

$$lp - m u_1 = 1$$

for some integers $l$ and $m$. Put now

$$\left.\begin{aligned} \boldsymbol{a}_1' &= l\boldsymbol{a}_1 - m\boldsymbol{b}_1 \\ \boldsymbol{a}_j' &= \boldsymbol{a}_j \qquad (2 \leqq j \leqq n), \end{aligned}\right\} \tag{2}$$

so that conversely

$$\left.\begin{aligned} \boldsymbol{a}_1 &= p\boldsymbol{a}_1' + m u_2 \boldsymbol{a}_2' + \cdots + m u_n \boldsymbol{a}_n' \\ \boldsymbol{a}_j &= \boldsymbol{a}_j' \qquad (2 \leqq j \leqq n). \end{aligned}\right\} \tag{3}$$

Let $\mathsf{M}_2$ be the lattice with basis $\boldsymbol{a}_j'$.

Then $M_1$ has index $p$ in $M_2$, so in particular

$$d(M_2) = p^{-1} d(M_1) \leqq \tfrac{1}{2} d(M_1). \tag{4}$$

But now, by (2), a basis of $M_2$ is in $\Lambda$ and so $M_2$ is entirely contained in $\Lambda$. We may now repeat the argument. If $M_2$ does not coincide with $\Lambda$ there is a third lattice $M_3$ which is in $\Lambda$ and contains $M_2$ as a sublattice. And so on. Now, by (4),

$$d(M_r) \leqq \tfrac{1}{2} d(M_{r-1}) \cdots \leqq (\tfrac{1}{2})^{r-1} d(M_1).$$

If

$$d(M_r) < (\eta/n)^n,$$

where $\eta$ is defined in (iii) of the enunciation of the Theorem, then, by Theorem II, $M_r$ would contain a point $\boldsymbol{d} \neq \boldsymbol{o}$ with

$$|d_j| < \eta/n \qquad (1 \leqq j \leqq n)$$

contrary to hypothesis. Hence the chain of lattices $M_1, \ldots, M_r, \ldots$ must have a last, $M_R$; and $M_R$ then coincides with $\Lambda$.

**III.5. Lattice constants.** We must now introduce a number of new definitions relating to lattices and points sets. The new concepts will be subjected to a searching analysis in Chapters IV and V; here we just prove enough to show their use and to enable applications of MINKOWSKI's theorem to be made.

Let $\mathscr{S}$ be any point set. If a lattice $\Lambda$ has no points in $\mathscr{S}$ other than $\boldsymbol{o}$ (if $\boldsymbol{o}$ is in $\mathscr{S}$), then we say that $\Lambda$ is admissible for $\mathscr{S}$ or $\mathscr{S}$-admissible. We call the infimum (greatest lower bound) of $d(\Lambda)$ for all $\Lambda$-admissible lattices the lattice constant of $\mathscr{S}$ and write

$$\Delta(\mathscr{S}) = \inf d(\Lambda) \qquad (\Lambda \text{ is } \mathscr{S}\text{-admissible}).$$

If there are no $\mathscr{S}$-admissible lattices then we say that $\mathscr{S}$ is of infinite type, and write $\Delta(\mathscr{S}) = \infty$; otherwise $\mathscr{S}$ is of finite type and $0 \leqq \Delta(\mathscr{S}) < \infty$. An $\mathscr{S}$-admissible lattice $\Lambda$ with $d(\Lambda) = \Delta(\mathscr{S})$ is said to be critical. Critical lattices play a very prominent role in Chapter V. Of course in general there is no reason why a general set $\mathscr{S}$ should have critical lattices at all.

Our definitions do not quite correspond with those of MAHLER (1946d, e). He is usually concerned with closed sets $\mathscr{S}$ and says that $\Lambda$ is $\mathscr{S}$-admissible if no interior point of $\mathscr{S}$ except $\boldsymbol{o}$ belongs to $\Lambda$, that is if $\Lambda$ is admissible in our sense for the set of interior points of $\mathscr{S}$. Our usage is a compromise between MAHLER's and that proposed by ROGERS (1952a).

**III.5.2.** The definition of $\Delta(\mathscr{S})$ may be stood on its head: $\Delta(\mathscr{S})$ is the greatest number $\Delta$ such that every lattice $\Lambda$ with $d(\Lambda) < \Delta$ has a point other than $\boldsymbol{o}$ in $\mathscr{S}$. The discussion of § 4 of Chapter I shows that many of the results of Chapter II may be interpreted as giving the value of $\Delta(\mathscr{S})$ for certain regions $\mathscr{S}$. Take for example the statement that if $f(\boldsymbol{x}) = f_{11}x_1^2 + 2f_{12}x_1x_2 + f_{22}x_2^2$ is a definite quadratic form and $D = f_{11}f_{22} - f_{12}^2$, then there are integers $\boldsymbol{u} = (u_1, u_2) \neq \boldsymbol{o}$ such that $f(\boldsymbol{u}) \leqq (4D/3)^{\frac{1}{2}}$, with equality only for forms equivalent to $f_{11}(x_1^2 + x_1x_2 + x_2^2)$ (Theorem II of Chapter II). This is equivalent to the statement that the 2-dimensional set

$$\mathscr{D}: \quad X_1^2 + X_2^2 < 1 \tag{1}$$

has lattice constant $\Delta(\mathscr{D}) = (\frac{3}{4})^{\frac{1}{2}}$ and that the critical lattices are precisely those with a base $\boldsymbol{b}_1 = (b_{11}, b_{21})$, $\boldsymbol{b}_2 = (b_{12}, b_{22})$ such that

$$(b_{11}x_1 + b_{12}x_2)^2 + (b_{21}x_1 + b_{22}x_2)^2 = x_1^2 + x_1x_2 + x_2^2 \tag{2}$$

identically. The reader will have no difficulty in making the translation for himself (cf. Lemma 4 of Chapter I). We can also make a geometrical interpretation of (2). Put

$$b_{11} = \cos\vartheta, \quad b_{21} = \sin\vartheta,$$
$$b_{12} = \cos\psi, \quad b_{22} = \sin\psi.$$

Then (2) is true provided that

$$\cos\vartheta\cos\psi + \sin\vartheta\sin\psi = \tfrac{1}{2},$$

that is provided that

$$\vartheta - \psi = \pm\pi/3.$$

Hence the critical lattice has as basis two points at angular distance $\pi/3$ on $X_1^2 + X_2^2 = 1$. A further point on $X_1^2 + X_2^2 = 1$ is $\boldsymbol{b}_1 - \boldsymbol{b}_2$, as is clear from (2). It is readily verified that the six points $\pm\boldsymbol{b}_1, \pm\boldsymbol{b}_2, \pm(\boldsymbol{b}_1 - \boldsymbol{b}_2)$ are the vertices of a regular hexagon inscribed in $X_1^2 + X_2^2 = 1$.

**III.5.3.** In this and in the next section we shall use MINKOWSKI's convex body Theorem II to evaluate or estimate $\Delta(\mathscr{S})$ for various sets $\mathscr{S}$. Theorem II is directly applicable when $\mathscr{S}$ is symmetric and convex, since it asserts that then

$$\Delta(\mathscr{S}) \geqq 2^{-n} V(\mathscr{S}). \tag{1}$$

This applies for example to the circular disc $\mathscr{D}$: $X_1^2 + X_2^2 < 1$ and gives $\Delta(\mathscr{D}) \geqq \pi/4 = 0.785\ldots$, which may be compared with the exact value $(\frac{3}{4})^{\frac{1}{2}} = 0.866\ldots$ obtained above.

Even if our region $\mathscr{S}$ is not convex or symmetric, we may obtain estimates for $\Delta(\mathscr{S})$ below if a convex symmetric body $\mathscr{T}$ is inscribable in it. Clearly $\Delta(\mathscr{S}) \geqq \Delta(\mathscr{T})$ if $\mathscr{T}$ is a subset of $\mathscr{S}$, since every $\mathscr{S}$-admissible lattice is automatically $\mathscr{T}$-admissible. Hence

$$\Delta(\mathscr{S}) \geqq \Delta(\mathscr{T}) \geqq 2^{-n} V(\mathscr{T}).$$

Consider for example the region

$$\mathscr{S}: \quad |X_1 \dots X_n| < 1.$$

This contains the convex symmetric region

$$\mathscr{T}: \quad |X_1| + \cdots + |X_n| < n$$

by the inequality of the arithmetic and geometric means. Now $\mathscr{T}$ is convex and symmetric, since it is defined by homogeneous linear inequalities, and its volume is

$$2^n n^n / n!.$$

Hence

$$\Delta(\mathscr{S}) \geqq n^n / n!.$$

We shall later obtain a rather better estimate than this (Chapter IX, § 8). We note the translation into the theory of forms: Let

$$L_j(\boldsymbol{x}) = \sum_{1 \leqq i \leqq n} c_{ji} x_i$$

be real linear forms in the $n$ variables $\boldsymbol{x} = (x_1, \dots, x_n)$ with $\det(c_{ij}) \neq 0$. Then there exists an integral $\boldsymbol{u} \neq \boldsymbol{o}$ such that

$$\left| \prod_j L_j(\boldsymbol{u}) \right| \leqq \frac{n!}{n^n} |\det(c_{ij})|.$$

MINKOWSKI's convex body theorem also permits the evaluation of $\Delta(\mathscr{S})$ for sets $\mathscr{S}$ which are not symmetric in $\boldsymbol{o}$. We reproduce here, with his kind permission, Professor MAHLER's elegant treatment of the simplex, hitherto unpublished[1]. Let $\mathscr{S}$ be an open simplex in $n$-dimensional space containing $\boldsymbol{o}$. If the faces of $\mathscr{S}$ are given by the equations

$$L_j(\boldsymbol{x}) = 1 \qquad (0 \leqq j \leqq n),$$

where the $L_j(\boldsymbol{x})$ are linear forms, then $\mathscr{S}$ is the set of points satisfying

$$L_j(\boldsymbol{x}) < 1 \qquad (0 \leqq j \leqq n).$$

There is one non-trivial relation between the linear forms, say

$$\sum_{0 \leqq j \leqq n} \alpha_j L_j(\boldsymbol{x}) = 0$$

identically in $\boldsymbol{x}$, where the $\alpha_j$ are real numbers, and without loss of generality

$$\alpha_0 > 0.$$

[1] It is given, however, in his mimeographed lecture course, Boulder (Colorado), U.S.A., 1950, together with other interesting results about non-symmetric sets.

If, say, $\alpha_1 \leqq 0$, then $\mathscr{S}$ would contain the infinite ray of points $\boldsymbol{x}$ satisfying

$$L_0(\boldsymbol{x}) \leqq 0 \quad L_j(\boldsymbol{x}) = 0 \qquad (j \neq 0, 1);$$

which is impossible, since $\mathscr{S}$ is a simplex. Hence

$$\alpha_j > 0 \qquad (0 \leqq j \leqq n).$$

We may suppose without loss of generality that

$$\alpha_0 = 1 = \min_j \alpha_j, \tag{2}$$

and then

$$L_0(\boldsymbol{x}) = -\sum_{1 \leqq j \leqq n} \alpha_j L_j(\boldsymbol{x}), \tag{3}$$

where

$$\alpha_j \geqq 1. \tag{4}$$

We show that

$$\Delta(\mathscr{S}) = 2^{-n} V(\mathscr{C}), \tag{5}$$

where $V(\mathscr{C})$ is the volume of the parallelopiped

$$\mathscr{C}: |L_j(\boldsymbol{x})| < 1 \qquad (1 \leqq j \leqq n).$$

In the first place, if $\Lambda$ is a lattice with $d(\Lambda) < 2^{-n} V(\mathscr{C})$, then there is a point $\boldsymbol{a} \neq \boldsymbol{o}$ of $\Lambda$ in $\mathscr{C}$. By taking $-\boldsymbol{a}$ instead of $\boldsymbol{a}$ if necessary, we may suppose that

$$L_0(\boldsymbol{a}) \leqq 0,$$

and then

$$L_j(\boldsymbol{a}) < 1 \qquad (0 \leqq j \leqq n);$$

so $\boldsymbol{a}$ is in $\mathscr{S}$. Hence $\Delta(\mathscr{S}) \geqq 2^{-n} V(\mathscr{C})$. On the other hand, we shall show that the lattice M of points $\boldsymbol{a}$ such that

$$L_j(\boldsymbol{a}) = u_j = \text{integer} \qquad (1 \leqq j \leqq n)$$

is admissible for $\mathscr{S}$.

If $\boldsymbol{a}$ is in $\mathscr{S}$, we must have $u_j \leqq 0$ $(1 \leqq j \leqq n)$, and $\min u_j \leqq -1$ if $\boldsymbol{a} \neq \boldsymbol{o}$. But then, by (4), we should have

$$L_0(\boldsymbol{a}) = -\sum \alpha_j u_j \geqq 1;$$

and so $\boldsymbol{a}$ is not in $\mathscr{S}$. Hence $\boldsymbol{o}$ is the only point of M in $\mathscr{S}$. Since $d(\mathsf{M}) = 2^{-n} V(\mathscr{C})$, this completes the proof of (5). We note that $2^{-n} V(\mathscr{C}) = |d_0|^{-1}$, where $d_0$ is the determinant of the $n$ forms $L_1, \ldots, L_n$. By (3) and (4), $d_0$ is the least in absolute value of the determinants of selections of $n$ out of the $n+1$ forms $L_0, \ldots, L_n$.

Estimates of $\Delta(\mathscr{S})$ for non-convex sets $\mathscr{S}$ may be obtained from Theorem I instead of Theorem II. Let $\mathscr{R}$ be any set such that all the differences

$$\boldsymbol{x}_1 - \boldsymbol{x}_2, \quad \boldsymbol{x}_1 \in \mathscr{R}, \ \boldsymbol{x}_2 \in \mathscr{R} \tag{6}$$

lie in $\mathscr{S}$. Then

$$\Delta(\mathscr{S}) \geqq V(\mathscr{R}),$$

since by Theorem I if $d(\Lambda) < V(\mathscr{R})$ there exist two points $\boldsymbol{x}_1, \boldsymbol{x}_2 \in \mathscr{R}$ such that $\boldsymbol{x}_1 - \boldsymbol{x}_2 \in \Lambda$; and by hypothesis $\boldsymbol{x}_1 - \boldsymbol{x}_2 \in \mathscr{S}$. Of course if $\mathscr{T}$ is a convex symmetric set inscribed in $\mathscr{S}$ we could take $\mathscr{R} = \frac{1}{2}\mathscr{T}$: but then we get just the same estimate $\Delta(\mathscr{S}) \geqq 2^{-n} V(\mathscr{T})$ as by the use of Theorem II. However MORDELL and MULLENDER found suitable sets $\mathscr{R}$ in the case they were treating such that $V(\mathscr{R})$ was greater than $2^{-n} V(\mathscr{T})$ for any convex symmetric inscribed $\mathscr{T}$. The increases are usually comparatively small and obtained at the expense of some complication. We refer the reader to MULLENDER (1948a) and the literature quoted there for further information.

In Chapter VI are obtained upper estimates for $\Delta(\mathscr{S})$ in terms of $V(\mathscr{S})$ which are valid for all sets (Minkowski-Hlawka Theorem and related topics).

**III.6. A method of MORDELL.** In this section we develop a method of MORDELL for finding $\Delta(\mathscr{S})$ precisely for point sets $\mathscr{S}$ which may or may not be convex. The method applies primarily to star bodies. This class of sets is defined by the properties that the origin is an inner point and any radius vector meets the boundary either not at all or in precisely one point: in other words, if $\boldsymbol{x}$ is any vector other than $\boldsymbol{o}$, then either $t\boldsymbol{x} \in \mathscr{S}$ for all $t \geqq 0$ or there exists a $t_0$ such that $t\boldsymbol{x}$ is an inner point of $\mathscr{S}$, a boundary point of $\mathscr{S}$ or not in $\mathscr{S}$ according as $t < t_0$, $t = t_0$ or $t > t_0$. We now have the rather trivial

LEMMA 4. *Let $\mathscr{S}$ be a star body and suppose that a constant $\Delta_0$ exists with the following two properties.*

*(i) every lattice $\Lambda$ with $d(\Lambda) = \Delta_0$ has a point other than $\boldsymbol{o}$ in or on the boundary of $\mathscr{S}$.*

*(ii) there exist lattices $\Lambda_c$ with $d(\Lambda_c) = \Delta_0$ having no points other than $\boldsymbol{o}$ in the interior of $\mathscr{S}$.*

*Then $\Delta(\mathscr{S}) = \Delta_0$. If further, $\mathscr{S}$ is open*[1]*, then the critical lattices are just the $\Lambda_c$.*

For suppose, if possible that $\mathsf{M}$ is an $\mathscr{S}$-admissible lattice with $d(\mathsf{M}) < \Delta_0$. Let $\gamma > 1$ be defined by $\gamma^n d(\mathsf{M}) = \Delta_0$. Then the lattice $\gamma \mathsf{M}$ of points $\gamma \boldsymbol{x}$, $\boldsymbol{x} \in \mathsf{M}$ has clearly no points in or on the boundary of $\mathscr{S}$, contrary to (i). Hence $\Delta(\mathscr{S}) \geqq \Delta_0$. On the other hand $(1+\varepsilon)\Lambda_c$ has no points in $\mathscr{S}$ for any $\varepsilon > 0$, where $\Lambda_c$ is one of the lattices given in (ii).

[1] i.e. does not contain any of its boundary points. MINKOWSKI and following him MAHLER define a star body to be closed. We depart from their nomenclature.

Hence $\Delta(\mathscr{S}) \leqq (1+\varepsilon)^n \Delta_0$, so $\Delta(\mathscr{S}) = \Delta_0$. The truth of the last sentence of the lemma is now obvious.

When the description of star-bodies by distance-functions is introduced in the next chapter, Lemma 4 will fall into place as part of a wider theory.

MORDELL'S method of finding $\Delta(\mathscr{S})$ for a given star-body $\mathscr{S}$ may now be described. First one must make an intelligent guess $\Delta_0$ at $\Delta(\mathscr{S})$: in particular so that (ii) of Lemma 4 is true. If $\Delta_0$ has been correctly chosen, then it may be possible to verify (i) and to find all the $\Lambda_c$ in (ii) by the following general procedure, of which the details naturally vary widely from case to case. We suppose for simplicity that $\mathscr{S}$ is open. Let $\mathsf{M}$ be any $\mathscr{S}$-admissible lattice with $d(\mathsf{M}) = \Delta_0$. Then if $\mathscr{T}_j$ $(1 \leqq j \leqq r)$ is any collection of closed convex symmetric sets each of volume

$$V(\mathscr{T}_j) = 2^n \Delta_0 \quad (1 \leqq j \leqq r),$$

there must be points $\boldsymbol{p}_j \neq \boldsymbol{o}$ of $\mathsf{M}$ in $\mathscr{T}_j$ for $1 \leqq j \leqq r$. Since $\mathsf{M}$ is $\mathscr{S}$-admissible, the $\boldsymbol{p}_j$ must lie in $\mathscr{R}_j$, the set of points of $\mathscr{T}_j$ which are not in $\mathscr{S}$. We may now use the hypothesis that the $\boldsymbol{p}_j$ are in a lattice $\mathsf{M}$ of determinant $\Delta_0$ to obtain further points of $\mathsf{M}$. Since these cannot lie in $\mathscr{S}$, this gives further information about the $\boldsymbol{p}_j$. In the end it may be possible to show that $\mathsf{M}$ is one of a set of lattices $\Lambda_c$, all of which have points on the boundary of $\mathscr{S}$. Lemma 4 shows that $\Delta(\mathscr{S}) = \Delta_0$. Of course the power of the method depends on a suitable choice of the $\mathscr{T}_j$.

MORDELL'S method is at its best in dealing with 2-dimensional regions, since for these it is easier to grasp the geometry of the figure. Before giving some concrete examples we must therefore study the geometry of a 2-dimensional lattice more closely.

**III.6.2.** Throughout § 6.2 we denote by $\Lambda$ a 2-dimensional lattice. We regard vectors as coordinates of points on a 2-dimensional euclidean plane, and use the normal geometric language to discuss their relations. By distance we mean the usual euclidean distance. For later reference we formulate our conclusions as lemmas.

We say that a point $\boldsymbol{u}$ of a (not necessarily 2-dimensional) lattice is primitive if it is not of the shape $\boldsymbol{u} = k\boldsymbol{u}_1$, where $\boldsymbol{u}_1 \in \Lambda$ and $k > 1$ is an integer.

LEMMA 5. *Let* $\boldsymbol{u}$ *be a primitive point of the 2-dimensional lattice* $\Lambda$. *Then the points of* $\Lambda$ *lie on lines* $\pi_r$ $(r = 0, \pm 1, \ldots)$ *which are parallel to* $\boldsymbol{o}\,\boldsymbol{u}$ *and at a perpendicular distance*

$$r\,d(\Lambda)/|\boldsymbol{u}|$$

*from it*[1]. *Each line* $\pi_r$ *contains infinitely many points of* $\Lambda$ *and these are spaced at a distance* $|\boldsymbol{u}|$.

[1] As before $|\boldsymbol{u}| = (u_1^2 + u_2^2)^{\frac{1}{2}}$, that is the distance from $\boldsymbol{o}$ to $\boldsymbol{u}$.

This is just a re-statement in geometrical language of what is known already. Since $\boldsymbol{u}$ is primitive, there is a point $\boldsymbol{v}$ which with $\boldsymbol{u}$ forms a basis for $\Lambda$ (Chapter I, Theorem I, Corollary 3). Hence

$$\det(\boldsymbol{u}, \boldsymbol{v}) = \pm d(\Lambda),$$

that is the perpendicular distance from $\boldsymbol{v}$ on the line through $\boldsymbol{o}$ and $\boldsymbol{u}$ is $d(\Lambda)/|\boldsymbol{u}|$. But now $\Lambda$ is just the set of points

$$r\boldsymbol{v} + s\boldsymbol{u} \qquad (r, s \text{ integers}).$$

Clearly the points with $r$ fixed but $s$ varying lie on a line $\pi_r$ with the required properties.

LEMMA 6. *Let $\boldsymbol{u}, \boldsymbol{v}$ be points of the 2-dimensional lattice $\Lambda$ such that $\boldsymbol{o}, \boldsymbol{u}, \boldsymbol{v}$ are not collinear. Then a necessary and sufficient condition that $\boldsymbol{u}, \boldsymbol{v}$ be a basis for $\Lambda$ is that the closed*[1] *triangle $\boldsymbol{ouv}$ should contain no points of $\Lambda$ other than the vertices.*

The condition is clearly necessary, by Lemma 5, so we must prove it sufficient. If there are no points of $\Lambda$ in the triangle $\boldsymbol{ouv}$ other than the vertices, then the same must be true of the triangles with vertices

$$-\boldsymbol{u}, \boldsymbol{o}, \boldsymbol{v}-\boldsymbol{u} \tag{1}$$

and

$$-\boldsymbol{v}, \boldsymbol{u}-\boldsymbol{v}, \boldsymbol{o}, \tag{2}$$

since, for example if $\boldsymbol{x}$ is a point of $\Lambda$ in (1), then $\boldsymbol{x}+\boldsymbol{u}$ is a point of $\Lambda$ in triangle $\boldsymbol{ouv}$. Similarly there can be no points of $\Lambda$ in the images of our first three triangles in the origin, since $-\boldsymbol{x}$ is in $\Lambda$ if $\boldsymbol{x}$ is. Hence there is no point of $\Lambda$ in the hexagon $\mathscr{H}$ with vertices $\pm\boldsymbol{u}$, $\pm\boldsymbol{v}$, $\pm(\boldsymbol{v}-\boldsymbol{u})$ except $\boldsymbol{o}$ and the vertices. By Theorem II

$$d(\Lambda) \geqq \tfrac{1}{4} V(\mathscr{H}) = \tfrac{3}{4} |\det(\boldsymbol{u}, \boldsymbol{v})|.$$

But

$$|\det(\boldsymbol{u}, \boldsymbol{v})| = I\, d(\Lambda),$$

where the integer $I$ is the index of the points $\boldsymbol{u}, \boldsymbol{v}$ in $\Lambda$ (Chapter I, § 2.2); and so $I = 1$, as required.

The analogue of Lemma 6 does not hold in space of dimension $>2$.

LEMMA 7. *Let $\mathscr{Q}$ be an open parallelogram with $\boldsymbol{o}$ as centre of area $4d(\Lambda)$, which contains no other point of $\Lambda$ than $\boldsymbol{o}$. Then $\Lambda$ has a basis consisting of the mid-point of one of the sides of $\mathscr{Q}$ and a point on one of the other pair of parallel sides.*

[1] i.e. the sides are counted as belonging to the triangle.

After a suitable transformation of coordinates, we may suppose that $\mathcal{Q}$ is the parallelogram

$$\mathcal{Q}: \quad |X_1| < 1, \quad |X_2| < 1$$

and that $d(\Lambda) = 1$. By Theorem III there is certainly a point of $\Lambda$ other than $\boldsymbol{o}$ in $|X_1| \leqq 1$, $|X_2| < 1$, and so $\Lambda$ must contain a point

$$\boldsymbol{u} = (1, u_2) \qquad |u_2| < 1.$$

Similarly, $\Lambda$ must contain a point

$$\boldsymbol{v} = (v_1, 1) \qquad |v_1| < 1.$$

But now, since $d(\Lambda) = 1$, the index of $(\boldsymbol{u}, \boldsymbol{v})$ in $\Lambda$ is

$$I = |\det(\boldsymbol{u}, \boldsymbol{v})| = 1 - u_2 v_1.$$

But $I$ is an integer and $|u_2 v_1| < 1$. Hence $I = 1$ and either $u_2 = 0$ or $v_1 = 0$.

LEMMA 8. *Let $\Lambda$ be a lattice of determinant $d(\Lambda)$ which has two points other than* $\mathbf{o}$ *in the closed parallelogram with vertices* $\mathbf{o}, \boldsymbol{a}, \mathbf{b}, \boldsymbol{a}+\mathbf{b}$ *and volume (area) $d(\Lambda)$. Then either*

*(i) the two points are collinear with* $\mathbf{o}$,

*or*

*(ii) one of the points is* $\boldsymbol{a}$ *and the other is on the line-segment* $\mathbf{b}, \boldsymbol{a}+\mathbf{b}$,

*or*

*(iii) one of the points is* $\mathbf{b}$ *and the other is on* $\boldsymbol{a}, \boldsymbol{a}+\mathbf{b}$.

For the points $\boldsymbol{p}, \boldsymbol{q}$, say, are of the type

$$\boldsymbol{p} = \pi_1 \boldsymbol{a} + \pi_2 \boldsymbol{b}, \qquad \boldsymbol{q} = \varkappa_1 \boldsymbol{a} + \varkappa_2 \boldsymbol{b}$$

where

$$0 \leqq \pi_j \leqq 1, \quad 0 \leqq \varkappa_j \leqq 1 \qquad (j = 1, 2).$$

The index $I$ of $\boldsymbol{p}, \boldsymbol{q}$ in $\Lambda$ is

$$I = |\pi_1 \varkappa_2 - \pi_2 \varkappa_1|.$$

Hence $\pi_1 \varkappa_2 - \pi_2 \varkappa_1 = 0$ or $\pm 1$; which gives the three alternatives quoted.

**III.6.3.** We first illustrate MORDELL'S method with an example where the amount of subsidiary argument required is a minimum.

Let $\mathcal{K}$ be the cross-shaped 2-dimensional region defined by

$$\min\{|x_1|, |x_2|\} < 1, \quad \max\{|x_1|, |x_2|\} < \tfrac{3}{2}.$$

We shall show that

$$\Delta(\mathcal{K}) = 2$$

and that the only critical lattices of $\mathcal{K}$ are those with the following bases:

$$\Lambda_1 \text{ basis } (1, 1) \quad \text{and} \quad (1, -1)$$
$$\Lambda_2 \text{ basis } (\tfrac{3}{2}, -\tfrac{1}{2}) \text{ and } (-\tfrac{1}{2}, \tfrac{3}{2})$$
$$\Lambda_3 \text{ basis } (\tfrac{3}{2}, \tfrac{1}{2}) \quad \text{and} \quad (\tfrac{1}{2}, \tfrac{3}{2}).$$

It is readily verified that these lattices are $\mathcal{K}$-admissible and have determinant 2. Hence by Lemma 4, it is enough to show that any $\mathcal{K}$-admissible lattice $\Lambda$ with $d(\Lambda) = 2$ must be one of $\Lambda_1$, $\Lambda_2$, $\Lambda_3$.

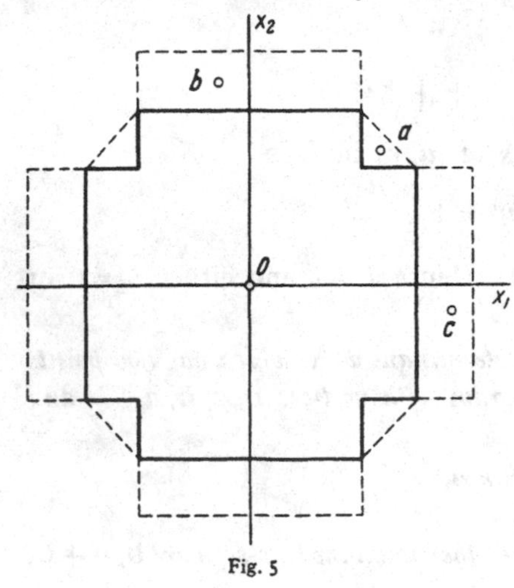

Fig. 5

From now on we suppose that

$$d(\Lambda) = 2: \quad \Lambda \text{ is } \mathcal{K}\text{-admissible.}$$

The convex symmetric octagon

$$\mathcal{S}_1: \quad |x_1| < \tfrac{3}{2}, \quad |x_2| < \tfrac{3}{2}, \quad |x_1| + |x_2| < \tfrac{5}{2}$$

has area

$$\tfrac{17}{2} > 2^2 d(\Lambda),$$

and so contains a point $\boldsymbol{a} \neq \boldsymbol{o}$ of $\Lambda$. The only points of $\mathcal{S}_1$ not in $\mathcal{K}$ are the four triangles with $|x_1| \geq 1$, $|x_2| \geq 1$ (see Fig. 5). Hence, by symmetry, we may suppose that $\Lambda$ contains a point $\boldsymbol{a} = (a_1, a_2)$ with

$$\boldsymbol{a}: \qquad 1 \leq a_1 < \tfrac{3}{2}, \quad 1 \leq a_2 < \tfrac{3}{2}, \quad a_1 + a_2 < \tfrac{5}{2}. \tag{1}$$

By Theorem III there is a point $\boldsymbol{b} \neq \boldsymbol{o}$ of $\Lambda$ in

$$|x_1| < 1 \quad |x_2| \leq 2.$$

On taking $-\boldsymbol{b}$ instead of $\boldsymbol{b}$ if necessary and using the fact that $\boldsymbol{b}$ is not in $\mathcal{K}$, we may assume that, the coordinates of $\boldsymbol{b}$ satisfy

$$\boldsymbol{b}: \qquad |b_1| < 1, \quad \tfrac{3}{2} \leq b_2 \leq 2. \tag{2}$$

Similarly there is a point $\boldsymbol{c}$ of $\Lambda$ satisfying

$$\boldsymbol{c}: \qquad \tfrac{3}{2} \leq c_1 \leq 2, \quad |c_2| < 1. \tag{3}$$

Now we show that $\boldsymbol{a}, \boldsymbol{b}$ is a basis for $\Lambda$. We have

$$\det(\boldsymbol{a}, \boldsymbol{b}) = a_1 b_2 - a_2 b_1 > 1 \cdot \tfrac{3}{2} - \tfrac{3}{2} \cdot 1 = 0$$

and

$$\det(\boldsymbol{a}, \boldsymbol{b}) < \tfrac{3}{2} \cdot 2 + \tfrac{3}{2} \cdot 1 = \tfrac{9}{2},$$

so

$$\det(\boldsymbol{a}, \boldsymbol{b}) = 2 \quad \text{or} \quad 4,$$

since $\det(\boldsymbol{a}, \boldsymbol{b})$ is an integral multiple of $d(\Lambda)$. Suppose first, if possible, that $\det(\boldsymbol{a}, \boldsymbol{b}) = 4$, so that the index of $\boldsymbol{a}, \boldsymbol{b}$ in $\Lambda$ is 2. For any integer $k > 1$ the points $k^{-1}\boldsymbol{a}$, $k^{-1}\boldsymbol{b}$ clearly lie in $\mathscr{K}$ and so are not in the $\mathscr{K}$-admissible lattice $\Lambda$: that is $\boldsymbol{a}$ and $\boldsymbol{b}$ are primitive points of $\Lambda$. We show now that $\tfrac{1}{2}(\boldsymbol{b} - \boldsymbol{a})$ is in $\Lambda$. Since $\boldsymbol{a}$ is primitive there is a basis $\boldsymbol{a}, \boldsymbol{d}$ where, say, $\det(\boldsymbol{a}, \boldsymbol{d}) = d(\Lambda) = 2$. Than $\boldsymbol{b} = u\boldsymbol{a} + v\boldsymbol{d}$ for some integers $u, v$; and indeed $v = 2$ since $\det(\boldsymbol{a}, \boldsymbol{b}) = 4 = 2\det(\boldsymbol{a}, \boldsymbol{d})$. Then $u$ is odd since $\boldsymbol{b}$ is primitive, so $\tfrac{1}{2}(\boldsymbol{b} - \boldsymbol{a})$ is in $\Lambda$ as asserted. But $\tfrac{1}{2}(\boldsymbol{b} - \boldsymbol{a})$ is clearly in $\mathscr{K}$, so we have a contradiction. Hence we can only have

$$\det(\boldsymbol{a}, \boldsymbol{b}) = 2 = d(\Lambda). \tag{4}$$

This gives the estimate

$$b_1 \geqq -\tfrac{1}{2}, \tag{5}$$

since otherwise we should have the contradiction

$$2 = a_1 b_2 - a_2 b_1 > 1 \cdot \tfrac{3}{2} + 1 \cdot \tfrac{1}{2}.$$

Similarly

$$\det(\boldsymbol{a}, \boldsymbol{c}) = -2 = -d(\Lambda) \tag{6}$$

and

$$c_2 \geqq -\tfrac{1}{2}. \tag{7}$$

Since $\boldsymbol{a}, \boldsymbol{b}$ is a basis for $\Lambda$ we have

$$\boldsymbol{c} = s\boldsymbol{a} + r\boldsymbol{b}$$

for some integers $r, s$. On substituting this in (6) and using (4) we obtain $r = -1$ and so

$$\boldsymbol{b} + \boldsymbol{c} = s\boldsymbol{a} \tag{8}$$

i.e.

$$b_1 + c_1 = s a_1, \quad b_2 + c_2 = s a_2. \tag{9}$$

But

$$\tfrac{1}{2} < b_1 + c_1 < 3, \quad 1 \leqq a_1 < \tfrac{3}{2}$$

by (1), (2), (3); so there are only the two possibilities

$$s = 1 \quad \text{or} \quad s = 2.$$

First case $s = 1$. From (1), (2), (3) and (9) we have

$$b_1 < 0, \quad c_2 < 0. \tag{10}$$

From (4), (5), (6) we have

$$\det(\boldsymbol{c}, \boldsymbol{b}) = 2$$

that is

$$c_1 b_2 - c_2 b_1 = 2.$$

But $c_1 \geqq \frac{3}{2}$, $b_2 \geqq \frac{3}{2}$ by (2) and (3); and $0 > b_1 \geqq -\frac{1}{2}$, $0 > c_2 \geqq -\frac{1}{2}$ by (5), (7) and (10). Hence (8) can hold only if

$$c_1 = b_2 = \tfrac{3}{2}, \quad c_2 = b_1 = -\tfrac{1}{2},$$

which gives the lattice $\Lambda_2$.

Second case $s = 2$. By (1), (2), (3) and (9) we now have

$$b_1 \geqq 0, \quad c_2 \geqq 0. \tag{11}$$

We now consider the lattice-point

$$(d_1, d_2) = \boldsymbol{d} = (\boldsymbol{b} - \boldsymbol{a}) = \tfrac{1}{2}(\boldsymbol{b} - \boldsymbol{c}).$$

By (2), (3) and (11) we have

$$0 \geqq 2d_1 = b_1 - c_1 \geqq -2,$$
$$0 \leqq 2d_2 = b_2 - c_2 \leqq 2.$$

Since $\boldsymbol{d}$ cannot be in $\mathscr{K}$ we must have $d_1 = -1$, $d_2 = +1$; that is $c_1 = b_2 = 2$, $b_1 = c_2 = 0$. This gives $a_1 = a_2 = 1$. Hence $\Lambda = \Lambda_1$.

In the proof we have made use of the symmetry of the figure. Since $\Lambda_1$ remains unchanged under transformation of $\mathscr{K}$ into itself, but $\Lambda_2$ and $\Lambda_3$ may be interchanged, we have shown that $\Lambda$ is one of $\Lambda_1, \Lambda_2, \Lambda_3$, as required.

**III.6.4.** As a second example of MORDELL's method we take the disc

$$\mathscr{D}: \quad x_1^2 + x_2^2 < 1,$$

which we have already discussed by other means (§ 5.2). We take $\Delta_0 = (\frac{3}{4})^{\frac{1}{2}}$ in Lemma 4. The lattices $\Lambda_c$ certainly exist; since they can be taken to be the lattices with a basis consisting of two of the vertices of an inscribed regular hexagon. We shall show that if $d(\Lambda) = (\frac{3}{4})^{\frac{1}{2}}$, then $\Lambda$ has a point other than $\boldsymbol{o}$ in $\mathscr{D}$ except when $\Lambda$ is a $\Lambda_c$.

There are certainly points of $\Lambda$ in the circle

$$x_1^2 + x_2^2 < 2,$$

since this has area $2\pi > 2^2 > 2^2 d(\Lambda)$. Since $\Lambda$ is $\mathscr{D}$-admissible, the point must lie in $1 \leqq x_1^2 + x_1^2 < 2$. After a suitable rotation of the coordinate system we may thus suppose without loss of generality that there is a point $\boldsymbol{p} = (p_1, p_2)$ in $\Lambda$ with

$$p_2 = -(\tfrac{3}{4})^{\frac{1}{2}}, \quad \tfrac{1}{2} \leqq p_1 < \tfrac{3}{2}.$$

But now, by Theorem III there is a point $\boldsymbol{q} = (q_1, q_2)$ other than $\boldsymbol{o}$ in the half-open parallelogram

$$\mathscr{Q}: \quad |x_1 + 3^{-\frac{1}{2}} x_2| \leqq 1, \quad |x_2| < \sqrt{\tfrac{3}{4}}$$

of area $2\sqrt{3}=4d(\Lambda)$ (see Fig. 6). The only portion of $\mathcal{Q}$ not contained in $\mathcal{D}$ is the curvilinear triangle $\mathcal{C}$ cut off by the arc of the circle between $\boldsymbol{a}=(1,0)$ and $\boldsymbol{b}=\left(\frac{1}{2},-\sqrt{\frac{3}{4}}\right)$ and the image of $\mathcal{C}$ in the origin. We may suppose without loss of generality that $\boldsymbol{q}$ is in $\mathcal{C}$.

Clearly both $\boldsymbol{p}$ and $\boldsymbol{q}$ are primitive, since, if either were of the shape $k\boldsymbol{u}$ with $\boldsymbol{u}\in\Lambda$ and integer $k>1$, then $\boldsymbol{u}$ would be in $\mathcal{D}$. Further $\boldsymbol{p}\neq\boldsymbol{q}$, since $p_2=-\sqrt{\frac{3}{4}}$ but $|q_2|<|\frac{3}{4}|$. We now apply Lemma 8. From what

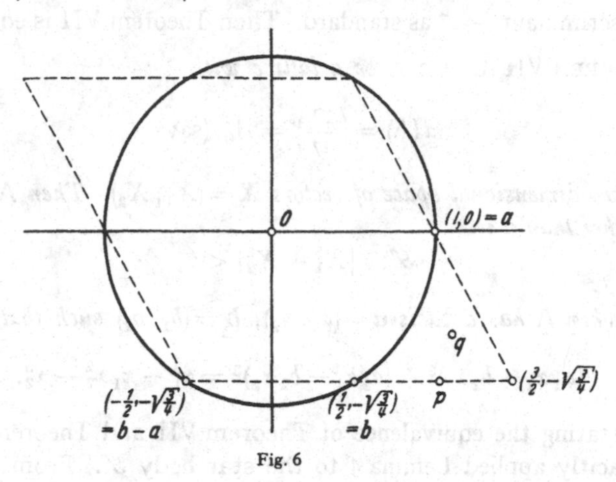


Fig. 6

has just been proved, $\boldsymbol{p},\boldsymbol{q},\boldsymbol{o}$ cannot be collinear. Hence either $\boldsymbol{p}=\boldsymbol{b}$ and $\boldsymbol{q}$ lies on the line-segment between $\boldsymbol{a}$ (inclusive) and $\boldsymbol{a}+\boldsymbol{b}$ (exclusive) or $\boldsymbol{q}=\boldsymbol{a}$ and $\boldsymbol{p}$ lies on the line-segment between $\boldsymbol{b}$ (inclusive) and $\boldsymbol{a}+\boldsymbol{b}$ (exclusive); and by symmetry we may suppose the second holds. Then $\boldsymbol{p}-\boldsymbol{q}=\boldsymbol{p}-\boldsymbol{a}$ lies between $\boldsymbol{b}-\boldsymbol{a}$ (inclusive) and $\boldsymbol{b}$ (exclusive). The only one of these points not in $\mathcal{D}$ is $\boldsymbol{b}-\boldsymbol{a}$. Hence also $\boldsymbol{p}=\boldsymbol{b}$. Hence $\Lambda$ is the lattice generated by $\boldsymbol{a}$ and $\boldsymbol{b}$. Since we made an arbitrary rotation of the coordinate system this completes the proof of the result stated.

**III.6.5.** As a final application of MORDELL's method we prove a result about binary cubic forms which fills a gap left in Chapter II, § 5. We use the same notation.

THEOREM VII. *If $f(x_1,x_2)$ be a binary cubic form of determinant $D<0$, there are integers $(u_1,u_2)\neq\boldsymbol{o}$ such that*

$$|f(\boldsymbol{u})|\leq\left|\frac{D}{23}\right|^{\frac{1}{4}}.$$

*The sign of equality is needed when and only when $f$ is equivalent to a multiple of*

$$x_1^3-x_1x_2^2-x_2^3.$$

This is the most important part of Theorem XI of Chapter II, which was left unproved. As already remarked, the form here is transformed into the form there by the substitution $x_1 \to x_1$, $x_2 \to -(x_1+x_2)$. We already noted that the exceptional form does require the sign of equality since it has $D=-23$ and represents only integers other than 0.

We must first express Theorem VII in a geometrical form. We saw in Chapter II that any two binary cubics with negative discriminant can be transformed into one another. It is convenient to take $X_1^3+X_2^3$ with discriminant $-27$ as standard. Then Theorem VII is equivalent to

THEOREM VII A. *Let $\Lambda$ be a lattice with*

$$d(\Lambda)=\left(\frac{23}{27}\right)^{\frac{1}{6}}=\Delta_0 \text{ (say)} \tag{1}$$

*in the two-dimensional space of vectors $\mathbf{X}=(X_1, X_2)$. Then $\Lambda$ contains a point other than $\mathbf{o}$ in*

$$\mathscr{S}: \ |X_1^3+X_2^3|<1, \tag{2}$$

*except when $\Lambda$ has a basis $\mathbf{a}=(a_1, a_2)$, $\mathbf{b}=(b_1, b_2)$ such that identically*

$$(a_1 x_1-b_1 x_2)^3+(a_2 x_1-b_2 x_2)^3=x_1^3-x_1 x_2^2-x_2^3. \tag{3}$$

In stating the equivalence of Theorem VII and Theorem VII A we have tacitly applied Lemma 4 to the star body $\mathscr{S}$. From now on we shall be concerned only with Theorem VII A. We use capital letters to denote points and coordinates, except that $\mathbf{o}$ is still the origin. Further, $\Lambda$ is a lattice with $d(\Lambda)$ given by (1) which has no point other than $\mathbf{o}$ in the set $\mathscr{S}$ defined by (2). The set $\mathscr{S}$ is shown in Fig. 7.

First, since $\Delta_0<1$, there is certainly a point $\mathbf{P}\neq\mathbf{o}$ of $\Lambda$ in the square

$$|X_1|<1, \quad |X_2|<1. \tag{4}$$

Since $\mathbf{P}$ does not lie in $\mathscr{S}$, either $\mathbf{P}$ or $-\mathbf{P}$ must lie in the first quadrant and we may suppose without loss of generality that

$$0 \leqq P_1<1, \quad 0 \leqq P_2<1. \tag{5}$$

From Fig. 7 (or from elementary algebra) we must have

$$P_1+P_2 \geqq 1. \tag{6}$$

Suppose, if possible, that there were two such points, $\mathbf{P}$ and $\mathbf{P}'$. Then their difference $\mathbf{P}''=\mathbf{P}-\mathbf{P}'$ satisfies (4). Hence on interchanging $\mathbf{P}$ and $\mathbf{P}'$ if need be, we may suppose that $\mathbf{P}''$ is in the first quadrant: of course it may coincide with $\mathbf{P}$ or $\mathbf{P}'$. Hence

$$\mathbf{P}=\mathbf{P}'+\mathbf{P}''.$$

But now, in the obvious notation, we have $P_1' + P_2' \geqq 1$, $P_1'' + P_2'' \geqq 1$, since neither $\boldsymbol{P}'$ nor $\boldsymbol{P}''$ is in $\mathscr{S}$. Hence we should have

$$P_1 + P_2 = (P_1' + P_2') + (P_1'' + P_2'') \geqq 2,$$

in contradiction to (5). To sum up what we have proved so far: there is precisely one pair of points $\pm \boldsymbol{P} \in \Lambda$ other than $\boldsymbol{o}$ in the square

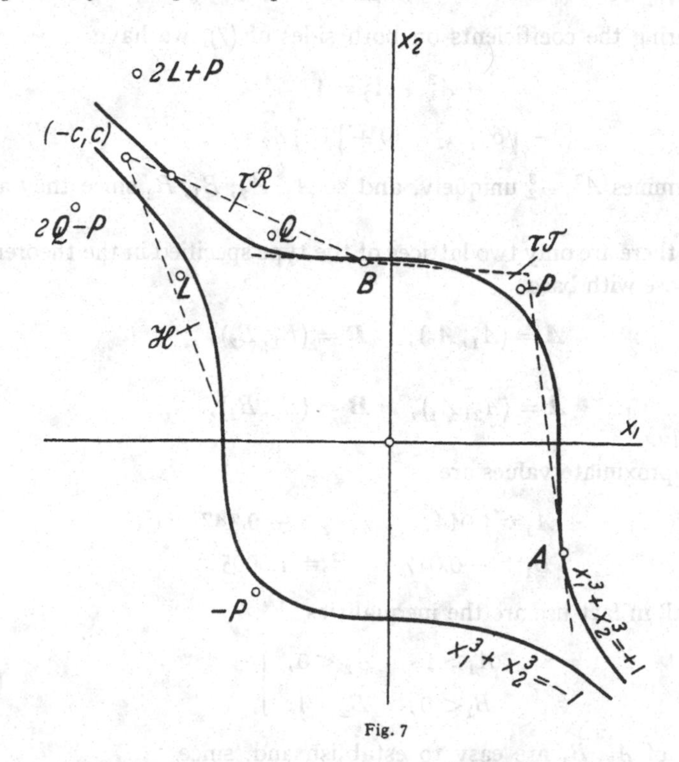


Fig. 7

$|X_1| < 1$, $|X_2| < 1$. We denote from now on by $\boldsymbol{P}$ the point of $\Lambda$ which satisfies (5) and (6).

We now examine more closely the lattices which satisfy (3). We must make use of the algebra developed in Chapter II, § 5. Let $A_1$, $B_1$, $A_2$, $B_2$ be any numbers such that identically

$$(A_1 x_1 - B_1 x_2)^3 + (A_2 x_1 - B_2 x_2)^3 = x_1^3 - x_1 x_2^2 - x_2^3 = f_0(\boldsymbol{x}) \text{ (say)}. \qquad (7)$$

On equating the hessians of both sides, we obtain

$$-9(A_1 B_2 - A_2 B_1)^2 (A_1 x_1 - B_1 x_2)(A_2 x_1 - B_2 x_2)$$
$$= \frac{1}{4}\left\{\left(\frac{\partial^2 f_0}{\partial x_1 \partial x_2}\right)^2 - \frac{\partial^2 f_0}{\partial x_1^2}\frac{\partial^2 f_0}{\partial x_2^2}\right\} = 3 x_1^2 + 9 x_1 x_2 + x_2^2.$$

The linear factors of both sides must coincide, and so, after interchanging $A_1$, $B_1$ and $A_2$, $B_2$ if need be we have

$$A_1 x_1 - B_1 x_2 = A_1 \left\{ x_1 + \frac{9-\sqrt{69}}{6}\, x_2 \right\},$$

$$A_2 x_1 - B_2 x_2 = A_2 \left\{ x_1 + \frac{9+\sqrt{69}}{6}\, x_2 \right\}.$$

On comparing the coefficients on both sides of (7), we have

$$A_1^3 + A_2^3 = 1$$

$$(9-\sqrt{69})\, A_1^3 + (9+\sqrt{69})\, A_2^3 = 0.$$

This determines $A_1^3$, $A_2^3$ uniquely, and so $A_1$, $A_2$, $B_1$, $B_2$ since they are all real.

Hence there are only two lattices of the type specified in the theorem, namely those with base

$$\boldsymbol{A} = (A_1, A_2), \quad \boldsymbol{B} = (B_1, B_2),$$

and

$$\tilde{\boldsymbol{A}} = (A_2, A_1), \quad \tilde{\boldsymbol{B}} = (B_2, B_1),$$

respectively.

The approximate values are

$$A_1 \doteqdot 1.014, \qquad A_2 \doteqdot -0.347$$

$$B_1 \doteqdot -0.017, \qquad B_2 \doteqdot 1.0005.$$

All we shall in fact use are the inequalities

$$\left.\begin{array}{ll} A_1 > 1, & A_2 < 0, \\ B_1 < 0, & B_2 > 1. \end{array}\right\} \tag{8}$$

The signs of $A_2$, $B_1$ are easy to establish and, since

$$A_1^3 + A_2^3 = B_1^3 + B_2^3 = 1,$$

by (7), the rest follows.

Comparison of discriminants on both sides of (7) gives

$$27(A_1 B_2 - A_2 B_1)^6 = 23,$$

and so

$$A_1 B_2 - A_2 B_1 = \pm \Delta_0,$$

where in fact the $+$ sign holds, but we do not use this information.

Let $\boldsymbol{X} = \boldsymbol{\tau x}$ be the transformation of the plane $\boldsymbol{X} = (X_1, X_2)$ into the plane $\boldsymbol{x} = (x_1, x_2)$ given by

$$X_1 = A_1 x_1 - B_1 x_2, \quad X_2 = A_2 x_1 - B_2 x_2.$$

Then the region $\boldsymbol{\tau}^{-1}\mathscr{S}$ of points $\boldsymbol{\tau}^{-1}\boldsymbol{X}$, $\boldsymbol{X}\in\mathscr{S}$ is given by

$$|x_1^3 - x_1 x_2^2 - x_2^3| < 1.$$

Further, $\boldsymbol{\tau}^{-1}\Lambda$ is a lattice of determinant

$$d(\boldsymbol{\tau}^{-1}\Lambda) = |\det(\boldsymbol{\tau})|^{-1} d(\Lambda) = 1 \tag{9}$$

(cf. Chapter I, § 3).

The region $\boldsymbol{\tau}^{-1}\mathscr{S}$ is shown in Fig. 8. The line $x_1 = 1$ touches $f_0(\boldsymbol{x}) = x_1^3 - x_1 x_2^2 - x_2^3 = 1$ at $x_2 = 0$ and meets it again at $x_2 = -1$. The line

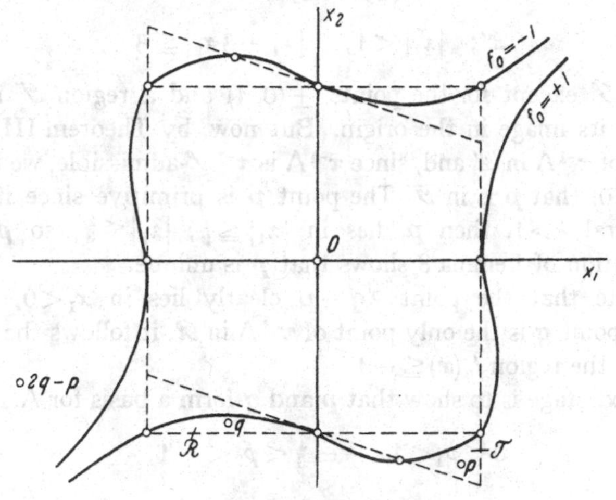


Fig. 8. $f_0 = x_1^3 - x_1 x_2^2 - x_2^3$

$x_2 = 1$ meets $x_1^3 - x_1 x_2^2 - x_2^3 = -1$ at $x_1 = 0, \pm 1$. Since no line meets a cubic curve in more than three points, it follows readily that the whole of the unit square

$$\mathscr{Q}: \quad |x_1| < 1, \quad |x_2| < 1$$

lies in $\boldsymbol{\tau}^{-1}\mathscr{S}$, except for a small region $\mathscr{R}$ in $x_1 < 0$, $x_2 < 0$ and the image $-\mathscr{R}$ of $\mathscr{R}$ in the origin.

Suppose first that $(1, 0) \in \boldsymbol{\tau}^{-1}\Lambda$. Since $d(\boldsymbol{\tau}^{-1}\Lambda) = 1$, there are points of $\boldsymbol{\tau}^{-1}\Lambda$ on the line $x_2 = 1$ spaced unit distance apart, by Lemma 5. Since none of these can lie in $\boldsymbol{\tau}^{-1}\mathscr{S}$ the only possibility is that $\boldsymbol{\tau}^{-1}\Lambda = \Lambda_0$, the lattice of points with integral co-ordinates. But then $\Lambda = \boldsymbol{\tau}\Lambda_0$, which is one of the exceptional lattices permitted by the theorem. Similarly, if $(0, 1) \in \boldsymbol{\tau}^{-1}\Lambda$, then $\Lambda = \boldsymbol{\tau}\Lambda_0$. Hence from now on we may assume that

$$(1, 0) \notin \boldsymbol{\tau}^{-1}\Lambda, \quad (0, 1) \notin \boldsymbol{\tau}^{-1}\Lambda. \tag{10}$$

By Lemma 7, either there is a point $\boldsymbol{q} \neq \boldsymbol{o}$ of $\Lambda$ in the square $\mathscr{Q}$, or $(1, 0) \in \Lambda$, or $(0, 1) \in \Lambda$. But the second and third alternatives have already been disposed of, and so, since $\boldsymbol{q}$ cannot lie in $\boldsymbol{\tau}^{-1}\mathscr{S}$, we may suppose that $\boldsymbol{q}$ is in $\mathscr{R}$. Further, $\boldsymbol{q}$ must be primitive, since if $\boldsymbol{q}=k\boldsymbol{q}_1$, with integral $k>1$ and $\boldsymbol{q}_1 \in \boldsymbol{\tau}^{-1}\Lambda$, then $\boldsymbol{q}_1$ would lie in $|x_1|<\frac{1}{2}$, $|x_2|<\frac{1}{2}$ and so certainly[1] in $\boldsymbol{\tau}^{-1}\mathscr{S}$, contrary to the hypothesis that $\boldsymbol{\tau}^{-1}\Lambda$ is $\boldsymbol{\tau}^{-1}\mathscr{S}$-admissible. Hence $\boldsymbol{q}$ is unique by Lemma 8.

We require another point of $\boldsymbol{\tau}^{-1}\Lambda$. The tangent to $f_0(\boldsymbol{x})=1$ at $(0, -1)$ is

$$-x_1-3x_2=3.$$

This meets $f_0(\boldsymbol{x})=1$ again at $\left(\frac{9}{25}, \frac{-28}{25}\right)$. Hence all of the parallelogram

$$\mathscr{Q}': \quad |x_1|<1, \qquad |x_1+3x_2|\leqq 3$$

lies in $\boldsymbol{\tau}^{-1}\mathscr{S}$ except for the points $\pm(0, 1)$ and a region $\mathscr{T}$ in $x_2<0$, $x_1>0$ and its image in the origin. But now, by Theorem III, there is a point $\boldsymbol{p}$ of $\boldsymbol{\tau}^{-1}\Lambda$ in $\mathscr{Q}'$ and, since $\boldsymbol{\tau}^{-1}\Lambda$ is $\boldsymbol{\tau}^{-1}\mathscr{S}$-admissible, we may suppose by (10) that $\boldsymbol{p}$ is in $\mathscr{T}$. The point $\boldsymbol{p}$ is primitive since if $\boldsymbol{p}=k\boldsymbol{p}_1$ with integral $k>1$, then $\boldsymbol{p}_1$ lies in $|x_1|\leqq\frac{1}{2}$, $|x_2|\leqq\frac{2}{3}$, so $\boldsymbol{p}_1\in\boldsymbol{\tau}^{-1}\mathscr{S}$. An application of Lemma 8 shows that $\boldsymbol{p}$ is unique.

We note that the point $2\boldsymbol{q}-\boldsymbol{p}$ clearly lies in $x_1<0$, $x_2>-1$. Since the point $\boldsymbol{q}$ is the only point of $\boldsymbol{\tau}^{-1}\Lambda$ in $\mathscr{R}$, it follows that $2\boldsymbol{q}-\boldsymbol{p}$ must lie in the region $f_0(\boldsymbol{x})\leqq -1$.

The next stage is to show that $\boldsymbol{p}$ and $\boldsymbol{q}$ form a basis for $\Lambda$. We have

$$0<p_1<1, \qquad -\tfrac{4}{3}<p_2<-1, \tag{11}$$

and

$$-1<q_1<0, \qquad -1<q_2<0. \tag{12}$$

Hence

$$\det(\boldsymbol{p},\boldsymbol{q})=p_1q_2-q_1p_2\begin{cases}<0\\ >-1.1-1.\tfrac{4}{3}>-3.\end{cases}$$

Since $\det(\boldsymbol{p},\boldsymbol{q})$ is a multiple of $\det(\boldsymbol{\tau}^{-1}\Lambda)=1$, the only possibilities are

$$\det(\boldsymbol{p},\boldsymbol{q})=-1$$

or

$$\det(\boldsymbol{p},\boldsymbol{q})=-2.$$

In the first case, $\boldsymbol{p},\boldsymbol{q}$ are a basis. Suppose, if possible, that $\det(\boldsymbol{p},\boldsymbol{q})=-2$. Since $\boldsymbol{p}$ is primitive, there is a basis $\boldsymbol{p},\boldsymbol{r}$, where $\det(\boldsymbol{p},\boldsymbol{r})=\pm d(\Lambda)=\pm 1$. Write

$$\boldsymbol{q}=u\boldsymbol{p}+v\boldsymbol{r}$$

where $u$ and $v$ are integers. Then

$$\det(\boldsymbol{p},\boldsymbol{q})=v\det(\boldsymbol{p},\boldsymbol{r}),$$

[1] For then $|f_0(\boldsymbol{q}_1)|\leqq|x_1|^3+|x_1||x_2|^2+|x_2|^3<\frac{3}{8}<1$. We shall not explicitly give such trivial estimations later.

so $v = \pm 2$. Now $u$ must be odd, since $\boldsymbol{q}$ is primitive, and so

$$\boldsymbol{t} = \tfrac{1}{2}(\boldsymbol{p} - \boldsymbol{q}) \in \boldsymbol{\tau}^{-1}\Lambda.$$

But then, by (11) and (12),

$$0 < t_1 < 1, \quad -\tfrac{2}{3} < t_2 < 0;$$

a trivial estimation shows that $|f_0(\boldsymbol{t})| < 1$, and so $\boldsymbol{t}$ would be in $\boldsymbol{\tau}^{-1}\mathscr{S}$, contrary to hypothesis.

To sum up: there is a basis $\boldsymbol{p} \in \mathscr{T}$, $\boldsymbol{q} \in \mathscr{R}$ of $\boldsymbol{\tau}^{-1}\Lambda$. The point $2\boldsymbol{q} - \boldsymbol{p}$ lies in $f_0(\boldsymbol{x}) \leqq -1$. There are no other points of $\tau^{-1}\Lambda$ in $\mathscr{T}$ or $\mathscr{R}$.

We must now translate our facts about $\boldsymbol{\tau}^{-1}\Lambda$ into facts about $\Lambda$. We write

$$\boldsymbol{A} = (A_1, A_2), \quad \boldsymbol{B} = (B_1, B_2).$$

The region $\boldsymbol{\tau}\mathscr{R}$ is bounded by the curve

$$X_1^3 + X_2^3 = 1,$$

the transform of $f_0(\boldsymbol{x}) = 1$, and the line-segment joining the points

$$\boldsymbol{\tau}(0, -1) = \boldsymbol{B}, \quad \boldsymbol{\tau}(-1, -1) = -\boldsymbol{A} + \boldsymbol{B},$$

and so is roughly as shown in Fig. 7. The point

$$\boldsymbol{Q} \text{ (say)} = \boldsymbol{\tau}\boldsymbol{q}$$

lies in $\boldsymbol{\tau}\mathscr{R}$.

Similarly $\boldsymbol{\tau}\mathscr{T}$ is bounded by $X_1^3 + X_2^3 = 1$ and the tangents at $\boldsymbol{\tau}(0, -1) = \boldsymbol{B}$ and at $\boldsymbol{\tau}(1, 0) = \boldsymbol{A}$. We now show that $\boldsymbol{\tau}\mathscr{T}$ lies in

$$0 < X_1 < 1, \quad 0 < X_2 < 1. \tag{13}$$

Indeed, since $B_2 > 1$, the tangent to $X_1^3 + X_2^3 = 1$ at $\boldsymbol{B}$ has negative gradient and so meets $X_1^3 + X_2^3 = 1$ again at a point in (13). Since $\boldsymbol{\tau}\mathscr{T}$ lies below this tangent, its points satisfy $X_2 < 1$. Similarly, since $A_1 > 1$, the points of $\boldsymbol{\tau}\mathscr{T}$ satisfy $X_1 < 1$. They clearly satisfy $X_1 > 0$, $X_2 > 0$.

But now we saw earlier that there is only one point, $\boldsymbol{P}$, of $\Lambda$ in (13). Since $\boldsymbol{\tau}\boldsymbol{p}$ is in $\boldsymbol{\tau}\mathscr{T}$ we must have

$$\boldsymbol{\tau}\boldsymbol{p} = \boldsymbol{P}.$$

To sum up the results of our translation: there is precisely one point $\boldsymbol{Q} \in \Lambda$ in $\boldsymbol{\tau}\mathscr{R}$. This point $\boldsymbol{Q}$ together with the unique point $\boldsymbol{P}$ of $\Lambda$ in (13) form a basis for $\Lambda$. The point $2\boldsymbol{Q} - \boldsymbol{P}$ lies in $X_1^3 + X_2^3 \leqq -1$.

Let $\mathscr{H}$ be the mirror image of $\boldsymbol{\tau}\mathscr{R}$ in $X_1 + X_2 = 0$. By symmetry there is precisely one point $\boldsymbol{L}$, say, of $\Lambda$ in $\mathscr{H}$: this point together with $-\boldsymbol{P}$ forms a basis for $\Lambda$, and the point $2\boldsymbol{L} + \boldsymbol{P}$ lies in $X_1^3 + X_2^3 \geqq 1$.

But now every point of the triangle $\boldsymbol{o}\boldsymbol{L}\boldsymbol{Q}$ is in one of the regions $\mathscr{S}, \boldsymbol{\tau}\mathscr{R}$ and $\mathscr{H}$. By hypothesis there is no point of $\Lambda$ in $\mathscr{S}$, and we proved that $\boldsymbol{Q}, \boldsymbol{L}$ are the only points of $\Lambda$ in $\boldsymbol{\tau}\mathscr{R}, \mathscr{H}$ respectively. Hence $\boldsymbol{Q}, \boldsymbol{L}$ forms a basis of $\Lambda$ by Lemma 6.

We have three bases $\boldsymbol{P}, \boldsymbol{Q}$: $\boldsymbol{Q}, \boldsymbol{L}$ and $\boldsymbol{L}, -\boldsymbol{P}$ for $\Lambda$ and must study their relations. Now

$$\det(\boldsymbol{P}, \boldsymbol{Q}) = \det(\boldsymbol{Q}, \boldsymbol{L}) = \det(\boldsymbol{L}, -\boldsymbol{P}) = d(\Lambda),$$

since the determinants are $\pm d(\Lambda)$ and are clearly positive. Write

$$\boldsymbol{P} = u\boldsymbol{Q} + v\boldsymbol{L}.$$

Then

$$\det(\boldsymbol{P}, \boldsymbol{Q}) = v \det(\boldsymbol{L}, \boldsymbol{Q}),$$
$$\det(\boldsymbol{P}, \boldsymbol{L}) = u \det(\boldsymbol{P}, \boldsymbol{Q}).$$

Hence

$$\boldsymbol{P} = \boldsymbol{Q} - \boldsymbol{L}.$$

We have now reached a contradiction, since

$$2\boldsymbol{Q} - \boldsymbol{P} = 2\boldsymbol{L} + \boldsymbol{P},$$

and this point has been shown to lie both in $X_1^3 + X_2^3 \geqq 1$ and in $X_1^3 + X_2^3 \leqq -1$. The contradiction shows that there are no $\mathscr{S}$-admissible lattices with $d(\Lambda) = \Delta_0$ except those mentioned in the enunciation of the theorem.

We have shown rather more. Let the line joining $\boldsymbol{B}$ and $-\boldsymbol{A}+\boldsymbol{B}$ (which forms part of the boundary of $\boldsymbol{\tau}\mathscr{R}$) meet $X_1 + X_2 = 0$ in the point $(-c, c)$. Then it is clear that our argument shows that there is a point of every lattice $\Lambda$ with $d(\Lambda) \leqq \Delta_0$ in the bounded region

$$|X_1^3 + X_2^3| < 1, \quad \max\{|X_1|, |X_2|\} \leqq c,$$

except when $\Lambda$ is one of the two critical lattices. That is, $|X_1^3 + X_2^3| < 1$ is boundedly reducible and indeed fully reducible in the sense of Chapter V, § 7.

**III.7. Representation of integers by quadratic forms**[1]. In this section we digress to present a number of results in the arithmetic theory of quadratic forms which can be proved very simply by the methods of the geometry of numbers. The principle tool is the following lemma.

LEMMA 9. *Let $n, m, k_1, \ldots, k_m$ be positive integers and $a_{ij}$ ($1 \leqq i \leqq m$, $1 \leqq j \leqq n$) be integers. The set $\Lambda$ of points $\boldsymbol{u}$ with integral co-ordinates*

[1] This section is not used later.

*satisfying the congruences*[1]

$$\sum_{1\leq j\leq n} a_{ij}u_j \equiv 0 \quad (k_i) \qquad (1\leq i\leq m)$$

*is a lattice with the determinant*

$$d(\Lambda) \leqq k_1 \ldots k_m.$$

That $\Lambda$ is a lattice follows, for example, from Theorem VI. Two points $\boldsymbol{u}$ and $\boldsymbol{v}$ of the lattice $\Lambda_0$ of all integral vectors are in the same class with respect to $\Lambda$ if and only if

$$\sum_j a_{ij}u_j \equiv \sum_j a_{ij}v_j \quad (k_i) \qquad (1\leq i\leq m).$$

Hence the index $I$ of $\Lambda$ in $\Lambda_0$, that is the number of classes, is at most $\prod_i k_i$, so

$$d(\Lambda) = I\, d(\Lambda_0) \leqq \prod_i k_i$$

(compare Lemma 1 of Chapter I).

**III.7.2.** As a first example we show that every prime number $p \equiv 1\,(4)$ is the sum of the squares of two integers. For then, as is well known, there is an integer $i$ such that

$$i^2 + 1 \equiv 0 \quad (p).$$

The set of integers $(u_1, u_2)$ such that

$$u_2 \equiv i\,u_1 \quad (p) \tag{1}$$

is, by Lemma 9, a lattice $\Lambda$ of determinant $d(\Lambda) \leqq p$. Hence by MINKOWSKI's convex body Theorem II there is certainly a point of $\Lambda$ in the disc

$$\mathscr{D}: \quad x_1^2 + x_2^2 < 2p$$

of area $V(\mathscr{D}) = 2\pi p > 2^2 d(\Lambda)$. Hence there are integers $u_1, u_2$ not both 0 satisfying (1) and

$$u_1^2 + u_2^2 < 2p.$$

But (1) implies

$$u_1^2 + u_2^2 \equiv u_1^2(1 + i^2) \equiv 0 \quad (p),$$

and so $u_1^2 + u_2^2 = p$, as required. The method is readily extended to show that a positive integer is the sum of two squares provided that it is not divisible by a prime $p \equiv -1\,(4)$.

**III.7.3.** As a second example, we shall show that every positive integer $m$ is of the shape

$$m = u_1^2 + u_2^2 + u_3^2 + u_4^2$$

[1] By $a \equiv b\ (k)$ we mean that $a - b$ is divisible by $k$.

with integral $u_1, u_2, u_1, u_4$. We may suppose without loss of generality that $m$ is not divisible by a square other than 1, so

$$m = p_1 \dots p_g,$$

with distinct primes $p_1, \dots, p_g$. We now show that to every prime $p$ there exist integers $a_p, b_p$ such that

$$a_p^2 + b_p^2 + 1 \equiv 0 \quad (p). \tag{1}$$

Indeed when $p$ is odd the numbers

$$a^2 \qquad (0 \leqq a < \tfrac{1}{2}p), \tag{2}$$

and

$$-1 - b^2 \qquad (0 \leqq b < \tfrac{1}{2}p) \tag{3}$$

are each a set of $\frac{1}{2}(p+1)$ integers which are incongruent modulo $p$. Since there are only $p$ classes modulo $p$, there must be some integer $c$ which is congruent to an element of each set (2) and (3), that is $a_p^2 \equiv c \equiv -1 - b_p^2$, so $a_p^2 + b_p^2 + 1 \equiv 0$. If $p = 2$, then $a_2 = 1$, $b_2 = 0$ will do.

We now consider (cf. DAVENPORT 1947b) the lattice of integral $\boldsymbol{u} = (u_1, \dots, u_4)$ which satisfy the $2g$ congruences

$$\left.\begin{aligned} u_1 &\equiv a_p u_3 + b_p u_4 \quad (p) \\ u_2 &\equiv b_p u_3 - a_p u_4 \quad (p) \end{aligned}\right\} \tag{4}$$

for $p = p_1, \dots, p_g$. By Lemma 9, these form a lattice $\Lambda$ of determinant

$$d(\Lambda) \leqq p_1^2 \dots p_g^2 = m^2.$$

Hence there is a lattice-point other than $\boldsymbol{o}$ in the set

$$x_1^2 + x_2^2 + x_3^2 + x_4^2 < 2m$$

of volume

$$\tfrac{1}{2}\pi^2 (2m)^2 > 2^4 m^2 \geqq 2^4 d(\Lambda).$$

If $\boldsymbol{u}$ is this point, then

$$0 < u_1^2 + u_2^2 + u_3^2 + u_4^2 < 2m$$

and, by (1) and (4),

$$u_1^2 + u_2^2 + u_3^2 + u_4^2 \equiv (a_p^2 + b_p^2 + 1)\, u_3^2 + (a_p^2 + b_p^2 + 1)\, u_4^2 \equiv 0 \quad (p)$$

for $p = p_1, \dots, p_g$; that is $m$ divides $u_1^2 + \dots + u_4^2$. This proves the result.

**III.7.4.** A famous theorem of LEGENDRE states that a ternary quadratic form $f(x_1, x_2, x_3)$ with rational coefficients represents 0 if obviously necessary congruence conditions are satisfied. Following DAVENPORT and MARSHALL HALL (1948a) and MORDELL (1951a) we verify this in a particular case, to which indeed the general case may be reduced by simple arguments.

Let

$$f(\boldsymbol{x}) = a_1 x_1^2 + a_2 x_2^2 + a_3 x_3^2,$$

where $a_1, a_2, a_3$ are square-free integers no two of which have a common factor, so $a_1 a_2 a_3$ is square-free. We show that there exist integers $\boldsymbol{u} \neq \boldsymbol{o}$ such that $f(\boldsymbol{u}) = 0$ provided that the following two conditions are satisfied

(i) there are integers $A_1, A_2, A_3$ such that

$$a_1 + A_3^2 a_2 \equiv 0 \ (a_3), \quad a_2 + A_1^2 a_3 \equiv 0 \ (a_1), \quad a_3 + A_2^2 a_1 \equiv 0 \ (a_2)$$

and

(ii) there are integers $v_1, v_2, v_3$ not all even such that

$$a_1 v_1^2 + a_2 v_2^2 + a_3 v_3^2 \equiv 0 \quad (2^{2+\lambda}),$$

where $\lambda = 1$ or 0 according as $a_1 a_2 a_3$ is even or odd.

Let

$$|a_1 a_2 a_3| = 2^\lambda p_1 \dots p_g$$

where $p_1, \dots, p_g$ are distinct odd primes and $\lambda = 1$ or 0. We shall take for $\Lambda$ the integral vectors $\boldsymbol{u} = (u_1, u_2, u_3)$ satisfying the following congruence conditions.

(I) Let $p$ be one of $p_1, \dots, p_g$. By symmetry we may suppose that $a_3 \equiv 0 \ (p)$. We impose the condition

$$u_2 \equiv A_3 u_1 \quad (p).$$

Then

$$a_1 u_1^2 + a_2 u_2^2 + a_3 u_3^2 \equiv a_1 u_1^2 + a_2 u_2^2 \equiv (a_1 + a_2 A_3^2) u_1^2 \equiv 0 \quad (p).$$

(II$_\alpha$) Suppose $\lambda = 0$, so $a_1, a_2, a_3$ are all odd. Now

$$v^2 \equiv 0 \quad \text{or} \quad 1 \quad (2^2)$$

for any integer $v$. In condition (ii) precisely one of $v_1, v_2, v_3$ must be even, say $v_3$. Then

$$0 \equiv a_1 v_1^2 + a_2 v_2^2 + a_3 v_3^2 \equiv a_1 + a_2 \quad (2^2).$$

We impose the two congruences

$$\left.\begin{aligned} u_1 &\equiv u_2 \quad (2), \\ u_3 &\equiv 0 \quad (2). \end{aligned}\right\}$$

Then

$$a_1 u_1^2 + a_2 u_2^2 + a_3 u_3^2 \equiv a_1 u_1^2 + a_2 u_2^2 \equiv 0 \quad (2^2).$$

(II$_\beta$) Suppose $\lambda = 1$, so one of $a_1, a_2, a_3$ is even, say $a_3$. Then $a_1 v_1^2 + a_2 v_2^2$ is even, so $v_1, v_2$ are both even or both odd. If $v_1, v_2$ were even then

$$a_3 v_3^2 = - a_1 v_1^2 - a_2 v_2^2$$

would be divisible by $2^2$, so $v_3$ would be even. Hence $v_1, v_2$ in (ii) must be odd, and

$$0 \equiv a_1 v_1^2 + a_2 v_2^2 + a_3 v_3^2 \equiv a_1 + a_2 + a_3 v_3^2, \quad (2^3)$$

since $v^2 \equiv 1$ $(2^3)$ if $v$ is odd. We impose the two conditions

$$\left.\begin{array}{ll} u_1 \equiv u_2 & (2^2), \\ u_3 \equiv v_3 u_1 & (2). \end{array}\right\}$$

Then it is readily verified that

$$a_1 u_1^2 + a_2 u_2^2 + a_3 u_3^2 \equiv 0. \quad (2^3)$$

In any case the lattice $\Lambda$ of integers $\boldsymbol{u}$ has determinant

$$d(\Lambda) \leqq 2^{\lambda+2} p_1 \dots p_g = 4|a_1 a_2 a_3|,$$

and the congruence conditions imply that

$$a_1 u_1^2 + a_2 u_2^2 + a_3 u_3^2 \equiv 0 \qquad (\text{mod } 2^{\lambda+2} p_1 \dots p_g = 4|a_1 a_2 a_3|).$$

But now, by MINKOWSKI's convex body theorem, there is a lattice point not $\boldsymbol{o}$ in the ellipsoid

$$\mathscr{E}: \quad |a_1| x_1^2 + |a_2| x_2^2 + |a_3| x_3^2 < 4|a_1 a_2 a_3|$$

of volume

$$V(\mathscr{E}) = \frac{\pi}{3} \cdot 2^5 |a_1 a_2 a_3| > 2^3 d(\Lambda).$$

If $\boldsymbol{u} \neq \boldsymbol{o}$ is the lattice point in $\mathscr{E}$, we must have $a_1 u_1^2 + a_2 u_2^2 + a_3 u_3^2 = 0$, since it is divisible by $4 a_1 a_2 a_3$; and

$$|a_1 u_1^2 + a_2 u_2^2 + a_3 u_3^2| \leqq |a_1| u_1^2 + |a_2| u_2^2 + |a_3| u_3^2 < 4|a_1 a_2 a_3|.$$

We conclude with a couple of remarks. An obviously necessary condition for solubility of $a_1 u_1^2 + a_2 u_2^2 + a_3 u_3^2 = 0$ is that $a_1, a_2, a_3$ should not all have the same sign. We did not use this at all. Hence this condition must be derivable from the others. The reader might verify that this can be done by means of the law of quadratic reciprocity.

In the second place we have not merely shown the existence of a solution, but we have shown that there is one which satisfies

$$|a_1| u_1^2 + |a_2| u_2^2 + |a_3| u_3^2 < 4|a_1 a_2 a_3|.$$

The right-hand side here may be improved to $2^{\frac{5}{3}}|a_1 a_2 a_3|$ by the use of the precise Theorem III of Chapter II instead of Theorem II, as the reader can easily verify.

Chapter IV

# Distance-Functions

**IV.1. Introduction.** In this chapter we introduce a number of concepts which are useful tools in all that follows.

**IV.1.2.** A distance-function $F(\boldsymbol{x})$ of variable vector $\boldsymbol{x}$ is any function which is

(i) non-negative, i.e. $F(\boldsymbol{x}) \geqq 0$,

(ii) continuous,

and

(iii) has the homogeneity-property that

$$F(t\boldsymbol{x}) = tF(\boldsymbol{x}) \qquad (t \geqq 0, \text{ real}).$$

The set $\mathscr{S}$ defined by

$$\mathscr{S}: \ F(\boldsymbol{x}) < 1 \tag{1}$$

turns out to be a star-body in the sense already introduced in the last chapter: that is, the origin $\boldsymbol{o}$ is an inner point of $\mathscr{S}$ and a radius vector

$$t\boldsymbol{x}_0 \qquad (0 \leqq t < \infty)$$

either lies entirely in $\mathscr{S}$ [which happens when $F(\boldsymbol{x}_0) = 0$] or there is a real number $t_0 = \{F(\boldsymbol{x}_0)\}^{-1} > 0$ such that $t\boldsymbol{x}_0$ is an interior point of, a boundary point of or outside of $\mathscr{S}$ according as $t < t_0$, $t = t_0$ or $t > t_0$. In § 2 we examine this relationship and show that conversely every star-body $\mathscr{S}$ determines a distance function $F(\boldsymbol{x})$ such that the set (1) is the set of interior points of $\mathscr{S}$. Since many, though not all, of the point-sets of interest in the geometry of numbers are star-bodies, the concept of distance-function plays an important rôle.

Most of the problems considered in Chapter II relate to star-bodies; and then it is easy to write down the corresponding distance functions. For example if $f(\boldsymbol{x})$ is a positive definite or semi-definite[1] form, the set

$$f(\boldsymbol{x}) < 1$$

corresponds to the distance-function

$$F(\boldsymbol{x}) = \{f(\boldsymbol{x})\}^{1/r},$$

where $r$ is the degree of $f(\boldsymbol{x})$. Again, if $f(\boldsymbol{x})$ is an indefinite form of degree $r$ and $k > 0$ is a number, then the set

$$-1 < f(\boldsymbol{x}) < k$$

[1] By semi-definite we mean that $f(\boldsymbol{x}) \geqq 0$ for all $\boldsymbol{x}$ but $f(\boldsymbol{x}) = 0$ for some $\boldsymbol{x} \neq \boldsymbol{o}$.

corresponds to the distance function

$$F(\boldsymbol{x}) = \begin{cases} k^{-1/r}|f(\boldsymbol{x})|^{1/r} & \text{if} \quad f(\boldsymbol{x}) \geqq 0, \\ |f(\boldsymbol{x})|^{1/r} & \text{if} \quad f(\boldsymbol{x}) \leqq 0. \end{cases}$$

The reader will readily verify that both the functions just defined are in fact distance-functions. One advantage of introducing distance-functions is that some of the ideas of Chapter II can be carried over to all star-bodies. A simple example of a 2-dimensional set which is not a star-body is

$$0 < x_1 x_2 < 1.$$

Clearly star-bodies $\mathscr{S}$ which are symmetric, i.e. have the property that $-\boldsymbol{x} \in \mathscr{S}$ when $\boldsymbol{x} \in \mathscr{S}$ correspond to distance-functions which are symmetric in the sense that

$$F(-\boldsymbol{x}) = F(\boldsymbol{x}).$$

K. Mahler (1950a) and C. A. Rogers (1952a) have investigated a wider class of sets which Rogers calls star-sets and which include the closed star-bodies as a sub-class. A star-set is a closed set such that $t\boldsymbol{x} \in \mathscr{S}$ whenever $0 \leqq t \leqq 1$ and $\boldsymbol{x} \in \mathscr{S}$. They are important in connection with certain problems ("bounded reducibility" cf. Chapter V, § 7) and we shall mention them again; but we refer the reader to the original memoirs for the details.

**IV.1.3.** Convex sets $\mathscr{K}$ are important as Minkowski's convex-body theorem shows. It turns out that the convex sets which have the origin $\boldsymbol{o}$ as an interior point are precisely the star-bodies whose distance-function satisfies the inequality

$$F(\boldsymbol{x} + \boldsymbol{y}) \leqq F(\boldsymbol{x}) + F(\boldsymbol{y}).$$

This we prove in § 3. We call such distance-functions convex.

In § 4 we show that an $n$-dimensional convex set $\mathscr{K}$ has a tac-(hyper)plane[1] at every point $\boldsymbol{a}$ on the boundary of $\mathscr{K}$; that is a (hyper)plane

$$\pi: \quad p_1 x_1 + \cdots + p_n x_n = k$$

which passes through $\boldsymbol{a}$ and is such that $\mathscr{K}$ lies entirely on one side of or in $\pi$; say

$$p_1 x_1 + \cdots + p_n x_n \leqq k \qquad (\text{all } \boldsymbol{x} \in \mathscr{K}).$$

Clearly if there is a tangent plane to $\mathscr{K}$ at $\boldsymbol{a}$, then it is the only tac-plane. But tac-planes exist even when tangent planes do not, and they do

[1] We use the words tac-plane and plane for tac-hyperplane and hyperplane. When $n = 2$ the corresponding thing is called a tac-line. The term supportplane (German: Stützebene) is sometimes used.

not need to be unique: for example when $\boldsymbol{a}$ is a corner of the square $|x_1|<1$, $|x_2|<1$.

In discussing tac-planes it is convenient to introduce the polar body of a convex body; a notion which we shall in any case require in Chapters VIII and XI. Any plane $\pi$ not passing through the origin can be put in the form

$$\pi: \quad y_1 x_1 + \cdots + y_n x_n = 1,$$

and so may be represented as a point $\boldsymbol{y} = (y_1, \ldots, y_n)$ in $n$-dimensional space. It turns out that the points $\boldsymbol{y}$ corresponding to planes $\pi$ which do not meet[1] $\mathscr{K}$ themselves form a convex set $\mathscr{K}^*$, say, the polar of $\mathscr{K}$. Further, the relationship between $\mathscr{K}$ and $\mathscr{K}^*$ is reciprocal in the sense that $\mathscr{K}$ may be obtained[2] from $\mathscr{K}^*$ in the same way as $\mathscr{K}^*$ was obtained from $\mathscr{K}$.

An example of a pair of polar bodies are the generalized cube

$$\mathscr{C}: \quad \max |x_j| \leqq 1$$

and the generalized octahedron

$$\mathscr{D}: \quad \sum |y_j| \leqq 1.$$

It is easy to see that a plane $\sum y_j x_j = 1$ for fixed $\boldsymbol{y}$ can contain a point of the interior of $\mathscr{C}$ only if $\boldsymbol{y}$ is not in $\mathscr{D}$; and vice versa. We discuss polar sets in § 4.

There is a rich theory of convex sets but we do not prove more than is relevant to the geometry of numbers. For the rest the reader is referred to the report of BONNESEN and FENCHEL (1934a) or EGGLESTON's tract (1958a).

**IV.2. General distance-functions.** We set up now the relationship between distance functions and star-bodies sketched in § 1.2.

THEOREM I. A. *If $F(\boldsymbol{x})$ is any distance function then the set*

$$\mathscr{S}: \quad F(\boldsymbol{x}) < 1$$

*is an open star-body. The boundary of $\mathscr{S}$ is the set of points $\boldsymbol{x}$ with $F(\boldsymbol{x}) = 1$ and points with $F(\boldsymbol{x}) > 1$ are exterior to $\mathscr{S}$ (that is, have a neighbourhood which does not meet $\mathscr{S}$).*

B. *Conversely any star-body $\mathscr{T}$ determines a unique distance-function $F(\boldsymbol{x})$. If $\mathscr{S}$ is the set of interior points of $\mathscr{T}$ then $\mathscr{S}$ is related to $F(\boldsymbol{x})$ in the way described in* A.

---

[1] We say that two point-sets meet if they have a point or points in common.

[2] Strictly speaking the set $\mathscr{K}^{**}$ obtained from $\mathscr{K}^*$ coincides with $\mathscr{K}$ except possibly on the boundary. The distance-functions of $\mathscr{K}$ and $\mathscr{K}^{**}$ are thus the same.

We note first that two distinct star-bodies $\mathscr{T}_1$ and $\mathscr{T}_2$ determine the same distance-function $F(\boldsymbol{x})$ if they have the same set of interior points, but a distance-function defines precisely one open star-body, namely $F(\boldsymbol{x})<1$ and one closed star-body, namely $F(\boldsymbol{x})\leqq 1$. Distinct distance-functions $F_1$, $F_2$ always determine distinct star-bodies. For then $F_1(\boldsymbol{x}_0)\neq F_2(\boldsymbol{x}_0)$ for some $\boldsymbol{x}_0$, say $F_1(\boldsymbol{x}_0)<F_2(\boldsymbol{x}_0)$; and then there is a $t$ such that

$$F_1(t\boldsymbol{x}_0)<1<F_2(t\boldsymbol{x}_0);$$

so $t\boldsymbol{x}_0$ is in one star-body but not the other.

The proof of Part A of the theorem is nearly trivial. If $F(\boldsymbol{x}_0)<1$, then, by the continuity of $F(\boldsymbol{x})$, there is a neighbourhood

$$|\boldsymbol{x}-\boldsymbol{x}_0|<\eta$$

of $\boldsymbol{x}_0$ which lies in $\mathscr{S}$; so $\boldsymbol{x}_0$ is an interior point of $\mathscr{S}$. Here we have used the standard notation

$$|\boldsymbol{x}| = (x_1^2+\cdots+x_n^2)^{\frac{1}{2}}.$$

Similarly, if $F(\boldsymbol{x}_0)>1$, then there is a neighbourhood of $\boldsymbol{x}_0$ which does not meet $\mathscr{S}$. Finally, if $F(\boldsymbol{x}_0)=1$, then every neighbourhood of $\boldsymbol{x}_0$ contains points $t\boldsymbol{x}_0$ both with $t>1$ and $t<1$, for which $F(t\boldsymbol{x}_0)>1$, $F(t\boldsymbol{x}_0)<1$ respectively: and so $\boldsymbol{x}_0$ is a boundary point of $\mathscr{S}$.

It remains to prove B. If $\mathscr{T}$ is any star-body, we define a function $F(\boldsymbol{x})$ as follows:

($\alpha$) $F(\boldsymbol{x})=0$ if $t\boldsymbol{x}\in\mathscr{T}$ for all $t>0$. In particular $F(\boldsymbol{o})=0$.

($\beta$) If $t\boldsymbol{x}$ is not in $\mathscr{T}$ for all $t>0$ then, by the definition of a star-body, there is a $t_0=t_0(\boldsymbol{x})>0$ such that $t\boldsymbol{x}$ is interior to or exterior to $\mathscr{S}$ according as $t<t_0$ or $t>t_0$; and $t_0\boldsymbol{x}$ is on the boundary of $\mathscr{S}$. We put

$$F(\boldsymbol{x}) = \{t_0(\boldsymbol{x})\}^{-1}.$$

Clearly, if $F(\boldsymbol{x})$ is a distance function, then it is related to the set $\mathscr{S}$ of interior points of $\mathscr{T}$ in the way described. Further, it follows trivially from the construction of $F(\boldsymbol{x})$ that it satisfies two of the defining properties of a distance function, namely $F(\boldsymbol{x})\geqq 0$ and $F(t\boldsymbol{x})=tF(\boldsymbol{x})$ for all $t>0$. It remains only to show that $F(\boldsymbol{x})$ is continuous.

We show first that $F(\boldsymbol{x})$ is continuous at $\boldsymbol{o}$. By the definition of a star-body, the origin $\boldsymbol{o}$ is an inner point of $\mathscr{T}$, so there is an $\eta>0$ such that the sphere

$$|\boldsymbol{x}|\leqq\eta$$

is contained in $\mathscr{T}$. Hence, if $\boldsymbol{x}_0\neq\boldsymbol{o}$, the vector

$$t'\boldsymbol{x}_0, \quad t'=\eta/|\boldsymbol{x}_0|$$

is certainly in $\mathscr{T}$, so

$$F(\boldsymbol{x}_0)\leqq\eta^{-1}|\boldsymbol{x}_0|. \tag{1}$$

Since $\eta$ is independent of $\boldsymbol{x}_0$, this proves the continuity of $F(\boldsymbol{x})$ at the origin.

We now prove continuity at a point $\boldsymbol{x}_0 \neq \boldsymbol{o}$. Let $\varepsilon > 0$ be arbitrarily small. The point

$$\boldsymbol{x}_1 = \{F(\boldsymbol{x}_0) + \varepsilon\}^{-1} \boldsymbol{x}_0 \tag{2}$$

is an interior point of $\mathscr{T}$ by the definition of $F$; and so there is a neighbourhood

$$|\boldsymbol{x} - \boldsymbol{x}_1| < \eta_1, \tag{3}$$

which lies in $\mathscr{T}$, that is, (3) implies

$$F(\boldsymbol{x}) \leqq 1. \tag{4}$$

Write

$$\boldsymbol{x} = \{F(\boldsymbol{x}_0) + \varepsilon\}^{-1} \boldsymbol{y}, \qquad \eta_1 = \{F(\boldsymbol{x}_0) + \varepsilon\}^{-1} \eta_2.$$

Then (3) is equivalent to

$$|\boldsymbol{y} - \boldsymbol{x}_0| < \eta_2 \tag{5}$$

and, by the homogeneity property of $F(\boldsymbol{x})$, which we have already proved, the inequality (4) is equivalent to

$$F(\boldsymbol{y}) \leqq F(\boldsymbol{x}_0) + \varepsilon. \tag{6}$$

We have thus found a neighbourhood (5) of $\boldsymbol{x}_0$ in which (6) holds. It remains to find a neighbourhood of points $\boldsymbol{y}$ in which

$$F(\boldsymbol{y}) \geqq F(\boldsymbol{x}_0) - \varepsilon. \tag{7}$$

If $F(\boldsymbol{x}_0) \leqq \varepsilon$, then (7) is true for all $\boldsymbol{y}$, since $F(\boldsymbol{y}) \geqq 0$. Otherwise one considers the point

$$\boldsymbol{x}_2 = \{F(\boldsymbol{x}_0) - \varepsilon\}^{-1} \boldsymbol{x}_0.$$

This is an exterior point of $\mathscr{T}$ and then the argument goes as before. This completes the proof of the theorem.

There is the trivial corollary of which we leave the proof to the reader.

COROLLARY. *Let $F_1(\boldsymbol{x})$ and $F_2(\boldsymbol{x})$ be distance functions. The star-body $F_1(\boldsymbol{x}) < 1$ is a subset of the star-body $F_2(\boldsymbol{x}) < 1$ if and only if*

$$F_2(\boldsymbol{x}) \leqq F_1(\boldsymbol{x}) \tag{8}$$

*for all $\boldsymbol{x}$.*

We record for later reference two results, the first of which we have already proved.

LEMMA 1. *For every distance-function $F(\boldsymbol{x})$ there is a constant $C$ such that*

$$F(\boldsymbol{x}) \leqq C |\boldsymbol{x}|$$

*for all $\boldsymbol{x}$.*

LEMMA 2. *A necessary and sufficient condition that the star-body* $F(\boldsymbol{x})<1$ *be bounded is that* $F(\boldsymbol{x})\neq 0$ *if* $\boldsymbol{x}\neq\boldsymbol{o}$. *There is then a constant* $c>0$ *such that*

$$F(\boldsymbol{x})\geqq c|\boldsymbol{x}| \tag{9}$$

*for all* $\boldsymbol{x}$.

We proved Lemma 1 above with $C=\eta^{-1}$, at least when $\boldsymbol{x}\neq\boldsymbol{o}$; and it is trivial when $\boldsymbol{x}=\boldsymbol{o}$. If there is a $\boldsymbol{x}_0\neq\boldsymbol{o}$ with $F(\boldsymbol{x}_0)=0$ then $t\boldsymbol{x}_0$ lies in $F(\boldsymbol{x})<1$ for all $t>0$, so $F(\boldsymbol{x})<1$ cannot be bounded; which proves half of Lemma 2. Suppose conversely that $F(\boldsymbol{x})\neq 0$ if $\boldsymbol{x}\neq\boldsymbol{o}$. The function $F(\boldsymbol{x})$ is continuous on the surface of the sphere $|\boldsymbol{x}|=1$; and so attains its minimum, say, at $\boldsymbol{x}_0$. Then $F(\boldsymbol{x}_0)>0$, by hypothesis. Put $F(\boldsymbol{x}_0)=c$. Then $F(\boldsymbol{x}_0)\geqq c$ if $|\boldsymbol{x}|=1$; and so (9) holds by homogeneity. This completes the proof of Lemma 2. We note that if (9) holds, then $F(\boldsymbol{x})<1$ is entirely in the sphere $|\boldsymbol{x}|<c^{-1}$.

The following trivial corollary rids Lemma 1 of its dependence on the particular distance-function $|\boldsymbol{x}|$.

COROLLARY. *Let* $F_1(\boldsymbol{x}), F_2(\boldsymbol{x})$ *be distance-functions and let* $F_1(\boldsymbol{x})<1$ *be a bounded set (i.e.* $F_1(\boldsymbol{x})=0$ *only for* $\boldsymbol{x}=\boldsymbol{o}$*). Then there is a constant* $C$ *such that*

$$F_2(\boldsymbol{x})\leqq C\,F_1(\boldsymbol{x})$$

*for all* $\boldsymbol{x}$.

*If, further,* $F_2(\boldsymbol{x})<1$ *is bounded, then there is a* $c>0$ *such that*

$$C\,F_1(\boldsymbol{x})\geqq F_2(\boldsymbol{x})\geqq c\,F_1(\boldsymbol{x}).$$

The second part of the corollary may be picturesquely summed up in the slogan "for qualitative purposes there is only one bounded star-body".

**IV.3. Convex sets.** A set $\mathscr{K}$ is convex if

$$t\boldsymbol{x}+(1-t)\boldsymbol{y}\qquad(0<t<1) \tag{1}$$

is in $\mathscr{K}$ whenever $\boldsymbol{x}$ and $\boldsymbol{y}$ are in $\mathscr{K}$. It is said to be strictly convex if the points (1) are all interior points of $\mathscr{K}$.

We show first that if $\boldsymbol{x}_1,\ldots,\boldsymbol{x}_r$ are any points of $\mathscr{K}$ and

$$t_j\geqq 0,\quad \sum t_j=1\qquad(1\leqq j\leqq r),$$

then

$$t_1\boldsymbol{x}_1+\cdots+t_r\boldsymbol{x}_r\in\mathscr{K}. \tag{2}$$

This is true for $r=2$ by the definition of convexity, and it is true for $r>2$ by induction, since we may suppose that $t_1\neq 1$, and then

$$t_1\boldsymbol{x}_1+\cdots+t_r\boldsymbol{x}_r=t_1\boldsymbol{x}_1+(1-t_1)\boldsymbol{y},$$

where

$$y = \frac{t_2}{1-t_1} \boldsymbol{x}_2 + \cdots + \frac{t_r}{1-t_1} \boldsymbol{x}_r \in \mathscr{K},$$

since it is of the shape (2) with $r-1$ summands.

Almost immediate consequences are

LEMMA 3. *A convex set $\mathscr{K}$ in $n$-dimensional space either lies entirely in a hyperplane*

$$\pi: \quad p_1 x_1 + \cdots + p_n x_n = k$$

*or it has interior points.*

LEMMA 4. *A convex set $\mathscr{K}$ with a volume $V(\mathscr{K})$ such that $0 < V(\mathscr{K}) < \infty$ is bounded.*

For if $\mathscr{K}$ does not lie in a hyperplane it contains $n+1$ points

$$\boldsymbol{x}_1, \ldots, \boldsymbol{x}_{n+1}$$

which do not lie in a hyperplane. The points $\sum t_j \boldsymbol{x}_j$ with $t_j \geqq 0$, $\sum t_j = 1$ are just the points of the simplex with vertices $\boldsymbol{x}_1, \ldots, \boldsymbol{x}_{n+1}$. The whole simplex must be contained in $\mathscr{K}$, and since a simplex has interior points this proves Lemma 3.

In Lemma 4, we note that $\mathscr{K}$ cannot lie in a hyperplane if $V(\mathscr{K}) > 0$; so we may suppose without loss of generality after a change of origin that $\boldsymbol{o}$ is an interior point of $\mathscr{K}$. There is then a number $\eta > 0$ such that all the vectors

$$\eta \boldsymbol{e}_j = (\overbrace{0, \ldots, 0}^{j-1}, \eta, \overbrace{0, \ldots, 0}^{n-j}) \qquad (1 \leqq j \leqq n)$$

are in $\mathscr{K}$. If $\boldsymbol{a} = (a_1, \ldots, a_n)$ be any other point of $\mathscr{K}$, we shall show that

$$\max_{1 \leqq j \leqq n} |a_j| \leqq \eta^{-n+1} (n!) \, V(\mathscr{K}).$$

If, say, $a_1 \neq 0$, then the whole of the simplex with vertices $\boldsymbol{o}, \boldsymbol{a}, \eta \boldsymbol{e}_2, \ldots, \eta \boldsymbol{e}_n$ is contained in $\mathscr{K}$ and has volume

$$(n!)^{-1} \cdot \eta^{n-1} |a_1|.$$

Since this can be at most $V(\mathscr{K})$, the result follows.

Finally we prove

THEOREM II. *A convex body $\mathscr{K}$ of which $\boldsymbol{o}$ is an interior point is a star-body. The corresponding distance function $F(\boldsymbol{x})$ satisfies the inequality*

$$F(\boldsymbol{x} + \boldsymbol{y}) \leqq F(\boldsymbol{x}) + F(\boldsymbol{y}) \tag{3}$$

*for all $\boldsymbol{x}$ and $\boldsymbol{y}$.*

*Conversely if $F(\boldsymbol{x})$ is a distance function for which (3) holds, then the star-body*

$$F(\boldsymbol{x}) < 1 \tag{4}$$

*is convex.*

The converse is trivial. If $F(\boldsymbol{x})<1$, $F(\boldsymbol{y})<1$ and $0<t<1$, then the inequality (3) applied to $t\boldsymbol{x}$ and $(1-t)\boldsymbol{y}$ gives

$$\begin{aligned} F\{t\boldsymbol{x}+(1-t)\boldsymbol{y}\} &\leqq F(t\boldsymbol{x})+F\{(1-t)\boldsymbol{y}\} \\ &= tF(\boldsymbol{x})+(1-t)F(\boldsymbol{y}) \\ &< t+(1-t) \\ &= 1. \end{aligned}$$

It remains, then, only to verify the direct assertion of Theorem II. We define a function $F(\boldsymbol{x})$ as follows:

$$F(\boldsymbol{x}) = \inf t^{-1}, \tag{5}$$

where the infimum is taken over all $t$ such that

$$t>0, \quad t\boldsymbol{x}\in\mathscr{K}. \tag{6}$$

Since $\boldsymbol{o}$ is an interior point of $\mathscr{K}$, there certainly do exist $t$ satisfying (6). It follows at once from the definition that $F(\boldsymbol{x})\geqq 0$, $F(\boldsymbol{o})=0$; and that $F(s\boldsymbol{x})=sF(\boldsymbol{x})$ for all $s\geqq 0$. Thus $F(\boldsymbol{x})$ will be a distance-function if we can prove continuity. We first prove the functional inequality (3) and then deduce continuity from (3).

Let $\boldsymbol{x},\boldsymbol{y}$ be any two vectors and $s,t$ any two positive numbers such that

$$s\boldsymbol{x}\in\mathscr{K}, \quad t\boldsymbol{y}\in\mathscr{K}. \tag{7}$$

Then

$$rs\boldsymbol{x}+(1-r)t\boldsymbol{y}\in\mathscr{K}$$

if $0<r<1$. We choose $r$ so that this point is multiple of $\boldsymbol{x}+\boldsymbol{y}$, i.e.

$$rs=(1-r)t; \quad r=t/(s+t).$$

Then

$$\frac{st}{s+t}(\boldsymbol{x}+\boldsymbol{y})\in\mathscr{K};$$

so

$$F(\boldsymbol{x}+\boldsymbol{y})\leqq\frac{s+t}{st}=s^{-1}+t^{-1}.$$

Hence

$$F(\boldsymbol{x}+\boldsymbol{y})\leqq F(\boldsymbol{x})+F(\boldsymbol{y})$$

since $F(\boldsymbol{x}), F(\boldsymbol{y})$ are the infima of $s^{-1}, t^{-1}$ over $s,t$ respectively which satisfy (7).

The function $F(\boldsymbol{x})$ is continuous at $\boldsymbol{o}$ by the same argument as was used for distance functions. Since $\boldsymbol{o}$ is an interior point there is a neighbourhood

$$|\boldsymbol{x}|\leqq\eta, \quad \eta>0$$

of $\boldsymbol{o}$ contained in $\mathscr{K}$, and so

$$F(\boldsymbol{x})\leqq\eta^{-1}|\boldsymbol{x}|.$$

The continuity at a general point $\boldsymbol{x}_0$ is now immediate. We have

$$F(\boldsymbol{x}_0+\boldsymbol{y}) \leqq F(\boldsymbol{x}_0)+F(\boldsymbol{y}),$$

and

$$F(\boldsymbol{x}_0) \leqq F(\boldsymbol{x}_0+\boldsymbol{y})+F(-\boldsymbol{y}).$$

Hence

$$|F(\boldsymbol{x}_0+\boldsymbol{y})-F(\boldsymbol{x}_0)| \leqq \max_{\pm} F(\pm \boldsymbol{y}) \leqq \eta^{-1}|\boldsymbol{y}| < \varepsilon$$

for any given $\varepsilon>0$, provided that $|\boldsymbol{y}|<\eta\varepsilon$.

Finally we must verify that the set

$$F(\boldsymbol{x})<1$$

is in fact the set of interior points of $\mathscr{K}$. A point $\boldsymbol{x}$ with $F(\boldsymbol{x})<1$ is certainly in $\mathscr{K}$ since, by the definition of $F(\boldsymbol{x})$, there is a $t>1$ such that $t\boldsymbol{x}\in\mathscr{K}$; and so

$$\boldsymbol{x}=t^{-1}(t\boldsymbol{x})+(1-t^{-1})\,\boldsymbol{o}$$

is in $\mathscr{K}$ by convexity. Since $F$ is continuous, the set $F(\boldsymbol{x})<1$ is open; and so all its points are inner points of $\mathscr{K}$. Conversely, if $\boldsymbol{x}$ is an inner point of $\mathscr{K}$, there is a $t>1$ such that $t\boldsymbol{x}\in\mathscr{K}$, and so $F(\boldsymbol{x})<1$ by the definition of $F(\boldsymbol{x})$. From the definition of $F$, no point $\boldsymbol{x}$ with $F(\boldsymbol{x})>1$ can belong to $\mathscr{K}$. Points with $F(\boldsymbol{x})=1$ may or may not belong to $\mathscr{K}$ but, since $F(\boldsymbol{x})$ is a distance-function, they must be boundary-points of $\mathscr{K}$.

For later reference we enunciate formally a result we have just proved:

COROLLARY. *Let $F(\boldsymbol{x})$ be a non-negative function of the vector $\boldsymbol{x}$ which satisfies the two conditions*

$$F(t\boldsymbol{x})=tF(\boldsymbol{x}) \quad \text{if} \quad t>0,$$

$$F(\boldsymbol{x}+\boldsymbol{y}) \leqq F(\boldsymbol{x})+F(\boldsymbol{y}),$$

*and which is continuous at $\boldsymbol{o}$. Then $F(\boldsymbol{x})$ is continuous for all $\boldsymbol{x}$; and so is a distance-function.*

**IV.3.2.** The next lemma is an essential preliminary to the treatment of polar bodies and tac-planes.

LEMMA 5[1]. *Let $\mathscr{K}_1, \mathscr{K}_2$ be a closed convex sets having no point in common. Then there is a hyperplane*

$$\pi\colon\ p_1x_1+\cdots+p_nx_n=k$$

*which separates $\mathscr{K}_1$ and $\mathscr{K}_2$: that is all the points of $\mathscr{K}_1$ are on the opposite side of $\pi$ from those of $\mathscr{K}_2$.*

[1] Proof given is valid only if at least one of $\mathscr{K}_1, \mathscr{K}_2$ is closed (as otherwise there need be no minimum distance).

Consider the distance $|\boldsymbol{x}_1-\boldsymbol{x}_2|$ when $\boldsymbol{x}_1, \boldsymbol{x}_2$ run through the points of $\mathcal{K}_1, \mathcal{K}_2$ respectively. Since $\mathcal{K}_1$ and $\mathcal{K}_2$ are closed, this distance attains its infinum at some points $\boldsymbol{x}_j' \in \mathcal{K}_j$ $(j=1, 2)$; and $\boldsymbol{x}_1' \neq \boldsymbol{x}_2'$ since $\mathcal{K}_1$ and $\mathcal{K}_2$ have no points in common. We show that the hyperplane $\pi$ which bisects perpendicularly the line-segment $\boldsymbol{x}_1'\boldsymbol{x}_2'$ will do what was required. After a suitable rotation of the co-ordinate system and a possible change of origin we may suppose that

$$\boldsymbol{x}_1' = (-\eta, 0, \ldots, 0), \qquad \boldsymbol{x}_2' = (\eta, 0, \ldots, 0)$$

for some $\eta > 0$. The plane $\pi$ is then

$$\pi: \quad x_1 = 0.$$

Suppose, if possible, that there is a point $\boldsymbol{z}$ in $\mathcal{K}_1$ with $z_1 \geqq 0$. By convexity, the point

$$\boldsymbol{z}_t = (1-t)\boldsymbol{x}_1' + t\boldsymbol{z} \qquad (0 < t < 1)$$

is in $\mathcal{K}_1$. The distance $|\boldsymbol{z}_t - \boldsymbol{x}_2'|$ is given by

$$\begin{aligned} |\boldsymbol{z}_t - \boldsymbol{x}_2'|^2 &= (2\eta - t\eta - tz_1)^2 + \sum_{2\leqq j\leqq n} (tz_j)^2 \\ &= 4\eta^2 - 4(\eta + z_1)\eta t + O(t^2) < 4\eta^2, \end{aligned}$$

if $t$ is small enough and strictly positive. This contradicts the definition of $\boldsymbol{x}_1'$ and $\boldsymbol{x}_2'$. The contradiction shows that $\boldsymbol{z}$ cannot in fact exist, and so proves the lemma.

COROLLARY. *If $\mathcal{K}$ is a convex closed set and $\boldsymbol{a}$ a point not in $\mathcal{K}$, there is a hyperplane separating $\mathcal{K}$ and $\boldsymbol{a}$.*

For we may put $\mathcal{K}_1 = \mathcal{K}$ and take $\mathcal{K}_2$ to be the set consisting of $\boldsymbol{a}$ alone.

**IV.3.3.** In introducing the polar set of a given convex set $\mathcal{K}$, we confine attention to the case when $\mathcal{K}$ is bounded and can be described by a distance function; that is $\boldsymbol{o}$ is an inner point and $0 < V(\mathcal{K}) < \infty$ by Lemmas 3 and 4 and Theorem II. If the reader is interested he will have no difficulty in extending the results to the other cases using Lemma 2.

We write

$$\boldsymbol{p}\boldsymbol{a} = p_1 a_1 + \cdots + p_n a_n$$

for the scalar product of two vectors $\boldsymbol{p}$ and $\boldsymbol{a}$.

THEOREM III. *Let $F(\boldsymbol{x})$ be the distance-function associated with a bounded convex set. For all vectors $\boldsymbol{y}$ let*

$$F^*(\boldsymbol{y}) = \sup_{\boldsymbol{x} \neq 0} \frac{\boldsymbol{x}\boldsymbol{y}}{F(\boldsymbol{x})}. \tag{1}$$

*Then $F^*(\boldsymbol{y})$ is the distance-function associated with a bounded convex set. The relationship is reciprocal in the sense that*

$$F(\boldsymbol{x}) = \sup_{\boldsymbol{y} \neq \boldsymbol{o}} \frac{\boldsymbol{x}\boldsymbol{y}}{F^*(\boldsymbol{y})}. \tag{2}$$

The functions $F$ and $F^*$, or the convex sets associated with them, are said to be polar to each other.

We must first show that $F^*$ is well-defined. Since the body $F(\boldsymbol{x}) < 1$ is bounded, we have $F(\boldsymbol{x}) \neq 0$ if $\boldsymbol{x} \neq \boldsymbol{o}$ by Lemma 2, and indeed there is a constant $c > 0$ such that $F(\boldsymbol{x}) \geqq c|\boldsymbol{x}|$. Since $\boldsymbol{x}\boldsymbol{y} \leqq |\boldsymbol{x}||\boldsymbol{y}|$, it follows that

$$F^*(\boldsymbol{y}) \leqq c^{-1}|\boldsymbol{y}|. \tag{3}$$

Immediate consequences of the definition are that

$$F^*(t\boldsymbol{y}) = tF^*(\boldsymbol{y}) \quad \text{if} \quad t > 0, \tag{4}$$

and

$$F^*(\boldsymbol{y}) > 0 \quad \text{if} \quad \boldsymbol{y} \neq \boldsymbol{o}. \tag{5}$$

Now if $\boldsymbol{y}_1, \boldsymbol{y}_2$ are any vectors, we have

$$\left.\begin{aligned} F^*(\boldsymbol{y}_1 + \boldsymbol{y}_2) = \sup_{\boldsymbol{x}} \frac{\boldsymbol{x}(\boldsymbol{y}_1 + \boldsymbol{y}_2)}{F(\boldsymbol{x})} &\leqq \sup_{\boldsymbol{x}} \frac{\boldsymbol{x}\boldsymbol{y}_1}{F(\boldsymbol{x})} + \sup_{\boldsymbol{x}} \frac{\boldsymbol{x}\boldsymbol{y}_2}{F(\boldsymbol{x})} \\ &= F^*(\boldsymbol{y}_1) + F^*(\boldsymbol{y}_2). \end{aligned}\right\} \tag{6}$$

But now (3), (4), (5) and (6) show that $F^*(\boldsymbol{y})$ is the distance-function of a convex set, by Theorem II and its Corollary. This convex set is bounded because of (5) and Lemma 2.

It remains only to prove (2); and here we need the convexity of $F(\boldsymbol{x})$, which we have not yet seriously used. If $\boldsymbol{x} = \boldsymbol{o}$, then (2) is trivial, so let $\boldsymbol{x}_0 \neq 0$ be fixed. From (1) we have

$$F(\boldsymbol{x})\, F^*(\boldsymbol{y}) \geqq \boldsymbol{x}\boldsymbol{y} \tag{7}$$

for all $\boldsymbol{x}$ and $\boldsymbol{y}$: and so certainly

$$F(\boldsymbol{x}_0) \geqq \sup_{\boldsymbol{y}} \frac{\boldsymbol{x}_0\boldsymbol{y}}{F^*(\boldsymbol{y})}. \tag{8}$$

Let $\varepsilon > 0$. Then by Lemma 5 Corollary there is a hyperplane $\pi$ separating $\boldsymbol{x}_0$ from the set of $\boldsymbol{x}$ such that

$$F(\boldsymbol{x}) \leqq (1 - \varepsilon)\, F(\boldsymbol{x}_0). \tag{9}$$

Since $\pi$ does not pass through the origin, it may be written in the shape

$$\pi: \quad \boldsymbol{x}\boldsymbol{y}_0 = 1. \tag{10}$$

Then $F(\boldsymbol{x}) \geqq (1 - \varepsilon)\, F(\boldsymbol{x}_0)$ for all points $\boldsymbol{x}$ on $\pi$, since $\pi$ does not meet (9); hence

$$F^*(\boldsymbol{y}_0) \leqq \frac{1}{(1 - \varepsilon)\, F(\boldsymbol{x}_0)}, \tag{11}$$

since one need clearly only consider the $\boldsymbol{x}$ with $\boldsymbol{x}\boldsymbol{y}=1$ in (1), by homogeneity, if $\boldsymbol{y}\neq\boldsymbol{o}$. Further,

$$\boldsymbol{x}_0\boldsymbol{y}_0>1, \tag{12}$$

since $\boldsymbol{x}_0$ is on the other side of $\pi$ from the origin, which is a point of (9). From (11) and (12) we have

$$\sup_{\boldsymbol{y}}\frac{\boldsymbol{x}_0\boldsymbol{y}}{F^*(\boldsymbol{y})}\geqq\frac{\boldsymbol{x}_0\boldsymbol{y}_0}{F^*(\boldsymbol{y}_0)}>(1-\varepsilon)\,F(\boldsymbol{x}_0). \tag{13}$$

The required result (2) now follows from (8) and (13), since $\varepsilon$ is arbitrarily small.

This concludes the proof of the theorem. The reader will be able to verify readily that the sets $F(\boldsymbol{x})<1$ and $F^*(\boldsymbol{y})<1$ are related to each other in the way described in § 1.3.

We have at once the

COROLLARY 1

$$F(\boldsymbol{x})\,F^*(\boldsymbol{y})\geqq\boldsymbol{x}\boldsymbol{y}$$

*for all* $\boldsymbol{x}, \boldsymbol{y}$ *For any* $\boldsymbol{y}_0\neq\boldsymbol{o}$ *there is an* $\boldsymbol{x}_0\neq\boldsymbol{o}$ *such that*

$$F(\boldsymbol{x}_0)\,F^*(\boldsymbol{y}_0)=\boldsymbol{x}_0\boldsymbol{y}_0; \tag{14}$$

*and vice versa.*

We have already noted the first inequality, which is an immediate consequence of the definition. By symmetry it is enough to show the existence of $\boldsymbol{x}_0$, given $\boldsymbol{y}_0$. The set $\mathscr{B}$ of points $\boldsymbol{x}$ with $F(\boldsymbol{x})=1$ is bounded; and it is closed since $F(\boldsymbol{x})$ is continuous. Hence the continuous function $\boldsymbol{x}\boldsymbol{y}_0$ attains its upper bound, say at $\boldsymbol{x}_0$. But we have already seen that the upper bound is $F^*(\boldsymbol{y}_0)$, so (14) must hold.

We also shall need later

COROLLARY 2. *Let* $\mathscr{K}_1$, $\mathscr{K}_2$ *be convex sets with non-zero volume having the origin as inner point and with respective polars* $\mathscr{K}_1^*$ *and* $\mathscr{K}_2^*$. *If* $\mathscr{K}_1$ *contains* $\mathscr{K}_2$ *then* $\mathscr{K}_2^*$ *contains* $\mathscr{K}_1^*$.

Let the corresponding distance functions be $F_1(\boldsymbol{x})$, $F_2(\boldsymbol{x})$, $F_1^*(\boldsymbol{x})$, $F_2^*(\boldsymbol{x})$. Then $F_2(\boldsymbol{x})\geqq F_1(\boldsymbol{x})$ by Theorem I Corollary. The definition (1) of the polar distance-function then gives immediately $F_2^*(\boldsymbol{y})\leqq F_1^*(\boldsymbol{y})$ for all $\boldsymbol{y}$.

The following corollary links polar distance-functions with the polar lattices and transformations introduced in Chapter I, § V.

COROLLARY 3. *Let* $F(\boldsymbol{x}), F^*(\boldsymbol{y})$ *be a pair of mutually polar convex distance-functions. Let* $\boldsymbol{\tau}$ *be a homogeneous linear transformation and* $\boldsymbol{\tau}^*$ *its polar transformation. Then* $F(\boldsymbol{\tau}\boldsymbol{x})$ *and* $F^*(\boldsymbol{\tau}^*\boldsymbol{y})$ *are mutually polar.*

For by the definition of $\boldsymbol{\tau}^*$ we have $\boldsymbol{\tau}\boldsymbol{x}\boldsymbol{\tau}^*\boldsymbol{y}=\boldsymbol{x}\boldsymbol{y}$ for all $\boldsymbol{x}, \boldsymbol{y}$. The truth of the corollary now follows from (1) and (2).

**IV.3.4.** A hyperplane $\pi$ through a point $\boldsymbol{x}_0$ on the boundary of a convex set $\mathscr{K}$ is said to be a tac-plane to $\mathscr{K}$ at $\boldsymbol{x}_0$ if no interior point of $\mathscr{K}$ is in $\pi$. The following Theorem IV is an almost immediate consequence of the results of § 3.3. We shall need Theorem IV in the next chapter, but § 3.3 only in Chapter VIII.

THEOREM IV. *Let $\mathscr{K}$ be any convex body with volume $V(\mathscr{K})$ such that $0 < V(\mathscr{K}) < \infty$. Then at every point $\boldsymbol{x}_0$ on the boundary of $\mathscr{K}$ there is at least one tac-plane. There are precisely two tac-planes to $\mathscr{K}$ parallel to any given hyperplane $\pi$.*

We may suppose that $\boldsymbol{o}$ is an interior point of $\mathscr{K}$. Let $F(\boldsymbol{x})$ be the corresponding distance function. Then $F(\boldsymbol{x}_0) = 1$. By Corollary 1 to Theorem III there is a $\boldsymbol{y}_0 \neq \boldsymbol{o}$ with

$$\boldsymbol{x}_0 \boldsymbol{y}_0 = F(\boldsymbol{x}_0)\, F^*(\boldsymbol{y}_0) = F^*(\boldsymbol{y}_0). \tag{1}$$

The plane

$$\pi' : \quad \boldsymbol{x}\boldsymbol{y}_0 = F^*(\boldsymbol{y}_0) \tag{2}$$

thus passes through $\boldsymbol{x}_0$. By the Corollary 1 to Theorem III we have

$$\boldsymbol{x}\boldsymbol{y}_0 \leqq F(\boldsymbol{x})\, F^*(\boldsymbol{y}_0),$$

so $F(\boldsymbol{x}) \geqq 1$ for all points of $\pi'$. Hence $\pi'$ contains no interior point of $\mathscr{K}$, so is a tac-plane.

Any plane (2) for fixed $\boldsymbol{y}_0$ is a tac-plane at some point $\boldsymbol{x}_0$. For by Corollary 1 to Theorem III there is an $\boldsymbol{x}_0$ such that (1) holds.

Hence if $\boldsymbol{y}_0$ is any vector, the two planes

$$\boldsymbol{x}\boldsymbol{y}_0 = F^*(\boldsymbol{y}_0) \tag{3}$$

and

$$\boldsymbol{x}\boldsymbol{y}_0 = -F^*(-\boldsymbol{y}_0) \tag{4}$$

are both tac-planes. It is clear that they are the only tac-planes parallel to $\boldsymbol{x}\boldsymbol{y}_0 = 0$. The origin lies between the hyperplanes (3) and (4), and hence so does the whole of the interior of $\mathscr{K}$.

**IV.3.5.** In Chapter IX we shall need the following result.

LEMMA 6. *Let $\mathscr{K}_1$ and $\mathscr{K}_2$ be open convex sets in n-dimensional space with*

$$0 < V(\mathscr{K}_j) < \infty \qquad (j = 1, 2).$$

*Suppose that $\mathscr{K}_1$ and $\mathscr{K}_2$ have no points in common but that $\boldsymbol{a}$ is a boundary point of both $\mathscr{K}_1$ and $\mathscr{K}_2$. Then there is a hyperplane through $\boldsymbol{a}$ which does not meet either $\mathscr{K}_1$ or $\mathscr{K}_2$ (and so is a tac-plane to both $\mathscr{K}_1$ and $\mathscr{K}_2$).*

The proof follows that of Theorem III. We may suppose without loss of generality that $\boldsymbol{o}$ is an inner point of $\mathscr{K}_1$. Let $\boldsymbol{b}$ be an inner point of $\mathscr{K}_2$. Then $\mathscr{K}_1$ and $\mathscr{K}_2$ may be described by distance-functions:

$$\mathscr{K}_1:\ F_1(\boldsymbol{x}) < 1,$$
$$\mathscr{K}_2:\ F_2(\boldsymbol{x}-\boldsymbol{b}) < 1.$$

For $\frac{1}{2} < t < 1$ let $\mathscr{K}_j^t$ $(j=1, 2)$ be given by

$$\mathscr{K}_1^t:\ F_1(\boldsymbol{x}) \leqq t,$$
$$\mathscr{K}_2^t:\ F_2(\boldsymbol{x}-\boldsymbol{b}) \leqq t,$$

so that $\mathscr{K}_j^t$ is a closed subset of $\mathscr{K}_j$. By Lemma 5 there is a plane $\pi^t$ separating $\mathscr{K}_1^t$ and $\mathscr{K}_2^t$. Since $\pi^t$ does not pass through the point $\boldsymbol{o}\in\mathscr{K}_1^t$, it has an equation

$$\sum_{1\leqq j\leqq n} p_{jt}\, x_j = 1.$$

Since $t > \frac{1}{2}$, the set $\mathscr{K}_1^t$ contains a neighbourhood $|\boldsymbol{x}| \leqq \eta$ of the origin, where $\eta > 0$. Since no points of this neighbourhood lie on $\pi^t$, we have

$$|p_{jt}| \leqq \eta^{-1} \qquad (1 \leqq j \leqq n). \tag{1}$$

Since $\boldsymbol{b}$ is on the opposite side of $\pi^t$ from $\boldsymbol{o}$, we have

$$\sum_{1\leqq j\leqq n} p_{jt}\, b_j > 1. \tag{2}$$

By (1) and WEIERSTRASS's compactness theorem, there exist $p_j'$ which are the limits of $p_{jt}$ as $t$ tends to 1 through a sequence of values $t_1 < t_2 < \cdots < t_m < \cdots$ which is the same for each $j$. By (2) not all the $p_j'$ are 0. The plane $\pi'$ defined by

$$\sum_j p_j' x_j = 1$$

clearly has all the properties required.

**IV.3.6.** The results of the rest of this § 3 will not be required until Chapter VIII, but it is convenient to give them here. They show that any two symmetric convex sets $\mathscr{K}_1$ and $\mathscr{K}_2$ with finite non-zero volumes behave similarly.

For more precise results, generalisations to convex sets which are not symmetric, and references to the literature, see for example BAMBAH (1955a), and for an interesting application see MAHLER (1955a, b).

A closed "generalized parallelopiped" in $n$-dimensional space with $\boldsymbol{o}$ as centre is the set of all points

$$\boldsymbol{x} = t_1\boldsymbol{x}_1 + \cdots + t_n\boldsymbol{x}_n \tag{1}$$

where $\boldsymbol{x}_1, \ldots, \boldsymbol{x}_n$ are fixed linearly independent vectors and $t_1, \ldots, t_n$ run through all real numbers in

$$\max |t_j| \leqq 1. \tag{2}$$

A closed "generalised octahedron" with $\boldsymbol{o}$ as centre is similarly the set of all vectors (1), where $t_1, \ldots, t_n$ run through all numbers in

$$\sum |t_j| \leqq 1. \tag{3}$$

We first prove the following refinement[1] of a result of MAHLER (1939b).

THEOREM V. *Let $\mathscr{K}$ be any closed symmetric convex set with volume $V(\mathscr{K})$ such that $0 < V(\mathscr{K}) < \infty$. Then there exist points $\pm \boldsymbol{x}_1, \ldots, \pm \boldsymbol{x}_n \in \mathscr{K}$ such that $\mathscr{K}$ is contained in the parallelopiped $\mathscr{C}$ with faces $\pm \pi_J (1 \leqq J \leqq n)$, where $\pi_J$ is the hyperplane through the points $\boldsymbol{x}_J \pm \boldsymbol{x}_j$ $(j \neq J)$. Further, the generalized octahedron $\mathscr{D}$ with vertices $\pm \boldsymbol{x}_j$ $(1 \leqq j \leqq n)$ is contained in $\mathscr{K}$.*

The last sentence is in any case trivial by convexity. We take for $\boldsymbol{x}_1, \ldots, \boldsymbol{x}_n$ points of $\mathscr{K}$ such that the volume of $\mathscr{D}$ is a maximum. Such a choice is possible since $\mathscr{K}$ is closed and bounded. If $\mathscr{K}$ were not contained in $\mathscr{C}$, there would be a point $\boldsymbol{y}$ on the opposite side of the origin from one of the faces $\pm \pi_J$, say on the opposite side of $\pi_n$. Then the generalized octahedron with vertices $\pm \boldsymbol{x}_1, \ldots, \pm \boldsymbol{x}_{n-1}, \pm \boldsymbol{y}$ would have greater volume than $\mathscr{D}$, contrary to construction.

COROLLARY 1.

$$V(\mathscr{K}) \leqq V(\mathscr{C}) \leqq n!\, V(\mathscr{K}),$$
$$V(\mathscr{K}) \geqq V(\mathscr{D}) \geqq (n!)^{-1} V(\mathscr{K}).$$

For the left-hand inequalities are trivial, and the right-hand ones follow from them and $V(\mathscr{C}) \leqq n!\, V(\mathscr{D})$.

COROLLARY 2. *Let $\mathscr{K}$, $\mathscr{L}$ be any two closed symmetric convex sets of finite non-zero volume. Then there is a homogeneous linear transformation $\boldsymbol{\tau}$ of the variables such that*

$$n^{-1} \boldsymbol{\tau} \mathscr{L} \subset \mathscr{K} \subset n \boldsymbol{\tau} \mathscr{L}$$

*and*

$$(n!)^{-1} V(\mathscr{K}) \leqq V(\boldsymbol{\tau} \mathscr{L}) \leqq (n!)\, V(\mathscr{K}).$$

Let $\boldsymbol{x}_1, \ldots, \boldsymbol{x}_n$ be the points of the theorem for $\mathscr{K}$ and let $\boldsymbol{y}_1, \ldots, \boldsymbol{y}_n$ be the corresponding points for $\mathscr{L}$. We determine $\boldsymbol{\tau}$ by the equations

$$\boldsymbol{\tau} \boldsymbol{y}_j = \boldsymbol{x}_j \qquad (1 \leqq j \leqq n).$$

Then the $\mathscr{C}$, $\mathscr{D}$ of the theorem are the same for $\mathscr{K}$ and $\boldsymbol{\tau}\mathscr{L}$. The stated results are now trivial, since $n^{-1}\mathscr{C} \subset \mathscr{D}$.

---

[1] Suggested by Professor C. A. ROGERS, who disclaims originality. The same method proves a corresponding result for non-symmetric bodies in which the inscribed and circumscribed bodies are both simplexes (cf. MAHLER 1950a). There are also results about inscribed and circumscribed ellipsoids (JOHN 1948a).

**IV.3.7.** As an application of the methods and results of § 3.6 we prove the following result about the volumes and lattice constants (Chapter III, § 5.1) of polar convex sets. We again denote the lattice constant of a set $\mathscr{S}$ by $\Delta(\mathscr{S})$.

THEOREM VI. *Let $\mathscr{K}$ and $\mathscr{K}^*$ be bounded symmetrical convex sets which are mutually polar. Then*

$$\frac{4^n}{(n!)^2} \leqq V(\mathscr{K})\, V(\mathscr{K}^*) \leqq 4^n$$

*and*

$$\Phi^2 \leqq \Delta(\mathscr{K})\, \Delta(\mathscr{K}^*) \leqq 1,$$

*where $\Phi$ is the lattice constant of the octahedron $\sum |x_j| \leqq 1$.*

The first pair of inequality is MAHLER's (1939a, b) and the proof of the second pair is practically identical. When $n=2$ MAHLER (1948a) has determined the best possible inequalities namely

$$\tfrac{1}{2} \leqq \Delta(\mathscr{K})\, \Delta(\mathscr{K}^*) \leqq \tfrac{3}{4},$$

equality on the left-hand side being necessary when $\mathscr{K}$ is a square and on the right when $\mathscr{K}$ is a circle. For related inequalities and references to later work see BAMBAH (1954c and 1955a).

We now prove the theorem for the lattice constants. The proof for the volumes is similar. Let $\boldsymbol{\tau}$ be any homogeneous linear transformation and $\boldsymbol{\tau}^*$ its polar transformation, so

$$\det(\boldsymbol{\tau}) \det(\boldsymbol{\tau}^*) = 1. \tag{1}$$

The bodies $\boldsymbol{\tau}\mathscr{K}$, $\boldsymbol{\tau}^*\mathscr{K}^*$ are mutually polar by Theorem III, Corollary 3. Since

$$\Delta(\boldsymbol{\tau}\mathscr{K}) = |\det(\boldsymbol{\tau})|\, \Delta(\mathscr{K}),$$

it follows from (1) that

$$\Delta(\boldsymbol{\tau}\mathscr{K})\, \Delta(\boldsymbol{\tau}^*\mathscr{K}^*) = \Delta(\mathscr{K})\, \Delta(\mathscr{K}^*).$$

Hence neither the hypotheses nor the conclusion of the theorem are affected if $\mathscr{K}$ is subjected to a homogeneous linear transformation and $\mathscr{K}^*$ to the polar transformation.

Suppose first that $\mathscr{K} = \mathscr{K}_0$ is a parallelopiped. After the application of a suitable homogeneous linear transformation we may suppose without loss of generality that $\mathscr{K}_0$ is the unit cube

$$|x_j| \leqq 1 \qquad (1 \leqq j \leqq n).$$

We saw already in § 1.3 that $\mathscr{K}_0^*$ is the generalized octahedron

$$\sum_j |y_j| \leqq 1.$$

Hence

$$\Delta(\mathscr{K}_0)\,\Delta(\mathscr{K}_0^*) = \Delta(\mathscr{K}_0^*) = \Phi \tag{2}$$

by the definition of $\Phi$.

Now consider a general $\mathscr{K}$, which we may suppose without loss of generality to be closed. Let $\mathscr{C}$ and $\mathscr{D}$ be the parallelopiped and octahedron given by Theorem V so that

$$\mathscr{C} \supset \mathscr{K} \supset \mathscr{D}. \tag{3}$$

The polar of the parallelopiped $\mathscr{C}$ is an octahedron $\mathscr{C}^*$ which is inscribed in $\mathscr{K}^*$ by Theorem III, Corollary 2. Similarly the polar of the octahedron $\mathscr{D}$ is a parallelopiped $\mathscr{D}^*$ and

$$\mathscr{D}^* \supset \mathscr{K}^* \supset \mathscr{C}^*. \tag{4}$$

We now show that

$$\Delta(\mathscr{D}) \geqq \Phi\,\Delta(\mathscr{C}), \tag{5}$$

where $\Phi$ is given in the enunciation. By Theorem V we have

$$V(\mathscr{D}) \geqq (n!)^{-1}\,V(\mathscr{C}). \tag{6}$$

But every octahedron may be transformed into any other by a homogeneous linear transformation, and so the ratio $\Delta(\mathscr{D})/V(\mathscr{D})$ is the same for all octahedra $\mathscr{D}$. In particular, taking $\mathscr{D}$ to be $\sum |x_j| < 1$, we have

$$\frac{\Delta(\mathscr{D})}{V(\mathscr{D})} = \frac{n!\,\Phi}{2^n}.$$

Similarly,

$$\frac{\Delta(\mathscr{C})}{V(\mathscr{C})} = \frac{1}{2^n},$$

and (5) follows from (6).

But now from (3) and (4) we have

$$\Delta(\mathscr{K})\,\Delta(\mathscr{K}^*) \geqq \Delta(\mathscr{D})\,\Delta(\mathscr{C}^*) \geqq \Phi\,\Delta(\mathscr{C})\,\Delta(\mathscr{C}^*) = \Phi^2,$$

on applying (2) with $\mathscr{K}_0 = \mathscr{C}$. Similarly

$$\Delta(\mathscr{K})\,\Delta(\mathscr{K}^*) \leqq \Delta(\mathscr{C})\,\Delta(\mathscr{D}^*) \leqq \Phi^{-1}\Delta(\mathscr{D})\,\Delta(\mathscr{D}^*) = 1,$$

on applying (2) with $\mathscr{K}_0 = \mathscr{D}^*$.

**IV.4. Distance-functions and lattices.** In the further study of the relationship between star-bodies (and in particular convex bodies) and lattices it is convenient to work with distance-functions rather than the star-bodies themselves. We write

$$F(\Lambda) = \inf_{\substack{\boldsymbol{a}\in\Lambda \\ \boldsymbol{a}\neq \boldsymbol{o}}} F(\boldsymbol{a}), \tag{1}$$

for any distance function $F$ and lattice $\Lambda$. In the language of § 5.1 of Chapter III the lattice $\Lambda$ is admissible for the star-body $F(\boldsymbol{x})<k$ if $k \leqq F(\Lambda)$ but not if $k>F(\Lambda)$.

We have

$$F(t\Lambda) = |t|\, F(\Lambda) \tag{2}$$

for any real number $t \neq 0$, where $t\Lambda$ is the set of $t\boldsymbol{a}$, $\boldsymbol{a} \in \Lambda$. If $t>0$ this follows from the property $F(t\boldsymbol{x}) = tF(\boldsymbol{x})$ of distance-functions, and if $t<0$ from the further observation that $\Lambda$ contains $-\boldsymbol{a}$ if it contains $\boldsymbol{a}$. In particular,

$$\frac{\{F(t\Lambda)\}^n}{d(t\Lambda)} = \frac{\{F(\Lambda)\}^n}{d(\Lambda)}, \tag{3}$$

where $n$ is the dimension of the space. We sum up the properties of $F(\Lambda)$ in the following theorem, which links our present point of view with that of § 5.1 of Chapter III.

THEOREM VII. *For any distance function $F$ write*

$$\delta(F) = \sup_{\Lambda} \frac{\{F(\Lambda)\}^n}{d(\Lambda)} \tag{4}$$

*over all lattices $\Lambda$. Then $\delta(F)<\infty$. Further,*

$$\delta(F) = \{\Delta(\mathscr{S})\}^{-1}, \tag{5}$$

*where $\Delta(\mathscr{S})$ is the lattice constant of the star-body*

$$\mathscr{S}:\ F(\boldsymbol{x}) < 1 .$$

*If $\Delta(\mathscr{S}) = \infty$, then* (5) *is to be interpreted as $\delta(F) = 0$.*

If $\delta(F) \neq 0$, then the supremum in (4) may be confined to lattices $\Lambda$ with $F(\Lambda)>0$, and then, by homogeneity, to those lattices with $F(\Lambda)=1$. Such lattices are admissible for $\mathscr{S}$ by definition, and so they have $d(\Lambda) \geqq \Delta(\Lambda)$. This shows that

$$\delta(F) \leqq \{\Delta(\mathscr{S})\}^{-1}. \tag{6}$$

On the other hand, if $\Lambda$ is $\mathscr{S}$-admissible then $F(\Lambda) \geqq 1$, and since there are $\mathscr{S}$-admissible lattices $\Lambda$ with $d(\Lambda)$ arbitrarily close to $\Delta(\mathscr{S})$, by the definition of $\Delta(\mathscr{S})$, we must have

$$\delta(F) \geqq \sup_{\Lambda \text{ is } \mathscr{S}\text{-admissible}} \{d(\Lambda)\}^{-1} \geqq \{\Delta(\mathscr{S})\}^{-1}. \tag{7}$$

Then (5) follows from (6) and (7).

If $\delta(F) = 0$, then $F(\Lambda) = 0$ for every $\Lambda$. From (7) this can hold only if there are no $\mathscr{S}$-admissible lattices, i.e. if $\Delta(\mathscr{S}) = \infty$. Conversely if $\Delta(\mathscr{S}) = \infty$ and $\Lambda$ is any lattice, the lattice $t\Lambda$ is not $\mathscr{S}$-admissible, for any $t>0$:

$$tF(\Lambda)=F(t\Lambda)<1.$$

Hence $F(\Lambda)=0$ on letting $t\to\infty$.

We note also the rather trivial

LEMMA 7. *Suppose that the distance-function $F(\boldsymbol{x})$ vanishes only for $\boldsymbol{x}=\boldsymbol{o}$. Then every lattice $\Lambda$ contains a point $\boldsymbol{a}\neq\boldsymbol{o}$ such that $F(\Lambda)=F(\boldsymbol{a})$. In particular, $F(\Lambda)>0$.*

For by Lemma 2 there is then a number $c>0$ such that

$$F(\boldsymbol{x})\geqq c|\boldsymbol{x}|.$$

Hence

$$F(\boldsymbol{x})\leqq F(\Lambda)+1 \tag{8}$$

implies that

$$|\boldsymbol{x}|\leqq c^{-1}\{F(\Lambda)+1\}. \tag{9}$$

But now by Lemma 1 of Chapter III there are only a finite number of points of $\Lambda$ for which $|\boldsymbol{x}|$ is less than a given bound, and so there are only a finite number of points $\boldsymbol{x}$ of $\Lambda$ satisfying (9). If we take $\boldsymbol{a}\neq\boldsymbol{o}$ to be one of those points for which $F(\boldsymbol{a})$ is least, then $\boldsymbol{a}$ enjoys the properties required.

# Chapter V

# MAHLER's compactness theorem

**V.1. Introduction.** So far we have been concerned with one lattice at a time. In this chapter we are concerned with properties of sets of lattices. We first must define what is meant by two lattices $\Lambda$ and $\mathsf{M}$ being near to each other; and this is done by means of homogeneous linear transformations. A homogeneous linear transformation $\boldsymbol{X}=\boldsymbol{\tau x}$ of $n$-dimensional euclidean space into itself is said to be near to identity transformation if the coefficients $\tau_{ij}$ in

$$X_i=\sum_{1\leqq j\leqq n}\tau_{ij}x_j \qquad (1\leqq i\leqq n)$$

are near those of the identity transformation, that is if

$$|\tau_{ii}-1| \qquad (1\leqq i\leqq n)$$

and

$$|\tau_{ij}| \qquad (1\leqq i\leqq n,\ 1\leqq j\leqq n,\ i\neq j)$$

are all small. The lattice $\mathsf{M}$ is thought of as near to $\Lambda$ if it is of the shape $\boldsymbol{\tau}\Lambda$ where $\boldsymbol{\tau}$ is near the identity transformation, and where $\boldsymbol{\tau}\Lambda$ denotes the set of $\boldsymbol{\tau a}$, $\boldsymbol{a}\in\Lambda$. Roughly speaking, $\mathsf{M}$ is near to $\Lambda$ if it can be obtained from $\Lambda$ by a small deformation of the underlying space. Convergence of a sequence of lattices $\Lambda_r$ to a lattice $\Lambda'$ may then be defined in the obvious way.

MINKOWSKI (1904a and 1907a) already used the idea of the continuous variation of lattices to show that a bounded convex set

$$\mathscr{S}: \; F(\boldsymbol{x}) < 1, \tag{1}$$

where $F(\boldsymbol{x})$ is the corresponding distance function, always has a critical lattice $\Lambda_c$ in the sense of § 5.1 of Chapter III; that is

$$F(\Lambda_c) = \inf_{\substack{\boldsymbol{a} \in \Lambda_c \\ \neq \boldsymbol{o}}} F(\boldsymbol{a}) \geqq 1, \tag{2}$$

and $d(\Lambda_c)$ is a minimum:

$$d(\Lambda_c) = \Delta(\mathscr{S}) = \inf_{F(\Lambda) \geq 1} d(\Lambda).$$

A critical lattice $\Lambda_c$ has the property that if it is slightly distorted to a lattice $\Lambda$ with $d(\Lambda) < d(\Lambda_c)$ then $F(\Lambda) < 1$; that is $\Lambda$ has a point other than $\boldsymbol{o}$ in $\mathscr{S}$. From this, MINKOWSKI obtained important properties of critical lattices and so gave an explicit process for finding $\Delta(\mathscr{S})$ for convex bodies $\mathscr{S}$, at least in 3-dimensional space. This was generalized and put on a more satisfactory basis by MAHLER (1946a), who gives general conditions under which a sequence of lattices $\Lambda_r$ should contain a convergent subsequence. In this way he showed that any star body $F(\boldsymbol{x}) < 1$ has a critical lattice if only there are any lattices $\Lambda$ with $F(\Lambda) > 0$. In an important sequence of papers, MAHLER (1946a, b, c, d, e) extended much of MINKOWSKI's work on critical lattice to general star bodies and made other applications of his compactness criteria. He has also [MAHLER (1949b)] considered the critical lattices of sets which are not star bodies, but we do not go into this here.

In this chapter we first consider the properties of homogeneous linear transformations which are needed for the treatment of convergence. Then we prove MAHLER's general criterion for a sequence of lattices $\Lambda_r$ to contain a convergent subsequence. After that, we study the properties of critical lattices of sets $\mathscr{S}$ taking in turn general star bodies, bounded star bodies, convex sets and spheres. As the sets become more specialized, there is more and more precise information about the critical lattices. Finally in § 10 we give an application to a problem in the theory of Dophantine approximation.

**V.2. Linear transformations.** Convergence for lattices will be defined in terms of homogeneous linear transformations, already introduced in Chapter I, § 3. We operate in $n$-dimensional space with some fixed euclidean coordinate system. If $\tau_{ij}$ is a set of $n^2$ real numbers, we denote by $\boldsymbol{\tau x}$ the transformation of our space into itself given by the equations

$$X_i = \sum_{1 \leq j \leq n} \tau_{ij} x_j \quad (1 \leq i \leq n),$$

where $\boldsymbol{X} = \boldsymbol{\tau x}$. We write $\det(\boldsymbol{\tau}) = \det(\tau_{ij})$. If $\det(\boldsymbol{\tau}) = 0$, the transformation $\boldsymbol{\tau}$ is singular: otherwise it is non-singular and possesses an inverse, which we denote by $\boldsymbol{\tau}^{-1}$. By $\boldsymbol{\sigma} + \boldsymbol{\tau}$, where $\boldsymbol{\sigma}$ and $\boldsymbol{\tau}$ are transformations, we mean the transformation

$$(\boldsymbol{\sigma} + \boldsymbol{\tau})\boldsymbol{x} = \boldsymbol{\sigma x} + \boldsymbol{\tau x};$$

and by $\boldsymbol{\sigma\tau}$ we mean the transformation

$$(\boldsymbol{\sigma\tau})\boldsymbol{x} = \boldsymbol{\sigma}(\boldsymbol{\tau x}).$$

If $\boldsymbol{\sigma}, \boldsymbol{\tau}$ correspond to the matrices of coefficients $\sigma_{ij}$, and $\tau_{ij}$, then the coefficients of $\boldsymbol{\sigma} + \boldsymbol{\tau}$ and $\boldsymbol{\sigma\tau}$ are clearly

$$\sigma_{ij} + \tau_{ij}$$

and

$$\sum_{1 \leq k \leq n} \sigma_{ik} \tau_{kj}, \tag{1}$$

respectively. We denote the identical transformation

$$X_i = x_i \quad (1 \leq i \leq n)$$

by $\boldsymbol{\iota}$.

We require a measure of the size of the coefficients of the matrix of a transformation $\boldsymbol{\tau}$. We write

$$\|\boldsymbol{\tau}\| = n \max |\tau_{ij}|.$$

Clearly

$$\left.\begin{aligned} \|-\boldsymbol{\tau}\| &= \|\boldsymbol{\tau}\|, \\ \|\boldsymbol{\sigma} + \boldsymbol{\tau}\| &\leq \|\boldsymbol{\sigma}\| + \|\boldsymbol{\tau}\|. \end{aligned}\right\} \tag{2}$$

Further,

$$\|\boldsymbol{\sigma\tau}\| \leq \|\boldsymbol{\sigma}\| \|\boldsymbol{\tau}\| \tag{3}$$

since the coefficients of $\boldsymbol{\sigma\tau}$ are given by (1). Further, if $\boldsymbol{X} = \boldsymbol{\tau x}$ we have trivially

$$\max_i |X_i| \leq \|\boldsymbol{\tau}\| \max_i |x_i|. \tag{4}$$

From this it follows crudely that

$$|\boldsymbol{\tau x}| \leq n^{\frac{1}{2}} \|\boldsymbol{\tau}\| |\boldsymbol{x}|, \tag{5}$$

since

$$\max_i |x_i| \leq |\boldsymbol{x}| \leq n^{\frac{1}{2}} \max_i |x_i|$$

for all $\boldsymbol{x}$.

We shall also need to use the fact that if $\boldsymbol{\tau}$ is near to the identical transformation $\boldsymbol{\iota}$, then $\boldsymbol{\tau}^{-1}$ exists and is also near to $\boldsymbol{\iota}$. This statement is made more precise in the following lemma.

LEMMA 1. *Let $\boldsymbol{\tau}=\boldsymbol{\iota}+\boldsymbol{\sigma}$ be a homogeneous linear transformation with*

$$\|\boldsymbol{\sigma}\|<1. \tag{6}$$

*Then $\boldsymbol{\tau}$ is nonsingular and*

$$\boldsymbol{\rho}=\boldsymbol{\iota}-\boldsymbol{\tau}^{-1} \tag{7}$$

*satisfies*

$$\|\boldsymbol{\rho}\|\leqq\frac{\|\boldsymbol{\sigma}\|}{1-\|\boldsymbol{\sigma}\|}. \tag{8}$$

We note first that if $\boldsymbol{\rho}$ exists, the inequality (8) follows at once from (2) and (3). We have

$$\boldsymbol{\rho}=\boldsymbol{\tau}^{-1}\boldsymbol{\sigma}=\boldsymbol{\sigma}-\boldsymbol{\rho}\boldsymbol{\sigma};$$

so

$$\|\boldsymbol{\rho}\|\leqq\|\boldsymbol{\sigma}\|+\|\boldsymbol{\rho}\boldsymbol{\sigma}\|\leqq\|\boldsymbol{\sigma}\|+\|\boldsymbol{\rho}\|\,\|\boldsymbol{\sigma}\|,$$

as required.

It remains to show that $\boldsymbol{\tau}$ is nonsingular; and for this it is convenient to use another characterization of $\|\boldsymbol{\tau}\|$. Put

$$F_1(\boldsymbol{x})=n^{-1}\sum_j|x_j|,\quad F_2(\boldsymbol{x})=\max_j|x_j|. \tag{9}$$

Then $F_1(\boldsymbol{x})$ and $F_2(\boldsymbol{x})$ are convex symmetric distance-functions vanishing only at $\boldsymbol{o}$, and

$$F_1(\boldsymbol{x})\leqq F_2(\boldsymbol{x}) \tag{10}$$

for all $\boldsymbol{x}$. Then clearly

$$\|\boldsymbol{\tau}\|=\sup_{\boldsymbol{x}\neq\boldsymbol{o}}\frac{F_2(\boldsymbol{\tau x})}{F_1(\boldsymbol{x})} \tag{11}$$

for all homogeneous transformations $\boldsymbol{\tau}$. For any $\boldsymbol{x}$ we have, by (10), (11), that

$$\begin{aligned}F_1(\boldsymbol{x})=F_1(\boldsymbol{\tau x}-\boldsymbol{\sigma x})&\leqq F_1(\boldsymbol{\tau x})+F_1(\boldsymbol{\sigma x})\\&\leqq F_1(\boldsymbol{\tau x})+F_2(\boldsymbol{\sigma x})\leqq F_1(\boldsymbol{\tau x})+\|\boldsymbol{\sigma}\|\,F_1(\boldsymbol{x}),\end{aligned}$$

the last line by (11) with $\boldsymbol{\sigma}$ for $\boldsymbol{\tau}$. In particular, since $\|\boldsymbol{\sigma}\|<1$ by hypothesis, we have $F_1(\boldsymbol{\tau x})=0$ only when $\boldsymbol{x}=\boldsymbol{o}$: that is $\boldsymbol{\tau x}=\boldsymbol{o}$ only when $\boldsymbol{x}=\boldsymbol{o}$: so $\boldsymbol{\tau}$ is nonsingular. This concludes the proof.

Our choice of $\|\boldsymbol{\tau}\|$ to represent the "size" of $\boldsymbol{\tau}$ is somewhat arbitrary. If $F$ is the distance-function of a symmetric convex bounded body, an alternative would be to use

$$\|\boldsymbol{\tau}\|_F=\sup_{\boldsymbol{x}\neq\boldsymbol{o}}\frac{F(\boldsymbol{\tau x})}{F(\boldsymbol{x})}. \tag{12}$$

The reader will have no difficulty in verifying that (2), (3) and Lemma 1 continue to hold when $\|\ \|_F$ is substituted for $\|\ \|$. Since we have used $|\boldsymbol{x}|$ to denote the size of the vector $\boldsymbol{x}$, it might have been more tidy to use $||\boldsymbol{\tau}||_{F_0}$, where $F_0(\boldsymbol{x})=|\boldsymbol{x}|$, to measure the size of $\boldsymbol{\tau}$. We have chosen $||\boldsymbol{\tau}||$ because of its simpler expression in terms of the $\tau_{ij}$. The choice of $||\ ||$ instead of some $||\ ||_F$ is, for all essential purposes, irrelevant, since it follows from Lemma 2, Corollary of Chapter IV that

$$0 < c_1' \leqq \frac{||\boldsymbol{\tau}||_F}{||\boldsymbol{\tau}||} \leqq c_2' < \infty,$$

where $c_1'$ and $c_2'$ are numbers depending on the particular function $F$, but not on $\boldsymbol{\tau}$.

We shall also need later two lemmas relating to distance functions and linear transformations.

LEMMA 2. *Let $F(\boldsymbol{x})$ be a distance function such that $F(\boldsymbol{x})=0$ only for $\boldsymbol{x}=\boldsymbol{o}$, and let $\boldsymbol{\tau}$ be a linear transformation. Then there is a number $c_1$ depending only on $F$ and $\boldsymbol{\tau}$ such that*

$$F(\boldsymbol{\tau x}) \leqq c_1 F(\boldsymbol{x})$$

*for all $\boldsymbol{x}$.*

For

$$F_1(\boldsymbol{x}) = F(\boldsymbol{\tau x})$$

is clearly a distance function. The result now follows at once from Lemma 2 Corollary of Chapter IV. If $\boldsymbol{\tau}$ is non-singular we may apply Lemma 2 with $\boldsymbol{\tau}^{-1}$ instead of $\boldsymbol{\tau}$ and obtain the

COROLLARY. *If $\boldsymbol{\tau}$ is non-singular there is a constant $c_2$ such that*

$$F(\boldsymbol{x}) \leqq c_2 F(\boldsymbol{\tau x}).$$

LEMMA 3. *Let $F(\boldsymbol{x})$ be a distance function such that $F(\boldsymbol{x})=0$ only for $\boldsymbol{x}=\boldsymbol{o}$. Then to every $\varepsilon$ in $0<\varepsilon<1$ there is an $\eta=\eta(\varepsilon)>0$, depending only on $F$ and $\varepsilon$, such that*

$$1-\varepsilon \leqq \frac{F(\boldsymbol{\tau x})}{F(\boldsymbol{x})} \leqq 1+\varepsilon \tag{13}$$

*for all homogeneous linear transformations $\boldsymbol{\tau}$ such that*[1]

$$||\boldsymbol{\tau}-\boldsymbol{\iota}|| < \eta \tag{14}$$

*and all $\boldsymbol{x}$.*

By Lemma 2 of Chapter IV there exists a number $c>0$ such that

$$F(\boldsymbol{x}) \geqq c|\boldsymbol{x}| \tag{15}$$

for all $\boldsymbol{x}$. Since $F(\boldsymbol{x})$ is continuous in the sphere $|\boldsymbol{x}|\leqq 2$, there exists a number $\eta_1$ in $0<\eta_1<1$ such that

$$|F(\boldsymbol{x}_1)-F(\boldsymbol{x}_2)| < c\varepsilon,$$

whenever

$$|\boldsymbol{x}_1-\boldsymbol{x}_2| < \eta_1; \quad |\boldsymbol{x}_1| \leqq 2, \ |\boldsymbol{x}_2| \leqq 2.$$

[1] As before, $\boldsymbol{\iota}$ denotes the identical transformation.

In particular, this is true when $|\boldsymbol{x}_2|=1$; and so, by homogeneity,

$$|F(\boldsymbol{x}_1)-F(\boldsymbol{x}_2)|<c\,\varepsilon\,|\boldsymbol{x}_2|, \tag{16}$$

whenever

$$|\boldsymbol{x}_1-\boldsymbol{x}_2|<\eta_1|\boldsymbol{x}_2|. \tag{17}$$

But now, by (5) and (14),

$$|\boldsymbol{\tau x}-\boldsymbol{x}|=|(\boldsymbol{\tau}-\boldsymbol{\iota})\,\boldsymbol{x}|\leqq n^{\frac{1}{2}}\|\boldsymbol{\tau}-\boldsymbol{\iota}\|\,|\boldsymbol{x}|<\eta_1|\boldsymbol{x}|;$$

provided that $n^{\frac{1}{2}}\eta<\eta_1$; which we may suppose.

But then from (15) and (16) with $\boldsymbol{x}_1=\boldsymbol{\tau x}$, $\boldsymbol{x}_2=\boldsymbol{x}$, we have

$$|F(\boldsymbol{\tau x})-F(\boldsymbol{x})|<\varepsilon F(\boldsymbol{x}),$$

which is equivalent to (13).

**V.3. Convergence of lattices.** If $\Lambda$ is a lattice and $\boldsymbol{\tau}$ a non-singular homogeneous transformation, we saw already in Chapter I, § 3 that the set of $\boldsymbol{\tau a}$, $\boldsymbol{a}\in\Lambda$ is a lattice $\boldsymbol{\tau}\Lambda$ with determinant

$$d(\boldsymbol{\tau}\Lambda)=|\det(\boldsymbol{\tau})|\,d(\Lambda). \tag{1}$$

If $\mathsf{M}$ is any other lattice, it may be put in the shape

$$\mathsf{M}=\boldsymbol{\tau}\Lambda$$

for some non-singular homogeneous transformation $\boldsymbol{\tau}$, and indeed in infinitely many ways. For if $\boldsymbol{a}_1,\ldots,\boldsymbol{a}_n$, $\boldsymbol{b}_1,\ldots,\boldsymbol{b}_n$ are bases for $\Lambda$ and $\mathsf{M}$ respectively, there is a uniquely defined homogeneous linear transformation $\boldsymbol{\tau}$ such that

$$\boldsymbol{b}_i=\boldsymbol{\tau a}_i \qquad (1\leqq i\leqq n);$$

and then

$$\mathsf{M}=\boldsymbol{\tau}\Lambda.$$

We say that a sequence of lattices $\Lambda_r$ $(1\leqq r<\infty)$ tends to the lattice $\Lambda'$ if there exist homogeneous linear transformations $\boldsymbol{\tau}_r$ such that

$$\Lambda_r=\boldsymbol{\tau}_r\Lambda' \tag{2}$$

and

$$\|\boldsymbol{\tau}_r-\boldsymbol{\iota}\|\to 0 \qquad (r\to\infty), \tag{3}$$

where $\boldsymbol{\iota}$ is the identity transformation and $\|\boldsymbol{\tau}\|$ is as defined in § 2. We write then

$$\Lambda_r\to\Lambda'.$$

From (1) and (3) we have immediately

$$d(\Lambda_r)\to d(\Lambda').$$

If $\boldsymbol{\alpha}$ is any non-singular homogeneous linear transformation it is also immediate that

$$\boldsymbol{\alpha}\Lambda_r \to \boldsymbol{\alpha}\Lambda'.$$

Indeed

$$\boldsymbol{\alpha}\Lambda_r = \boldsymbol{\alpha}\boldsymbol{\tau}_r\boldsymbol{\alpha}^{-1}\{\boldsymbol{\alpha}\Lambda'\}$$

and

$$\boldsymbol{\alpha}\boldsymbol{\tau}_r\boldsymbol{\alpha}^{-1} - \boldsymbol{\iota} = \boldsymbol{\alpha}(\boldsymbol{\tau}_r - \boldsymbol{\iota})\boldsymbol{\alpha}^{-1},$$

so

$$\|\boldsymbol{\alpha}\boldsymbol{\tau}_r\boldsymbol{\alpha}^{-1} - \boldsymbol{\iota}\| \leqq \|\boldsymbol{\alpha}\|\,\|\boldsymbol{\alpha}^{-1}\|\,\|\boldsymbol{\tau}_r - \boldsymbol{\iota}\| \to 0,$$

by (3) of § 2.

LEMMA 4. *A necessary and sufficient condition that the sequence of lattices $\Lambda_r$ $(1 \leqq r < \infty)$ tends to $\Lambda'$ is that there exist bases*

$$\boldsymbol{b}_1^r, \ldots, \boldsymbol{b}_n^r,$$

*and*

$$\boldsymbol{b}_1', \ldots, \boldsymbol{b}_n'$$

*of $\Lambda_r$, $\Lambda'$ respectively, such that*

$$\boldsymbol{b}_j^r \to \boldsymbol{b}_j' \quad (1 \leqq j \leqq n) \quad (r \to \infty). \tag{4}$$

The last limit is meant, of course, in the sense of the ordinary convergence of vectors: $|\boldsymbol{b}_j^r - \boldsymbol{b}_j'| \to 0$.

The proof of Lemma 4 is almost trivial. Suppose first that $\Lambda_r \to \Lambda'$ and let $\boldsymbol{\tau}_r$ be the transformation satisfying (2) and (3). Choose any basis $\boldsymbol{b}_j'$ for $\Lambda'$ and put

$$\boldsymbol{b}_j^r = \boldsymbol{\tau}_r \boldsymbol{b}_j' \quad (1 \leqq j \leqq n;\ 1 \leqq r < \infty). \tag{5}$$

Then by (5) of § 2 and (3), we have

$$|\boldsymbol{b}_j^r - \boldsymbol{b}_j'| = |(\boldsymbol{\tau}_r - \boldsymbol{\iota})\boldsymbol{b}_j'| \leqq n^{\frac{1}{2}} \|\boldsymbol{\tau}_r - \boldsymbol{\iota}\|\,|\boldsymbol{b}_j'| \to 0 \quad (r \to \infty).$$

Suppose conversely that the bases are given satisfying (4). We may define $\boldsymbol{\tau}_r$ uniquely by (5). Then clearly $\|\boldsymbol{\tau}_r - \boldsymbol{\iota}\| \to 0$.

The following criterion is rather less trivial.

THEOREM I. *A necessary and sufficient condition that $\Lambda_r \to \Lambda'$ is that the following two conditions be both satisfied:*

*(i) if $\boldsymbol{a}' \in \Lambda'$, there are points $\boldsymbol{a}^r \in \Lambda_r$ for $r = 1, 2, \ldots$ such that*

$$\boldsymbol{a}^r \to \boldsymbol{a}' \quad (r \to \infty). \tag{6}$$

*(ii) if $\boldsymbol{c}$ is not in $\Lambda'$, there is a number $\eta > 0$ and an integer $r_0 > 0$, both depending on $\boldsymbol{c}$, such that*

$$|\boldsymbol{a}^r - \boldsymbol{c}| > \eta \tag{7}$$

*for all $\boldsymbol{a}^r \in \Lambda_r$ with $r \geqq r_0$.*

It is quite straightforward that (i) and (ii) are satisfied when $\Lambda_r \to \Lambda'$. In (i) we have only to put

$$\boldsymbol{a}^r = \boldsymbol{\tau}_r \boldsymbol{a}',$$

where the $\boldsymbol{\tau}_r$ are the transformations such that

$$\Lambda_r = \boldsymbol{\tau}_r \Lambda', \qquad \|\boldsymbol{\tau}_r - \boldsymbol{\iota}\| \to 0.$$

Then, as before,

$$\|\boldsymbol{a}^r - \boldsymbol{a}'\| \leqq n^{\frac{1}{2}} \|\boldsymbol{\tau}_r - \boldsymbol{\iota}\| \, |\boldsymbol{a}'| \to 0 \quad (r \to \infty).$$

To prove (ii), we note that there certainly is an $\eta_1 > 0$ such that

$$|\boldsymbol{a}' - \mathbf{c}| > \eta_1 \tag{8}$$

for all $\boldsymbol{a}' \in \Lambda'$. Put

$$\eta = \tfrac{1}{2}\eta_1. \tag{9}$$

Suppose, if possible, that there is a point $\boldsymbol{a}^r \in \Lambda_r$ such that

$$|\boldsymbol{a}^r - \mathbf{c}| \leqq \eta. \tag{10}$$

Then

$$|\boldsymbol{a}^r| \leqq |\mathbf{c}| + \eta. \tag{11}$$

By the definition of $\boldsymbol{\tau}_r$ we have

$$\boldsymbol{a}^r = \boldsymbol{\tau}_r \boldsymbol{a}' \tag{12}$$

for some $\boldsymbol{a}' \in \Lambda$. Then

$$\boldsymbol{a}^r - \boldsymbol{a}' = \boldsymbol{\rho}_r \boldsymbol{a}^r, \tag{13}$$

where

$$\boldsymbol{\rho}_r = \boldsymbol{\iota} - \boldsymbol{\tau}_r^{-1}.$$

Now

$$\|\boldsymbol{\rho}_r\| \to 0 \qquad (r \to \infty)$$

by Lemma 1 and since $\|\boldsymbol{\tau}_r - \boldsymbol{\iota}\| \to 0$. Hence by (5) of § 2 and (11), we have

$$|\boldsymbol{a}^r - \boldsymbol{a}'| \leqq n^{\frac{1}{2}} \|\boldsymbol{\rho}_r\| \, |\boldsymbol{a}^r| \leqq n^{\frac{1}{2}} \|\boldsymbol{\rho}_r\| \{|\mathbf{c}| + \eta\} < \eta$$

for all $r$ greater than some $r_0$. From this and (9) and (10) we have

$$|\boldsymbol{a}' - \mathbf{c}| \leqq 2\eta = \eta_1.$$

This is in contradiction to (8). Hence statement (ii) of the theorem is true.

We must now show that if (i) and (ii) of the theorem are true then $\Lambda_r \to \Lambda'$. We require a lemma of some independent interest.

LEMMA 5. *Let* $\mathbf{c}_1, \ldots, \mathbf{c}_n$ *be linearly independent points of a lattice* $\Lambda$ *but not a basis. Then* $\Lambda$ *contains a point*

$$\boldsymbol{d} = \vartheta_1 \mathbf{c}_1 + \cdots + \vartheta_n \mathbf{c}_n,$$

*where* $\vartheta_1, \ldots, \vartheta_n$ *are numbers such that*

$$\tfrac{1}{4} \leqq \max_j |\vartheta_j| \leqq \tfrac{1}{2}. \tag{14}$$

We first prove Lemma 5. Since $\boldsymbol{c}_1, \ldots, \boldsymbol{c}_n$ is not a basis, there certainly exist points

$$\boldsymbol{a} = \alpha_1 \boldsymbol{c}_1 + \cdots + \alpha_n \boldsymbol{c}_n$$

in $\Lambda$ for which $\alpha_1, \ldots, \alpha_n$ are not integers. We may suppose without loss of generality that

$$|\alpha_j| \leqq \tfrac{1}{2} \qquad (1 \leqq j \leqq n).$$

Let $t$ be the least non-negative integer such that

$$2^t \max_j |\alpha_j| \geqq \tfrac{1}{4}.$$

Then

$$2^t \max_j |\alpha_j| \leqq \tfrac{1}{2},$$

and

$$\boldsymbol{d} = 2^t \boldsymbol{a}$$

will do what is required. A slight refinement of the argument, which is left to the reader, shows that the $\frac{1}{4}$ in (14) may be replaced by $\frac{1}{3}$ but by no larger number.

We now revert to the proof of Theorem I. Suppose that $\Lambda_r$ and $\Lambda'$ satisfy (i) and (ii). Let $\boldsymbol{b}_1', \ldots, \boldsymbol{b}_n'$ be any basis for $\Lambda'$. By (i) there exist sequences of points

$$\boldsymbol{b}_j^r \to \boldsymbol{b}_j' \qquad (1 \leqq j \leqq n,\ \boldsymbol{b}_j^r \in \Lambda_r). \tag{15}$$

We show that $\boldsymbol{b}_j^r$ $(1 \leqq j \leqq n)$ is actually a basis for $\Lambda_r$ except, possibly, for a finite number of $r$. For if $\boldsymbol{b}_1^r, \ldots, \boldsymbol{b}_n^r$ is not a basis for $\Lambda_r$, let

$$\boldsymbol{d}_r = \vartheta_{1r} \boldsymbol{b}_1^r + \cdots + \vartheta_{nr} \boldsymbol{b}_n^r \tag{16}$$

be a point of $\Lambda$ with

$$\tfrac{1}{4} \leqq \max_j |\vartheta_{jr}| \leqq \tfrac{1}{2} \tag{17}$$

which is given by Lemma 5. Since the $\vartheta_{jr}$ are bounded, they contain a convergent subsequence by a classical theorem of WEIERSTRASS (cf. § 1.2 of Chapter III), say

$$\lim_{t \to \infty} \vartheta_{j r_t} = \vartheta_j', \tag{18}$$

where

$$r_1 < r_2 < \cdots < r_t < \cdots$$

is an increasing sequence of integers. Then

$$\boldsymbol{d}' \text{ (say)} = \sum_j \vartheta_j' \boldsymbol{b}_j' = \lim_{t \to \infty} \boldsymbol{d}_{r_t},$$

by (15), (16) and (18). Hence $\boldsymbol{d}' \in \Lambda'$ by (ii) of the enunciation of the theorem. This is a contradiction since

$$\tfrac{1}{4} \leqq \max_j |\vartheta_j'| \leqq \tfrac{1}{2},$$

by (14) and (15), and since $\boldsymbol{b}_j'$ $(1 \leqq j \leqq n)$ was defined to be a basis for $\Lambda'$. The contradiction shows that $\boldsymbol{b}_j^r$ is a basis for $\Lambda_r$ except for a finite number of $r$. If the $\boldsymbol{b}_j^r$ are changed for these exceptional $r$ so that $\boldsymbol{b}_j^r$ $(1 \leqq j \leqq r)$ is a basis for $\Lambda_r$ for all $r$ this does not affect the limits (15). Hence the criterion is certainly sufficient by Lemma 4.

**V.3.2.** In Chapter X we shall need the notion of a neighbourhood of a lattice, and we shall mention it again in passing briefly in §9 of this chapter.

A set $\mathfrak{L}$ of lattices $\Lambda$ is said to be a neighbourhood of the lattice $\mathsf{M}$ if it contains all lattices

$$\Lambda = \boldsymbol{\tau} \mathsf{M} \tag{1}$$

with

$$\|\boldsymbol{\tau} - \boldsymbol{\iota}\| < \eta$$

for some $\eta > 0$ depending on the particular neighbourhood. The neighbourhood $\mathfrak{L}$ may contain other lattices $\Lambda$ than those given by (1) and (2); but there is some $\eta > 0$ such that it contains all these. If $\boldsymbol{\alpha}$ is any non-singular homogeneous transformation we show that the set $\boldsymbol{\alpha}\mathfrak{L}$ of lattices $\boldsymbol{\alpha}\Lambda$, $\Lambda \in \mathfrak{L}$ is a neighbourhood of $\boldsymbol{\alpha}\mathsf{M}$. Indeed $\boldsymbol{\alpha}\mathfrak{L}$ contains all lattices

$$\mathsf{N} = \boldsymbol{\sigma}(\boldsymbol{\alpha}\mathsf{M})$$

with

$$\|\boldsymbol{\sigma} - \boldsymbol{\iota}\| < \{\|\boldsymbol{\alpha}\| \, \|\boldsymbol{\alpha}^{-1}\|\}^{-1}\eta;$$

since then

$$\mathsf{N} = \boldsymbol{\alpha}\Lambda$$

where

$$\Lambda = \boldsymbol{\alpha}^{-1}\boldsymbol{\sigma}\boldsymbol{\alpha}\mathsf{M}:$$

and then

$$\|\boldsymbol{\alpha}^{-1}\boldsymbol{\sigma}\boldsymbol{\alpha} - \boldsymbol{\iota}\| \leqq \|\boldsymbol{\alpha}^{-1}\| \, \|\boldsymbol{\alpha}\| \, \|\boldsymbol{\sigma} - \boldsymbol{\iota}\| < \eta$$

as in §3.1.

Clearly the sequence $\Lambda_r$ $(1 \leqq r < \infty)$ of lattices tends to $\mathsf{M}$ if and only if every neighbourhood of $\mathsf{M}$ contains all but a finite number of the $\Lambda_r$.

Although we nowhere use it, we note that it is in fact possible to introduce explicitly a metric into the space of all lattices. Let $\Lambda$ and $\mathsf{M}$ be two lattices and let

$$\mu = \inf \|\boldsymbol{\sigma} - \boldsymbol{\iota}\|, \quad \nu = \inf \|\boldsymbol{\tau} - \boldsymbol{\iota}\|,$$

where the infima are over all non-singular $\sigma$ and $\tau$ such that

$$\Lambda = \sigma M \qquad M = \tau \Lambda.$$

Put

$$D(M, \Lambda) = D(\Lambda, M) = \max\{\log(1+\mu), \log(1+\nu)\}.$$

Then we have the triangle inequality

$$D(\Lambda, N) \leqq D(\Lambda, M) + D(M, N);$$

since if

$$\Lambda = (\iota + \rho_1) M, \qquad M = (\iota + \rho_2) N;$$

then

$$\Lambda = (\iota + \rho_3) N,$$

where

$$\|\rho_3\| = \|\rho_1 + \rho_2 + \rho_1\rho_2\| \leqq \|\rho_1\| + \|\rho_2\| + \|\rho_1\|\,\|\rho_2\|;$$

and so

$$\log(1 + \|\rho_3\|) \leqq \log(1 + \|\rho_1\|) + \log(1 + \|\rho_2\|).$$

The neighbourhood defined above is the one associated with this metric, since if

$$\Lambda = \sigma M$$

with

$$\|\sigma - \iota\| < \eta < 1;$$

then

$$M = \sigma^{-1}\Lambda,$$

where

$$\|\sigma^{-1} - \iota\| \leqq \frac{\|\sigma - \iota\|}{1 - \|\sigma - \iota\|} < \frac{\eta}{1-\eta},$$

and so

$$D(\Lambda, M) < \frac{\eta}{1-\eta}.$$

**V.3.3.** The continuity of the distance-function $F(\boldsymbol{x})$ of the vector $\boldsymbol{x}$ is reflected as a semi-continuity of the function

$$F(\Lambda) = \inf_{\substack{\boldsymbol{a} \in \Lambda \\ \neq \boldsymbol{o}}} F(\boldsymbol{a}) \tag{1}$$

of the lattice $\Lambda$ considered in § 4 of Chapter IV. For certain later applications it is useful to allow the distance-function $F$ and the lattice $\Lambda$ to vary simultaneously.

THEOREM II. *Let $\Lambda_r$ $(1 \leqq r < \infty)$ be a sequence of lattices tending to the lattice $\Lambda'$. Let $F_r(\boldsymbol{x})$ $(1 \leqq r < \infty)$ be a sequence of distance functions which converge uniformly to the distance-function $F'(\boldsymbol{x})$ in the unit sphere $|\boldsymbol{x}| < 1$. Then*

$$F'(\Lambda') \geqq \limsup_{r \to \infty} F_r(\Lambda_r). \tag{2}$$

The proof is very simple. Since $F_r(t\boldsymbol{x}) = tF_r(\boldsymbol{x})$ for $t>0$, the convergence of $F_r(\boldsymbol{x})$ to $F'(\boldsymbol{x})$ is uniform in any bounded set of points; in particular, since the distance-function $F'(\boldsymbol{x})$ is continuous by definition, if $\boldsymbol{a}_r$ is any sequence of points converging to a point $\boldsymbol{a}'$, we have

$$\lim_{r\to\infty} F_r(\boldsymbol{a}_r) = F'(\boldsymbol{a}').$$

But by Theorem I, every point $\boldsymbol{a}' \neq \boldsymbol{o}$ of $\Lambda'$ is the limit of points $\boldsymbol{a}_r \neq \boldsymbol{o}$ of $\Lambda_r$. Hence

$$F'(\boldsymbol{a}') = \lim_{r\to\infty} F_r(\boldsymbol{a}_r) \geqq \limsup_{r\to\infty} F_r(\Lambda_r),$$

since $F_r(\boldsymbol{a}_r) \geqq F_r(\Lambda_r)$. The result (2) now follows from the definition (1).

The sign of equality need not hold in (2) even when $F_r = F'$ for all $r$, but we defer giving an example until § 10.5. However, much more than Theorem II is true if $F'(\boldsymbol{x}) = 0$ only for $\boldsymbol{x} = \boldsymbol{o}$, i.e. if the set $F'(\boldsymbol{x}) < 1$ is bounded (Lemma 2 of Chapter IV).

COROLLARY. *Suppose that the hypotheses of Theorem II hold and that the only point $\boldsymbol{x}$ such that $F'(\boldsymbol{x}) = 0$ is $\boldsymbol{x} = \boldsymbol{o}$. Then*

$$\lim_{r\to\infty} F_r(\Lambda_r)$$

*exists and is equal to $F'(\Lambda')$.*

The proof is similar to that of Lemma 3. By Lemma 2 of Chapter IV, there is a $c>0$ such that

$$F'(\boldsymbol{x}) \geqq c|\boldsymbol{x}| \tag{3}$$

for all $\boldsymbol{x}$. Let $\varepsilon>0$ be arbitrarily small. By the uniformity of the convergence of $F_r(\boldsymbol{x})$, there is an $r_0$ such that

$$|F_r(\boldsymbol{x}) - F'(\boldsymbol{x})| < c\varepsilon \tag{4}$$

for all $r \geqq r_0$ and all $\boldsymbol{x}$ with $|\boldsymbol{x}| = 1$. Hence for all $\boldsymbol{x}$ whatsoever and $r \geqq r_0$, we have

$$|F_r(\boldsymbol{x}) - F'(\boldsymbol{x})| < c\varepsilon|\boldsymbol{x}| \leqq \varepsilon F'(\boldsymbol{x});$$

so

$$1-\varepsilon < \frac{F_r(\boldsymbol{x})}{F'(\boldsymbol{x})} < 1+\varepsilon. \tag{5}$$

Now let $\Lambda_r = \boldsymbol{\tau}_r\Lambda'$, where $\boldsymbol{\tau}_r$ are homogeneous linear transformations such that

$$\|\boldsymbol{\tau}_r - \boldsymbol{\iota}\| \to 0 \qquad (r\to\infty)$$

in the language of § 2. Then

$$1-\varepsilon < \frac{F'(\boldsymbol{\tau}_r\boldsymbol{x})}{F'(\boldsymbol{x})} < 1+\varepsilon \tag{6}$$

for all $r$ greater than some $r_1$, by Lemma 3. Hence by (6) and (5) with $\tau_r x$ for $x$ we have

$$(1-\varepsilon)^2 < \frac{F_r(\tau_r x)}{F'(x)} < (1+\varepsilon)^2$$

for all $r > \max(r_0, r_1)$. But now $\Lambda_r$ is just the set of $\tau_r x$ with $x \in \Lambda'$, and so[1]

$$(1-\varepsilon)^2 \leqq \frac{F_r(\Lambda_r)}{F'(\Lambda')} \leqq (1+\varepsilon)^2.$$

Since $\varepsilon$ is arbitrarily small, this proves the corollary.

**V.3.4.** An almost immediate consequence of Theorem II, Corollary is the following result, which shows that no bounded star body can have successive minima in the sense of Chapter II, § 4.

LEMMA 6. *Let $F(x)$ be an $n$-dimensional distance function which vanishes only when $x = o$ and let $\eta$ be any number for which*

$$0 < \eta < \delta(F) = \sup_{\Lambda} \frac{\{F(\Lambda)\}^n}{d(\Lambda)}. \tag{1}$$

*Then there exists a lattice $M_\eta$ such that*

$$\{F(M_\eta)\}^n = \eta\, d(M_\eta).$$

After Theorem VI we shall be able to replace the second $<$ in (1) by $\leqq$.

Suppose that $\eta$ satisfies (1). Then there exists a lattice N, such that

$$\{F(N)\}^n > \eta\, d(N). \tag{2}$$

Let $b_1, \ldots, b_n$ be any basis for N; and for $0 < \varepsilon < 1$ let $N_\varepsilon$ be the lattice with basis

$$\varepsilon b_1, b_2, \ldots, b_n.$$

Then

$$d(N_\varepsilon) = \varepsilon d(N)$$

and

$$F(N_\varepsilon) \leqq F(\varepsilon b_1) = \varepsilon F(b_1).$$

Hence

$$\frac{\{F(N_\varepsilon)\}^n}{d(N_\varepsilon)} \leqq \varepsilon^{n-1} \frac{\{F(b_1)\}^n}{d(N)} \to 0 \qquad (\varepsilon \to 0). \tag{3}$$

But now, by Theorem II Corollary, $F(N_\varepsilon)$ is a continuous function of $\varepsilon$. Hence by (2) and (3) we may put $M_\eta = N_\varepsilon$ for an appropriate value of $\varepsilon$.

**V.3.5.** For the sake of completeness we enunciate the following lemma, which interprets the uniformity of the convergence of $F_r(x)$ to

[1] Note that $F'(\Lambda') \neq 0$ by Lemma 7 of Chapter IV.

$F'(\boldsymbol{x})$ in terms of the corresponding star bodies

$$\mathscr{S}_r: \quad F_r(\boldsymbol{x}) < 1 \tag{1}$$

and

$$\mathscr{S}': \quad F'(\boldsymbol{x}) < 1. \tag{2}$$

Since we do not use the lemma we do not give the proof, but the reader should have no difficulty in constructing one along the lines of the proof of Theorem I of Chapter IV.

LEMMA 7. *A necessary and sufficient condition that the sequence of distance functions $F_r(\boldsymbol{x})$ tend to the distance-function $F'(\boldsymbol{x})$ uniformly in $|\boldsymbol{x}| \leqq 1$ is that the bodies $\mathscr{S}_r$ and $\mathscr{S}'$ defined by (1) and (2) have the following properties:*

*(i) If $\mathbf{c}$ is an (inner) point of $\mathscr{S}'$, then there exists an $\eta > 0$ and an integer $r_0$ (depending on c) such that all points $\boldsymbol{x}$ of the neighbourhood $|\boldsymbol{x} - \mathbf{c}| < \eta$ belong to $\mathscr{S}_r$ for all $r$ greater than $r_0$.*

*(ii) If $\mathbf{c}$ is an exterior point to $\mathscr{S}'$ (i.e. $F'(\boldsymbol{x}) > 1$) then there is an $\eta > 0$ and $r_0$ such that no point $\boldsymbol{x}$ of the neighbourhood $|\boldsymbol{x} - \mathbf{c}| < \eta$ belongs to $\mathscr{S}_r$ for any $r > r_0$.*

**V.4. Compactness for lattices.** In this section we are concerned with conditions under which an infinite sequence $\Lambda_r$ of lattices should contain a subsequence $\mathsf{M}_t = \Lambda_{r_t}$ which converges to a lattice $\mathsf{M}'$, not necessarily belonging to the sequence.

The simplest such condition is when every lattice of the sequence has a basis every point of which lies in some fixed sphere

$$|\boldsymbol{x}| \leqq R \tag{1}$$

and $d(\Lambda_r)$ is bounded below by a positive constant, say

$$d(\Lambda_r) \geqq \varkappa > 0 \qquad (\text{all } r). \tag{2}$$

Since all the lattices have bases in (1) we may by WEIERSTRASS' compactness theorem, find a subsequence of lattices $\mathsf{M}_t = \Lambda_{r_t}$ with bases $\boldsymbol{b}_1^t, \ldots, \boldsymbol{b}_n^t$ in (1) such that all the limits

$$\lim_{t \to \infty} \boldsymbol{b}_j^t = \boldsymbol{b}_j'$$

exist. By (2) we have

$$|\det(\boldsymbol{b}_1', \ldots, \boldsymbol{b}_n')| = \lim_{t \to \infty} |\det(\boldsymbol{b}_1^t, \ldots, \boldsymbol{b}_n^t)| = \lim_{t \to \infty} d(\mathsf{M}_t) \geqq \varkappa > 0:$$

and so $\boldsymbol{b}_1', \ldots, \boldsymbol{b}_n'$ are linearly independent. Hence there exists a lattice $\mathsf{M}'$ with basis $\boldsymbol{b}_1', \ldots, \boldsymbol{b}_n'$ and, by Lemma 4,

$$\mathsf{M}_t \to \mathsf{M}' \qquad (t \to \infty).$$

A slight extension of this idea gives the following theorem which however turns out not to be very useful. We give it partly for historical interest and partly because the lemma on which it depends will be used later.

THEOREM III. *Let* $\Lambda_r$ $(1 \leqq r < \infty)$ *be an infinite sequence of n-dimensional lattices enjoying the following two properties:*

*(i) there exists an R such that every* $\Lambda_r$ *has n linearly independent points in the sphere*

$$|\boldsymbol{x}| \leqq R.$$

*(ii) there exists a* $\varkappa > 0$ *such that*

$$d(\Lambda_r) \geqq \varkappa$$

*for all r.*

*Then* $\Lambda_r$ *contains a subsequence of lattices* $\mathsf{M}_t$ *for which*

$$\mathsf{M}' = \lim_{t \to \infty} \mathsf{M}_t$$

*exists.*

The proof of Theorem III depends on the following lemma due to MAHLER (1938a) and rediscovered by WEYL (1942a).

LEMMA 8. *Let* $F(\boldsymbol{x})$ *be any symmetric convex distance function and* $\boldsymbol{a}_1, \ldots, \boldsymbol{a}_n$ *be n linearly independent points of a lattice* $\Lambda$. *Then there exists a basis* $\boldsymbol{b}_1, \ldots, \boldsymbol{b}_n$ *of* $\Lambda$ *for which*

$$F(\boldsymbol{b}_j) \leqq \max\left[F(\boldsymbol{a}_j), \tfrac{1}{2}\{F(\boldsymbol{a}_1) + \cdots + F(\boldsymbol{a}_j)\}\right].$$

Before proving Lemma 8 we show that Theorem III follows from it by applying it to the convex function $F(\boldsymbol{x}) = |\boldsymbol{x}|$ and to the $n$ linearly independent points $\boldsymbol{a}_1^r, \ldots, \boldsymbol{a}_n^r$ of $\Lambda_r$ given by (i) of the theorem. Then Lemma 8 shows that $\Lambda_r$ has a basis $\boldsymbol{b}_1^r, \ldots, \boldsymbol{b}_n^r$ with

$$|\boldsymbol{b}_j^r| \leqq \max\left[|\boldsymbol{a}_j^r|, \tfrac{1}{2}\{|\boldsymbol{a}_1^r| + \cdots + |\boldsymbol{a}_j^r|\}\right] \leqq nR/2.$$

We have thus reduced Theorem III to the trivial case discussed at the beginning but with $nR/2$ instead of $R$.

It remains to prove Lemma 8. By Theorem I of Chapter I there is a basis $\boldsymbol{c}_1, \ldots, \boldsymbol{c}_n$ of $\Lambda$ such that

$$\left.\begin{aligned} \boldsymbol{a}_1 &= v_{11}\boldsymbol{c}_1, \\ \boldsymbol{a}_2 &= v_{21}\boldsymbol{c}_1 + v_{22}\boldsymbol{c}_2, \\ &\cdots\cdots\cdots\cdots \\ \boldsymbol{a}_n &= v_{n1}\boldsymbol{c}_1 + \cdots + v_{nn}\boldsymbol{c}_n, \end{aligned}\right\} \tag{3}$$

where the $v_{ij}$ are integers and $v_{ii} \neq 0$. We shall take $\boldsymbol{b}_j$ of the shape

$$\boldsymbol{b}_j = \boldsymbol{c}_j + t_{j,j-1}\boldsymbol{a}_{j-1} + \cdots + t_{j1}\boldsymbol{a}_1 \in \Lambda, \tag{4}$$

where the $t_{ji}$ are numbers to be determined. Clearly $\boldsymbol{b}_1, \ldots, \boldsymbol{b}_n$ is a basis for $\Lambda$ for any set of numbers $t_{ji}$ such that $\boldsymbol{b}_j \in \Lambda$.

We distinguish two cases for each $j$. If $v_{jj} = \pm 1$, we put $\boldsymbol{b}_j = \pm \boldsymbol{a}_j$. This certainly has the shape (4) and

$$F(\boldsymbol{b}_j) = F(\boldsymbol{a}_j).$$

Otherwise $|v_{jj}| \geqq 2$. On solving (3) for the $\boldsymbol{c}_j$ we have

$$\boldsymbol{c}_j = v_{jj}^{-1} \boldsymbol{a}_j + k_{j,j-1} \boldsymbol{a}_{j-1} + \cdots + k_{j1} \boldsymbol{a}_1, \tag{5}$$

where $k_{ji}$ are certain real numbers. Choose $t_{ji}$ in (4) to be integers such that

$$|k_{ji} + t_{ji}| \leqq \tfrac{1}{2}.$$

Then $\boldsymbol{b}_j \in \Lambda$ and

$$\boldsymbol{b}_j = l_{jj} \boldsymbol{a}_j + l_{j,j-1} \boldsymbol{a}_{j-1} + \cdots + l_{j1} \boldsymbol{a}_1, \tag{6}$$

where

$$|l_{jj}| = |v_{jj}^{-1}| \leqq \tfrac{1}{2}$$

and

$$|l_j| = |k_{ji} + t_{ji}| \leqq \tfrac{1}{2} \qquad (i < j).$$

Then by the convexity symmetry and homogeneity of $F(\boldsymbol{x})$ we have

$$\left.\begin{aligned} F(\boldsymbol{b}_j) &\leqq F(l_{jj} \boldsymbol{a}_j) + \cdots + F(l_{j1} \boldsymbol{a}_1) \\ &= |l_{jj}| F(\boldsymbol{a}_j) + \cdots + |l_{j1}| F(\boldsymbol{a}_1) \\ &\leqq \tfrac{1}{2}\{F(\boldsymbol{a}_j) + \cdots + F(\boldsymbol{a}_1)\}. \end{aligned}\right\} \tag{7}$$

This concludes the proof of the lemma.

When $F(\boldsymbol{x})$ is the usual euclidean distance, an argument due to REMAK (1938a) gives a sharper result. See also VAN DER WAERDEN (1956a).

When

$$F(\boldsymbol{a}_1) \leqq F(\boldsymbol{a}_2) \leqq \cdots \leqq F(\boldsymbol{a}_n), \tag{8}$$

Lemma 8 gives

$$F(\boldsymbol{b}_j) \leqq \max\left(1, \frac{n}{2}\right) F(\boldsymbol{a}_j).$$

**V.4.2.** We owe to MAHLER (1946d, e) a criterion for the existence of a convergent subsequence of lattices in a sequence of lattices, which is much more fertile of applications than Theorem III, and which may be said to have completely transformed the subject. MAHLER proved his criterion by using the theory of successive minima[1] of a sphere to show that it is equivalent to that of Theorem III. We shall give

[1] Not to be confused with the "successive minima" discussed in Chapter II which are quite different.

MAHLER's argument[1] when we discuss successive minima in Chapter VIII, but here we give a direct treatment due to CHABAUTY (1950a), who shows that it generalizes significantly to a more general situation (subgroups of locally compact topological groups). MAHLER'S criterion is expressed in

THEOREM IV. *Let* $\Lambda_r$ *be any infinite sequence of lattices satisfying the following two conditions*

*(i)* $d(\Lambda_r) \leqq K$ *for all lattices* $\Lambda_r$, *where* $K$ *is independent of* $r$.

*(ii)* $|\Lambda_r| \geqq \varkappa > 0$ *for all* $r$ *where* $\varkappa$ *is independent of* $r$ *and, as usual,*

$$|\Lambda| = \inf_{\substack{\boldsymbol{a} \in \Lambda \\ \neq \boldsymbol{o}}} |\boldsymbol{a}|.$$

*Then* $\Lambda_r$ *contains a subsequence* $\mathsf{M}_r = \Lambda_{r_t}$ *which converges to a limit* $\mathsf{M}'$.

We prove[2] Theorem IV by induction. The result of the $j$-th stage $(1 \leqq j \leqq n)$ will be the following statement:

$\mathfrak{S}_j$: There exist $j$ linearly independent points $\boldsymbol{a}_1, \ldots, \boldsymbol{a}_j$ and a subsequence $\mathsf{N}_t = \mathsf{N}_t^j \, (1 \leqq t < \infty)$ of $\Lambda_r$ which satisfies the following conditions:

$\mathfrak{S}_j'$: Each point $\boldsymbol{a}_i$ $(1 \leqq i \leqq j)$ is the limit of points

$$\boldsymbol{a}_i^t \in \mathsf{N}_t = \mathsf{N}_t^j \tag{1}$$

$\mathfrak{S}_j''$: Suppose that $t_1 < t_2 < \cdots$ is any increasing sequence of integers and there exist points $\boldsymbol{c}_{t_s} \in \mathsf{N}_{t_s}$ such that

$$\lim_{s \to \infty} \boldsymbol{c}_{t_s} = \gamma_1 \boldsymbol{a}_1 + \cdots + \gamma_j \boldsymbol{a}_j \tag{2}$$

with real $\gamma_1, \ldots, \gamma_j$. Then $\gamma_1, \ldots, \gamma_j$ must be integers.

Before continuing the proof we note that the statement $\mathfrak{S}_n$ implies that the lattices $\mathsf{M}_t = \mathsf{N}_t^n$ converge to the lattice $\mathsf{M}'$ with basis $\boldsymbol{a}_1, \ldots, \boldsymbol{a}_n$; the parts $\mathfrak{S}_n'$ and $\mathfrak{S}_n''$ of corresponding respectively to (i) and (ii) of the Theorem I. Hence it suffices to prove $\mathfrak{S}_n$.

We do not give a separate proof of $\mathfrak{S}_1$ since that is a simple version of the deduction of $\mathfrak{S}_{j+1}$ from $\mathfrak{S}_j$. For the rest of this section we shall assume therefore that $\mathfrak{S}_j$ holds for some $j$ in $1 \leqq j < n$ and will deduce $\mathfrak{S}_{j+1}$. The sequence $\mathsf{N}_t^{j+1}$ will be a subsequence of the sequence $\mathsf{N}_t = \mathsf{N}_t^j$, and the points $\boldsymbol{a}_1, \ldots, \boldsymbol{a}_j$ will be the same in $\mathfrak{S}_j$ and $\mathfrak{S}_{j+1}$.

A non-singular homogeneous linear transformation of the variables does not affect either statement $\mathfrak{S}_j$ or the hypotheses of the theorem, though it will in general replace the $K$ and $\varkappa$ in (i) and (ii) by different numbers. Hence we may suppose without loss of generality that

$$\boldsymbol{a}_i = \boldsymbol{e}_i = (\overbrace{0, \ldots, 0}^{i-1}, 1, \overbrace{0, \ldots, 0}^{n-i}) \qquad (1 \leqq i \leqq j). \tag{3}$$

---

[1] The reader may prefer, instead of studying the proof here, to turn to §§ 1, 2 of Chapter VIII, which are independent of the intervening matter.

[2] I now prefer the proof given by CHABAUTY (1950a) to the version given here.

Define the number $\psi$ by

$$(\tfrac{3}{4})^j \psi^{n-j} = K, \tag{4}$$

where $K$ is the number occurring in hypothesis (i) of the theorem. By Theorem III of Chapter III, each lattice $\mathsf{N}_t$ contains a point $\boldsymbol{x} \neq \boldsymbol{o}$ with[1]

$$\left.\begin{array}{ll} |x_i| \leqq \frac{3}{4} & (1 \leqq i \leqq j) \\ |x_i| \leqq \psi & (j+1 \leqq i \leqq n). \end{array}\right\} \tag{5}$$

Let $\boldsymbol{c}^t$ be one of the finite number of points of $\mathsf{N}_t$ other than $\boldsymbol{o}$ in (5) for which

$$\max_{j+1 \leqq i \leqq n} |x_i| \tag{6}$$

is a minimum. Since the $\boldsymbol{c}^t$ are in the bounded set (5), they contain a convergent subsequence, say

$$\boldsymbol{c}^{t_s} \to \boldsymbol{a}_{j+1}, \tag{7}$$

where

$$t_1 < t_2 < \cdots.$$

Write

$$\boldsymbol{a}_{j+1} = (A_1, \ldots, A_n), \tag{8}$$

so that clearly

$$\left.\begin{array}{ll} |A_i| \leqq \frac{3}{4} & (1 \leqq i \leqq j) \\ |A_i| \leqq \psi & (j+1 \leqq i \leqq n). \end{array}\right\} \tag{9}$$

Suppose first, if possible, that $A_{j+1} = \cdots = A_n = 0$, so that $\boldsymbol{a}_{j+1}$ is linearly dependent on $\boldsymbol{a}_1, \ldots, \boldsymbol{a}_j$. We are assuming statement $\mathfrak{S}_j$ to be already established. Hence by (7) we could apply $\mathfrak{S}_j''$ with $\gamma_i = A_i$ $(1 \leqq i \leqq j)$ and it would follow from (9) that $A_1 = \cdots = A_j = 0$, and so

$$\lim_{s \to \infty} \boldsymbol{c}^{t_s} = \boldsymbol{o}.$$

This contradicts hypothesis (ii) of the theorem, since $\boldsymbol{c}^{t_s} \in \mathsf{N}_{t_s} = \Lambda_r$ for some $r$ and $\boldsymbol{c}^{t_s} \neq \boldsymbol{o}$. Hence the vectors $\boldsymbol{a}_1, \ldots, \boldsymbol{a}_{j+1}$ are linearly independent. We put $\Gamma_s = \mathsf{N}_{t_s}$, and will show that the statement $\mathfrak{S}_{j+1}$ now holds for $\mathsf{N}_s^{j+1} = \Gamma_s$.

The statement $\mathfrak{S}_{j+1}'$ is trivially true. So far as $\boldsymbol{a}_{j+1}$ is concerned, $\mathfrak{S}_{j+1}'$ follows from (7); and so far as the remaining $\boldsymbol{a}_i$ $(1 \leqq i \leqq j)$ are concerned, $\mathfrak{S}_{j+1}'$ follows from $\mathfrak{S}_j'$ since $\Gamma_s$ is a subsequence of $\mathsf{N}_t$.

It remains to prove $\mathfrak{S}_{j+1}''$. Suppose, if possible, that there is an increasing sequence of integers

$$s_1 < s_2 < \cdots < s_m < \cdots \tag{10}$$

and vectors

$$\boldsymbol{d}^{s_m} \in \Gamma_{s_m}$$

[1] The only property of $\frac{3}{4}$ we use is $\frac{1}{2} < \frac{3}{4} < 1$.

such that

$$\left.\begin{aligned}\lim_{t\to\infty} \boldsymbol{d}^{s_m} &= \boldsymbol{d} \text{ (say)}\\ &= \delta_1 \boldsymbol{a}_1 + \cdots + \delta_{j+1} \boldsymbol{a}_{j+1},\end{aligned}\right\} \tag{11}$$

where $\delta_1, \ldots, \delta_{j+1}$ are not all integers. By $\mathfrak{S}'_{j+1}$, which we have already proved, we may add integer multiples of $\boldsymbol{a}_1, \ldots, \boldsymbol{a}_{j+1}$ to the right-hand side of (11), after appropriate modification of the sequence $\boldsymbol{d}^{s_m}$. Hence we may suppose in the first place that

$$|\delta_{j+1}| \leqq \tfrac{1}{2} \tag{12}$$

and in the second place, by (3), that

$$|d_i| \leqq \tfrac{1}{2} < \tfrac{3}{4} \quad (1 \leqq i \leqq j), \tag{13}$$

where, as usual, $\boldsymbol{d} = (d_1, \ldots, d_n)$. From (8) and (12) we have

$$\left.\begin{aligned}\max_{j+1\leqq i\leqq n} |d_i| &= |\delta_{j+1}| \max_{j+1\leqq i\leqq n} |A_i|\\ &\leqq \tfrac{1}{2} \max_{j+1\leqq i\leqq n} |A_i|\end{aligned}\right\} \tag{14}$$

$$< \psi. \tag{15}$$

We now show that this in contradiction with the definition of the vectors $\boldsymbol{c}^t$ as the vectors $\boldsymbol{x}$ of $\mathsf{N}_t$ in (5) other than $\boldsymbol{o}$ for which (6) is as small as possible. Since $\boldsymbol{c}^{t_r} \to \boldsymbol{a}_{j+1}$, we have

$$\lim_{r\to\infty} \max_{j+1\leqq i\leqq n} |c_{i t_r}| = \max_{j+1\leqq i\leqq n} |A_i|, \tag{16}$$

where

$$\boldsymbol{c}^t = (c_{1t}, \ldots, c_{nt}).$$

By (13) and (15) the vector $\boldsymbol{d}^{s_m}$ certainly lies in the region defined by (5) when $m$ is large enough. Further, $\boldsymbol{d}^{s_m} \in \mathsf{N}_T$, where $T = t_{s_m}$. But now, by (14) and (16), the function (6) is certainly greater for $\boldsymbol{c}^T$ than it is for $\boldsymbol{d}^{s_m}$ when $m$ is large enough, which contradicts the choice of $\boldsymbol{c}^T$. The contradiction shows that if (11) holds then $\delta_1, \ldots, \delta_{j+1}$ are all integers; that is the statement $\mathfrak{S}''_{j+1}$ holds.

This ends the deduction of $\mathfrak{S}_{j+1}$ from $\mathfrak{S}_j$, and so concludes the proof of Theorem IV.

We note a form which is often useful in applications and which does not depend on the use of the special distance-function $|\boldsymbol{x}|$.

COROLLARY. *Let $F(\boldsymbol{x})$ be any distance function and let $\Lambda_r$ be any infinite sequence of lattices satisfying the two conditions*

*(i) $d(\Lambda_r) \leqq K$ for all $r$, where $K$ is independent of $r$.*

*(ii) $F(\Lambda_r) \geqq \varkappa > 0$ for all $r$, where $\varkappa$ is independent of $r$ and, as usual,*

$$F(\Lambda) = \inf_{\substack{\boldsymbol{a}\in\Lambda\\ \neq \boldsymbol{o}}} F(\boldsymbol{a}).$$

*Then $\Lambda_r$ contains a convergent subsequence.*

For by Lemma 1 of Chapter IV there is a $C>0$ such that $F(\boldsymbol{x}) \leqq C|\boldsymbol{x}|$, and so

$$|\Lambda_r| \geqq C^{-1} F(\Lambda_r) \geqq C^{-1}\varkappa > 0.$$

**V.4.3.** An almost immediate consequence (cf. MAHLER 1949a) of Theorem IV is

THEOREM V. *Let $\mathscr{S}$ be any open set. Let $\mathscr{S}_1, \mathscr{S}_2, \ldots, \mathscr{S}_r, \ldots$ be a sequence of open subsets of $\mathscr{S}$ such that*

*(i) $\mathscr{S}_r$ is contained in $\mathscr{S}_t$ if $r<t$,*

*(ii) the origin is an inner point of $\mathscr{S}_1$,*

*(iii) every point $\boldsymbol{x}$ of $\mathscr{S}$ is in $\mathscr{S}_r$ for some $r$.*

*Then*

$$\Delta(\mathscr{S}) = \lim_{r\to\infty} \Delta(\mathscr{S}_r).$$

We recall that $\Delta(\mathscr{S})$ is the lower bound of $d(\Lambda)$ over $\mathscr{S}$-admissible lattices $\Lambda$, i.e. $\Lambda$ having only $\boldsymbol{o}$ in $\mathscr{S}$. Clearly

$$\Delta(\mathscr{S}_r) \leqq \Delta(\mathscr{S}) \tag{1}$$

for all $r$. Suppose that

$$\liminf_{r\to\infty} \Delta(\mathscr{S}_r) < \Delta(\mathscr{S}). \tag{2}$$

Then there is an increasing sequence of integers $r_1 < r_2 < \cdots$ and lattices $\Lambda_{r_t}$ such that

$$\lim_{t\to\infty} d(\Lambda_{r_t}) < \Delta(\mathscr{S});$$

and $\Lambda_{r_t}$ is $\mathscr{S}_{r_t}$-admissible. By (ii) and Theorem IV we may extract a convergent sequence of lattices from the $\Lambda_{r_t}$, so that without loss of generality

$$\lim_{t\to\infty} \Lambda_{r_t} = \Lambda'; \quad d(\Lambda') < \Delta(\mathscr{S}).$$

Hence there is a point $\boldsymbol{p} \neq \boldsymbol{o}$ of $\Lambda'$ in $\mathscr{S}$. By (iii) then $\boldsymbol{p}$ is in $\mathscr{S}_R$ for some $R$. By (i) and since $\mathscr{S}_r$ is open by hypothesis, there is a neighbourhood

$$|\boldsymbol{x} - \boldsymbol{p}| < \eta \tag{3}$$

every point of which is in $\mathscr{S}_r$ for all $r \geqq R$.

But now

$$\boldsymbol{p} = \lim_{r\to\infty} \boldsymbol{p}', \quad \boldsymbol{p}' \in \Lambda_r$$

by Theorem I. Hence $\boldsymbol{x} = \boldsymbol{p}'$ satisfies (3) for all $r$ greater than some $r_1$. For $r > \max(R, r_1)$ this means that $\boldsymbol{p}_r$ is in $\mathscr{S}$ contrary to our assumption. The contradiction arose from the assumption that (2) is true. Hence the theorem is true by (1).

When the $\mathcal{S}$ and $\mathcal{S}_r$ are star-bodies, Theorem V follows fairly immediately[1] from Theorem II but we shall in fact apply Theorem V when $\mathcal{S}$ is not a star-body in Chapter VIII. The proof of Theorem V gives also the following corollary which is a trivial consequence of the theorem when the $\mathcal{S}$ and $\mathcal{S}_r$ are star-bodies, but which is valid when they are not.

COROLLARY. *Suppose that for* $1 \leqq r < \infty$ *there is an* $\mathcal{S}_r$*-admissible lattice* $\Lambda_r$ *with*

$$d(\Lambda_r) = \Delta_1,$$

*for some number* $\Delta_1$. *Then there is an* $\mathcal{S}$*-admissible lattice* $\Lambda$ *with* $d(\Lambda) = \Delta_1$.

**V.5. Critical lattices.** Let $F(\boldsymbol{x})$ be a distance-function. It may well be that $F(\Lambda) = 0$ for every lattice $\Lambda$, in which case we say, following MAHLER, that the distance function and its associated star-body are of "infinite type". An example of a distance function of infinite type in two dimensions is

$$F(\boldsymbol{x}) = |x_1^2 x_2|^{\frac{1}{3}}.$$

Any lattice $\Lambda$ of determinant $d(\Lambda) = d$ contains a point $\boldsymbol{a} = (a_1, a_2) \neq \boldsymbol{o}$ with

$$|a_1| \leqq \varepsilon, \quad |a_2| \leqq d/\varepsilon,$$

where $\varepsilon > 0$ is arbitrarily small, by MINKOWSKI's convex body Theorem II of Chapter III. Then

$$F(\boldsymbol{a}) \leqq |\varepsilon d|^{\frac{1}{3}}$$

is arbitrarily small, so $F(\Lambda) = 0$. It is not always possible to decide whether a distance function is of finite type or not, for example, this is not known in the case of the 5-dimensional distance functions

$$F(\boldsymbol{x}) = |x_1^2 + x_2^2 + x_3^2 + x_4^2 - x_5^2|^{\frac{1}{2}}$$

and

$$F(\boldsymbol{x}) = |x_1^2 + x_2^2 + x_3^2 - x_4^2 - x_5^2|^{\frac{1}{2}}.$$

The problem whether these functions are of infinite type or not is equivalent to the problem whether all indefinite quadratic forms in 5 variables represent arbitrarily small values (including 0) or not for integer values of the variables (cf. § 3 of Chapter I). A classical theorem of MEYER says that if the coefficients of the form are rational then it represents 0. Recently DAVENPORT and more recently B. J. BIRCH have developed an attack on this problem but it appears to work only for

[1] When the $\mathcal{S}$ and $\mathcal{S}_r$ are star-bodies, say, with distance-functions $F(\boldsymbol{x})$ and $F_r(\boldsymbol{x})$, the hypotheses of Theorem V imply that, for each $x$, $F_r(\boldsymbol{x})$ tends monotonely to $F(\boldsymbol{x})$. Since $F_r(\boldsymbol{x})$ and $F(\boldsymbol{x})$ are continuous, this convergence must be uniform; and so Theorem II applies.

indefinite forms in more variables than 5 [see DAVENPORT (1956a) and later work of DAVENPORT and BIRCH]. The results of Chapters VI and X sometimes permit one to decide whether a given distance function $F(\boldsymbol{x})$ is of finite type or not but beyond that very little is known. For another unsolved problem of this type with important implications see CASSELS and SWINNERTON-DYER (1950a).

Most of the investigations in the geometry of numbers are concerned with distance-functions $F$ of finite type, i.e. not of infinite type. Then

$$\delta(F) = \sup_{\Lambda} \frac{\{F(\Lambda)\}^n}{d(\Lambda)} \tag{1}$$

s strictly positive. Then by Theorem VI of Chapter IV,

$$0 < \delta(F) < \infty \tag{2}$$

and

$$\delta(F)\,\Delta(\mathscr{S}) = 1, \tag{3}$$

where $\Delta(\mathscr{S})$ is the lattice constant of the set

$$\mathscr{S}:\ F(\boldsymbol{x}) < 1. \tag{4}$$

We recollect that a critical lattice for $\mathscr{S}$ is a lattice $\Lambda$ which is $\mathscr{S}$-admissible and which has determinant $d(\Lambda) = \Delta(\mathscr{S})$ (Chapter III, §5). A general theorem of MAHLER states that a set $\mathscr{S}$ of the type (4) always has critical lattices if it has admissible lattices.

THEOREM VI. *Let the distance-function $F(\boldsymbol{x})$ be of finite type. Then there exist lattices $\Lambda$ such that*

$$F(\Lambda) = 1, \quad d(\Lambda) = \{\delta(F)\}^{-1} = \Delta(\mathscr{S}),$$

*where $\delta(F)$ is defined in* (1) *and $\Delta(\mathscr{S})$ is the lattice constant of the region defined by* (4).

The proof of Theorem VI is now quite simple. By the definition of $\Delta(\mathscr{S})$, there exists a sequence of lattices $\Lambda_r$ such that

$$F(\Lambda_r) \geqq 1, \quad d(\Lambda_r) \to \Delta(\mathscr{S}). \tag{5}$$

We may now apply Theorem IV Corollary 1, its conditions (i) and (ii) being satisfied by (5). Hence there exists a convergent subsequence, and so, after a change of notation, we may suppose that

$$\Lambda_r \to \Lambda'$$

for some lattice $\Lambda'$. By (5) we have

$$d(\Lambda') = \lim_{r\to\infty} d(\Lambda_r) = \Delta(\mathscr{S}).$$

By (5) and Theorem II, we have

$$F(\Lambda') \geqq \limsup_{r\to\infty} F(\Lambda_r) \geqq 1.$$

If $F(\Lambda')>1$ there would exist a real number $\vartheta<1$ such that

$$F(\vartheta\Lambda') \geqq 1, \quad d(\vartheta\Lambda') = \vartheta^n d(\Lambda') < \Delta(\mathscr{S});$$

in contradiction to the definition of $\Delta(\mathscr{S})$ as a lower bound. Hence $F(\Lambda')=1$. This concludes the proof of the theorem.

In evaluating $\Delta(\mathscr{S})$ for star-bodies $\mathscr{S}$ we may therefore confine attention to critical lattices.

There is an alternative formulation of Theorem VI which does not need to distinguish between the two cases $\delta(F)=0$ and $\delta(F)>0$:

COROLLARY. *For every distance-function $F(\boldsymbol{x})$ in $n$-dimensional space there is a lattice $\mathsf{M}$ such that*

$$d(\mathsf{M}) = 1$$

*and*

$$\{F(\mathsf{M})\}^n = \delta(F) = \sup_{\Lambda} \frac{\{F(\Lambda)\}^n}{d(\Lambda)},$$

For if $\delta(F)=0$, any lattice $\mathsf{M}$ with $d(\mathsf{M})=1$ will do. Otherwise $\mathsf{M}=\vartheta\Lambda'$ will do, where $\Lambda'$ is a critical lattice and $\vartheta$ is chosen so that $d(\mathsf{M})=1$.

**V.5.2.** It would be natural to assume that every critical lattice $\Lambda$ for a star-body

$$\mathscr{S}: \quad F(\boldsymbol{x}) < 1$$

should contain a point $\boldsymbol{a}$ with $F(\boldsymbol{a})=1$, but in fact this is not the case even in 2 dimensions. Here we construct a counter-example using the phenomenon of successive minima discussed in § 4 of Chapter II. Write

$$F_0(\boldsymbol{x}) = |x_1 x_2|^{\frac{1}{2}}. \tag{1}$$

Theorem IV of Chapter I when translated into our present language implies that

$$\{F_0(\Lambda)\}^2 \leqq d(\Lambda)/8^{\frac{1}{2}} \tag{2}$$

except when $\Lambda$ is a lattice $\Lambda_c$ with basis

$$\boldsymbol{a}_1 = (a_{11}, a_{21}), \quad \boldsymbol{a}_2 = (a_{12}, a_{22}) \tag{3}$$

such that

$$(u_1 a_{11} + u_2 a_{12})(u_1 a_{21} + u_2 a_{22}) = k(u_1^2 + u_1 u_2 - u_2^2) \tag{4}$$

identically in $u_1, u_2$ for some number $k$; in which case

$$\{F_0(\Lambda_c)\}^2 = d(\Lambda_c)/5^{\frac{1}{2}}. \tag{5}$$

In particular

$$\delta(F_0) = 5^{-\frac{1}{2}}. \tag{6}$$

Now consider the distance-function

$$F_1(\boldsymbol{x}) = F_0(\boldsymbol{x})\left[1 + \frac{|x_1 x_2|}{100\{|x_1| + |x_2|\}^2}\right]; \tag{7}$$

so that

$$F_0(\boldsymbol{x}) \leqq F_1(\boldsymbol{x}) \leqq \frac{401}{400} F_0(\boldsymbol{x}). \tag{8}$$

From (8) and (2) or (5) we have

$$\{F_1(\Lambda)\}^2 \leqq \left(\frac{401}{400}\right)^2 d(\Lambda)/8^{\frac{1}{2}} \tag{9}$$

if $\Lambda$ is not a $\Lambda_c$; and

$$\{F_1(\Lambda_c)\}^2 \geqq d(\Lambda_c)/5^{\frac{1}{2}} \tag{10}$$

respectively. Since

$$8^{-\frac{1}{2}}\left(\frac{401}{400}\right)^2 < 5^{-\frac{1}{2}},$$

a critical lattice for $F_1(\boldsymbol{x}) < 1$ is necessarily a $\Lambda_c$.

We show now that equality holds in (10). After a possible interchange of $x_1$ and $x_2$ we may suppose that

$$u_1 a_{11} + u_2 a_{12} = a_{11}(u_1 + \omega u_2)$$
$$u_1 a_{21} + u_2 a_{22} = a_{21}(u_1 + \psi u_2),$$

where

$$2\omega = 1 + 5^{\frac{1}{2}}, \quad 2\psi = 1 - 5^{\frac{1}{2}}, \quad k = a_{11} a_{21},$$

on factorising the right-hand side of (4). Here

$$\omega\psi = -1.$$

Since

$$\omega^2 = \omega + 1, \quad \psi^2 = \psi + 1,$$

we have

$$\omega^t = u_1^{(t)} + u_2^{(t)}\omega; \quad \psi^t = u_1^{(t)} + u_2^{(t)}\psi;$$

for every positive integer $t$ and certain integers $u_1^{(t)}, u_2^{(t)}$. Hence

$$\boldsymbol{y}^t \text{ (say)} = (a_{11}\omega^t, a_{21}\psi^t) \in \Lambda_c.$$

But now, since $\omega\psi = -1$, we have

$$F_1(\boldsymbol{y}^t) = |a_{11} a_{21}|^{\frac{1}{2}}\left[1 + \frac{|a_{11} a_{21}|}{100\{|a_{11}|\omega^t + a_{21}\omega^{-t}\}^2}\right] \to |a_{11} a_{21}|^{\frac{1}{2}} \qquad (t \to \infty)$$
$$= k^{\frac{1}{2}}.$$

Hence

$$\{F_1(\Lambda_c)\}^2 \leqq k = d(\Lambda_c)/5^{\frac{1}{2}}.$$

This with (10) gives

$$\{F_1(\Lambda_c)\}^2 = d(\Lambda_c)/5^{\frac{1}{2}}.$$

But now if $\boldsymbol{a} \neq \boldsymbol{o}$ is a point of $\Lambda_c$, we have trivially

$$\{F_1(\boldsymbol{a})\}^2 > \{F_0(\boldsymbol{a})\}^2 \geqq d(\Lambda_c)/5^{\frac{1}{2}}.$$

In particular, if $k=1$, so that $d(\Lambda_c) = 5^{\frac{1}{2}} = \Delta(\mathscr{S}_1)$, where $\mathscr{S}_1$ is the region $F_1(\boldsymbol{x}) < 1$, there are no points $\boldsymbol{a}$ of $\Lambda_c$ on the boundary $F_1(\boldsymbol{x}) = 1$ of $\Lambda_c$.

By an ingenious argument, again using the phenomenon of successive minima, ROGERS (1947c) has constructed a distance-function $F(\boldsymbol{x})$ such that the critical lattice of the unbounded star-body $F(\boldsymbol{x}) < 1$ has only one pair of points $\pm\boldsymbol{a}$ with $F(\pm\boldsymbol{a}) = 1$. All other points $\boldsymbol{b} \neq \boldsymbol{o}$ of $\Lambda$ satisfy $F(\boldsymbol{b}) \geqq t$ for some explicitly given $t > 1$. This is in striking contrast with the results we shall prove in § 6 about bounded star-bodies.

**V.6. Bounded star-bodies.** For bounded star-bodies a great deal is known about critical lattices. [See in particular MAHLER (d, e) and for an extremely detailed treatment of the 2-dimensional case MAHLER (a, b, c).] In contrast to the negative result of § 5.2 we now have

THEOREM VII. *Every critical lattice $\Lambda$ of a bounded star body $\mathscr{S}$ has $n$ linearly independent points on the boundary of $\mathscr{S}$.*

For suppose not. Then there exists a basis $\boldsymbol{b}_1, \ldots, \boldsymbol{b}_n$ of $\Lambda$ such that any point

$$\boldsymbol{p} = u_1 \boldsymbol{b}_1 + \cdots + u_n \boldsymbol{b}_n \qquad (u_1, \ldots, u_n, \text{ integers}) \tag{1}$$

of $\Lambda$ on the boundary of $\mathscr{S}$ has $u_n = 0$. Since $\mathscr{S}$ is bounded, there exists a number $Y$ such that if a point

$$y_1 \boldsymbol{b}_1 + \cdots + y_n \boldsymbol{b}_n$$

with real $y_1, \ldots, y_n$ is in or on the boundary of $\mathscr{S}$, then certainly

$$|y_i| \leqq Y \qquad (1 \leqq i \leqq n).$$

Now let $\varepsilon$ be a number in, say,

$$0 < \varepsilon < \tfrac{1}{2},$$

and let $\Lambda_\varepsilon$ be the lattice with basis

$$\boldsymbol{b}_1, \ldots, \boldsymbol{b}_{n-1} \quad \text{and} \quad (1-\varepsilon)\boldsymbol{b}_n.$$

Consider a point

$$\boldsymbol{p}_\varepsilon = u_1 \boldsymbol{b}_1 + \cdots + u_{n-1} \boldsymbol{b}_{n-1} + u_n (1-\varepsilon) \boldsymbol{b}_n \tag{2}$$

of $\Lambda_\varepsilon$, where $u_1, \ldots, u_n$ are integers. If $u_n=0$, then $\boldsymbol{p}_\varepsilon$ is in $\Lambda$; and so is either on the boundary of $\mathscr{S}$ or outside $\mathscr{S}$. If

$$\max_{1\leqq j\leqq n} |u_j| > 2Y,$$

then certainly $\boldsymbol{p}_\varepsilon$ is outside $\mathscr{S}$. We need therefore consider only the points with

$$\max |u_j| \leqq 2Y, \quad u_n \neq 0. \tag{3}$$

But now for these $u_j$ the corresponding point $\boldsymbol{p}$ given by (1) is an exterior point of $\mathscr{S}$, since $u_n \neq 0$; that is some whole neighbourhood of $\boldsymbol{p}$ lies outside $\mathscr{S}$. Hence $\boldsymbol{p}_\varepsilon$ cannot be in $\mathscr{S}$ for all $\varepsilon$ smaller than some $\varepsilon_0$, which may depend in the first place on $u_1, \ldots, u_n$. But there are only a finite number of $u_1, \ldots, u_n$ to consider, by (3), and hence $\Lambda_\varepsilon$ is $\mathscr{S}$-admissible if $\varepsilon$ is small enough. But now

$$d(\Lambda_\varepsilon) = (1-\varepsilon)\, d(\Lambda) = (1-\varepsilon)\, \Delta(\mathscr{S}),$$

since $\Lambda$ was assumed to be critical. But this contradicts the definition of $\Delta(\mathscr{S})$ as the lower bound of the determinants of admissible lattices.

It is only exceptionally that there can be as few as $n$ pairs of linearly independent points $\pm \boldsymbol{a}_j$ $(1\leqq j\leqq n)$ of a critical lattice on the boundary of $\mathscr{S}$. Rather surprisingly, it is possible, however, at least when $n=2$, for a star-body to have a continuous infinity of critical lattices each with only $n$ pairs of points on the boundary, see OLLERENSHAW (1945a).

COROLLARY. *Suppose that* $\pm \boldsymbol{a}_j$ $(1\leqq j\leqq n)$ *are the only points of* $\Lambda$ *on the boundary of* $\mathscr{S}$. *Then there exists an* $\varepsilon_0$ *such that all points*

$$\boldsymbol{a}_n + \varepsilon_1 \boldsymbol{a}_1 + \cdots + \varepsilon_{n-1} \boldsymbol{a}_{n-1} \tag{4}$$

*with*

$$\max_{1\leqq j\leqq n-1} |\varepsilon_j| \leqq \varepsilon_0 \tag{5}$$

*are either in or on the boundary of* $\mathscr{S}$.

For $\boldsymbol{a}_1, \ldots, \boldsymbol{a}_n$ are linearly independent by the theorem; and so there exists a basis $\boldsymbol{b}_1, \ldots, \boldsymbol{b}_n$ such that

$$\boldsymbol{a}_i = v_{i1} \boldsymbol{b}_1 + \cdots + v_{ii} \boldsymbol{b}_i \qquad (1 \leqq i \leqq n) \tag{6}$$

with integers $v_{ij}$ and $v_{ii} \neq 0$. Let $\Lambda_\eta$ be the lattice with basis

$$\boldsymbol{b}_1, \ldots, \boldsymbol{b}_{n-1}, \boldsymbol{b}_n^\eta,$$

where

$$\boldsymbol{b}_n^\eta = \boldsymbol{b}_n + \eta_1 \boldsymbol{b}_1 + \cdots + \eta_{n-1} \boldsymbol{b}_{n-1}, \tag{7}$$

and $\eta_1, \ldots, \eta_{n-1}$ are small real numbers. As in the proof of the theorem, if $\max |\eta_j|$ is small enough, the only points of $\Lambda_\eta$ which can lie in or on the boundary of $\mathscr{S}$ are $\pm \boldsymbol{a}_1, \ldots, \pm \boldsymbol{a}_{n-1}$ (which are unchanged by the substitution of $b_n^\eta$ for $b_n$) and $\pm a_n^\eta$, where

$$a_n^\eta \text{ (say)} = v_{n1} b_1 + \cdots + v_{n,n-1} b_{n-1} + v_{nn} b_n^\eta. \tag{8}$$

But

$$d(\Lambda_\eta) = |\det(b_1, \ldots, b_{n-1}, b_n^\eta)| = |\det(b_1, \ldots, b_n)| = d(\Lambda) = \Delta(\mathscr{S}).$$

Hence either $a_n^\eta$ is in $\mathscr{S}$, when there is nothing more to prove, or $\Lambda_n^\eta$ is critical, and then $a_n^\eta$ is on the boundary of $\mathscr{S}$ by the theorem. Since every vector of the shape (4) can be put in the shape (8), where $\max|\eta_j|$ is small if $\max|\varepsilon_j|$ is small, this proves the corollary.

**V.6.2.** For the continued study of the points of a critical lattice on the boundary of a bounded star-body, we need an estimate of $\det(a_1, \ldots, a_n)$ in terms of

$$|a_j| \qquad (1 \leqq j \leqq n),$$

where $a_1, \ldots, a_n$ are any $n$-dimensional vectors. For our present purposes any estimate, however crude, would suffice, but, since we shall later need a more precise estimate, we prove it here.

LEMMA 9 (HADAMARD). *Let $a_1, \ldots, a_n$ be $n$-dimensional vectors. Then*

$$|\det(a_1, \ldots, a_n)| \leqq |a_1| \ldots |a_n|.$$

We note that the simple example

$$a_j = e_j = (\overbrace{0, \ldots, 0}^{j-1}, 1, \overbrace{0, \ldots, 0}^{n-j})$$

shows that $\leqq$ cannot in general be improved to $<$. The inequality is the $n$-dimensional analogue of the fact that the volume of a parallelopiped is at most the product of the length of the sides.

If the determinant is 0 there is nothing to prove. Hence we may suppose that $a_1, \ldots, a_n$ are linearly independent. We construct a sequence of vectors $c_j$ $(1 \leqq j \leqq n)$ such that

$$c_i c_j = 0 \qquad (i \neq j) \tag{1}$$

(scalar product of two vectors), and

$$a_i = t_{i1} c_1 + \cdots + t_{i,i-1} c_{i-1} + c_i \tag{2}$$

for some real numbers $t_{ij}$. Indeed if $c_1 = a_1$ and the $c_i$ are defined recursively by

$$c_i = a_i - \sum_{j<i} (a_i c_j) |c_j|^{-2} c_j,$$

it is readily verified that the $c_i$ have the required properties. By (1) and (2) we have

$$|a_i|^2 = a_i a_i = t_{i1}^2 |c_1|^2 + \cdots + t_{i,i-1}^2 |c_{i-1}|^2 + |c_i|^2 \geqq |c_i|^2, \tag{3}$$

and

$$\det(a_1, \ldots, a_n) = \det(c_1, \ldots, c_n). \tag{4}$$

On the other hand, on regarding the $\boldsymbol{c}_1, \ldots, \boldsymbol{c}_n$ in $\det(\boldsymbol{c}_1, \ldots, \boldsymbol{c}_n)$ first as rows and then as columns and multiplying the two determinants together, we have[1]

$$\{\det(\boldsymbol{c}_1, \ldots, \boldsymbol{c}_n)\}^2 = \det\{\boldsymbol{c}_i \boldsymbol{c}_j\} = \prod |\boldsymbol{c}_i|^2, \tag{5}$$

by (1). The required inequality now follows from (3), (4) and (5).

**V.6.3.** We may now show that, in principle, the evaluation of $\Delta(\mathscr{S})$ for a bounded $n$-dimensional star-body $\mathscr{S}$ may be reduced to a finite set of ordinary minimal problems. Except for convex bodies, for which see § 7, this is hardly in practice a fruitful approach, though it might well be adaptable to machine computation.

We may suppose without loss of generality that $\mathscr{S}$ is defined by

$$\mathscr{S}: \quad F(\boldsymbol{x}) < 1, \tag{1}$$

where $F(\boldsymbol{x})$ is a distance-function. By Lemmas 1 and 2 of Chapter IV, there are numbers $c > 0$ and $C$ such that

$$c|\boldsymbol{x}| \leqq F(\boldsymbol{x}) \leqq C|\boldsymbol{x}|. \tag{2}$$

In particular, a lattice $\Lambda$ admissible for $\mathscr{S}$ has no points in the sphere

$$|\boldsymbol{x}| < C^{-1},$$

and so has

$$d(\Lambda) \geqq 2^{-n} C^{-n} V_n, \tag{3}$$

by MINKOWSKI's convex body Theorem II of Chapter III, where $V_n$ is the volume of the unit sphere $|\boldsymbol{x}| < 1$.

Now let $\Lambda$ be a critical lattice, so that there are (at least) $n$ linearly independent points $\boldsymbol{a}_1, \ldots, \boldsymbol{a}_n$ of $\Lambda$ on the boundary $F(\boldsymbol{x}) = 1$ of $\mathscr{S}$. Then by (2) we have

$$|\boldsymbol{a}_j| \leqq c^{-1} \qquad (1 \leqq j \leqq n), \tag{4}$$

and so by HADAMARD's Lemma 9 we have

$$|\det(\boldsymbol{a}_1, \ldots, \boldsymbol{a}_n)| \leqq c^{-n}. \tag{5}$$

Hence in the language of Chapter I the index $I$ of $\boldsymbol{a}_1, \ldots, \boldsymbol{a}_n$ in $\Lambda$ is

$$I = \frac{|\det(\boldsymbol{a}_1, \ldots, \boldsymbol{a}_n)|}{d(\Lambda)} \leqq \left(\frac{2C}{c}\right)^n V_n^{-1} = I_0. \tag{6}$$

Hence by the corollaries to Theorem I of Chapter I, there is a basis $\boldsymbol{b}_1, \ldots, \boldsymbol{b}_n$ of $\Lambda$ such that

$$\boldsymbol{a}_i = v_{i1} \boldsymbol{b}_1 + \cdots + v_{ii} \boldsymbol{b}_i, \tag{7}$$

---

[1] Alternatively one may observe that, by (1), $\left|\sum_i x_i \boldsymbol{c}_i\right|^2 = \sum x_i^2 |\boldsymbol{c}_i|^2$ and compare determinants.

where the $v_{ij}$ are integers,

$$0 \leqq v_{ij} < v_{ii} \qquad (j < i), \tag{8}$$

and

$$0 < \prod v_{ii} = I \leqq I_0. \tag{9}$$

There are thus only a finite set of possibility for the integers $v_{ij}$. For each set of integers $v_{ij}$, the points $\boldsymbol{a}_i$ on the boundary determine the $\boldsymbol{b}_i$, by (6). The $\boldsymbol{a}_i$ are to be chosen so as to make

$$d(\Lambda) = \frac{|\det(\boldsymbol{a}_1, \ldots, \boldsymbol{a}_n)|}{v_{11} \cdots v_{nn}}$$

a minimum, subject to no points of $\Lambda$ being in $\mathscr{S}$ and, in particular, subject to (3). Then $\Delta(\mathscr{S})$ is clearly the minimum of $d(\Lambda)$ over the $\Lambda$ so obtained and over all of the finite number of choices for the $v_{ij}$.

We now verify that if $\Lambda$ is a lattice constructed with $n$ points $\boldsymbol{a}_1, \ldots, \boldsymbol{a}_n$ on the boundary and satisfying (3), (5), (6), (7), (8), (9), and if $\boldsymbol{d}$ were any point of $\Lambda$ in $\mathscr{S}$, then $\boldsymbol{d}$ has the shape

$$\boldsymbol{d} = u_1 \boldsymbol{b}_1 + \cdots + u_n \boldsymbol{b}_n,$$

where bounds can be given for the integers $u_j$. Indeed then $|\boldsymbol{d}| \leqq c^{-1}$; and so for each integer $j$ we have

$$|\det(\boldsymbol{a}_1, \ldots, \boldsymbol{a}_{j-1}, \boldsymbol{d}, \boldsymbol{a}_{j+1}, \ldots, \boldsymbol{a}_n)| \leqq c^{-n}$$

by (4) and HADAMARD'S Lemma 9. Hence, if (3) is true, the index of $\boldsymbol{a}_1, \ldots, \boldsymbol{a}_{j-1}, \boldsymbol{d}, \boldsymbol{a}_{j+1}, \ldots, \boldsymbol{a}_n$ in $\Lambda$ is at most $I_0$ for $j = 1, 2, \ldots, n$: and it is easily verified that this gives bounds for the $u_j$. It is thus, in principle, a finite problem to find $\Delta(\mathscr{S})$.

The lattice constants of a great many 2-dimensional bounded star-bodies have been evaluated. There is a partial list in KELLER (1954a) to which may be added among others the bodies discussed by OLLERENSHAW (1945a, b, 1953g). The treatment of bounded non-convex body in more than 2 dimensions by such methods seems inevitably laborious. Perhaps the only cases worked out are those of N. MULLINEUX (1951a).

**V.6.4.** In the evaluation of $\Delta(\mathscr{S})$ for a given star set $\mathscr{S}$ it is usually best to combine the techniques just introduced with those discussed in Chapter III. We consider an instructive example due to N. MULLINEUX which we shall have occasion to discuss further in § 7.

LEMMA 10. *Let $k$ be an arbitrary positive number and put*

$$D = (k^2 + 4k)^{\frac{1}{2}},$$

*and*

$$g = \tfrac{1}{2}(k + 2 + D),$$

*so that*

$$g^{-1} = \tfrac{1}{2}(k + 2 - D).$$

*Let $\mathscr{S}$ be the 2-dimensional star-body defined by*

$$-1 < x_1 x_2 < k, \quad |x_1 + x_2| < D.$$

*Then*

$$\Delta(\mathscr{S}) = D.$$

*The only critical lattices have bases of one of the two following kinds*

*(i) the point* $(1, -1)$ *and any point on* $x_1 + x_2 = D$,

*(ii) the points*

$$\boldsymbol{p} = (-g^{-1}t, gt^{-1}), \quad \boldsymbol{q} = (-t, t^{-1})$$

*where $t$ is any number in the range*

$$1 < t < g.$$

We must first verify that the lattices defined above are $\mathscr{S}$-admissible. This is certainly true for (i). We now verify it for (ii). It is readily verified that the line $x_1 + x_2 = D$ meets $x_1 x_2 = -1$ in the points

$$(-g^{-1}, g), \quad (g, -g^{-1}).$$

Hence the points $\boldsymbol{p}$ and $\boldsymbol{q}$ above do lie on the portion of the boundary of $\mathscr{S}$ given by $x_1 x_2 = -1$. The point

$$\boldsymbol{r} = \boldsymbol{p} - \boldsymbol{q} = \{\tfrac{1}{2}(-k + D)t, \tfrac{1}{2}(k + D)t^{-1}\} = (r_1, r_2)$$

lies on

$$r_1 r_2 = k.$$

Further,

$$0 < r_1 + r_2 < D,$$

since $1 < t < g$ and

$$\tfrac{1}{2}(-k + D)t + \tfrac{1}{2}(k + D)t^{-1}$$

equals $D$ both for $t = 1$ and for $t = g$. Hence a lattice of type (ii) has six points $\pm\boldsymbol{p}, \pm\boldsymbol{q}, \pm\boldsymbol{r}$ on the boundary of $\mathscr{S}$. There can be no further point of the lattice in $\mathscr{S}$, since it is easy to verify that every point of $\mathscr{S}$ except $\pm\boldsymbol{r}$ lies either strictly between the (infinite) line $\lambda$ through $\boldsymbol{p}$ and $\boldsymbol{r}$ and its image $-\lambda$ in $\boldsymbol{o}$; or strictly between the line $\mu$ through $\boldsymbol{q}, \boldsymbol{r}$ and its image $-\mu$ in $\boldsymbol{o}$; for example the line $\lambda$ meets $x_1 x_2 = -1$ and $x_1 x_2 = k$ respectively apart from $\boldsymbol{p}, \boldsymbol{r}$ in the points

$$(gt, -g^{-1}t^{-1}), \quad \{\tfrac{1}{2}(k + D)t, \tfrac{1}{2}(-k + D)t^{-1}\};$$

and for both of these $|x_1 + x_2| > D$.

For later use we note that the whole of the line-segment joining $\boldsymbol{p}, \boldsymbol{r}$ must lie in $\mathscr{S}$ except the end points, since a line can meet a hyperbola $x_1 x_2 = -1$ or $x_1 x_2 = k$ in at most two points. Hence the whole of the closed parallelogram with vertices at $\boldsymbol{o}, \boldsymbol{p}, \boldsymbol{r}$ and $-\boldsymbol{q}$ must lie in $\mathscr{S}$ except for $\boldsymbol{p}, \boldsymbol{r}$ and $-\boldsymbol{q}$.

We are now in a position to prove Lemma 10. Let M be a critical lattice. Suppose, if possible, that there is no point of M on the portion

$$x_1 x_2 = -1 \quad |x_1 + x_2| < D$$

of the boundary of $\mathscr{S}$. Then the set of points[1]

$$\mathsf{M}_\varepsilon: \quad \{(1-\varepsilon) x_1 + \varepsilon x_2, \varepsilon x_1 + (1-\varepsilon) x_2\}, \qquad (x_1, x_2) \in \mathsf{M},$$

for small enough $\varepsilon$, will also be $\mathscr{S}$-admissible since

$$\{(1-\varepsilon) x_1 + \varepsilon x_2\} + \{\varepsilon x_1 + (1-\varepsilon) x_2\} = x_1 + x_2$$

and

$$\{(1-\varepsilon) x_1 + \varepsilon x_2\}\{\varepsilon x_1 + (1-\varepsilon) x_2\} = x_1 x_2 + (\varepsilon - \varepsilon^2)(x_1 - x_2)^2 \geqq x_1 x_2.$$

Since

$$d(\mathsf{M}_\varepsilon) = (1 - 2\varepsilon)\, d(\mathsf{M}) < \Delta(\mathscr{S}),$$

this contradicts the hypothesis that M is critical. Hence there is a point $\boldsymbol{q} = (q_1, q_2)$ on the boundary $x_1 x_2 = -1$ of $\mathscr{S}$; and, by symmetry, we may suppose that

$$-q_1 \geqq 1 \geqq q_2 > 0.$$

Suppose first that $\boldsymbol{q} \neq (-1, 1)$. Then

$$(q_1, q_2) = (-t, t^{-1})$$

for some $t$ in $1 < t < g$. Let us identify this $\boldsymbol{q}$, with the $\boldsymbol{q}$ of the lattice $\Lambda$ introduced earlier, and let $\boldsymbol{p}, \boldsymbol{r}$ have the meanings introduced then. Since $\Lambda$ is admissible and M is critical, we have

$$d(\mathsf{M}) \leqq d(\Lambda).$$

The line $\lambda$ of points $\boldsymbol{x}$ with

$$\det(\boldsymbol{x}, \boldsymbol{q}) = d(\Lambda)$$

passes through $\boldsymbol{p}$ and $\boldsymbol{r}$, so the line

$$\det(\boldsymbol{x}, \boldsymbol{q}) = d(\mathsf{M}) \tag{1}$$

must either coincide with it or lie between it and the line through $\boldsymbol{o}$ and $\boldsymbol{q}$. But now $\boldsymbol{q}$ is a primitive point of M, since $r^{-1}\boldsymbol{q}$ $\mathscr{S}$ for any

[1] This argument becomes more transparent on introducing temporarily the co-ordinates $y_1 = \frac{1}{2}(x_1 + x_2)$, $y_2 = \frac{1}{2}(x_1 - x_2)$.

integer $r > 1$; and so there are points of M on the line (1) and at a distance $|\boldsymbol{q}|$ apart. Hence there must be a point of M other than $\boldsymbol{o}$ and $-\boldsymbol{q}$ in the closed parallelogram with vertices at $\boldsymbol{o}$, $-\boldsymbol{q}$, $\boldsymbol{p}$ and $\boldsymbol{r}$. But we have already seen that the only points of this parallelogram which are not in $\mathscr{S}$ are the vertices $\boldsymbol{p}$, $\boldsymbol{r}$ and $-\boldsymbol{q}$. Hence either $\boldsymbol{p}$ or $\boldsymbol{r}$ is in M; and in both cases then M coincides with $\Lambda$.

There remains the possibility that $\boldsymbol{q} = (-1, 1)$. If the definition of $\boldsymbol{p}$ and $\boldsymbol{r}$ is extended in the obvious way to $t = 1$, the situation remains the same, except that now the whole line-segment joining $\boldsymbol{p}$ and $\boldsymbol{r}$ is part of the boundary $x_1 + x_2 = D$ of $\mathscr{S}$. Hence we may deduce only that M has a basis consisting of $(-1, 1)$ and some point on $x_1 + x_2 = D$.

For this type of proof compare OLLERENSHAW (1945b).

For later use we note that we have also proved the

COROLLARY 1. *The only critical lattices for*

$$-1 < x_1 x_2 < k, \qquad |x_1 + x_2| \leqq D$$

*are those of type (ii), where now $t$ is allowed also to take the value* 1.

For the other lattices of type (i) have a point on $-1 < x_1 x_2 < D$, $|x_1 + x_2| = D$. Here our usage differs from that of MAHLER (1946a), since he calls a lattice admissible for a set $\mathscr{S}$ if it has no points other than $\boldsymbol{o}$ in the interior of $\mathscr{S}$. Thus MAHLER calls the lattice of type (i) admissible (and so critical) for the set of the corollary.

Lemma 10 may be regarded as a more precise version of Theorem IV of Chapter II. To make the connection more clear we prove

COROLLARY 2. *If $k$ is an integer, the critical lattices of type (ii) are admissible for*

$$-1 < x_1 x_2 < k.$$

For the general point of a lattice of type (ii) is

$$\boldsymbol{x} = u_1 \boldsymbol{p} + u_2 \boldsymbol{r},$$

where $u_1$, $u_2$ are integers. Then

$$x_1 x_2 = (u_1 p_1 + u_2 r_1)(u_1 p_2 + u_2 r_2) = -u_1^2 + k u_1 u_2 + k u_2^2.$$

We showed in § 4.4 of Chapter II that $-u^2 + k u_1 u_2 + k u_2^2$ does not take any values strictly between $-1$ and $+k$ when $k$ is a positive integer and $u_1$, $u_2$ are integers not both 0.

**V.7. Reducibility.** It may happen that if $\mathscr{S}_1$ is a star-body, there is some star-body $\mathscr{S}_2$ which is properly contained in $\mathscr{S}_1$ but which has the same lattice constant: $\Delta(\mathscr{S}_2) = \Delta(\mathscr{S}_1)$. We say then that $\mathscr{S}_1$ is reducible. If no such $\mathscr{S}_2$ exists, then $\mathscr{S}_1$ is said to be irreducible. Criteria for the reducibility of a bounded star-body have been given by MAHLER (1946a) and ROGERS (1947a). Later, ROGERS (1952a) gave a most ingenious example of a reducible star-body which does not contain an irreducible star-body of the same lattice constant: but he was able to show that if a rather wider class of point sets, which he calls "star sets", is considered, then every bounded reducible star set contains an irreducible star set. Convex 2-dimensional sets were considered in great detail by MAHLER (1947a). Mrs. OLLERENSHAW (1953b) has shown that the $n$-dimensional unit cube is irreducible for all $n$ and that the unit sphere is irreducible at least for $n \leqq 5$. She shows further that a 3-dimensional cylinder is irreducible if its 2-dimensional base is irreducible.

We refer the reader to the papers quoted for the general theory. The following lemma shows in a simple case the sort of ideas involved in the proof that a star-body is irreducible.

LEMMA 11. *The star-body*

$$\mathscr{D}: \quad x_1^2 + x_2^2 < 1$$

*is irreducible.*

For suppose $\mathscr{S}$ is a star-body strictly contained in $\mathscr{D}$. Then there is a point $\boldsymbol{p}$ on the boundary of $\mathscr{D}$ which is not on the boundary of $\mathscr{S}$. But now (§ 6.4 of Chapter III) there is a critical lattice $\Lambda$ of $\mathscr{D}$ having points at $\pm\boldsymbol{p}$. The only other points of $\Lambda$ on the boundary of $\mathscr{D}$ are the points $\pm\boldsymbol{q}$, $\pm\boldsymbol{r}$ which, together with $\pm\boldsymbol{p}$, are at the vertices of a regular hexagon. Since $\mathscr{S} \subset \mathscr{D}$, the lattice $\Lambda$ must be admissible for $\mathscr{S}$. But now the only points of $\Lambda$ on the boundary of $\mathscr{S}$ can be $\pm\boldsymbol{q}$ and $\pm\boldsymbol{r}$. These points clearly do not satisfy the criterion of Theorem VII, Corollary. Hence $\Lambda$ is not critical for $\mathscr{S}$, that is

$$\Delta(\mathscr{S}) < d(\Lambda) = \Delta(\mathscr{D}).$$

Since $\mathscr{S}$ is any star-body contained in $\mathscr{D}$, this proves the lemma.

A similar proof shows that MULLINEUX's star-body $\mathscr{S}$ defined in Lemma 10 is irreducible. Again, if $\boldsymbol{p}$ is a point on the boundary of $\mathscr{S}$ then, apart from a finite number of exceptional $\boldsymbol{p}$, there is a critical lattice for $\mathscr{S}$ which has only three pairs of points $\pm\boldsymbol{p}$, $\pm\boldsymbol{q}$ and $\pm\boldsymbol{r}$ on the boundary of $\mathscr{S}$; and the points $\pm\boldsymbol{q}$, $\pm\boldsymbol{r}$ cannot be the only points on the boundary of a critical lattice of any set $\mathscr{T}$ contained in $\mathscr{S}$. The finite number of exceptional points $\boldsymbol{p}$ for which such a lattice does not exist cannot affect the argument, since if $\mathscr{T}$ is properly contained in $\mathscr{S}$ there are infinitely many boundary points of $\mathscr{S}$ which are not boundary points of $\mathscr{T}$.

**V.7.2.** If $\mathscr{S}$ is an unbounded star set but there is a bounded star set $\mathscr{T}$ contained in $\mathscr{S}$ such that $\Delta(\mathscr{T}) = \Delta(\mathscr{S})$, then $\mathscr{S}$ is said to be boundedly reducible. Corollary 2 of Lemma 10 shows that the 2-dimensional star-body

$$\mathscr{S}_k: \; -1 < x_1 x_2 < k \tag{1}$$

is boundedly reducible when $k$ is a positive integer, since $\Delta(\mathscr{S}_k) = \Delta(\mathscr{T}_k)$, where $\mathscr{T}_k$ is MULLINEUX's set

$$\mathscr{T}_k: \quad -1 < x_1 x_2 < k, \qquad |x_1 + x_2| < (k^2 + 4k)^{\frac{1}{2}}. \tag{2}$$

On the other hand, $\mathscr{S}_k$ is not boundedly reducible for every $k$. Thus we saw in § 4.4 of Chapter II that the critical lattices M for $\mathscr{S}_{\frac{1}{10}}$ are admissible for $|x_1 x_2| < \frac{11}{10}$, and so have no points on $x_1 x_2 = -1$. But then, precisely as in the proof of Lemma 10, M cannot be critical for a bounded set $\mathscr{T}$ contained in $\mathscr{S}_{\frac{1}{10}}$, since the lattice $\mathsf{M}_\varepsilon$ of points

$$\{(1-\varepsilon)\, x_1 + \varepsilon\, x_2, \; \varepsilon\, x_1 + (1-\varepsilon)\, x_2\}, \qquad (x_1, x_2) \in \mathsf{M}$$

would be admissible for $\mathscr{T}$ for sufficiently small $\varepsilon$.

The proof of Theorem VII of Chapter III shows that the 2-dimensional star-body

$$|x_1^3 + x_2^3| < 1$$

is boundedly reducible, since the proof used only a bounded portion of the set. MAHLER (1946a) has developed criteria for sets of certain types to be boundedly reducible if their critical lattices are known. Bounded reducibility is further discussed by DAVENPORT and ROGERS (1950a). DAVENPORT and ROGERS introduce the concept of full reducibility. If $\mathscr{T}$ is a set contained in the set $\mathscr{S}$ and $\Delta(\mathscr{T}) = \Delta(\mathscr{S})$ then clearly every lattice critical for $\mathscr{S}$ is also critical for $\mathscr{T}$, but in general $\mathscr{T}$ might have more critical lattices. For example when $k$ is a positive integer the sets defined in (1) and (2) have the same lattice constant, but the critical lattices of $\mathscr{T}_k$ of the type (i) of the enunciation of Lemma 10 will in general have points in $\mathscr{S}_k$. On the other hand, the set

$$\mathscr{T}_k': \quad -1 < x_1 x_2 < k, \qquad |x_1 + x_2| \leqq (k^2 + 4k)^{\frac{1}{2}}$$

has no more critical lattices then $\mathscr{S}_k$ by Lemma 10 Corollary 1. If an unbounded set $\mathscr{S}$ contains a bounded set $\mathscr{T}$ with the same lattice constant and no more critical lattices then $\mathscr{S}$ is said by DAVENPORT and ROGERS (1950a) to be fully reducible[1]. They, following MAHLER, use the concept to show that lattices of certain types have infinitely many points in certain regions. We shall be discussing this from a rather different point of view later in Chapter X. We do not discuss bounded and full reducibility

[1] Their definition is not quite the same as ours since they use MAHLER's definition of an admissible lattice. But it is not difficult to see that it is equivalent to ours.

further but refer the reader to the papers quoted. The following example illustrates the connection with the existence of infinitely many lattice points in sets.

LEMMA 12. *Let $k$ be a positive integer and $\Lambda$ a lattice with*

$$d(\Lambda) \leqq (k^2+4k)^{\frac{1}{2}}.$$

*Then there are infinitely many points of $\Lambda$ in*

$$\overline{\mathscr{S}_k}:\quad -1 \leqq x_1 x_2 \leqq k. \tag{3}$$

*There are infinitely many points of $\Lambda$ in*

$$\mathscr{S}_k:\quad -1 < x_1 x_2 < k, \tag{4}$$

*except when $\Lambda$ is critical for $\mathscr{S}_k$.*

If $\Lambda$ contains a point $(0, x_2)$ with $x_2 \neq 0$, it contains all the points $(0, r x_2)$ $(r = 1, 2, 3, \ldots)$ and so the lemma is trivially true. Otherwise it suffices to show that for every $\varepsilon > 0$ there is a point $(x_1, x_2)$ of $\Lambda$ in $\overline{\mathscr{S}_k}$ for which $|x_1| \leqq \varepsilon$; and that this point is in $\mathscr{S}_k$ unless $\Lambda$ is critical for $\mathscr{S}_k$.

Let $t$ be any positive number. Then the lattice $\Lambda_t$ of points

$$(x_1, x_2) = (t X_1, t^{-1} X_2) \qquad (X_1, X_2) \in \Lambda \tag{5}$$

has the same determinant as $\Lambda$. Hence by Lemma 10, Corollary 1 there is a point of $\Lambda_t$ in

$$-1 \leqq x_1 x_2 \leqq k, \quad |x_1 + x_2| \leqq (k^2 + 4k)^{\frac{1}{2}}; \tag{6}$$

and indeed in $\mathscr{S}_k$ unless $\Lambda_t$ is critical for $\mathscr{S}_k$. But now the region (6) is bounded, so all the points of (6) satisfy

$$|x_1| \leqq \gamma$$

for some number $\gamma$ which depends only on $k$. Hence, by (5), the original lattice $\Lambda$ contains a point $(X_1, X_2) \neq \boldsymbol{o}$ such that

$$-1 \leqq X_1 X_2 \leqq k, \quad |X_1| \leqq \gamma t^{-1}.$$

Further, $\Lambda$ is critical for $\mathscr{S}$ if and only if $\Lambda_t$ is. Since $\gamma t^{-1}$ is arbitrarily small when $t$ is a arbitrarily large, this proves the result.

**V.8. Convex bodies.** For convex bodies stronger results than Theorem VII hold about the lattice points of a critical lattice on the boundary. The following theorem of SWINNERTON-DYER (1953a) generalised an old result of KORKINE and ZOLOTAREFF for spheres.

THEOREM VIII. *Let $\mathscr{K}$ be a bounded open symmetric convex set in $n$ dimensions and let $\Lambda$ be a critical lattice for $\mathscr{K}$. Then $\Lambda$ has at least $\frac{1}{2}n(n+1)$ pairs of points $\pm\boldsymbol{a}$ on the boundary of $\mathscr{K}$.*

We reproduce SWINNERTON-DYER's elegant proof. Let $\boldsymbol{b}_1, \ldots, \boldsymbol{b}_n$ be a basis for $\Lambda$ and let $\Lambda'$ be a lattice with the basis $\boldsymbol{b}'_j$ $(1 \leqq j \leqq n)$, where

$$\boldsymbol{b}'_j - \boldsymbol{b}_j = \eta \sum_{1 \leqq i \leqq n} a_{ji} \boldsymbol{b}_i, \tag{1}$$

and the $a_{ij}$ and $\eta$ are real numbers to be determined later. Let $\pm \boldsymbol{p}_1, \ldots, \pm \boldsymbol{p}_N$ be the only points of $\Lambda$ on the boundary of $\mathscr{K}$ and let $\pm \boldsymbol{p}'_1, \ldots, \pm \boldsymbol{p}'_N$ be the points of $\Lambda'$ which correspond to them in an obvious way. Let $\pi_1, \ldots, \pi_N$ be tac-planes to $\mathscr{K}$ at $\boldsymbol{p}_1, \ldots, \boldsymbol{p}_N$ (Theorem IV of Chapter IV). If there is more than one tac-plane, we choose one arbitrarily. We then impose on $\Lambda'$ the condition that $\boldsymbol{p}'_J$ lies in $\pi_J$ for $1 \leqq J \leqq N$. By (1), and since $\boldsymbol{p}_J$ lies on $\pi_J$, this imposes a condition of the type

$$\sum_{\substack{1 \leqq i \leqq n \\ 1 \leqq j \leqq n}} a_{ij} t_{ij}^{(J)} = 0 \qquad (1 \leqq J \leqq N), \tag{2}$$

where the numbers $t_{ij}^{(J)}$ depend only on the point $\boldsymbol{p}_J$ and the choice of tac-plane $\pi_J$. We also impose the conditions

$$a_{ij} = a_{ji} \qquad (i \neq j). \tag{3}$$

The total number of linear conditions (2) and (3) imposed on the $n^2$ numbers $a_{ij}$ is $\frac{1}{2} n(n-1) + N$. Hence if $N < \frac{1}{2} n(n+1)$, there exists a set of real numbers $a_{ij}$ not all 0 satisfying (2) and (3). We select any one such solution and keep it fixed in what follows.

Since the points $\boldsymbol{p}_J$ lie on tac-planes to the open set $\mathscr{K}$, they do not lie in $\mathscr{K}$. When $|\eta|$ is small enough, there are no further points of $\Lambda'$ in $\mathscr{K}$ other than $\boldsymbol{o}$, by the argument of § 6.1. Hence $\Lambda'$ is admissible for $\mathscr{K}$. Since $\Lambda$ is critical, we must then have

$$d(\Lambda') = |\det(\boldsymbol{b}'_1, \ldots, \boldsymbol{b}'_n)| \geqq |\det(\boldsymbol{b}_1, \ldots, \boldsymbol{b}_n)| = d(\Lambda) = \Delta(\mathscr{K});$$

that is

$$1 \leqq \det \begin{pmatrix} 1 + a_{11}\eta & a_{12}\eta & a_{1n}\eta \\ a_{21}\eta & 1 + a_{22}\eta & a_{2n}\eta \\ \cdot & \cdot \cdot \cdot \cdot \cdot \cdot \cdot \cdot \cdot \cdot \cdot & \cdot \\ a_{n1}\eta & a_{n2}\eta & 1 + a_{nn}\eta \end{pmatrix}$$

$$= 1 + A_1\eta + A_2\eta^2 + \cdots + A_n\eta^n \text{ (say)}.$$

Since this must be true for all sufficiently small values of $|\eta|$, it follows that

$$A_1 = \sum_i a_{ii} = 0$$

and

$$A_2 = -\sum_{i<j} a_{ij} a_{ji} + \sum_{i<j} a_{ii} a_{jj} \geqq 0.$$

Hence on using the symmetry conditions (3) we have

$$0 \leqq 2A_2 - A_1^2 = - \sum_{\substack{1 \leqq i \leqq n \\ 1 \leqq j \leqq n}} a_{ij}^2 .$$

Hence $a_{ij} = 0$ for all $i$ and $j$; which is a contradiction. The contradiction arises from the assumption that there are fewer than $\frac{1}{2}n(n+1)$ pairs of points of $\Lambda$ on the boundary of $\mathscr{K}$. Hence the theorem is established.

**V.8.2.** For bounded symmetric convex star sets the considerations of § 6.3 about the maximum number of points of a critical lattice on the boundary and about their index may be made much more precise, as was shown already by MINKOWSKI. His results apply indeed not merely to critical but to all admissible lattices. We recollect that a body $\mathscr{K}$ is strictly convex if every point $t\boldsymbol{p} + (1-t)\boldsymbol{q}$ $(0<t<1)$ is an interior point of $\mathscr{K}$ whenever $\boldsymbol{p}$ and $\boldsymbol{q}$ are distinct points in or on the boundary of $\mathscr{K}$.

THEOREM IX. *Let $\Lambda$ be an admissible lattice for the convex symmetric open set $\mathscr{K}$. Then there are at most $\frac{1}{2}(3^n - 1)$ pairs of points $\pm\boldsymbol{a}$ of $\Lambda$ on the boundary of $\mathscr{K}$. If $\mathscr{K}$ is strictly convex, the number of pairs is at most $2^n - 1$.*

The proofs are very simple. Suppose first that $\mathscr{K}$ is strictly convex. Let $\boldsymbol{b}_1, \ldots, \boldsymbol{b}_n$ be any basis for $\Lambda$ and let

$$\boldsymbol{a} = u_1 \boldsymbol{b}_1 + \cdots + u_n \boldsymbol{b}_n$$

be a point of $\Lambda$ on the boundary of $\mathscr{K}$. Then not all of $u_1, \ldots, u_n$ are even, since otherwise $\frac{1}{2}\boldsymbol{a}$ would belong to $\Lambda$; and $\frac{1}{2}\boldsymbol{a}$ is certainly an inner point of $\mathscr{K}$. Let now

$$\boldsymbol{a}' = u_1' \boldsymbol{b}_1 + \cdots + u_n' \boldsymbol{b}_n,$$

if possible, be another point of $\Lambda$ on the boundary of $\mathscr{K}$ such that[1]

$$u_j' \equiv u_j \pmod 2 \qquad (1 \leqq j \leqq n).$$

Then $\frac{1}{2}(\boldsymbol{a} + \boldsymbol{a}') \in \Lambda$. By the strict convexity, $\frac{1}{2}(\boldsymbol{a} + \boldsymbol{a}')$ is an inner point of $\mathscr{K}$ and so must be $\boldsymbol{o}$, that is $\boldsymbol{a}' = -\boldsymbol{a}$. Hence the total number of boundary points is at most the number of residue classes for $(u_1, \ldots, u_n)$ modulo 2 excluding $(0, \ldots, 0)$, that is $2^n - 1$, as required.

When $K$ is not strictly convex one must work with congruences modulo 3; the details are left to the reader.

THEOREM X. *Let $\mathscr{K}$ be a convex symmetric open n-dimensional set and $\Lambda$ an admissible lattice for $\mathscr{K}$. If $\boldsymbol{a}_1, \ldots, \boldsymbol{a}_n$ are points of $\Lambda$ on the*

[1] The notation means that $u_j - u_j'$ is divisible by 2.

*boundary of $\mathscr{K}$ then their index $I$ satisfies*

$$I \leqq n!. \tag{1}$$

*There is inequality in (1) if $\mathscr{K}$ is strictly convex.*

If $\boldsymbol{a}_1, \ldots, \boldsymbol{a}_n$ are linearly dependent, then their index is 0 and there is nothing to prove. Otherwise, every point $\boldsymbol{c}$ of $\Lambda$ may be put in the shape

$$\boldsymbol{c} = v_1 \boldsymbol{a}_1 + \cdots + v_n \boldsymbol{a}_n, \tag{2}$$

where $v_1, \ldots, v_n$ are rational numbers. The sets of numbers $\boldsymbol{v}$ such that (2) is in $\Lambda$ clearly form a lattice $\mathsf{M}$ of determinant

$$d(\mathsf{M}) = \frac{d(\Lambda)}{|\det(\boldsymbol{a}_1, \ldots, \boldsymbol{a}_n)|} = I^{-1}.$$

Hence, by MINKOWSKI's convex body Theorem II of Chapter III, there is point $\boldsymbol{v} \neq \boldsymbol{o}$ of $\mathsf{M}$ such that

$$|v_1| + \cdots + |v_n| \leqq (n!/I)^{1/n}. \tag{3}$$

Let $F$ be the distance function associated with $\mathscr{K}$, so that

$$F(\boldsymbol{a}_j) = 1 \qquad (1 \leqq j \leqq n).$$

For the $\boldsymbol{c} \in \Lambda$ given by (2) and (3) we thus have by the convexity and symmetry of $\mathscr{K}$, that

$$F(\boldsymbol{c}) \leqq |v_1| F(\boldsymbol{a}_1) + \cdots + |v_n| F(\boldsymbol{a}_n) \leqq (n!/I)^{1/n}. \tag{4}$$

But $F(\boldsymbol{c}) \geqq 1$ since $\Lambda$ is admissible for $\mathscr{K}$ and so $I \leqq n!$ as required. If $I = n!$ and $\mathscr{K}$ is strictly convex we should have $F(\boldsymbol{c}) < 1$ unless both $\mathsf{M}$ is a critical lattice for $|v_1| + \cdots + |v_n| < 1$ and every point of $\mathsf{M}$ on the boundary has $n-1$ of the co-ordinates $v_1, \ldots, v_n$ equal to 0. But these two requirements are incompatible by SWINNERTON-DYER's Theorem VIII.

The[1] estimate for $I$ in Theorem X can usually be much improved and more information obtained about the relationship of $\boldsymbol{a}_1, \ldots, \boldsymbol{a}_n$ to a basis for the lattice. Thus for $n = 3$ we have

COROLLARY. *If $\mathscr{K}$ is strictly convex and $n = 3$, then $I = 1$ or 2. If $I = 2$, then $\frac{1}{2}(\boldsymbol{a}_1 + \boldsymbol{a}_2 + \boldsymbol{a}_3) \in \Lambda$.*

For $I \leqq 5$. If $I = 5$, then there are integers $u_1, u_2, u_3$ not all divisible by 5 such that

$$\boldsymbol{c} = \tfrac{1}{5}(u_1 \boldsymbol{a}_1 + u_2 \boldsymbol{a}_2 + u_3 \boldsymbol{a}_3) \in \Lambda.$$

[1] We do not use the rest of § 8.2 later but do refer to it at the end of § 8.5.

We may suppose that 5 does not divide $u_1$ and, by taking $2\boldsymbol{c}$ instead of $\boldsymbol{c}$ if necessary, that

$$u_1 \equiv \pm 1 \pmod 5.$$

Hence by adding appropriate integer multiples of $\boldsymbol{a}_1, \boldsymbol{a}_2, \boldsymbol{a}_3$ to $\boldsymbol{c}$ we may suppose, without loss of generality, that

$$u_1 = \pm 1, \quad |u_2| \leqq 2, \quad |u_3| \leqq 2.$$

But then by the strict convexity we should have

$$F(\boldsymbol{c}) < \tfrac{1}{5}F(\boldsymbol{a}_1) + \tfrac{2}{5}F(\boldsymbol{a}_2) + \tfrac{2}{5}F(\boldsymbol{a}_3) = 1;$$

a contradiction. Hence $I \neq 5$. Similarly $I \neq 3$.

Suppose now $I = 4$. Then there exists a base $\boldsymbol{b}_1, \boldsymbol{b}_2, \boldsymbol{b}_3$ for $\Lambda$ such that

$$\begin{aligned} \boldsymbol{a}_1 &= v_{11}\boldsymbol{b}_1, \\ \boldsymbol{a}_2 &= v_{21}\boldsymbol{b}_1 + v_{22}\boldsymbol{b}_2, \\ \boldsymbol{a}_3 &= v_{31}\boldsymbol{b}_1 + v_{32}\boldsymbol{b}_2 + v_{33}\boldsymbol{b}_3, \end{aligned}$$

where

$$0 \leqq v_{ij} < v_{ii} \quad (j < i),$$

and

$$v_{11}v_{22}v_{33} = 4.$$

Then $v_{11} = 1$, since otherwise $\frac{1}{2}\boldsymbol{a}_1 \in \Lambda$ and $F(\frac{1}{2}\boldsymbol{a}_1) < F(\boldsymbol{a}_1) = 1$. If $v_{22} \neq 1$, then either $\frac{1}{2}\boldsymbol{a}_2$ or $\frac{1}{2}(\boldsymbol{a}_1 + \boldsymbol{a}_2)$ is in $\Lambda$; and again we have a contradiction. Hence

$$v_{11} = v_{22} = 1; \quad \text{so} \quad v_{33} = 4.$$

If $v_{31}$ were even, we should have either $\frac{1}{2}\boldsymbol{a}_3$ or $\frac{1}{2}(\boldsymbol{a}_2 + \boldsymbol{a}_3)$ in $\Lambda$; so $v_{31}$ is odd. Similarly, $v_{32}$ is odd. Hence there is a point

$$\boldsymbol{c} = \tfrac{1}{4}(u_1\boldsymbol{a}_1 + u_2\boldsymbol{a}_2 + \boldsymbol{a}_3) \in \Lambda,$$

where $u_1, u_2$ are odd. By adding integer multiples of $\boldsymbol{a}_1$ and $\boldsymbol{a}_2$ to $\boldsymbol{c}$, we may suppose that $u_1 = \pm 1$, $u_2 = \pm 1$. But then

$$F(\boldsymbol{c}) < \tfrac{1}{4}\{F(\boldsymbol{a}_1) + F(\boldsymbol{a}_2) + F(\boldsymbol{a}_3)\} = \tfrac{3}{4} < 1.$$

Hence $I \neq 4$.

Finally, when $I = 2$ it follows, just as for $I = 4$, that the only possibility is $v_{11} = v_{22} = 1$, $v_{21} = 0$ and $v_{33} = 2$. Further, the argument that $v_{31}, v_{32}$ are both odd continues to hold. Hence $\frac{1}{2}(\boldsymbol{a}_1 + \boldsymbol{a}_2 + \boldsymbol{a}_3) \in \Lambda$.

**V.8.3.** When $\mathscr{K}$ is a bounded symmetrical strictly convex 2-dimensional set, the lower bound 3 for the number of pairs of points $\pm\boldsymbol{a}$ of a critical lattice on the boundary given by Theorem VIII coincides with the upper bound give by Theorem IX. We have indeed

THEOREM XI. A. *Let $\mathscr{K}$ be an open convex symmetrical 2-dimensional convex body. Then a critical lattice $\Lambda$ of $\mathscr{K}$ has six points $\pm\boldsymbol{p}, \pm\boldsymbol{q}, \pm\boldsymbol{r}$ on the boundary of $\mathscr{K}$ such that*

$$\boldsymbol{p}+\boldsymbol{q}+\boldsymbol{r}=\boldsymbol{o} \tag{1}$$

*and any two of $\boldsymbol{p}, \boldsymbol{q}, \boldsymbol{r}$ is a basis for $\Lambda$.*

B. *Further, if $\pm\boldsymbol{p}, \pm\boldsymbol{q}, \pm\boldsymbol{r}$ are any points on the boundary of $\mathscr{K}$ such that* (1) *holds, then the lattice* $\mathsf{M}$ *with basis $\boldsymbol{p}, \boldsymbol{q}$ is admissible for $\mathscr{K}$. There are no further points of* $\mathsf{M}$ *on the boundary, except when $\mathscr{K}$ is a parallelogram and two of $\boldsymbol{p}, \boldsymbol{q}, \boldsymbol{r}$ are mid-points of its sides.*

The first part of Theorem XI is an almost immediate consequence of the last three theorems. By Theorem VIII there are three pairs of points $\pm\boldsymbol{p}, \pm\boldsymbol{q}, \pm\boldsymbol{r}$ on the boundary of $\mathscr{K}$. By Theorem IX, the index of $\boldsymbol{p}, \boldsymbol{q}$ is 1 or 2. Since $\frac{1}{2}\boldsymbol{p}, \frac{1}{2}\boldsymbol{q}$ are (inner) points of $\mathscr{K}$, they cannot belong to $\Lambda$. Hence, if the index is 2, the point $\frac{1}{2}(\boldsymbol{p}+\boldsymbol{q})$ is in $\Lambda$. It is also in $\mathscr{K}$ or on the boundary of $\mathscr{K}$, the latter only if $\mathscr{K}$ is not strictly convex. If the index is 2, we may thus take $\frac{1}{2}(\boldsymbol{p}+\boldsymbol{q})=\boldsymbol{q}'$ instead of $\boldsymbol{q}$. The index of $\boldsymbol{p}$ and $\boldsymbol{q}'$ is 1. Hence without loss of generality the index of $\boldsymbol{p}$ and $\boldsymbol{q}$ is 1. Hence $\boldsymbol{r}=u\boldsymbol{p}+v\boldsymbol{q}$ for some integers $u$ and $v$, where $|u|\leqq 2$, $|v|\leqq 2$, since the indexes of $\boldsymbol{p}, \boldsymbol{r}$ and of $\boldsymbol{q}, \boldsymbol{r}$ are at most 2. Not both $u$ and $v$ can be even, since otherwise $\frac{1}{2}\boldsymbol{r}$ would be in $\Lambda$. If, say, $u=\pm 2$ is even, then $v=\pm 1$ is odd, and $\boldsymbol{r}'=\frac{1}{2}(\boldsymbol{r}+v\boldsymbol{q})=\frac{1}{2}u\boldsymbol{p}+v\boldsymbol{q}$ is in or on the boundary of $\mathscr{K}$. It must be on the boundary since $\Lambda$ is admissible. Hence by taking $\boldsymbol{r}'$ instead of $\boldsymbol{r}$ we may suppose, without loss of generality, that $|u|=|v|=1$. By changing the signs of $\boldsymbol{p}$ and $\boldsymbol{q}$, where necessary, we may suppose that $u=v=-1$, that is, that (1) holds. This proves A.

It remains to prove B. Suppose, if possible, that the point

$$\begin{aligned}\boldsymbol{c}&=u\boldsymbol{p}+v\boldsymbol{q}\\&=(v-u)\,\boldsymbol{q}+(-u)\,\boldsymbol{r}\\&=(u-v)\,\boldsymbol{p}+(-v)\,\boldsymbol{r}\end{aligned}$$

is in or on the boundary of $\mathscr{K}$ for some integers $u, v$. If, say, $|u|>|v|+1$, then the point

$$\boldsymbol{p}=u^{-1}\boldsymbol{c}-vu^{-1}\boldsymbol{q}$$

would be an inner point of $\mathscr{K}$, because we should have $|u^{-1}|+|vu^{-1}|<1$. Hence from the three expressions for $\boldsymbol{c}$ we deduce that

$$\begin{aligned}\big||u|-|v|\big|&\leqq 1,\\\big||u-v|-|u|\big|&\leqq 1,\\\big||u-v|-|v|\big|&\leqq 1.\end{aligned}$$

It is easy to see that the only integral solutions of those inequalities giving primitive lattice points distinct from $\pm \boldsymbol{p}$, $\pm \boldsymbol{q}$, $\pm \boldsymbol{r}$ are

$$\pm(u, v) = (2, 1), \ (1, 2) \quad \text{or} \quad (1, -1).$$

Hence after permuting $\boldsymbol{p}, \boldsymbol{q}, \boldsymbol{r}$ cyclically if need be, we may suppose that $\boldsymbol{c} = \boldsymbol{p} - \boldsymbol{q}$ is in or on the boundary of $\mathscr{K}$. Since now

$$\boldsymbol{p} = \tfrac{1}{2}\boldsymbol{c} - \tfrac{1}{2}\boldsymbol{r}, \quad \boldsymbol{q} = -\tfrac{1}{2}\boldsymbol{c} - \tfrac{1}{2}\boldsymbol{r},$$

the only possibility is that $\boldsymbol{c}$ is a boundary point.

We now show that $\mathscr{K}$ contains the whole parallelogram $\mathscr{P}$ of points

$$\boldsymbol{x} = \lambda \boldsymbol{p} + \mu \boldsymbol{q}$$

with

$$\max\{|\lambda|, |\mu|\} < 1.$$

Indeed

$$\boldsymbol{x} = \varrho \boldsymbol{c} + \sigma \boldsymbol{r},$$

where

$$|\varrho| + |\sigma| = \tfrac{1}{2}|\lambda - \mu| + \tfrac{1}{2}|\lambda + \mu| = \max\{|\lambda|, |\mu|\}.$$

But now the area $V(\mathscr{P})$ of $\mathscr{P}$ is

$$V(\mathscr{P}) = 4|\det(\boldsymbol{p}, \boldsymbol{q})| = 4d(\mathsf{M}).$$

On the other hand, by MINKOWSKI's convex body theorem, we have

$$V(\mathscr{K}) \leqq 4d(\mathsf{M}).$$

Since $\mathscr{K}$ includes $\mathscr{P}$, and since $\mathscr{K}$ is open, the only possibility is that $\mathscr{K}$ coincides with $\mathscr{P}$. This concludes the proof of the theorem.

Theorem XI gives one a ready criterion for finding the lattice constant of 2-dimensional convex star-bodies. It is easy to see that if $\boldsymbol{p}$ is a given point on the boundary of $\mathscr{K}$, then there is precisely one hexagon of boundary points $\pm \boldsymbol{p}$, $\pm \boldsymbol{q}$, $\pm \boldsymbol{r}$ for which (1) holds. The lattice constant of $\mathscr{K}$ is then the lower bound of $\det(\boldsymbol{p}, \boldsymbol{q})$ for these hexagons.

**V.8.4.** As an application of Theorem XI we prove

LEMMA 13. *Let $\mathscr{S}$ be a convex symmetric open hexagon. Then*

$$\Delta(\mathscr{S}) = \tfrac{1}{4} V(\mathscr{S}). \tag{1}$$

*The only critical lattice* $\mathsf{M}$ *is that which has points at the mid-points of all the sides of* $\mathscr{S}$.

By MINKOWSKI's convex body theorem,

$$\Delta(\mathscr{S}) \geqq \tfrac{1}{4} V(\mathscr{S}). \tag{2}$$

Let the vertices of $\mathscr{S}$ taken in counter-clockwise order be

$$\boldsymbol{a}, -\boldsymbol{b}, \boldsymbol{c}, -\boldsymbol{a}, \boldsymbol{b}, -\boldsymbol{c}.$$

Then the lattice M of the lemma has basis $\frac{1}{2}(\boldsymbol{a}-\boldsymbol{b})$ and $\frac{1}{2}(\boldsymbol{b}-\boldsymbol{c})$. It clearly contains also $\frac{1}{2}(\boldsymbol{c}-\boldsymbol{a})$. Hence, by Theorem XI, M is $\mathscr{S}$-admissible. We now show that

$$d(\mathsf{M}) = \tfrac{1}{4} V(\mathscr{S}). \tag{3}$$

On dissecting $\mathscr{S}$ into triangles with a vertex at $\boldsymbol{o}$, we have

$$-V(\mathscr{S}) = \det(\boldsymbol{a}, \boldsymbol{b}) + \det(\boldsymbol{b}, \boldsymbol{c}) + \det(\boldsymbol{c}, \boldsymbol{a}) = 4\det(\boldsymbol{u}, \boldsymbol{v})$$

on putting $\boldsymbol{b}=\boldsymbol{a}+2\boldsymbol{u}$, $\boldsymbol{c}=\boldsymbol{a}+2\boldsymbol{v}$. This proves (3). Then (1) follows from (2) and (3) since M is $\mathscr{S}$-admissible.

Now let $\Lambda$ be any critical lattice for $\mathscr{S}$. Then $d(\Lambda)=\frac{1}{4}V(\mathscr{S})$. If $\Lambda$ did not have a point on a particular side of $\mathscr{S}$ there would be a symmetric convex set larger than $\mathscr{S}$ which contained no point of $\mathscr{S}$ except $\boldsymbol{o}$; which would contradict MINKOWSKI's convex body theorem. Hence, by Theorem XI, $\Lambda$ has precisely 6 points $\pm\boldsymbol{p}$, $\pm\boldsymbol{q}$, $\pm\boldsymbol{r}$ on the boundary of $\mathscr{S}$; one on each side. If, say, the points $\pm\boldsymbol{p}$ are not the mid-points of their sides, then by rotating slightly the sides about $\pm\boldsymbol{p}$, leaving the other pairs of sides fixed, it would be possible to find a convex symmetric set $\mathscr{T}$ of volume $V(\mathscr{T})>V(\mathscr{S})$ containing no points of $\Lambda$ except $\boldsymbol{o}$; again contradicting MINKOWSKI's convex body theorem. Hence $\pm\boldsymbol{p}$, $\pm\boldsymbol{q}$, $\pm\boldsymbol{r}$ are the mid-points of their sides, and $\Lambda=\mathsf{M}$.

It would, of course, be possible directly to compute the determinants of all lattices having points $\boldsymbol{p}, \boldsymbol{q}, \boldsymbol{r}$ with $\boldsymbol{p}+\boldsymbol{q}+\boldsymbol{r}=\boldsymbol{o}$ on the boundary of $\mathscr{S}$ and to show that M gives a minimum.

**V.8.5.** MINKOWSKI (1904a) has extended the argument of Theorem XI to 3 dimensions and proved the following.

THEOREM XII. *To find the lattice constant $\Delta(\mathscr{K})$ of an open symmetrical convex set $\mathscr{K}$ in 3 dimensions it suffices to consider the minimum of the determinants of lattices generated by three points $\boldsymbol{a}_1, \boldsymbol{a}_2, \boldsymbol{a}_3$ on the boundary of $\Lambda$ and satisfying one of the following three conditions:*

(A) *the points $\boldsymbol{a}_1-\boldsymbol{a}_2$, $\boldsymbol{a}_2-\boldsymbol{a}_3$, $\boldsymbol{a}_3-\boldsymbol{a}_1$ are on the boundary of $\mathscr{K}$ and $-\boldsymbol{a}_1+\boldsymbol{a}_2+\boldsymbol{a}_3$, $\boldsymbol{a}_1-\boldsymbol{a}_2+\boldsymbol{a}_3$, $\boldsymbol{a}_1+\boldsymbol{a}_2-\boldsymbol{a}_3$ are outside $\mathscr{K}$.*

(B) *the points $\boldsymbol{a}_1+\boldsymbol{a}_2$, $\boldsymbol{a}_2+\boldsymbol{a}_3$, $\boldsymbol{a}_3+\boldsymbol{a}_1$ are on the boundary of $\mathscr{K}$ and $\boldsymbol{a}_1+\boldsymbol{a}_2+\boldsymbol{a}_3$ is outside $\mathscr{K}$.*

(C) *the points $\boldsymbol{a}_1+\boldsymbol{a}_2$, $\boldsymbol{a}_2+\boldsymbol{a}_3$, $\boldsymbol{a}_3+\boldsymbol{a}_1$ and $\boldsymbol{a}_1+\boldsymbol{a}_2+\boldsymbol{a}_3$ are on the boundary of $\mathscr{K}$.*

We refer the reader to the original paper for the proof. Alternatively the reader may construct a proof by combining the ideas of the proof of Theorem XI with those at the end of § 8.2. The corresponding result in 4-dimensional space, which is fairly complicated, has been found by K. H. WOLFF (1954a), who states that some of the auxiliary results are due to E. BRUNNGRABER (1944a).

MINKOWSKI (1904a) used Theorem XII to find the lattice constant of the octahedron

$$|x_1| + |x_2| + |x_3| < 1,$$

namely 19/108. The lattice constants of further convex 3-dimensional bodies have been determined by CHALK (1950a) and WHITWORTH (1948a and 1951a). In all cases a considerable amount of rather tedious detail is necessary.

**V.9. Spheres.** We now consider more particularly the $n$-dimensional spheres

$$\mathscr{D}_n:\quad |\boldsymbol{x}|^2 = x_1^2 + \cdots + x_n^2 < 1. \tag{1}$$

We denote the lattice constant of $\mathscr{D}_n$ by

$$\Gamma_n = \Delta(\mathscr{D}_n). \tag{2}$$

The value of $\Gamma_n$ is known for $1 \leqq n \leqq 8$, see Appendix A. We here find again $\Gamma_3$, which we already found in another context in Chapter II, Theorem III. From this the value of $\Gamma_4$ will follow almost at once by a general theorem of MORDELL in Chapter X.

We must first prove a result for spheres which is more precise than the mere application of Theorem X.

THEOREM XIII. *Let $\Lambda$ be a lattice admissible for $\mathscr{D}_n$: $|\boldsymbol{x}|^2 < 1$; and let $\boldsymbol{a}_1, \ldots, \boldsymbol{a}_n$ be points of $\Lambda$ on the boundary of $\mathscr{D}_n$. Then the index $I$ of $\boldsymbol{a}_1, \ldots, \boldsymbol{a}_n$ satisfies*

$$I \leqq \{d(\Lambda)\}^{-1} \leqq \{\Delta(\mathscr{D}_n)\}^{-1} = \Gamma_n^{-1}. \tag{3}$$

For $|\boldsymbol{a}_j| = 1$ $(1 \leqq j \leqq n)$, and so, by HADAMARD's Lemma 9, we have

$$|\det(\boldsymbol{a}_1, \ldots, \boldsymbol{a}_n)| \leqq |\boldsymbol{a}_1| \ldots |\boldsymbol{a}_n| = 1.$$

Since

$$I = \frac{|\det(\boldsymbol{a}_1, \ldots, \boldsymbol{a}_n)|}{d(\Lambda)},$$

the first half of (3) follows. The second half of (3) is a trivial consequence of the definition of $\Gamma_n$.

COROLLARY. *If $n = 3$ the index is 0 or 1.*

For $\mathscr{D}_n$ has volume $4\pi/3$, and so

$$\Gamma_3 \geqq \pi/6 > \tfrac{1}{2},$$

by MINKOWSKI's convex body Theorem II of Chapter III.

THEOREM XIV.

$$\Gamma_3 = 2^{-\frac{1}{2}}.$$

*A critical lattice for $\mathscr{D}_3$ has a basis $\boldsymbol{m}_1, \boldsymbol{m}_2, \boldsymbol{m}_3$ such that*

$$|u_1\boldsymbol{m}_1 + u_2\boldsymbol{m}_2 + u_3\boldsymbol{m}_3|^2 = u_1^2 + u_2^2 + u_3^2 + u_2u_3 + u_3u_1 + u_1u_2$$

*identically in $u_1, u_2$ and $u_3$.*

Let $\Lambda$ be a critical lattice for $\mathcal{D}_3$. By Theorem VIII there are at least $\frac{1}{2}n(n+1)=6$ pairs of points $\pm\boldsymbol{m}$ of $\Lambda$ on the boundary of $\mathcal{D}_3$ and by Theorem VII there is a linearly independent set of 3, say $\boldsymbol{m}_1$, $\boldsymbol{m}_2$, $\boldsymbol{m}_3$. By Theorem XIII, $\boldsymbol{m}_1, \boldsymbol{m}_2, \boldsymbol{m}_3$ is a basis for $\Lambda$. If

$$\boldsymbol{m} = u_1\boldsymbol{m}_1 + u_2\boldsymbol{m}_2 + u_3\boldsymbol{m}_3$$

is another point of $\Lambda$ on the boundary of $\mathcal{D}_3$, the only possible value for the $u_j$ are $0$, $\pm 1$ by Theorem XIII. There can be at most one such pair $\pm\boldsymbol{m}$ with $u_1u_2u_3\neq 0$. For if, say,

$$\boldsymbol{m} = u_1\boldsymbol{m}_1 + u_2\boldsymbol{m}_2 + u_3\boldsymbol{m}_3, \qquad u_1u_2u_3 \neq 0,$$
$$\boldsymbol{m}' = u_1'\boldsymbol{m}_1 + u_2'\boldsymbol{m}_2 + u_3'\boldsymbol{m}_3, \qquad u_1'u_2'u_3' \neq 0,$$

the index $|u_2u_3'-u_3u_2'|$ of $\boldsymbol{m}_1, \boldsymbol{m}, \boldsymbol{m}'$ is even, so must be 0. Similarly

$$u_2u_3' - u_2'u_3 = u_1u_3' - u_1'u_3 = u_1u_2' - u_1'u_2 = 0,$$

so $\boldsymbol{m}'=\pm\boldsymbol{m}$. Hence there must be at least one point $u_1\boldsymbol{m}_1+u_2\boldsymbol{m}_2+u_3\boldsymbol{m}_3$ with $u_1u_2u_3=0$ on the boundary of $\mathcal{D}_3$ other than $\pm\boldsymbol{m}_1$, $\pm\boldsymbol{m}_2$, $\pm\boldsymbol{m}_3$. We may suppose without loss of generality that it is

$$\boldsymbol{m}_4 = \boldsymbol{m}_1 - \boldsymbol{m}_2.$$

Then neither $\boldsymbol{m}_1+\boldsymbol{m}_2$ nor $\boldsymbol{m}_1+\boldsymbol{m}_2\pm\boldsymbol{m}_3$ can occur as boundary points, since they would give index 2 with $\boldsymbol{m}_3$ and $\boldsymbol{m}_4$. Hence at least two of the remaining possibilities

$$\boldsymbol{m}_1 \pm \boldsymbol{m}_3, \quad \boldsymbol{m}_2 \pm \boldsymbol{m}_3, \quad \boldsymbol{m}_1 - \boldsymbol{m}_2 \pm \boldsymbol{m}_3$$

must occur. Since $\boldsymbol{m}_1-\boldsymbol{m}_2+\boldsymbol{m}_3$ and $\boldsymbol{m}_1-\boldsymbol{m}_2-\boldsymbol{m}_3$ cannot both occur, we may suppose without loss of generality that

$$\boldsymbol{m}_5 = \boldsymbol{m}_2 - \boldsymbol{m}_3$$

occurs. Then $\boldsymbol{m}_2+\boldsymbol{m}_3$ and $\boldsymbol{m}_1-\boldsymbol{m}_2-\boldsymbol{m}_3$ do not occur, since they give index 2 with $\boldsymbol{m}_2$ and $\boldsymbol{m}_5$: and $\boldsymbol{m}_1+\boldsymbol{m}_3$ cannot occur, since it gives index 2 with $\boldsymbol{m}_4$ and $\boldsymbol{m}_5$. Hence the only possibilities for $\pm\boldsymbol{m}_6$ are

$$\boldsymbol{m}_3 - \boldsymbol{m}_1 \quad \text{or} \quad \boldsymbol{m}_1 - \boldsymbol{m}_2 + \boldsymbol{m}_3.$$

In the second of these cases take $\boldsymbol{m}_5$ instead of $\boldsymbol{m}_3$. Then without loss of generality

$$\boldsymbol{m}_6 = \boldsymbol{m}_3 - \boldsymbol{m}_1.$$

Write

$$f(u_1, u_2, u_3) = |u_1\boldsymbol{m}_1 + u_2\boldsymbol{m}_2 + u_3\boldsymbol{m}_3|^2,$$

where $u_1, u_2, u_3$ are variables, so $f(\boldsymbol{u})$ is a quadratic form. Then

$$f(1,0,0) = f(0,1,0) = f(0,0,1)$$
$$= f(1,0,-1) = f(0,1,-1) = f(1,-1,0) = 1.$$

Hence

$$f(u) = u_1^2 + u_2^2 + u_3^2 + u_2 u_3 + u_3 u_1 + u_1 u_2$$

with determinant $D(f) = \frac{1}{2}$, and so

$$\{\det(m_1, m_2, m_3)\}^2 = \tfrac{1}{2},$$

as required.

**V.9.2.** Let $\mathscr{S}$ be a star-body and $\Lambda$ an $\mathscr{S}$-admissible lattice. We say that $\Lambda$ is extreme for $\mathscr{S}$ if there is a neighbourhood $\mathfrak{L}$ of $\Lambda$, in the sense of § 3.2, in which every $\mathscr{S}$-admissible lattice M satisfies

$$d(\mathsf{M}) \geqq d(\Lambda).$$

Clearly a critical lattice is extreme; but an extreme lattice need not be critical. Some of the results proved already extend to extreme lattices, notably SWINNERTON-DYER's Theorem VIII.

The extreme lattices of $n$-dimensional spheres have been exhaustively studied. For example there are six distinct types of extreme lattice for the 6-dimensional sphere as was shown by BARNES (1957b). There is a general theorem of VORONOI (1907a) which helps to characterise the extreme lattices of an $n$-dimensional sphere (they are "perfect" and "eutactic"). BARNES (1957a) has given an extremely elegant proof of VORONOI's characterisation. Unfortunately we cannot discuss these points further here, so we refer the reader to the two papers by BARNES where there are further references to the copious literature.

**V.10. Applications to diophantine approximation**[1]. The theory of Diophantine approximation deals with the approximation of rational or irrational numbers by rational numbers with special properties. The geometry of numbers has many applications to Diophantine approximation. The author's recent Cambridge Tract [CASSELS (1957a)] deals with Diophantine approximation and we do not intend to repeat what was done there. We give however a theorem of DAVENPORT generalizing work of FURTWÄNGLER which is an interesting application of MAHLER'S compactness techniques.

First, we note an obvious consequence of MINKOWSKI's linear forms Theorem III of Chapter III. Let $\vartheta_1, \ldots, \vartheta_n$ be real numbers and $Q$ an integer. By Theorem III of Chapter III there exist $n+1$ integers $u_0, \ldots, u_n$, not all 0, such that

$$|u_0 \vartheta_j - u_j| < Q^{-1/n} \qquad (1 \leqq j \leqq n), \tag{1}$$

$$|u_0| \leqq Q; \tag{2}$$

since $u_0\vartheta_j - u_j$ $(1 \leqq j \leqq n)$ together with $u_0$ form $n+1$ linear forms in $u_0, \ldots, u_n$ with determinant 1. Were $u_0 = 0$, we should have $|u_j| < Q^{-1/n}$, so $u_j = 0$ $(1 \leqq j \leqq n)$. Hence $u_0 \neq 0$, and on replacing $u_0, \ldots, u_n$ by

[1] Not used later in book.

$-u_0, \ldots, -u_n$ if need be, we may suppose that

$$0 < u_0 < Q. \tag{2'}$$

Further, (1) may be written

$$\left| \vartheta_j - \frac{u_j}{u_0} \right| < \frac{1}{u_0 Q^{1/n}}, \tag{1'}$$

which shows that the $u_j/u_0$ are good rational approximations to the $\vartheta_j$, all with the same denominator $u_0$.

We may look at (1) and (2') from another point of view. On eliminating $Q$ we have

$$u_0 \left\{ \max_{1 \leq j \leq n} |u_0 \vartheta_j - u_j| \right\}^n < 1. \tag{3}$$

There are in fact infinitely many solutions $u_0 > 0, u_1, \ldots, u_n$ of (3). If all of $\vartheta_1, \ldots, \vartheta_n$ are rational, this is trivial since then there exist integers $v_0 > 0, v_1, \ldots, v_n$ such that

$$v_0 \vartheta_j = v_j \qquad (1 \leq j \leq n),$$

and then we may put

$$u_j = r v_j \qquad (0 \leq j \leq n),$$

where $r$ is any positive integer: and then the left-hand side of (3) is 0. Otherwise we may suppose that $\vartheta_1$ is irrational. Suppose that $R$ integral solutions $u_j^{(r)}$ $(0 \leq j \leq n, 1 \leq r \leq R)$ have already been found with $u_0^{(r)} > 0$. Since $\vartheta_1$ is irrational, we may choose $Q$ so large that

$$|u_0^{(r)} \vartheta_1 - u_1^{(r)}| > Q^{-1/n} \qquad (1 \leq r \leq R).$$

For this value of $Q$ the solution of (1) and (2') gives a solution of (3) which is clearly not identical with any of the earlier ones.

**V.10.2.** For different purposes one may be interested in different properties of the approximations $u_j/u_0$ to the $\vartheta_j$. For example, instead of

$$\max_{1 \leq j \leq n} |u_0 \vartheta_j - u_j|$$

we may wish to make

$$\sum_{1 \leq j \leq n} (u_0 \vartheta_j - u_j)^2 \tag{1}$$

or

$$\prod_{1 \leq j \leq n} |u_0 \vartheta_j - u_j| \tag{2}$$

small. Or again one may be interested in "asymmetric" inequalities, of the type

$$-k_0 u_0^{-1/n} \leq u_0 \vartheta_j - u_j \leq k_1 u_0^{-1/n} \qquad (1 \leq j \leq n), \tag{3}$$

where $k_0$ and $k_1$ are positive numbers. All these different problems may be brought into one general shape. Let $\Phi(x_1, \ldots, x_n)$ be a distance-function of $n$ variables. How small can[1]

$$u_0 \Phi^n(u_0\vartheta_1 - u_1, \ldots, u_0\vartheta_n - u_n)$$

be made for infinitely many sets of integers $u_0 > 0$ and $u_1, \ldots, u_n$? We write

$$D(\Phi : \vartheta_1, \ldots, \vartheta_n) = \liminf_{\substack{u_0 \to \infty \\ u_0, u_1, \ldots, u_n \text{ integers}}} u_0 \Phi^n(u_0\vartheta_1 - u_1, \ldots, u_0\vartheta_n - u_n) \quad (4)$$

and

$$D(\Phi) = \sup_{\vartheta_1, \ldots, \vartheta_n} D(\Phi : \vartheta_1, \ldots, \vartheta_n); \quad (5)$$

so that $D(\Phi)$ is the number we wish to estimate.

The non-negative function $F(x_0, \ldots, x_n)$ of $n+1$ real variables defined by

$$F^{n+1}(x_0, \ldots, x_n) = \left\{ \begin{array}{rl} x_0 \Phi^n(x_1, \ldots, x_n) & \text{if} \quad x_0 \geqq 0 \\ -x_0 \Phi^n(-x_1, \ldots, -x_n) & \text{if} \quad x_0 \leqq 0 \end{array} \right\} \quad (6)$$

is a distance-function when $\Phi$ is a distance function of $n$ variables: since it clearly has the three defining properties that it is non-negative, continuous and satisfies

$$F(t x_0, \ldots, t x_n) = t F(x_0, \ldots, x_n)$$

when $t > 0$. By definition, $F$ is symmetric:

$$F(-x_0, \ldots, -x_n) = F(x_0, \ldots, x_n). \quad (7)$$

It satisfies the identity

$$F(t^n x_0, t^{-1} x_1, \ldots, t^{-1} x_n) = F(x_0, \ldots, x_n) \quad (8)$$

for any $t > 0$, since

$$\Phi(t^{-1} x_1, \ldots, t^{-1} x_n) = t^{-1} \Phi(x_1, \ldots, x_n).$$

As in § 4 of Chapter IV we write

$$\delta(F) = \sup_{\Lambda} \frac{F^{n+1}(\Lambda)}{d(\Lambda)},$$

where the supremum is over all $(n+1)$-dimensional lattices, so that

$$\delta(F) = \{\Delta(\mathscr{S})\}^{-1},$$

where $\mathscr{S}$ is the $(n+1)$-dimensional star-body

$$\mathscr{S}: \quad F(x_0, \ldots, x_n) < 1.$$

DAVENPORT'S result may now be put in the following shape.

[1] By $\Phi^n$ is meant the $n$-th power of $\Phi$.

THEOREM XV. *Let $\Phi$ and $F$ be related as above. Then*

$$D(\Phi) \leqq \delta(F) \tag{9}$$

*always. If $\Phi(\boldsymbol{x}) = 0$ only for $\boldsymbol{x} = \boldsymbol{o}$, then*

$$D(\Phi) = \delta(F). \tag{10}$$

The first part of Theorem XV is due essentially to MAHLER and is related to the theory of automorphic bodies which we shall study in Chapter X. When $D(\Phi) = 0$, there is nothing to prove. Otherwise, let $c$ be any positive number such that

$$c < D(\Phi). \tag{11}$$

Then, by the definition of $D(\Phi)$, there are real numbers $\vartheta_1, \ldots, \vartheta_n$ and an integer $U_0$ such that

$$u_0 \Phi^n(u_0\vartheta_1 - u_1, \ldots, u_0\vartheta_n - u_n) \geqq c, \tag{12}$$

whenever $u_0, \ldots, u_n$ are integers and

$$u_0 \geqq U_0. \tag{13}$$

In particular, $\vartheta_1, \ldots, \vartheta_n$ are not all rational; and so there exists a number $\varkappa > 0$ such that

$$\max_{1 \leqq j \leqq n} |u_0\vartheta_j - u_j| \geqq \varkappa > 0 \tag{14}$$

for all integers $u_0, \ldots, u_n$ with

$$0 < u_0 \leqq U_0.$$

Clearly

$$\varkappa \leqq \tfrac{1}{2} < 1. \tag{15}$$

Let $\mathsf{M}_1$ be the $n+1$-dimensional lattice of points

$$(x_0, \ldots, x_n) = (u_0, u_0\vartheta_1 - u_1, \ldots, u_0\vartheta_n - u_n), \tag{16}$$

where $u_0, \ldots, u_n$ run through all integers. Clearly

$$d(\mathsf{M}_1) = 1. \tag{17}$$

The function

$$F_1(x_0, \ldots, x_n) = \max\left[F(x_0, \ldots, x_n), \frac{c^{1/(n+1)}}{\varkappa} \max_{1 \leqq j \leqq n} |x_j|\right] \tag{18}$$

is clearly an $(n+1)$-dimensional distance-function and

$$F_1(-\boldsymbol{x}) = F_1(\boldsymbol{x}) \tag{19}$$

by (7). We show now that

$$F_1^{n+1}(\mathsf{M}_1) \geqq c. \tag{20}$$

Consider a point (16) of $\mathsf{M}_1$, where, by (19), we may suppose that $u_0 \geqq 0$. If $u_0=0$ but not all of $u_1, \ldots, u_n$ are 0, then the second term of the outer maximum in (18) is

$$\frac{c^{1/(n+1)}}{\varkappa} \max_{1 \leqq j \leqq n} |u_j| \geqq \frac{c^{1/(n+1)}}{\varkappa} \geqq c^{1/(n+1)},$$

by (15). If $0 < u_0 \leqq U_0$, then the second term of the outer maximum in (18) is still $\geqq c^{1/(n+1)}$, by (14). If $u_0 \geqq U_0$, the first term of the outer maximum in (18) is $\geqq c^{1/(n+1)}$ by (12). Hence in any case,

$$F_1(\boldsymbol{x}) \geqq c^{1/(n+1)}$$

for all $\boldsymbol{x} \in \mathsf{M}_1$ except $\boldsymbol{o}$. This completes the proof of (20).

For positive integers $r = 1, 2, \ldots$ write more generally

$$F_r(x_0, \ldots, x_n) = \max\left[F(x_0, \ldots, x_n), \frac{c^{1/(n+1)}}{r\varkappa} \max_{1 \leqq j \leqq n} |x_j|\right]. \tag{21}$$

Then

$$F(\boldsymbol{x}) \leqq F_r(\boldsymbol{x}) \leqq F_1(\boldsymbol{x}) \tag{21'}$$

and

$$\lim_{r \to \infty} F_r(\boldsymbol{x}) = F(\boldsymbol{x}) \tag{22}$$

uniformly in any bounded set of points $\boldsymbol{x}$. We have the identity

$$F_r(x_0, \ldots, x_n) = F_1(r^n x_0, r^{-1} x_1, \ldots, r^{-1} x_n), \tag{23}$$

by (8).

Let $\mathsf{M}_r$ be the lattice

$$\mathsf{M}_r: \quad (r^{-n} x_0, r x_1, \ldots, r x_n), \qquad \boldsymbol{x} \in \mathsf{M}_1.$$

Clearly

$$d(\mathsf{M}_r) = d(\mathsf{M}_1) = 1 \tag{24}$$

and

$$F_r^{n+1}(\mathsf{M}_r) = F_1^{n+1}(\mathsf{M}_1) \geqq c, \tag{25}$$

by (17), (20) and (23). Consequently, by (21'), we have the weaker assertion

$$F_1^{n+1}(\mathsf{M}_r) \geqq c > 0 \quad (1 \leqq r < \infty). \tag{26}$$

By (24), (26) and Theorem IV Corollary, there exists a convergent subsequence of the $\mathsf{M}_r$, say

$$\mathsf{M}_{r_s} \to \mathsf{N}.$$

By (24) we have

$$d(\mathsf{N}) = 1. \tag{27}$$

Since (22) holds uniformly in any bounded set, we have

$$F^{n+1}(\mathsf{N}) \geqq \limsup_{s \to \infty} F_{r_s}^{n+1}(\mathsf{M}_{r_s}) \geqq c, \tag{28}$$

by (25) and Theorem II. Hence

$$\delta(F) = \sup_{\Lambda} \frac{F^{n+1}(\Lambda)}{d(\Lambda)} \geqq \frac{F^{n+1}(\mathsf{N})}{d(\mathsf{N})} \geqq c.$$

Since $c$ was any positive number smaller than $D(\Phi)$, this proves $\delta(F) \geqq D(\Phi)$, the first part of Theorem XV.

The second part of Theorem XV requires quite different techniques and uses the basis constructed in Theorem II of Chapter I. By the Corollary to Theorem VI, there is a lattice $\Lambda$ with

$$d(\Lambda) = 1 \tag{29}$$

and

$$F^{n+1}(\Lambda) = \delta(F). \tag{30}$$

We denote the $(n+1)$-dimensional vector $(x_0, \ldots, x_n)$ in which $x_j = 1$ but the remaining co-ordinates are 0 by

$$\boldsymbol{e}_j = \left(\overbrace{0, \ldots, 0}^{j}, 1, \overbrace{0, \ldots, 0}^{n-j}\right) \qquad (0 \leqq j \leqq n).$$

By Theorem II of Chapter I, with $\varepsilon = \frac{1}{2}$ and $n+1$ for $n$, there exists, for all sufficiently large numbers $N$, a basis $\boldsymbol{a}_0, \boldsymbol{a}_1, \ldots, \boldsymbol{a}_n$ of $\Lambda$ such that

$$|\boldsymbol{a}_j - N\boldsymbol{e}_j| < N^{\frac{1}{2}} \qquad (1 \leqq j \leqq n). \tag{31}$$

Then

$$\boldsymbol{a}_j = N \sum_{0 \leqq i \leqq n} t_{ji} \boldsymbol{e}_i \qquad (1 \leqq j \leqq n), \tag{32}$$

where

$$|t_{jj} - 1| \leqq N^{-\frac{1}{2}} \qquad (1 \leqq j \leqq n) \tag{33}$$

and

$$|t_{ji}| \leqq N^{-\frac{1}{2}} \qquad (1 \leqq j \leqq n,\ 0 \leqq i \leqq n,\ i \neq j). \tag{34}$$

Since $\boldsymbol{a}_0, \boldsymbol{a}_1, \ldots, \boldsymbol{a}_n$ are linearly independent, there are real numbers $\lambda_0, \lambda_1, \ldots, \lambda_n$ such that

$$\boldsymbol{e}_0 = \lambda_0 \boldsymbol{a}_0 + \lambda_1 \boldsymbol{a}_1 + \cdots + \lambda_n \boldsymbol{a}_n,$$

where we may suppose that

$$\lambda_0 \geqq 0,$$

on taking $-\boldsymbol{a}_0$ for $\boldsymbol{a}_0$ if necessary. Since $d(\Lambda) = 1$, we have now

$$\begin{aligned} \lambda_0 &= \lambda_0 |\det(\boldsymbol{a}_0, \ldots, \boldsymbol{a}_n)| \\ &= |\det(\boldsymbol{e}_0, \boldsymbol{a}_1, \ldots, \boldsymbol{a}_n)| \\ &= N^n \{1 + O(N^{-\frac{1}{2}})\}, \end{aligned}$$

where the constant implied by the $O$ depends only on $n$. We may thus write

$$\boldsymbol{a}_0 = \mu \boldsymbol{e}_0 + \vartheta_1 \boldsymbol{a}_1 + \cdots + \vartheta_n \boldsymbol{a}_n, \tag{35}$$

where $\vartheta_1, \ldots, \vartheta_n$ are certain real numbers, and

$$\mu = \lambda_0^{-1} = N^{-n}\{1 + O(N^{-\frac{1}{2}})\}. \tag{36}$$

Let $\delta'$ be any number such that

$$\delta' < \delta(F).$$

We wish to show that

$$\liminf_{u_0 \to \infty} u_0 \Phi^n(u_0\vartheta_1 - u_1, \ldots, u_0\vartheta_n - u_n) > \delta' \tag{37}$$

for the $\vartheta_1, \ldots, \vartheta_n$ we have just constructed; provided that $N$ is greater than some $N_0$ which may depend on $\delta'$ and the function $\Phi$. After the first part of Theorem XV, this will complete the proof of the theorem. If $\delta(F) = 0$ there is nothing to prove. Otherwise we may suppose without loss of generality that

$$0 < \delta' < \delta(F). \tag{37'}$$

To prove (37) we may clearly confine attention to integers $u_0, \ldots, u_n$, if any, for which

$$u_0 > 0, \quad u_0 \Phi^n(y_1, \ldots, y_n) \leqq \delta(F), \tag{38}$$

where we have put

$$y_j = u_0\vartheta_j - u_j \quad (1 \leqq j \leqq n). \tag{39}$$

So far we have not used the fact that $\Phi(\boldsymbol{x}) = 0$ only for $\boldsymbol{x} = \boldsymbol{o}$. By Lemma 2 of Chapter IV, this implies that

$$\Phi(\boldsymbol{x}) \geqq c|\boldsymbol{x}| \geqq c \max\{|x_1|, \ldots, |x_n|\}$$

or some $c > 0$. Hence, by (38), we have

$$u_0 \max_{1 \leqq j \leqq n} |y_j|^n \leqq c^{-n}\delta(F). \tag{40}$$

We now consider the point

$$\boldsymbol{Y} = u_0\boldsymbol{a}_0 - u_1\boldsymbol{a}_1 - \cdots - u_n\boldsymbol{a}_n$$

of $\Lambda$. By (35) and (39) this is of the shape

$$\boldsymbol{Y} = \mu u_0 \boldsymbol{e}_0 + \sum_{1 \leqq j \leqq n} y_j \boldsymbol{a}_j;$$

and so, by (32), has co-ordinates $(Y_0, \ldots, Y_n)$, where

$$Y_0 = \mu u_0 + N \sum_{1 \leqq j \leqq n} y_j t_{j0}, \tag{41}$$

$$Y_i = \phantom{\mu u_0 +} N \sum_{1 \leqq j \leqq n} y_j t_{ji} \quad (1 \leqq i \leqq n). \tag{42}$$

Let $\varepsilon$ be an arbitrarily small positive number to be determined later. By (33), (34) and Lemma 3, the inequality

$$\Phi\left(\sum_j t_{j1} y_j, \ldots, \sum_j t_{jn} y_j\right) \leqq (1+\varepsilon)\, \Phi(y_1, \ldots, y_n)$$

holds for all real numbers $y_1, \ldots, y_n$ whatsoever, provided that $N$ is greater than a number depending only on the number $\varepsilon$ and the function $\Phi$. Hence, by (42),

$$\Phi(Y_1, \ldots, Y_n) \leqq (1+\varepsilon)\, N\, \Phi(y_1, \ldots, y_n). \tag{43}$$

By (40) and (41), we have

$$0 < Y_0 \leqq \mu(1+\varepsilon)\, u_0 \qquad (\text{all } u_0 \geqq U_0) \tag{44}$$

for some $U_0$ which will depend, of course, on $N$. But now $\boldsymbol{Y} \in \Lambda$ and $F^{n+1}(\Lambda) = \delta(F)$, by hypothesis. Hence

$$\delta(F) \leqq Y_0\, \Phi^n(Y_1, \ldots, Y_n), \tag{45}$$

by the definition (6) of $F$. From (36), (37'), (43) and (44), we have

$$u_0\, \Phi^n(y_1, \ldots, y_n) \geqq (N^n \mu)^{-1} (1+\varepsilon)^{-n-1}\, \delta(F) > \delta' \qquad (\text{all } u_0 \geqq U_0),$$

provided that first $\varepsilon$ is chosen small enough, then $N$ is chosen large enough, and finally $U_0$ is chosen large enough. This concludes the proof of (37), and so of the theorem.

**V.10.3.** The condition that $\Phi(\boldsymbol{x}) = 0$ only for $\boldsymbol{x} = \boldsymbol{o}$ is necessary for the second part of Theorem XV. The case when $n = 2$ and

$$\Phi^2(x_1, x_2) = |x_1 x_2|$$

represents a fascinating problem of LITTLEWOOD. It is not in fact known whether there exist numbers $\vartheta_1$ and $\vartheta_2$ such that

$$\liminf_{u_0 \to \infty} u_0 |u_0 \vartheta_1 - u_1|\, |u_0 \vartheta_2 - u_2| > 0,$$

where $u_0, u_1, u_2$ are integers. The corresponding function $F(x_0, x_1, x_2)$ is given by

$$F^3(x_0, x_1, x_2) = |x_0 x_1 x_2|:$$

and for this we have DAVENPORT's result that

$$\delta(F) = 1/7,$$

which we shall prove in Chapter X. But it follows from work of CASSELS and SWINNERTON-DYER (1955a) and from DAVENPORT's results about the successive minima of $F$, that at least

$$D(\Phi) \leqq 1/9 \cdot 1.$$

There is a companion result to Theorem XV, also due to DAVENPORT, which relates to the approximation of a single linear form to 0. Here one is concerned with

$$D'(\Phi:\vartheta_1,\ldots,\vartheta_n) = \liminf_{\substack{\max|u_1|,\ldots,|u_n|\to\infty \\ u_0+u_1\vartheta_1+\cdots+u_n\vartheta_n\geq 0 \\ u_0,\ldots,u_n \text{ integers}}} |u_0+u_1\vartheta_1+\cdots+u_n\vartheta_n|\,\Phi^n(u_1,\ldots,u_n),$$

where the condition $u_0+u_1\vartheta_1+\cdots+u_n\vartheta_n\geq 0$ may clearly be omitted if $\Phi$ is symmetric. Then Theorem XV remains valid if $D(\Phi)$ is replaced by

$$D'(\Phi) = \sup_{\vartheta_1,\ldots,\vartheta_n} D'(\Phi:\vartheta_1,\ldots,\vartheta_n);$$

and the proof is substantially similar.

**V.10.4.** Note that we have not shown the existence in the second part of Theorem XV of $\vartheta_1,\ldots,\vartheta_n$ such that

$$\liminf_{u_0\to\infty} u_0\,\Phi^n(u_0\vartheta_1-u_1,\ldots,u_0\vartheta_n-u_n) = \delta(F):$$

and indeed in general such $\vartheta_1,\ldots,\vartheta_n$ do not exist[1]. When $n=1$, however, a $\vartheta_1$ does exist, as is easy to show. Here, of course, the only possibility for the distance function $\Phi(x_1)$ of one variable is

$$\Phi(x_1) = \begin{cases} k\,x_1 & \text{if} \quad x_1\geq 0 \\ -l\,x_1 & \text{if} \quad x_1\leq 0, \end{cases}$$

where $k$ and $l$ are positive constants. As in the proof of the second part of Theorem XV, we consider a lattice $\Lambda$ with

$$d(\Lambda)=1, \qquad F^2(\Lambda)=\delta(F).$$

Let

$$\boldsymbol{a}=(a_0,a_1), \qquad \boldsymbol{b}=(b_0,b_1)$$

be a basis for $\Lambda$, where without loss of generality

$$b_1>0 \quad a_0b_1-a_1b_0=d(\Lambda)=1. \tag{1}$$

Put

$$\vartheta=\vartheta_1=a_1/b_1. \tag{2}$$

After Theorem XV it is enough to show that

$$\liminf_{u_0\to\infty} u_0\,\Phi(u_0\vartheta+u_1)\geq\delta(F).$$

As in the proof of Theorem XV, it is enough to consider value of $u_0$ and $u_1$, such that

$$u_0\,|u_0\vartheta+u_1|\leq c^{-2}\delta(F), \tag{3}$$

where $c$ is a constant such that $\Phi(x_1)\geq c|x_1|$ for all $x_1$.

[1] For example when $n=2$ and $\Phi^2(x_1,x_2)=x_1^2+x_2^2$, as one may show by "isolation" techniques. Cf. Chapter X.

We consider now the point

$$\boldsymbol{Y} = u_0 \boldsymbol{a} + u_1 \boldsymbol{b} = (u_0 a_0 + u_1 b_0, u_0 a_1 + u_1 b_1) = (Y_0, Y_1)$$

of $\Lambda$. By (1) and (2), we have

$$\Phi(Y_1) = b_1 \Phi(u_0 \vartheta + u_1). \tag{4}$$

But now, by (3), we have

$$\lim_{u_0 \to \infty} \frac{Y_0}{u_0} = \lim \left( a_0 + b_0 \frac{u_1}{u_0} \right) = a_0 - b_0 \vartheta = b_1^{-1}, \tag{5}$$

by (1) and (2). But

$$Y_0 \Phi(Y_1) \geqq \delta(F);$$

and so

$$\liminf u_0 \Phi(u_0 \vartheta + u_1) \geqq \delta(F)$$

by (4) and (5).

In particular, Theorem IV of Chapter II shows that

$$\liminf_{u_0 \to \infty} u_0 |u_0 \vartheta + u_1| \leqq 5^{-\frac{1}{2}}$$

for all $\vartheta$: and there exist numbers $\vartheta$ for which the sign of equality is required. Indeed the "successive minima" of Theorem IV of Chapter II correspond to a sequence of successive minima here. The original proofs of this used continued fractions, but there is a proof due to C. A. ROGERS which uses the isolation techniques which will be discussed in Chapter X and which is given in the author's Tract (CASSELS 1957a).

**V.10.5.** The proof of Theorem XV gives a simple case when inequality necessarily occurs in Theorem II, that is, when we have a convergent sequence of lattices,

$$\mathsf{M}_r \to \mathsf{M}'$$

and a distance function $F$ such that

$$F(\mathsf{M}') > \limsup_{r \to \infty} F(\mathsf{M}_r).$$

Let $F$ be the distance-function and $\mathsf{M}_r$ the lattices occurring in the first half of the proof. Then

$$F(\mathsf{M}_r) = 0$$

for all $r$, since $\mathsf{M}_r$ has points with $x_0 = 0$. On the other hand, we constructed a convergent subsequence $\mathsf{M}_{r_s}$ of the $\mathsf{M}_r$ such that

$$\mathsf{M}_{r_s} \to \mathsf{N},$$

where

$$F^{n+1}(\mathsf{N}) \geqq D(\Phi : \vartheta_1, \ldots, \vartheta_n).$$

The right-hand side here may well be strictly positive, as § 10.4 shows.

# Chapter VI

# The theorem of MINKOWSKI-HLAWKA

**VI.1. Introduction.** Hitherto we have been primarily concerned to estimate the lattice constant $\Delta(\mathscr{S})$ of a set $\mathscr{S}$ from below, that is to find numbers $\Delta_0$ such that every lattice $\Lambda$ with $d(\Lambda)<\Delta_0$ certainly has points other than $\boldsymbol{o}$ in $\mathscr{S}$. In this chapter we are concerned with estimates for $\Delta(\mathscr{S})$ from above; that is we wish to find numbers $\Delta_1$ such that there are certainly lattices $\Lambda$ with $d(\Lambda)=\Delta_1$ which have no points other than the origin in $\mathscr{S}$, i.e. are $\mathscr{S}$-admissible.

HLAWKA (1944a) showed that if $\mathscr{S}$ is any bounded $n$-dimensional set with a volume (content) $V$ in the sense of JORDAN[1] and if $\Delta_1>V$, then there is a lattice $\Lambda$ with $d(\Lambda)=\Delta_1$ which is admissible for $\mathscr{S}$. He showed, further, that if $\mathscr{S}$ is a bounded symmetric star-body, then it is enough that

$$\Delta_1 > V/2\zeta(n), \tag{1}$$

where

$$\zeta(n) = 1 + 2^{-n} + 3^{-n} + \cdots: \tag{2}$$

thereby confirming a conjecture of MINKOWSKI. These results were put in a wider setting by SIEGEL (1945a). Denote by $N_{\mathscr{S}}(\Lambda)=N(\Lambda)$ the number of points of $\Lambda$ other than $\boldsymbol{o}$ in a set $\mathscr{S}$; and by $P_{\mathscr{S}}(\Lambda)=P(\Lambda)$ the number of primitive[2] points of $\Lambda$ in $\mathscr{S}$. SIEGEL[3] gave a very natural way to define averages over the set of all lattices $\Lambda$ with a fixed determinant $d(\Lambda)=\Delta_1$. If $\psi(\Lambda)$ is any function of a lattice $\Lambda$, let us denote this average by

$$\underset{\Lambda}{\mathfrak{M}}\{\psi(\Lambda)\}. \tag{3}$$

SIEGEL showed that

$$\underset{\Lambda}{\mathfrak{M}}\{N_{\mathscr{S}}(\Lambda)\} = V(\mathscr{S})/\Delta_1, \tag{4}$$

and

$$\underset{\Lambda}{\mathfrak{M}}\{P_{\mathscr{S}}(\Lambda)\} = V(\mathscr{S})/\zeta(n)\,\Delta_1, \tag{5}$$

where $\mathscr{S}$ is any bounded set, not necessarily a star-body and not necessarily convex, which possesses a volume $V(\mathscr{S})$ in JORDAN's sense.

---

[1] This is rather more restrictive than the sense of LEBESGUE, but if the volume is defined in the sense of JORDAN it is also defined in that of LEBESGUE and equal to it. Let $\chi(\boldsymbol{x})$ be the characteristic function of $\mathscr{S}$, that is $\chi(\boldsymbol{x})=1$ if $\boldsymbol{x}\in S$ and $\chi(\boldsymbol{x})=0$ otherwise. Then $\mathscr{S}$ has a volume in the sense of JORDAN if $\chi(\boldsymbol{x})$ is integrable in the sense of RIEMANN, and the volume is equal to the integral of $\chi(\boldsymbol{x})$ over all space.

[2] That is points $\boldsymbol{a}\in\Lambda$ which are not of the form $\boldsymbol{a}=k\boldsymbol{b}$, where $\boldsymbol{b}\in\Lambda$ and $k>1$ is an integer.

[3] For a particularly simple exposition of SIEGEL's averaging process, see MACBEATH and ROGERS (1958a).

HLAWKA's theorems follow at once from (4) and (5). If $\Delta_1 > V(\mathcal{S})$, then, from the definition of the average, there must certainly by (4) be at least one lattice, say M, such that $N_{\mathcal{S}}(\mathsf{M}) \leqq \underset{\Lambda}{\mathfrak{M}}(N_{\mathcal{S}}(\Lambda)) < 1$. Since $N_{\mathcal{S}}(\mathsf{M})$ is an integer, we must have $N_{\mathcal{S}}(\mathsf{M}) = 0$, so M is $\mathcal{S}$-admissible. Similarly, if $\mathcal{S}$ is a symmetric star-body and $\Delta_1 > V(\mathcal{S})/2\zeta(n)$, then there must be some lattice N for which $P_{\mathcal{S}}(\mathsf{N}) < 2$. Since $\mathcal{S}$ is symmetric, points of N, other than the origin, occur in pairs, $\pm \boldsymbol{a}$, so $P_{\mathcal{S}}(\mathsf{N}) = 0$. Hence $\mathcal{S}$ contains no primitive points of N and, being a star-body, can contain no points of N at all other than $\boldsymbol{o}$.

The constant $\zeta(n)$ occurs in (5), roughly speaking, because the probability that a point of a lattice $\Lambda$ chosen at random should be primitive is $\{\zeta(n)\}^{-1}$. More precisely, the ratio of the number of primitive points of $\Lambda$ to the total number of points of $\Lambda$ in a large sphere $|\boldsymbol{x}| < R$ tends to $\{\zeta(n)\}^{-1}$ as $R \to \infty$.

When $\mathcal{S}$ is convex, improvements of the Minkowski-Hlawka theorem were obtained fairly soon after the original proof [see e.g. MAHLER (1947b), DAVENPORT and ROGERS (1947a) and LEKKERKERKER (1957a)]. However, even so, the smallest value of

$$Q(\mathcal{S}) = \frac{V(\mathcal{S})}{\Delta(\mathcal{S})} \tag{6}$$

is not known even for 2-dimensional symmetric convex sets: though the same conjecture was made independently by REINHARDT (1934a) and MAHLER (1947c) that it is attained when $\mathcal{S}$ is a certain "smoothed octagon", that is an octagon in which the corners are replaced by certain hyperbolic arcs.

Mrs. OLLERENSHAW (1953a) has given an example of a 2-dimensional non-convex symmetric star-body $\mathcal{S}$ for which $Q(\mathcal{S})$ is smaller than for the REINHARDT-MAHLER convex octagon and constructed from it a set which is not a star-body for which

$$Q = 1.3173\ldots.$$

It is not known whether this is the smallest possible value for a 2-dimensional set.

For a long time no improvement was obtained on the Minkowski-Hlawka theorem for general sets or for star-bodies. However, almost simultaneously, improvements were made by ROGERS (1955a, 1955b and 1956a) and SCHMIDT (1956a and 1956b). ROGERS's work depends on elaborate estimates of the average

$$\underset{\Lambda}{\mathfrak{M}}[\{N_{\mathcal{S}}(\Lambda)\}^k] \tag{7}$$

for positive integers $k$, where we have used the same notation as in (4). In a later paper ROGERS (1958a), using ideas of SCHMIDT combined with his own, shows that there is an absolute constant $C$ such that

$$Q(\mathscr{S}) = \frac{V(\mathscr{S})}{\Delta(\mathscr{S})} \geqq \frac{1}{2} n \log \frac{4}{3} - 2 \log n - C \tag{8}$$

for all symmetric sets[1], provided that the dimension $n$ is greater than some absolute constant $n_0$. We shall not discuss ROGERS's work further but refer the reader to the original memoires. SCHMIDT, on the other hand, uses an elegant device which is more effective than ROGERS's method for small dimensions but much less effective when the dimension is large. We shall discuss it more in detail in § 4.

The work just described can be generalized in several directions. In the first place, instead of operating with the number $N_{\mathscr{S}}(\Lambda)$ defined above, one may consider more generally

$$\sum_{\substack{\boldsymbol{a} \in \Lambda \\ \neq \boldsymbol{o}}} f(\boldsymbol{a}), \tag{9}$$

where $f(\boldsymbol{x})$ is some function defined at all points of space and which may be subjected to certain conditions (e.g. that it be non-negative or Riemann-integrable). If $f(\boldsymbol{x})$ is the characteristic function of $\mathscr{S}$, then the sum (9) is just $N_{\mathscr{S}}(\Lambda)$. Again, one may confine the sum in (9) to primitive points of $\Lambda$, when there is an analogue of $P_{\mathscr{S}}(\Lambda)$. In fact most of the work so far described has dealt with generalisations of this kind. Again, it was shown by MACBEATH and ROGERS (1955a) that the Minkowski-Hlawka theorem extends to more general sets of points than lattices. It is enough for $\Lambda$ to be any set of points such that the ratio of the number of points of the set $\Lambda$ in the sphere $|\boldsymbol{x}| < R$ to the volume of the sphere should tend to a finite non-zero limit $d$ as $R \to \infty$. Indeed (4) continues to hold with a modified definition of the mean $\mathfrak{M}$ and with $\Delta_1 = d^{-1}$.

Finally, we observe that MAHLER's Theorem V Corollary of Chapter III often permits the results of this chapter to be extended to unbounded sets $\mathscr{S}$ on taking $\mathscr{S}_r$ to be the set of points of $\mathscr{S}$ in the sphere $|\boldsymbol{x}| < r$.

**VI.1.2.** In this book we shall not consider any of these generalizations in detail. In § 3 we shall prove the Minkowski-Hlawka Theorem in its original formulation, that is, the existence of a lattice $\Lambda$ admissible for a symmetric star-body $\mathscr{S}$ with finite volume $V(\mathscr{S})$ and with determinant arbitrarily near to $V(\mathscr{S})$. We shall use an averaging argument, but the type of average will be chosen to facilitate the proof, not for

[1] Professor ROGERS tells me that Dr. SCHMIDT has obtained an improvement of (8) which is in course of publication in *Acta Mathematica*.

any deeper reason[1]. Then in § 4 we shall give an improvement of the Minkowski-Hlawka theorem using SCHMIDT's ideas but not carrying the detail quite so far as he does.

The arguments of §§ 3, 4 depend on a thorough investigation of the properties of sublattices of prime index in a lattice and this is carried out in § 2. These investigations further enable one to prove the result conjectured by ROGERS that if $\mathscr{S}$ is a symmetric star-body and $md(\Lambda) < \Delta(\mathscr{S})$ for some integer $m$ and some lattice $\Lambda$, then $\mathscr{S}$ contains at least $m$ pairs of points $\pm \boldsymbol{a} \in \Lambda$ other than $\boldsymbol{o}$. This we do in § 5.

In § 6 we give an entirely different generalization of the Minkowski-Hlawka Theorem which applies only in 2 dimensions. We show namely that certain sets $\mathscr{S}$ of infinite volume (= area) are of finite type, that is, possess admissible lattices. The proof depends on a generalization of a theorem of MARSHALL HALL (1947a) due to the author (CASSELS 1956a).

We do not use the contents of this chapter later in the book.

**VI.2. Sublattices of prime index.** An important tool in the work of both ROGERS and SCHMIDT is the existence of sublattices of a given lattice with certain special properties. We shall use the definition and properties of an index introduced in Chapter I.

LEMMA 1. *Let $p$ be a prime number and $\Lambda$ an $n$-dimensional lattice. Let $\boldsymbol{a}_1, \ldots, \boldsymbol{a}_R$ be any points of $\Lambda$ which are not of the shape $p\boldsymbol{a}$, $\boldsymbol{a} \in \Lambda$ and let $k_1, \ldots, k_R$ be real numbers. Then there is a lattice $\mathsf{M}$ of index $p$ in $\Lambda$ such that*

$$\sum_{\boldsymbol{a}_r \in \mathsf{M}} k_r \leqq \frac{p^{n-1}-1}{p^n-1} \sum_{1 \leqq r \leqq R} k_r. \tag{1}$$

Let $\boldsymbol{b}_1, \ldots, \boldsymbol{b}_n$ be a basis for $\Lambda$. Let $c_1, \ldots, c_n$ be integers and

$$0 \leqq c_j < p \qquad (1 \leqq j \leqq n), \tag{2}$$

$$(c_1, \ldots, c_n) \neq (0, \ldots, 0). \tag{3}$$

Let $\mathsf{M}(c_1, \ldots, c_n)$ be the lattice of points $u_1\boldsymbol{b}_1 + \cdots + u_n\boldsymbol{b}_n$, where $u_1, \ldots, u_n$ are integers, such that

$$u_1 c_1 + \cdots + u_n c_n \equiv 0 \quad (p).$$

Clearly $\mathsf{M}(c_1, \ldots, c_n)$ is of index $p$. There are $p^n - 1$ such lattices and we now show that a point $\boldsymbol{a}_r$ belongs to precisely $p^{n-1} - 1$ of them.

[1] Other averaging processes have been used. For a particularly brief proof of Theorem II using one of them, see CASSELS (1953a). It has been shown by ROGERS (1955a) that many of the averaging processes that can be used to prove the Minkowski-Hlawka Theorem are essentially equivalent to SIEGEL's.

We have

$$\boldsymbol{a}_r = v_{r1}\boldsymbol{b}_1 + \cdots + v_{rn}\boldsymbol{b}_n,$$

where $v_{r1}, \ldots, v_{rn}$ are integers not all divisible by $p$, by hypothesis. Without loss of generality, $v_{r1}$ is not divisible by $p$. The congruence

$$v_{r1}c_1 + \cdots + v_{rn}c_n \equiv 0 \quad (p) \tag{4}$$

then determines $c_1$ uniquely if $c_2, \ldots, c_n$ are given subject to (2). In particular, (4) gives $c_1 = 0$ if already $c_2 = \cdots = c_n = 0$; contrary to (3). But $c_2, \ldots, c_n$ may be given any other of the $p^{n-1} - 1$ possible sets of values subject to (2). Hence the average of the left-hand side of (1) over all lattices $\mathsf{M} = \mathsf{M}(c_1, \ldots, c_n)$ is given by the right-hand side, and so (1) must be true for at least one of them.

We have at once the

COROLLARY 1. *Let $p$ be a prime number and let $\boldsymbol{a}_1, \ldots, \boldsymbol{a}_p$ be $p$ points of $\Lambda$ none of which is of the shape $p\boldsymbol{b}$, $\boldsymbol{b} \in \Lambda$. Then there is a lattice $\mathsf{M}$ of index $p$ in $\Lambda$ which contains none of $\boldsymbol{a}_1, \ldots, \boldsymbol{a}_p$.*

For we may put $k_r = 1$ for $1 \leqq r \leqq p$. For the lattice $\mathsf{M}$ of the theorem we have

$$\sum_{\boldsymbol{a}_r \in \mathsf{M}} 1 \leqq \frac{p^{n-1} - 1}{p^n - 1} p < 1.$$

The number $p$ of points in the corollary cannot be replaced by $p + 1$. It is easy to see that if $\boldsymbol{a}_1, \boldsymbol{a}_2$ are any two points of $\Lambda$, then at least one of the $p + 1$ points

$$\boldsymbol{a}_1, \quad \boldsymbol{a}_2 + r\boldsymbol{a}_1 \quad (0 \leqq r \leqq p - 1)$$

is in each sublattice of index $p$.

More generally we have the following corollary, due to SCHMIDT in essence.

COROLLARY 2. *Suppose that the number $R$ of points $\boldsymbol{a}_r$ satisfies*

$$R < \frac{p^{m+1} - 1}{p - 1}$$

*for some integer $m$. Then there is a lattice $\mathsf{M}$ of index $p$ in $\Lambda$ such that*

$$\sum_{\boldsymbol{a}_r \in \mathsf{M}} k_r \leqq \frac{p^{m-1} - 1}{p^m - 1} \sum_{1 \leqq r \leqq R} k_r. \tag{5}$$

*(i.e. $n$ in (1) may be replaced by $m$).*

If the dimension $n$ of the space is $\leqq m$ the result follows at once since

$$\frac{p^{m-1} - 1}{p^m - 1} \geqq \frac{p^{n-1} - 1}{p^n - 1} \qquad \text{if } m \geqq n.$$

When $n > m$ we use induction on the dimension $n$. We say that two vectors $\boldsymbol{a}$ and $\boldsymbol{a}'$ of $\Lambda$, neither of the shape $p\boldsymbol{b}$, $\boldsymbol{b} \in \Lambda$, are proportional mod $p$ if there is an integer $u$ and a vector $\mathbf{c}$ of $\Lambda$ such that

$$\boldsymbol{a} = u\boldsymbol{a}' + p\boldsymbol{c}. \tag{6}$$

Clearly $u$ is prime to $p$. The relationship is a symmetric one between $\boldsymbol{a}$ and $\boldsymbol{a}'$, since there is an integer $v$ such that $uv \equiv 1(p)$; and then

$$v\boldsymbol{a} = \boldsymbol{a}' + p\boldsymbol{c}'$$

or some $\mathbf{c}' \in \Lambda$. Further, if $\boldsymbol{a}$ proportional both to $\boldsymbol{a}'$ and $\boldsymbol{a}''$, then $\boldsymbol{a}'$ s proportional to $\boldsymbol{a}''$. We thus have a subdivision into classes or "rays". The number of rays is clearly

$$\frac{p^n - 1}{p - 1}.$$

Since we are now supposing that $n > m$, at least one of these rays must contain no members of the set $\boldsymbol{a}_r$ $(1 \leqq r \leqq R)$. If $\boldsymbol{c}$ is in this ray, it is of the shape $\mathbf{c} = w\boldsymbol{b}$ where $\boldsymbol{b}$ is primitive and $w$ is an integer prime to $p$. Hence the primitive point $\boldsymbol{b}$ is in the ray, and we may suppose that $\boldsymbol{b} = \boldsymbol{b}_1$, where $\boldsymbol{b}_1, \ldots, \boldsymbol{b}_n$ is a basis for $\Lambda$. Then every point $\boldsymbol{a}_r$ is of the shape

$$\boldsymbol{a}_r = v_{r1}\boldsymbol{b}_1 + \cdots + v_{rn}\boldsymbol{b}_n,$$

where by the construction of $\boldsymbol{b}_1$, at least one of $v_{r2}, \ldots, v_{rn}$ is not divisible by $p$. Hence if we make $\boldsymbol{a}_r$ correspond to the vector

$$\tilde{\boldsymbol{a}}_r = (v_{r2}, \ldots, v_{rn})$$

in the $(n-1)$-dimensional lattice $\Lambda_0$ of points with integer coordinates, then $\tilde{\boldsymbol{a}}_r$ is not of the shape $p\tilde{\boldsymbol{b}}$, $\tilde{\boldsymbol{b}} \in \Lambda_0$. Since we are assuming that the corollary has already been proved for smaller values of $n$, there exist integers $c_2, \ldots, c_n$ such that

$$\sum_{c_2 v_{r2} + \cdots + c_n v_{rn} \equiv 0 \ (p)} k_r \leqq \frac{p^{m-1} - 1}{p^m - 1} \sum_{1 \leqq r \leqq R} k_r.$$

The lattice M of points

$$u_1\boldsymbol{b}_1 + \cdots + u_n\boldsymbol{b}_n$$

with

$$c_2 u_2 + \cdots + c_n u_n \equiv 0 \quad (p)$$

then does what is required.

**VI.2.2.** A refinement of the argument gives a rather more special result than Lemma 1 in which now the $k_r$ must be non-negative.

LEMMA 2. *Let $p$ be a prime-number and $\Lambda$ an $n$-dimensional lattice. Let $\boldsymbol{a}_0, \ldots, \boldsymbol{a}_R$ be any $R+1$ points of $\Lambda$ which are not of the shape $p\boldsymbol{b}$,*

$\boldsymbol{b} \in \Lambda$ *and let* $k_1, \ldots, k_R$ *be non-negative real numbers. Then there is a lattice* $\mathsf{M}$ *of index* $p$ *in* $\Lambda$ *such that*

$$\boldsymbol{a}_0 \notin \mathsf{M}$$

*and*

$$\sum_{\boldsymbol{a}_r \in \mathsf{M}} k_r \leq p^{-1} \sum_{1 \leq r \leq R} k_r. \tag{1}$$

We may choose a basis $\boldsymbol{b}_1, \ldots, \boldsymbol{b}_n$ for $\Lambda$ such that

$$\boldsymbol{a}_0 = v_0 \boldsymbol{b}_1,$$

where $v_0$ is some integer, which is not divisible by $p$ by hypothesis. For integers $c_j$ $(2 \leq j \leq n)$ with

$$0 \leq c_j < p \qquad (2 \leq j \leq n), \tag{2}$$

denote by $\mathsf{N}(c_2, \ldots, c_p)$ the lattice of points

$$u_1 \boldsymbol{b}_1 + \cdots + u_n \boldsymbol{b}_n,$$

where the integers $u_1, \ldots, u_n$ satisfy

$$u_1 + c_2 u_2 + \cdots + c_n u_n \equiv 0 \quad (p). \tag{3}$$

Clearly $\boldsymbol{a}_0 \notin \mathsf{N}(c_2, \ldots, c_p)$.

For $1 \leq r \leq R$, let

$$\boldsymbol{a}_r = v_{r1} \boldsymbol{b}_1 + \cdots + v_{rn} \boldsymbol{b}_n.$$

By hypothesis, not all of the integers $v_{r1}, \ldots, v_{rn}$ are divisible by $p$. If all of $v_{r2}, \ldots, v_{rn}$ are divisible by $p$, then $v_{r1}$ is not divisible by $p$: and so $\boldsymbol{a}_r$ does not belong to any $\mathsf{N}(c_2, \ldots, c_n)$. If, say, $v_{r2}$ is not divisible by $p$, the condition

$$v_{r1} + c_2 v_{r2} + \cdots + c_n v_{rn} \equiv 0 \quad (p)$$

is satisfied for precisely one value of $c_2$ if $c_3, \ldots, c_n$ are fixed; that is $\boldsymbol{a}_r$ belongs to precisely $p^{n-2}$ of the $p^{n-1}$ lattices $\mathsf{N}(c_2, \ldots, c_n)$. Hence if $\mathsf{M}$ runs through all the $p^{n-1}$ lattices $\mathsf{N}(c_2, \ldots, c_n)$ the average value of the left-hand side of (1) is

$$p^{-1} \Sigma' k_r,$$

where $\Sigma'$ denotes that the $r$ for which $v_{r2}, \ldots, v_{rn}$ are all divisible by $p$ must be omitted. Since $k_r \geq 0$ for all $r$, by hypothesis, this shows that at least one of the lattices $\mathsf{M} = \mathsf{N}(c_2, \ldots, c_n)$ satisfies (1).

**VI.3. The Minkowski-Hlawka Theorem.** Following ROGERS (1942b and 1951b) we now prove the following theorem of HLAWKA.

THEOREM I. *Let* $f(\boldsymbol{x})$ *be a Riemann-integrable function of the variables* $\boldsymbol{x} = (x_1, \ldots, x_n)$ *which vanishes outside a bounded set. Let* $\Delta_1 > 0$ *and* $\varepsilon > 0$

*be given. Then there is a lattice* M *of determinant* $\Delta_1$ *such that*

$$\Delta_1 \sum_{\substack{a \in M \\ \neq o}} f(a) < \int f(x)\, dx + \varepsilon, \tag{1}$$

*where*

$$dx = dx_1 \dots dx_n.$$

We may suppose that $f(x)$ vanishes outside the cube

$$\max_j |x_j| \leqq S \qquad (1 \leqq j \leqq n). \tag{2}$$

Let $p$ be a prime number and let $\eta > 0$ be determined by the equation

$$p\eta^n = \Delta_1. \tag{3}$$

We may choose $p$ so large that

$$p\eta > S. \tag{4}$$

Let $\Lambda$ be the lattice of points

$$\eta(u_1, \dots, u_n), \tag{5}$$

where $u_1, \dots, u_n$ are integers, so

$$d(\Lambda) = \eta^n. \tag{6}$$

Now

$$\eta^n \sum_{\substack{a \in \Lambda \\ a \neq o}} f(a) < \int f(x)\, dx + \tfrac{1}{2}\varepsilon \tag{7}$$

if $\eta$ is small enough, by the definition of Riemann integration; and so (7) is true when $p$ is large enough, by (3).

A point $a$ of $\Lambda$ other than $o$ for which $f(a) \neq 0$ lies in (2); and so cannot be of the shape $pb$, $b \in \Lambda$ by (4). Hence we may apply Lemma 1 where $a_1, \dots, a_R$ are all the points $a$ of $\Lambda$ other than $o$ at which $f(a) \neq 0$ and

$$k_r = f(a_r).$$

Then M has determinant

$$d(M) = p\, d(\Lambda) = p\eta^n = \Delta_1, \tag{8}$$

and

$$\sum_{\substack{a \in M \\ \neq o}} f(a) \leqq \frac{p^{n-1} - 1}{p^n - 1} \sum_{\substack{a \in \Lambda \\ a \neq o}} f(a). \tag{9}$$

Finally, (1) follows from (3), (7) and (9), when $p$ is chosen large enough.

As in § 1 we have the

COROLLARY. *Let* $\mathscr{S}$ *be a set with Jordan-volume* $V(\mathscr{S})$ *and let* $\Delta_1 > V(\mathscr{S})$. *Then there is a lattice* M *with* $d(M) = \Delta_1$ *which is admissible for* $\mathscr{S}$.

For let $f(\boldsymbol{x})$ be the characteristic function of $\mathscr{S}$, and choose $\varepsilon$ so that $\Delta_1 > V(\mathscr{S}) + \varepsilon$. The number of points of $\mathsf{M}$ other than $\boldsymbol{o}$ in $\mathscr{S}$ is then

$$\sum_{\substack{\boldsymbol{a}\in\mathsf{M}\\ \boldsymbol{a}\neq\boldsymbol{o}}} f(\boldsymbol{a}) < \Delta_1^{-1}\{V(\mathscr{S}) + \varepsilon\} < 1,$$

by (1). Since the number is an integer, it must be 0.

**VI.3.2.** The result corresponding to Theorem I in which only primitive points are summed over is:

THEOREM II. *Let $f(\boldsymbol{x})$, $\Delta_1$ and $\varepsilon$ be as in the enunciation of Theorem I. Then there exists a lattice $\mathsf{M}$ of determinant $d(\mathsf{M}) = \Delta_1$ such that*

$$\zeta(n)\,\Delta_1 \sum_{\boldsymbol{a}\in\mathsf{M}}{}^{*} f(\boldsymbol{a}) < \int f(\boldsymbol{x})\,d\boldsymbol{x} + \varepsilon,$$

*where the star (*) indicates that only primitive points are to be summed over.*

We only indicate briefly the modification required to the proof of Theorem I. In any case Theorem II is embraced in the generalization of Theorem I to point sets $\Lambda$ other than lattices due to MACBEATH and ROGERS (1955a), which was discussed in § 1. The exposition still follows ROGERS (1947b and 1951b).

In the first place, it is trivial that a point of $\mathsf{M}$ in the cube (2) of § 3.1 is a primitive point of $\mathsf{M}$ if and only if it is primitive as a point of $\Lambda$. Hence it is enough to show that

$$\lim_{\eta\to 0} \eta^n \sum_{\boldsymbol{a}\in\Lambda}{}^{*} f(\boldsymbol{a}) = \{\zeta(n)\}^{-1} \int f(\boldsymbol{x})\,d\boldsymbol{x}. \tag{1}$$

Now

$$\sum_{\substack{\boldsymbol{a}\in\Lambda\\ \boldsymbol{a}\neq\boldsymbol{o}}} f(\boldsymbol{a}) = \sum_{r=1}^{\infty} \sum_{\boldsymbol{a}\in\Lambda}{}^{*} f(r\boldsymbol{a}).$$

Hence by MÖBIUS' inversion formula [e.g. HARDY and WRIGHT (1938a) Chapter XVI], we have

$$\sum_{\boldsymbol{a}\in\Lambda}{}^{*} f(\boldsymbol{a}) = \sum_{r} \mu(r) \sum_{\substack{\boldsymbol{a}\in\Lambda\\ \boldsymbol{a}\neq\boldsymbol{o}}} f(r\boldsymbol{a}).$$

Hence

$$\eta^n \sum_{\boldsymbol{a}\in\Lambda}{}^{*} f(\boldsymbol{a}) = \sum_{r\geq 1} \frac{\mu(r)}{r^n}\,\sigma(r\eta),$$

where, for any $\xi > 0$, we have put

$$\sigma(\xi) = \xi^n \sum_{\substack{\boldsymbol{u}\text{ integral}\\ \boldsymbol{u}\neq\boldsymbol{o}}} f(\xi\boldsymbol{u}).$$

But now $\sigma(\xi)$ is bounded for all $\xi$, and

$$\lim_{\xi\to 0}\sigma(\xi) = \int f(\boldsymbol{x})\,d\boldsymbol{x}.$$

The result now follows on letting $p\to\infty$, so $\eta\to 0$, since

$$\sum_{r\geq 1}\frac{\mu(r)}{r^n} = \{\zeta(n)\}^{-1}.$$

As in § 1 we have the

COROLLARY (The "Minkowski-Hlawka Theorem"). *Let $\mathscr{S}$ be a bounded symmetric star-body with volume $V(\mathscr{S})$ and let $2\zeta(n)\,\Delta_1 > V(\mathscr{S})$. Then there is a lattice $\mathsf{M}$ with $d(\mathsf{M}) = \Delta_1$ which is admissible for $\mathscr{S}$.*

**VI.4. SCHMIDT's theorems.** We are now in a position to illustrate SCHMIDT's method of improving the corollaries to the last two theorems. We first give a simple example

LEMMA 3. *Let $\mathscr{S}$ be a symmetric star-body in $n$-dimensions with Jordan-volume $V(\mathscr{S})$ and let $\Delta_1$ be any number such that*

$$3\zeta(n)\,\Delta_1 > (1+2^{1-n})\,V(\mathscr{S}).$$

*Then there is a $\mathscr{S}$-admissible lattice $\mathsf{M}$ of determinant $\Delta_1$.*

Let $g(\boldsymbol{x})$ be the characteristic function of $\mathscr{S}$, and let

$$f(\boldsymbol{x}) = g(\boldsymbol{x}) + 2g(2\boldsymbol{x}),$$

so that

$$f(\boldsymbol{x}) = \begin{cases} 3 & \text{if } \boldsymbol{x}\in\tfrac{1}{2}\mathscr{S} \\ 1 & \text{if } \boldsymbol{x}\in\mathscr{S},\ \boldsymbol{x}\notin\tfrac{1}{2}\mathscr{S} \\ 0 & \text{otherwise} \end{cases}$$

and

$$\int f(\boldsymbol{x})\,d\boldsymbol{x} = (1+2^{1-n})\,V(\mathscr{S}).$$

Choose $\varepsilon$ so small that

$$3\zeta(n)\,\Delta_1 > (1+2^{1-n})\,\{V(\mathscr{S})+\varepsilon\}.$$

By Theorem II with $\Delta_1/2$ for $\Delta_1$ and this $\varepsilon$, there is a lattice $\Lambda$ with determinant

$$d(\Lambda) = \tfrac{1}{2}\Delta_1$$

such that

$$\sum_{\boldsymbol{a}\in\Lambda,\ \text{primitive}} f(\boldsymbol{a}) < 6.$$

Since $f(-\boldsymbol{x}) = f(\boldsymbol{x})$, by the symmetry of $\mathscr{S}$, there is thus no primitive point of $\Lambda$ for which $f(\boldsymbol{a}) = 3$, and so no point of $\Lambda$ at all in $\tfrac{1}{2}\mathscr{S}$ except $\boldsymbol{o}$. Further, there are at most two pairs of primitive points say $\pm\boldsymbol{a}_1$, $\pm\boldsymbol{a}_2$

of $\Lambda$ in $\mathscr{S}$. By Lemma 1 Corollary 1, there is a lattice M of index 2 which contains neither $a_1$ nor $a_2$. Since $a_1, a_2$ are not in $\frac{1}{2}\mathscr{S}$, the points $2a_1, 2a_2$ of M are not in $\mathscr{S}$. Hence M is $\mathscr{S}$-admissible. Since

$$d(\mathsf{M}) = 2d(\Lambda) = \Delta_1,$$

the lattice M does what is required.

**VI.4.2.** When $n=2$, the result of Lemma 3 is no stronger than Theorem II Corollary.

By further elaboration, SCHMIDT (1956a) improved Lemma 3 somewhat but for values of $n$ at all large Lemma 3 is weaker than the following Theorem III which applies to all Jordan-measurable bounded sets not merely symmetric star-bodies. To obtain results about symmetric sets, Theorem III should not be applied to $\mathscr{S}$ directly but, say, to the "half-set" $\mathscr{S}_1$, of points

$$\boldsymbol{x} \in \mathscr{S}, \quad x_1 \geqq 0.$$

Then

$$V(\mathscr{S}_1) = \tfrac{1}{2} V(\mathscr{S}),$$

and a lattice M is $\mathscr{S}_1$-admissible if and only if it is $\mathscr{S}$-admissible. There is thus an additional factor 2 for symmetric sets.

THEOREM III. *Let $\mathscr{S}$ be any bounded n-dimensional Jordan-measurable set of volume $V(\mathscr{S})$ and let $\Delta_1$ be any number such that*

$$(1 + 2^{1-n})(1 + 3^{1-n}) V(\mathscr{S}) < 2\Delta_1. \tag{1}$$

*Then there is a lattice M of determinant $\Delta_1$ having no points, except possibly $\boldsymbol{o}$, in $\mathscr{S}$.*

Let $g(\boldsymbol{x})$ be the characteristic function of $S$ and put

$$f(\boldsymbol{x}) = g(\boldsymbol{x}) + 2g(2\boldsymbol{x}) + 3g(3\boldsymbol{x}) + 6g(6\boldsymbol{x}). \tag{2}$$

Then

$$\begin{aligned} \int f(\boldsymbol{x})\, d\boldsymbol{x} &= (1 + 2\cdot 2^{-n} + 3\cdot 3^{-n} + 6\cdot 6^{-n}) \int g(\boldsymbol{x})\, d\boldsymbol{x} \\ &= (1 + 2^{1-n})(1 + 3^{1-n}) V(\mathscr{S}). \end{aligned}$$

By Theorem I there is thus a lattice $\Lambda$ of determinant

$$d(\Lambda) = \Delta_1/6, \tag{3}$$

such that

$$\sum_{\substack{\boldsymbol{a} \in \Lambda \\ \boldsymbol{a} \neq \boldsymbol{o}}} f(\boldsymbol{a}) < 12. \tag{4}$$

We shall construct a lattice M of index 6 in $\Lambda$ with the required properties.

We classify the points $\boldsymbol{a}$ of $\Lambda$ in $\mathscr{S}$, other than $\boldsymbol{o}$, into four types $\mathfrak{T}_1, \mathfrak{T}_2, \mathfrak{T}_3$ and $\mathfrak{T}_6$:

(i) $\boldsymbol{a}$ is in $\mathfrak{T}_1$ if it is not of either the shape $\boldsymbol{a}=2\boldsymbol{b}$ or $\boldsymbol{a}=3\boldsymbol{b}$ with $\boldsymbol{b}\in\Lambda$.

(ii) $\boldsymbol{a}$ is in $\mathfrak{T}_2$ if it is of the shape $\boldsymbol{a}=2\boldsymbol{b}$ but not of the shape $\boldsymbol{a}=3\boldsymbol{b}$, with $\boldsymbol{b}\in\Lambda$.

(iii) $\boldsymbol{a}$ is in $\mathfrak{T}_3$ if it is of the shape $\boldsymbol{a}=3\boldsymbol{b}$ but not of the shape $\boldsymbol{a}=2\boldsymbol{b}$, $\boldsymbol{b}\in\Lambda$.

(iv) $\boldsymbol{a}$ is in $\mathfrak{T}_6$ if it is of the shape $\boldsymbol{a}=6\boldsymbol{b}$, $\boldsymbol{b}\in\Lambda$.

Let $N_1, N_2, N_3, N_6$ be the numbers of lattice points in the corresponding classes. Then by (2) and (4) we have

$$N_1+3N_2+4N_3+12N_6<12, \tag{5}$$

since, for example, the coutribution to (4) of $\boldsymbol{a}\in\mathfrak{T}_6$ is

$$1+2+3+6=12.$$

In particular, by (5),

$$N_6=0.$$

Suppose, first, that $N_3>0$. We apply Lemma 2 with $p=2$, taking $\boldsymbol{a}_0$ to be one of the $N_3$ points in $\mathfrak{T}_3$ and $\boldsymbol{a}_1,\ldots,\boldsymbol{a}_R$ to be the remaining points in $\mathfrak{T}_3$ (if any) together with any points in $\mathfrak{T}_1$. The numbers $k_r$ of the lemma are taken as 1 if $\boldsymbol{a}_r\in\mathfrak{T}_1$ and 4 if $\boldsymbol{a}_r\in\mathfrak{T}_3$. Then, by Lemma 2, there is a lattice $\Gamma$ of index 2 which contains $N_1'$, $N_3'$ points of $\mathfrak{T}_1$, $\mathfrak{T}_3$ respectively, where

$$N_1'+4N_3'\leqq\tfrac{1}{2}(N_1+4N_3-4) \tag{6}$$

(the $-4$ being the contribution of $\boldsymbol{a}_0$, which is definitely lost). All the points of $\mathfrak{T}_2$ are, of course, in $\Gamma$. By (5) and (6) we have

$$2N_1'+3N_2+8N_3'+4\leqq 11.$$

Hence $N_3'=0$ and $N_1'+N_2\leqq\frac{7}{2}$, so $N_1'+N_2\leqq 3$. But now by Lemma 1, Corollary 1 there is a sublattice $\mathsf{M}$ of $\Gamma$ of index 3 which contains none of these $N_1'+N_2$ points. Then $\mathsf{M}$ does what is required.

We may thus suppose now that

$$N_3=0.$$

We now apply Lemma 1, Corollary 2 with $p=3$ to the points $\boldsymbol{a}_r$ with $k_r=1$ if $\boldsymbol{a}_r\in\mathfrak{T}_1$ and $k_r=3$ if $\boldsymbol{a}_r\in\mathfrak{T}_2$. Since there are at most

$$11<(3^3-1)/(3-1)$$

points $\boldsymbol{a}_r$, we may take $m=2$, so, in the notation of the corollary,

$$\frac{p^{m-1}-1}{p^m-1}=\frac{1}{4}.$$

Hence there is a sublattice $\Gamma$ of index 3 which contains $N_1'$, $N_2'$ points of $\mathfrak{T}_1$, $\mathfrak{T}_2$ respectively, where

$$N_1' + 3N_2' \leqq \frac{1}{4}(N_1 + 3N_2) \leqq \frac{11}{4} < 3.$$

Hence $N_2' = 0$ and $N_1' \leqq 2$. By Lemma 1, Corollary 1, there is a sublattice $\mathsf{M}$ of $\Gamma$ of index 2 which contains none of these $N_1'$ points. This lattice $\mathsf{M}$ does what is required.

Thus in every case we have constructed a lattice $\mathsf{M}$ of index 6 in $\Lambda$ which is admissible for $\mathscr{S}$. Since

$$d(\mathsf{M}) = 6d(\Lambda) = \Delta_1,$$

the lattice $\mathsf{M}$ has all the required properties.

As SCHMIDT remarks, Theorem III can be improved somewhat at the expense of further elaboration; but for large $n$ is weaker than ROGERS' results which we referred to in § 1 and which we cannot prove here. In particular the factor $(1+2^{1-n})(1+3^{1-n})$ on the left of (1) may be replaced by something smaller if $\mathscr{S}$ is a star-body, since then a point in $r^{-1}\mathscr{S}$ is automatically in $t^{-1}\mathscr{S}$ if $t \leqq r$.

**VI.5. A conjecture of Rogers.** We digress now from the general theme of the chapter to prove a result which was conjectured by ROGERS (1951 a), who compares it with the generalization of Theorem II of Chapter III from $m=1$ to $m>1$. It was proved by ROGERS when the number $m$ occurring in it is a prime and by SCHMIDT (1955 a) for all except a finite number[1] of $m$. It has been proved generally in a rather wider context by the author (CASSELS 1958 a). We do not use it later.

THEOREM IV. *Let $\mathscr{S}$ by a symmetric star-body and let $\Lambda$ be a lattice with*

$$m\,d(\Lambda) < \Delta(\mathscr{S}), \tag{1}$$

*where $m \geqq 1$ is an integer. Then $\mathscr{S}$ contains at least $m$ pairs $\pm\boldsymbol{a}$ of points of $\Lambda$ other than $\boldsymbol{o}$.*

Theorem IV is an immediate consequence of the following theorem in which the reference to star-bodies disappears.

THEOREM V. *Let $\boldsymbol{a}_1, \ldots, \boldsymbol{a}_R$ be primitive points of a lattice $\Lambda$ and let*

$$j_r \qquad (1 \leqq r \leqq R) \tag{2}$$

*be positive integers. Then there is a lattice of index at most*

$$j_1 + \cdots + j_R + 1 = J + 1 \text{ (say)} \tag{3}$$

[1] For all $m \leqq 10^7$ and all sufficiently large $m$, according to the review in Mathematical Reviews!

*which contains none of the points*

$$\pm i_r \boldsymbol{a}_r \qquad (1 \leq i_r \leq j_r, \ 1 \leq r \leq R). \tag{4}$$

We show first that Theorem V implies Theorem IV. Suppose that $\Lambda$ in Theorem IV contains fewer than $m$ pairs of points of $\mathscr{S}$. Since $\mathscr{S}$ is a star-body, the points of $\Lambda$ in $\mathscr{S}$ can be put in the shape (4), where the number of pairs is

$$J < m.$$

Hence by Theorem V there is a lattice M of index $\leq m$ in $\Lambda$ which contains none of these points, i.e. M is $\mathscr{S}$-admissible. Since

$$d(\mathsf{M}) \leq m\, d(\Lambda) < \Delta(\mathscr{S}),$$

by (1), this is a contradiction to the definition of $\Delta(\mathscr{S})$.

The proof of Theorem V depends on the following lemma, which gives the existence of primes with certain properties. It is due to SYLVESTER (1892a) and was rediscovered by SCHUR (1929a) who gave a rather simpler proof. The proof is in any case rather involved, so we do not give it here but refer the reader to the original papers.

LEMMA 4 (SYLVESTER). *Let $X$, $Y$ be integers and*

$$1 \leq X \leq Y.$$

*Then there is a prime number $p > X$ which divides one of the numbers*

$$Y+1, \ldots, Y+X.$$

We now prove Theorem V. Suppose first that $R=1$. Since $\boldsymbol{a}_1$ is primitive, it may be taken as part of a basis for $\Lambda$:

$$\boldsymbol{a}_1 = \boldsymbol{b}_1, \quad \boldsymbol{b}_2, \ldots, \boldsymbol{b}_n,$$

where $n$ is the dimension. Clearly the lattice M of points

$$u_1 \boldsymbol{b}_1 + \cdots + u_n \boldsymbol{b}_n,$$

where $u_1, \ldots, u_n$ are integers and

$$u_1 \equiv 0 \qquad (J+1),$$

does all that is required.

We now consider the case when $R>1$ and use induction on $J$. Without loss of generality

$$j_1 = \max_{1 \leq r \leq R} j_r. \tag{5}$$

Let $p$ be the prime given by SYLVESTER's Lemma 4 with

$$X = \min(j_1, j_2 + \cdots + j_R),$$
$$Y = \max(j_1, j_2 + \cdots + j_R).$$

Then

$$p > X \geqq j_r \qquad (2 \leqq r \leqq R). \tag{6}$$

Since $p$ divides one of the numbers $Y+1, \ldots, Y+X$, we have

$$\left[\frac{X}{p}\right] + \left[\frac{Y}{p}\right] < \left[\frac{X+Y}{p}\right],$$

that is

$$\left[\frac{j_1}{p}\right] + \left[\frac{j_2 + \cdots + j_R}{p}\right] < \left[\frac{j_1 + \cdots + j_R}{p}\right], \tag{7}$$

where for any real number $x$ we denote by $[x]$ in this proof the integer such that $[x] \leqq x < [x]+1$. By Lemma 2, there is a lattice $\Gamma$ of index $p$ which does not contain $\boldsymbol{a}_1$ and such that

$$p \sum_{\boldsymbol{a}_r \in \Gamma} j_r \leqq \sum_{2 \leqq r \leqq R} j_r;$$

that is

$$\sum_{\boldsymbol{a}_r \in \Gamma} j_r \leqq \left[\frac{j_2 + \cdots + j_R}{p}\right]. \tag{8}$$

By (6), if a point $i_r \boldsymbol{a}_r$ in (4) with $r > 1$ is in $\Gamma$, then $\boldsymbol{a}_r$ is in $\Gamma$. Since $\boldsymbol{a}_1$ is not in $\Gamma$, the only points (4) with $r = 1$ in $\Gamma$ are the

$$\pm i_1' (p \boldsymbol{a}_1) \qquad \left(1 \leqq i_1' \leqq \left[\frac{j_1}{p}\right]\right). \tag{9}$$

But now, by the hypothesis of the induction argument, there is a lattice $\mathsf{M}$ of index at most

$$1 + \left[\frac{j_1}{p}\right] + \sum_{\boldsymbol{a}_r \in \Gamma} j_r \leqq 1 + \left[\frac{j_1}{p}\right] + \left[\frac{j_2 + \cdots + j_R}{p}\right]$$

in $\Gamma$ which contains none of the points (4) at all. The index of $\mathsf{M}$ in $\Lambda$ is $p$ times the index of $\mathsf{M}$ in $\Gamma$; and so, by (7), is

$$\leqq p\left\{1 + \left[\frac{j_1}{p}\right] + \left[\frac{j_2 + \cdots + j_R}{p}\right]\right\} \leqq J < J+1.$$

This concludes the proof of Theorem V.

**VI.6. Unbounded star-bodies.** The results of §§ 3, 4 extend to unbounded star-bodies. For example we have

THEOREM VI. *Let $\mathscr{S}$ be a bounded or unbounded symmetric star-body. Then*

$$\Delta(\mathscr{S}) \leqq \{2\zeta(n)\}^{-1} V(\mathscr{S}). \tag{1}$$

When $\mathscr{S}$ is bounded this is just Theorem II, Corollary. When $\mathscr{S}$ is unbounded it follows from Theorem II, Corollary together with Theorem V, Corollary of Chapter V.

In the same way any of the other estimates of §§ 3, 4 may be extended to unbounded star-bodies $\mathscr{S}$, or indeed, to any open sets of finite volume of which the origin is an inner point.

**VI.6.2.** There certainly exist star-bodies $\mathscr{S}$ of finite type [i.e. with $\Delta(\mathscr{S})<\infty$] and infinite volume. A 2-dimensional example is

$$\mathscr{S}_1: \quad |x_1 x_2| < 1; \tag{1}$$

for which $\Delta(\mathscr{S}_1) = 5^{\frac{1}{2}}$, as we saw in Chapter II. More generally, in $n$-dimensions the body

$$|x_1 \dots x_n| < 1$$

is of finite type but infinite volume, since admissible lattices are given by the norm-forms of totally real algebraic fields of degree $n$ (see Chapter X). In general, in more than 2 dimensions it is very difficult to decide whether a given star-body is of finite type or not. Two 3-dimensional examples are discussed in CASSELS and SWINNERTON-DYER (1955a), for which a decision on this point would have interesting repercussions. In 2 dimensions however there do exist general criteria which we shall now discuss.

**VI.6.3.** From now on we put [1]

$$n = 2.$$

In an obvious sense, the body $\mathscr{S}_1$ defined in (1) of § 6.2 has two pairs of asymptotic arms, the asymptotes being the $x_1$ and $x_2$ axis. It is possible to inscribe in $\mathscr{S}$ arbitrarily narrow parallelograms with one pair of sides parallel to an asymptote and area 1, for example

$$|x_1| < \varepsilon, \quad |x_2| < \varepsilon^{-1}.$$

In a sense $\mathscr{S}_1$ is a limiting case, since if it is possible to inscribe in a star-body $\mathscr{S}$ parallelograms with centre the origin and arbitrarily large volume (area), then $\mathscr{S}$ is of infinite type by MINKOWSKI's convex body Theorem II of Chapter III. Roughly speaking, any star-body with a pair of arms wider than those of $\mathscr{S}_1$ is of infinite type. We now show that a 2-dimensional star-body may have any finite number of arms like those of $\mathscr{S}_1$ and still remain of finite type.

[1] It is customary to call 2-dimensional star-bodies "star domains" but we do not follow this usage. Similarly we may sometimes continue to speak of volume where area is more usual.

THEOREM VII. *Let*
$$f_0(x_1, x_2) = \varepsilon(x_1^2 + x_2^2)$$
*for some* $\varepsilon > 0$, *and let*
$$f_j(x_1, x_2) \qquad (1 \leqq j \leqq J)$$
*be any finite number of indefinite quadratic forms. Suppose that the distance-function* $F(x_1, x_2)$ *satisfies*
$$F^2(x_1, x_2) \geqq \min_{0 \leqq j \leqq J} |f_j(x_1, x_2)| \tag{2}$$
*for all* $(x_1, x_2)$. *Then the star-body*
$$\mathscr{S}: \quad F(x_1, x_2) < 1 \tag{3}$$
*is of finite type.*

The exponent 2 in (2) is dictated by reasons of homogeneity.

We shall deduce Theorem VII from the following generalization of a theorem of MARSHALL HALL (1947a) which is due to the author (CASSELS 1956a).

THEOREM VIII. *Let* $\beta_1, \ldots, \beta_K$ *be any real numbers. Then there exists a real number* $\alpha$ *such that*
$$|u| \, |(\alpha + \beta_k) u + v| > \frac{1}{8(K+1)^2} \qquad (1 \leqq k \leqq K) \tag{4}$$
*for all integers* $u \neq 0$ *and* $v$.

We first deduce Theorem VII from Theorem VIII, and then prove Theorem VIII in § 6.4. After a suitable rotation of the co-ordinate system, we may suppose without loss of generality that
$$f_j(1, 0) \neq 0 \qquad (1 \leqq j \leqq J);$$
and so
$$f_j(x_1, x_2) = \lambda_j (x_1 + \vartheta_j x_2)(x_1 + \varphi_j x_2) \qquad (1 \leqq j \leqq J) \tag{5}$$
for real numbers $\lambda_j, \vartheta_j, \varphi_j$ such that
$$\lambda_j \neq 0, \quad \vartheta_j \neq \varphi_j.$$
But now
$$|f_j(x_1, x_2)| \geqq \mu_j \min\{|x_2(x_1 + \vartheta_j x_2)|, |x_2(x_1 + \varphi_j x_2)|\}, \tag{6}$$
where
$$\mu_j = \tfrac{1}{2} |\lambda_j| \, |\vartheta_j - \varphi_j| > 0;$$
since if, for example
$$|x_1 + \vartheta_j x_2| \leqq |x_1 + \varphi_j x_2|,$$
then we have
$$|(\vartheta_j - \varphi_j) x_2| = |(x_1 + \vartheta_j x_2) - (x_1 + \varphi_j x_2)| \leqq 2 |x_1 + \varphi_j x_2|.$$

We apply Theorem VIII where the $\beta_1, \ldots, \beta_K$ are the $\vartheta_j, \varphi_j$ in some order, so $K = 2J$. Let $\alpha$ be the number given by Theorem VIII, so that

$$\left.\begin{aligned} |u\{(\alpha+\vartheta_j)u+v\}| \geqq \eta > 0 \\ |u\{(\alpha+\varphi_j)u+v\}| \geqq \eta > 0 \end{aligned}\right\} \tag{7}$$

or integers $u \neq 0$, $v$, where

$$\eta = \{8(2J+1)^2\}^{-1}.$$

Let $\Lambda$ be the lattice of points

$$(x_1, x_2) = R(\alpha u + v, u), \tag{8}$$

where $u, v$ run through all integer values and $R$ is a positive number yet to be chosen. If $u \neq 0$ we have, by (6) and (7)

$$|f_j\{R(\alpha u + v), Ru\}| \geqq \mu_j R^2 \eta. \tag{9}$$

If however $u = 0$ but $v \neq 0$ then, by (5),

$$|f_j(Rv, 0)| \geqq |\lambda_j| R^2. \tag{10}$$

Similarly

$$f_0(x_1, x_2) = \varepsilon(x_1^2 + x_2^2) \geqq \varepsilon R^2 \tag{11}$$

for all $(x_1, x_2) \in \Lambda$ other than $\boldsymbol{o}$, on distinguishing the two cases $u \neq 0$ and $u = 0$, $v \neq 0$ in (8). We may choose $R$ so large that the right-hand sides of (9), (10) and (11) are all not less than 1. Then for all $(x_1, x_2) \in \Lambda$ except $\boldsymbol{o}$, we have, by (2),

$$F^2(x_1, x_2) \geqq \min_{0 \leqq j \leqq J} |f_j(x_1, x_2)| \geqq 1;$$

that is $\Lambda$ is $\mathscr{S}$-admissible. This concludes the proof of Theorem VII.

**VI.6.4.** We now prove Theorem VIII which was enunciated in § 6.3. Write

$$\varkappa = \{2(K+1)\}^{\frac{1}{2}}. \tag{1}$$

We shall construct a sequence of open intervals $\mathscr{I}_{-1}, \mathscr{I}_0, \mathscr{I}_1, \ldots$ which enjoy the following three properties:

(i)$_m$ $\mathscr{I}_{m+1}$ is contained in $\mathscr{I}_m$.

(ii)$_m$ $\mathscr{I}_m$ is of length $\varkappa^{-2m-2}$.

(iii)$_m$ the inequality

$$u|(\alpha + \beta_k)u + v| > \tfrac{1}{2}\varkappa^{-4} \qquad (1 \leqq k \leqq K) \tag{2}$$

holds for all numbers $\alpha$ in $\mathscr{I}_m$ and for all integers $v$ and $u$ with

$$0 < u \leqq \varkappa^m. \tag{3}$$

If we can construct the $\mathscr{I}_m$ we shall have proved Theorem VIII, since there is a number $\alpha$ contained in all the intervals $\mathscr{I}_m$ and then (2) holds with this $\alpha$ for all integers $u>0$ and $v$.

We may take $\mathscr{I}_{-1}$ to be the interval $0<\alpha<1$, since there are no integers $u$ in (3) with $m=-1$. We thus assume that $\mathscr{I}_m$ has already been constructed and construct $\mathscr{I}_{m+1}$. By (ii)$_m$, the open interval $\mathscr{I}_m$ is the set of $\alpha$ satisfying

$$\alpha'<\alpha<\alpha'' \tag{4}$$

for some numbers $\alpha'$ and $\alpha''$ for which

$$\alpha''-\alpha'=\varkappa^{-2m-2}. \tag{5}$$

For each $k$ $(1\leqq k\leqq K)$, there is at most one fraction $v_k/u_k$ in its lowest terms such that

$$-\left(\frac{v_k}{u_k}+\beta_k\right)\in\mathscr{I}_m, \quad 0<u_k\leqq\varkappa^{m+1}, \tag{6}$$

since two fractions $v/u$ with $0<u\leqq\varkappa^{m+1}$ differ by at least $\varkappa^{-2m-2}$. By (iii)$_m$, we have

$$u_k>\varkappa^m \qquad (1\leqq k\leqq K). \tag{7}$$

Let $\mathscr{G}$ be the set of $\alpha$ such that

$$\alpha'+\tfrac{1}{2}\varkappa^{-2m-4}<\alpha<\alpha''-\tfrac{1}{2}\varkappa^{-2m-4}, \tag{8}$$

and

$$u_k|(\alpha+\beta_k)u_k+v_k|>\tfrac{1}{2}\varkappa^{-4} \tag{9}$$

for all $k$ in $1\leqq k\leqq K$ for which a $v_k/u_k$ of the type (6) exists. Then $\mathscr{G}$ consists of at most $K+1$ intervals. Their total length is

$$\begin{aligned}\alpha''-\alpha'-\varkappa^{-2m-4}-\sum_k\varkappa^{-4}u_k^{-2}\\ &\geqq\varkappa^{-2m-2}-(K+1)\varkappa^{-2m-4}\\ &=(K+1)\varkappa^{-2m-4},\end{aligned}$$

by (1), (5) and (7). We may therefore find in $\mathscr{G}$ an open interval $\mathscr{I}_{m+1}$ of length exactly $\varkappa^{-2m-4}$. Then $\mathscr{I}_{m+1}$ satisfies (i)$_m$ and (ii)$_{m+1}$, by construction. It remains only to verify (iii)$_{m+1}$. We may clearly suppose that $u$ and $v$ are coprime and that

$$\varkappa^m<u\leqq\varkappa^{m+1} \tag{10}$$

by (i)$_m$ and (iii)$_m$. If $v/u=v_k/u_k$ is a fraction of the type (6), then

$$u|(\alpha+\beta_k)u+v|>\tfrac{1}{2}\varkappa^{-4} \tag{11}$$

for all $\alpha\in\mathscr{I}_{m+1}$, by (9). Otherwise $-\left(\frac{v}{u}+\beta_k\right)$ is not in $\mathscr{I}_m$, and so

$$\left|\frac{v}{u}+(\alpha+\beta_k)\right|>\frac{1}{2}\varkappa^{-2m-4}$$

for all $\alpha\in\mathscr{I}_{m+1}$, by (8); then (11) follows, by (10). Thus $\mathscr{I}_{m+1}$ has all the required properties.

Chapter VII

# The quotient space

**VII.1. Introduction.** Before resuming the general study of the geometry of numbers, it is convenient to introduce here the concept of the quotient space of an $n$-dimensional space by a lattice. This concept plays an important rôle in the discussion of inhomogeneous problems in Chapter XI: but we shall also need it in Chapter VIII as it gives the most natural interpretation of MINKOWSKI's theorem about the successive minima of a convex body with respect to a lattice.

In § 2 we give the definition and most important properties of a quotient space. In § 3 we prove a result which will be basic for one topic in Chapter XI.

**VII.2. General properties.** Let $\Lambda$ be a lattice in $n$-dimensional euclidean space. Two points $\boldsymbol{y}_1, \boldsymbol{y}_2$ of the space are said to be congruent modulo $\Lambda$, written

$$\boldsymbol{y}_1 \equiv \boldsymbol{y}_2 \quad (\Lambda), \tag{1}$$

if the difference $\boldsymbol{y}_1 - \boldsymbol{y}_2$ is in $\Lambda$. This relationship is clearly symmetrical in $\boldsymbol{y}_1$ and $\boldsymbol{y}_2$. If

$$\boldsymbol{y}_1 \equiv \boldsymbol{y}_2 \quad (\Lambda), \qquad \boldsymbol{y}_2 \equiv \boldsymbol{y}_3 \quad (\Lambda),$$

then

$$\boldsymbol{y}_1 \equiv \boldsymbol{y}_3 \quad (\Lambda).$$

The points $\boldsymbol{y}$ may therefore be divided into classes $\mathfrak{y}$ so that two points $\boldsymbol{y}$ and $\boldsymbol{y}'$ are congruent if and only if they are in the same class. A class $\mathfrak{y}$ consists of all the points $\boldsymbol{y}_0 + \boldsymbol{a}$, where $\boldsymbol{y}_0$ is some fixed member of $\mathfrak{y}$ and $\boldsymbol{a}$ runs through all points of $\Lambda$.

If

$$\boldsymbol{y}' \equiv \boldsymbol{y} \quad (\Lambda), \qquad \boldsymbol{z}' \equiv \boldsymbol{z} \quad (\Lambda),$$

then clearly

$$\boldsymbol{y}' + \boldsymbol{z}' \equiv \boldsymbol{y} + \boldsymbol{z} \quad (\Lambda).$$

Hence there is no ambiguity in defining the sum $\mathfrak{y} + \mathfrak{z}$ of two classes as the class to which $\boldsymbol{y} + \boldsymbol{z}$ belongs when $\boldsymbol{y}, \boldsymbol{z}$ are any members of $\mathfrak{y}, \mathfrak{z}$ respectively.

Similarly, if $t$ is an integer, the definition of $t\mathfrak{y}$ as the class to which $t\boldsymbol{y}$ belongs when $\boldsymbol{y}$ is in $\mathfrak{y}$ is unambiguous. On the other hand, if $t$ is not an integer, it is not, in general, true that $t\boldsymbol{y}' \equiv t\boldsymbol{y}$ when $\boldsymbol{y}' \equiv \boldsymbol{y}$. Hence $t\mathfrak{y}$ for real numbers $t$ other than integers must be left undefined.

So far, of course, we have only followed the standard procedure for finding the quotient group of an abelian group (namely the additive group of all vectors) by a subgroup (namely the additive group of vectors in $\Lambda$). We shall say that the classes $\mathfrak{y}$ are points of the quotient space $\mathscr{R}/\Lambda$, where $\mathscr{R}$ will denote the original $n$-dimensional euclidean space.

**VII.2.2.** Let $F(\boldsymbol{x})$ be any distance function defined in $\mathscr{R}$ and put[1]

$$F(\mathfrak{y}) = \inf_{\boldsymbol{y}\in\mathfrak{y}} F(\boldsymbol{y}) \tag{1}$$

for $\mathfrak{y}\in\mathscr{R}/\Lambda$. This is the function which will be important in inhomogeneous problems (Chapter XI). Note that

$$F(\mathfrak{o}) = 0, \tag{2}$$

where $\mathfrak{o}$ is the class to which $\boldsymbol{o}$ belongs. For reference we enunciate the principal properties of $F(\mathfrak{x})$, $\mathfrak{x}\in\mathscr{R}/\Lambda$, in the following lemma.

LEMMA 1. *Let $F(\boldsymbol{x})$ be a distance function and let $F(\mathfrak{x})$ be defined, as above, for $\mathfrak{x}\in\mathscr{R}/\Lambda$. Then*

*(i) $F(t\mathfrak{x}) \leqq tF(\mathfrak{x})$ for integers $t \geqq 0$.*

*(ii) If $F(\boldsymbol{x})$ is convex, then so is $F(\mathfrak{x})$, in the sense that*

$$F(\mathfrak{x}+\mathfrak{y}) \leqq F(\mathfrak{x}) + F(\mathfrak{y})$$

*for all $\mathfrak{x}$, $\mathfrak{y}$.*

*(iii) If $F(\boldsymbol{x}) = 0$ only for $\boldsymbol{x} = 0$, then $F(\mathfrak{x}) = 0$ only for $\mathfrak{x} = \mathfrak{o}$. Further, for each $\mathfrak{y}\in\mathscr{R}/\Lambda$ there is a $\boldsymbol{y}\in\mathfrak{y}$ such that $F(\mathfrak{y}) = F(\boldsymbol{y})$.*

*(iv) If $F_1(\boldsymbol{x})$, $F_2(\boldsymbol{x})$ are two distance function and $F_1(\boldsymbol{x}) \leqq cF_2(\boldsymbol{x})$ for some number $c$ and all $\boldsymbol{x}\in\mathscr{R}$, then $F_1(\mathfrak{x}) \leqq cF_2(\mathfrak{x})$ for all $\mathfrak{x}\in\mathscr{R}/\Lambda$.*

Here (iv) is an immediate consequence of the definition (1). By the definition of a distance function, we have $F(t\boldsymbol{x}) = tF(\boldsymbol{x})$ for all real $t > 0$. Hence, if $t > 0$ is an integer, we have

$$F(t\mathfrak{x}) = \inf_{\boldsymbol{y}\in t\mathfrak{x}} F(\boldsymbol{y}) \leqq \inf_{\boldsymbol{x}\in\mathfrak{x}} F(t\boldsymbol{x}) = t \inf_{\boldsymbol{x}\in\mathfrak{x}} F(\boldsymbol{x}) = tF(\mathfrak{x}).$$

This establishes (i). The proof of (ii) is similar and may be left to the reader.

It remains to prove (iii). Let $\mathfrak{y}\in\mathscr{R}/\Lambda$ and let $\boldsymbol{y}_0\in\mathfrak{y}$, so that the general element of $\mathfrak{y}$ is $\boldsymbol{y}_0+\boldsymbol{a}$, $\boldsymbol{a}\in\Lambda$. By Lemma 2 of Chapter IV, there is a constant $c > 0$ such that $F(\boldsymbol{x}) \geqq c|\boldsymbol{x}|$ for all $\boldsymbol{x}$; and so

$$F(\boldsymbol{y}_0+\boldsymbol{a}) \geqq c|\boldsymbol{y}_0+\boldsymbol{a}| \geqq c\big||\boldsymbol{a}| - |\boldsymbol{y}_0|\big|.$$

In particular, if $F(\boldsymbol{a}+\boldsymbol{y}_0) \leqq F(\boldsymbol{y}_0)$, we have

$$|\boldsymbol{a}| \leqq |\boldsymbol{y}_0| + c^{-1}F(\boldsymbol{y}_0). \tag{3}$$

There are only a finite number of $\boldsymbol{a}\in\Lambda$ in (3). Hence there exists an $\boldsymbol{a}_0\in\Lambda$ such that $F(\boldsymbol{y}_0+\boldsymbol{a}_0) = \inf_{\boldsymbol{a}\in\Lambda} F(\boldsymbol{y}_0+\boldsymbol{a})$. By definition, $F(\mathfrak{y}) = F(\boldsymbol{y}_0+\boldsymbol{a}_0)$. Further, $F(\mathfrak{y}) = 0$ only if $F(\boldsymbol{y}_0+\boldsymbol{a}_0) = 0$, that is $\mathfrak{y} = \mathfrak{o}$.

[1] There should be no confusion with the usage of Chapter IV, since there the arguments were lattices; and here they are classes with respect to a lattice.

**VII.2.3.** Let $\mathfrak{y}_r$ $(1 \leqq r < \infty)$ be a sequence of elements of $\mathscr{R}/\Lambda$. We say that the sequence tends to $\mathfrak{y}' \in \mathscr{R}/\Lambda$ if

$$\lim_{r \to \infty} |\mathfrak{y}_r - \mathfrak{y}'| = 0, \tag{1}$$

where, in conformity with the notation of § 2.2, we have written

$$|\mathfrak{x}| = \inf_{\boldsymbol{x} \in \mathfrak{x}} |\boldsymbol{x}|. \tag{2}$$

LEMMA 2. *A necessary and sufficient condition that $\mathfrak{y}_r \to \mathfrak{y}'$ is that there exist elements $\boldsymbol{y}_r \in \mathfrak{y}_r$ and $\boldsymbol{y}' \in \mathfrak{y}'$ such that*

$$\boldsymbol{y}_r \to \boldsymbol{y}'. \tag{3}$$

Suppose, first, that the $\boldsymbol{y}_r$, $\boldsymbol{y}'$ exist such that (3) holds. Then

$$|\mathfrak{y}_r - \mathfrak{y}'| \leqq |\boldsymbol{y}_r - \boldsymbol{y}'|;$$

so (1) holds, that is $\mathfrak{y}_r \to \mathfrak{y}'$.

Suppose, now, that (1) holds. By Lemma 1 (iv) there exist $\boldsymbol{z}_r \in \mathfrak{y}_r - \mathfrak{y}'$ such that

$$|\boldsymbol{z}_r| = |\mathfrak{y}_r - \mathfrak{y}'|.$$

Let $\boldsymbol{y}'$ be any element of $\mathfrak{y}'$ and put $\boldsymbol{y}_r = \boldsymbol{y}' + \boldsymbol{z}_r$. Then the $\boldsymbol{y}_r$ clearly have all the properties required.

**VII.2.4.** Let

$$\boldsymbol{b}_1, \ldots, \boldsymbol{b}_n \tag{1}$$

be any basis for $\Lambda$. Then every point $\boldsymbol{x}$ of space can be put uniquely in the shape

$$\boldsymbol{x} = \xi_1 \boldsymbol{b}_1 + \cdots + \xi_n \boldsymbol{b}_n \tag{2}$$

for some real numbers $\xi_1, \ldots, \xi_n$; and $\boldsymbol{x} \in \Lambda$ if and only if $\xi_1, \ldots, \xi_n$ are integers. Hence to every vector $\boldsymbol{x}$ there is a unique $\boldsymbol{a} \in \Lambda$ such that

$$\boldsymbol{y} = \boldsymbol{x} - \boldsymbol{a} = \eta_1 \boldsymbol{b}_1 + \cdots + \eta_n \boldsymbol{b}_n, \tag{3}$$

where

$$0 \leqq \eta_j < 1. \tag{4}$$

In other words, every $\mathfrak{x} \in \mathscr{R}/\Lambda$ has precisely one representative $\boldsymbol{y} \in \mathfrak{x}$ in the half-open parallelopiped $\mathscr{P}$ defined by (3) and (4). We say that this parallelopiped is a fundamental parallelopiped for $\Lambda$. Different bases $\boldsymbol{b}_j$ in general give rise to different fundamental parallelopipeds.

An immediate consequence of Lemma 2 and the existence of a fundamental parallelopiped is

LEMMA 3. *The quotient space $\mathscr{R}/\Lambda$ is compact. That is, any sequence $\mathfrak{y}_r$ $(1 \leq r < \infty)$ of elements of $\mathscr{R}/\Lambda$ contains a convergent subsequence:*

$$\mathfrak{y}_{r_s} \to \mathfrak{y}'. \tag{5}$$

The fundamental parallelopiped $\mathscr{P}$ is not compact, since although it is bounded it is not closed. Let $\overline{\mathscr{P}}$ be its closure, that is the set of points (3) with $0 \leq \eta_j \leq 1$ $(1 \leq j \leq n)$. Let $\boldsymbol{y}_r$ be the representative of $\mathfrak{y}_r$ in $\mathscr{P}$. By WEIERSTRASS's compactness theorem (§ 1.3 of Chapter III), there is a convergent subsequence

$$\boldsymbol{y}_{r_s} \to \boldsymbol{y}',$$

where $\boldsymbol{y}' \in \overline{\mathscr{P}}$. Then (5) holds by Lemma 2, where $\boldsymbol{y}' \in \mathfrak{y}'$.

**VII.2.5.** We are now in a position to introduce a measure into the quotient space $\mathscr{R}/\Lambda$. Let $\mathsf{S}$ be any set of elements of $\mathscr{R}/\Lambda$. We call a set $\mathscr{S}$ of elements of $\mathscr{R}$ a set of representatives for $\mathsf{S}$ if (i) for each $\mathfrak{x} \in \mathsf{S}$ there is precisely one $\boldsymbol{x} \in \mathfrak{x}$ which belongs to $\mathscr{S}$ and (ii) each $\boldsymbol{x} \in \mathscr{S}$ belongs to an $\mathfrak{x} \in \mathsf{S}$. We say that $\mathsf{S}$ is measurable if at least one set $\mathscr{S}$ of representatives is measurable.

Let $\mathscr{S}_1$ be the set of elements $\boldsymbol{x} \in \mathscr{P}$ of the shape

$$\boldsymbol{x} = \boldsymbol{y} + \boldsymbol{u}, \quad \boldsymbol{y} \in \overline{\mathscr{S}}, \quad \boldsymbol{u} \in \Lambda,$$

where $\overline{\mathscr{S}}$ is any measurable set of representatives of $\mathsf{S}$ and $\mathscr{P}$ is a fundamental parallelopiped. By Theorem I Corollary of Chapter III, the set $\mathscr{S}_1$ is measurable, and

$$V(\mathscr{S}_1) = V(\mathscr{S}).$$

In particular, if $\mathscr{S}$, $\mathscr{S}'$ are any two measurable sets of representatives of $\mathsf{S}$, we have $V(\mathscr{S}) = V(\mathscr{S}')$. This common value will be denoted by

$$m(\mathsf{S})$$

and will be called the measure of $\mathsf{S}$.

Clearly the measure of the whole of the quotient space is the volume of the fundamental parallelopiped $\mathscr{P}$, that is $d(\Lambda)$.

Let $\boldsymbol{\tau}$ be any homogeneous mapping of $n$-dimensional space $\mathscr{R}$ onto itself. In a natural way, it gives a mapping of $\mathscr{R}/\Lambda$ into $\mathscr{R}/\boldsymbol{\tau}\Lambda$, which we may also denote by $\boldsymbol{\tau}$. If $m'$ is the measure defined in $\mathscr{R}/\boldsymbol{\tau}\Lambda$ in the way that $m$ is defined in $\mathscr{R}/\Lambda$, then clearly

$$m'(\boldsymbol{\tau}\mathsf{S}) = |\det(\boldsymbol{\tau})|\, m(\mathsf{S})$$

for any set $\mathsf{S}$ in $\mathscr{R}/\Lambda$.

**VII.3. The sum theorem**[1]. If $C$ and $D$ are two sets of points in the quotient space $\mathcal{R}/\Lambda$ we denote by $C+D$ the set of all points

$$\mathfrak{c}+\mathfrak{d}, \quad \text{where} \quad \mathfrak{c}\in C,\ \mathfrak{d}\in D.$$

This section is devoted to proving

THEOREM I. *Let $C$ and $D$ be non-empty sets in $\mathcal{R}/\Lambda$ with measures $m(C)$ and $m(D)$ respectively.*

*(i) If $m(C)+m(D)>d(\Lambda)$, then $C+D$ is the whole space $\mathcal{R}/\Lambda$.*

*(ii) If $m(C)+m(D)\leq d(\Lambda)$, then $m(C+D)\geq m(C)+m(D)$.*

This theorem is due to MACBEATH (1953a). It was discovered independently by KNESER (1955a), who first recognized its importance for the geometry of numbers. Theorem I is, in fact, now only part of a much wider theory, for which see KNESER (1956a) and the literature cited there. It falls into the same circle of ideas as the so-called "$\alpha+\beta$ hypothesis" about the densities of sequences of integers which was first proved by MANN. As all this is rather aside from the main theme of the book we do not discuss it further. It is convenient to prove Theorem I here but the application to the geometry of numbers will not be made until Chapter XI.

Part (i) of Theorem I is easy. Suppose that there is a point $\mathfrak{x}$ of $\mathcal{R}/\Lambda$ which does not belong to $C+D$. Then none of the points

$$\mathfrak{x}-\mathfrak{c}, \quad \mathfrak{c}\in C \tag{1}$$

can belong to $D$. We may denote the set (1) by $\mathfrak{x}-C$. Clearly

$$m(\mathfrak{x}-C)=m(C). \tag{2}$$

But $D$ and $\mathfrak{x}-C$ have no points in common, so

$$m(\mathfrak{x}-C)+m(D)\leq m(\mathcal{R}/\Lambda)=d(\Lambda). \tag{3}$$

Then $m(C)+m(D)\leq d(\Lambda)$, by (2) and (3). This proves (i).

In what follows we denote, as is conventional, by $C\cap D$ and $C\cup D$ the sets of points which belong to both $C$ and $D$ and to either $C$ or $D$ (or both) respectively. We note for further reference the identity

$$m(C\cap D)+m(C\cup D)=m(C)+m(D); \tag{4}$$

which becomes clear on noting that points of $C\cap D$ occur in two sets on each side of (4), but points of $C\cup D$ other than those of $C\cap D$ occur

[1] The results of § 3 will not be needed until Chapter XI.

in precisely one set on each side. Further, we show that

$$C+D \supset (C \cap D) + (C \cup D) \tag{5}$$

($\supset$ means "contains"). For let

$$\mathfrak{a} \in C \cap D, \quad \mathfrak{b} \in C \cup D.$$

Suppose $\mathfrak{b}$ belongs to $C$: then we may regard $\mathfrak{a}$ as belonging to $D$ since it belongs to both $C$ and $D$. Hence $\mathfrak{a}+\mathfrak{b}=\mathfrak{b}+\mathfrak{a} \in C+D$. Similarly, if $\mathfrak{b}$ belongs to $D$ we regard $\mathfrak{a}$ as belonging to $C$.

It follows from (4) and (5) that, if the conclusions of Theorem I are true when $C \cap D$, $C \cup D$ are read for $C$, $D$ respectively, then the conclusions are also true for $C$ and $D$ themselves. This is one of the principal ingredients of the proof. The other is provided by

LEMMA 4. *There is some* $\mathfrak{x} \in \mathscr{R}/\Lambda$ *such that*

$$d(\Lambda)\, m\{(C+\mathfrak{x}) \cap D\} = m(C)\, m(D).$$

Before proving Lemma 4 we complete the proof of Theorem I with its use. Let $C$, $D$ be two sets with

$$m(C) = \gamma\, d(\Lambda), \quad m(D) = \delta\, d(\Lambda)$$

and

$$\gamma + \delta \leqq 1.$$

If $\gamma = 0$, the conclusions of the theorem certainly hold, since $C$ is non-empty, by hypothesis, and if $\mathfrak{c} \in C$ the set $\mathfrak{c}+D$, which is contained in $C+D$, has measure $m(D) = m(C) + m(D)$. We may thus suppose without loss of generality that

$$0 < \gamma \leqq \delta, \quad \gamma + \delta \leqq 1. \tag{6}$$

Now let $\mathfrak{x}$ be given by Lemma 4, and put

$$C_1 = (C+\mathfrak{x}) \cap D, \quad D_1 = \{(C+\mathfrak{x}) \cup D\} - \mathfrak{x}.$$

Write

$$m(C_1) = \gamma_1\, d(\Lambda), \quad m(D_1) = \delta_1\, d(\Lambda),$$

so that

$$\gamma_1 + \delta_1 = \gamma + \delta,$$

and

$$\gamma_1 = \gamma\delta$$

by (4) applied to $C+\mathfrak{x}$ and $D$ and by Lemma 4 respectively. Further,

$$C + D \supset C_1 + D_1,$$

by (5) applied to $C+\mathfrak{x}$ and $D$. We may now repeat the process on $C_1, D_1$. In this way we get a sequence of sets $C_r, D_r$ with measures $\gamma_r d(\Lambda)$, $\delta_r d(\Lambda)$ respectively, such that

$$C+D \supset C_r + D_r, \tag{7}$$

and

$$\gamma_r + \delta_r = \gamma + \delta, \tag{8}$$

$$\gamma_r = \gamma_{r-1}\delta_{r-1}. \tag{9}$$

But now, by the argument used when $\gamma=0$, it is certainly true that

$$m(C_r + D_r) \geqq m(D_r) = \delta_r d(\Lambda). \tag{10}$$

It follows from (6), (8) with $r-1$ for $r$ and (9), that

$$\gamma_r \leqq \gamma_{r-1}(1-\gamma_{r-1});$$

and so

$$\gamma_r \to 0 \qquad (r\to\infty). \tag{11}$$

Hence

$$\delta_r \to \gamma + \delta \qquad (r\to\infty), \tag{12}$$

by (8). But

$$m(C+D) \geqq \delta_r d(\Lambda), \tag{13}$$

by (7) and (10). In letting $r\to\infty$ in (13) and using (12) we have

$$m(C+D) \geqq (\gamma+\delta)\, d(\Lambda) = m(C) + m(D)$$

as required.

It remains only to prove Lemma 3. We note, first, that

$$m\{(C+\mathfrak{x})\cap D\} \tag{14}$$

varies continuously with $\mathfrak{x}$. This is clearly true with the "well-behaved" sets $C$ and $D$ to which we will wish to apply Theorem I, but it is in fact true for all measurable $C$ and $D$, see for example A. Weil (1951a). In the second place, in an appropriate sense, to be explained more fully below, the average of (14) as $\mathfrak{x}$ runs through $\mathscr{R}/\Lambda$ is $m(C)\, m(D)/d(\Lambda)$. Perhaps the simplest way is to observe that we may introduce integration in $\mathscr{R}/\Lambda$ in the obvious way. Let $\varphi(\mathfrak{x})$ be a function defined in $\mathscr{R}/\Lambda$ and let $f(\boldsymbol{x})$ be the function in $\mathscr{R}$ such that

$$f(\boldsymbol{x}) = \varphi(\mathfrak{x}),$$

when $\boldsymbol{x}$ belongs to the class $\mathfrak{x}$. Then we write

$$\int_{\mathscr{R}/\Lambda} \varphi(\mathfrak{x})\, d\mathfrak{x} = \int_{\mathscr{P}} f(\boldsymbol{x})\, d\boldsymbol{x},$$

where $\mathscr{P}$ is a fundamental parallelopiped. Exactly as in § 2.5, one may show that this definition is independent of the choice of fundamental parallelopiped $\mathscr{P}$. Let $\varphi(\mathfrak{x}), \chi(\mathfrak{x})$ be the characteristic functions of $C, D$ respectively; so that

$$m\{(C+\mathfrak{x})\cap D\} = \int_{\mathscr{R}/\Lambda} \varphi(\mathfrak{y}+\mathfrak{x})\,\chi(\mathfrak{y})\,d\mathfrak{y}.$$

Then

$$\int_{\mathscr{R}/\Lambda} m\{(C+\mathfrak{x})\cap D\}\,d\mathfrak{x} = \int_{\mathscr{R}/\Lambda}\Big\{\int_{\mathscr{R}/\Lambda} \varphi(\mathfrak{y}+\mathfrak{x})\,\chi(\mathfrak{y})\,d\mathfrak{x}\Big\}\,d\mathfrak{y}. \tag{15}$$

But

$$\int_{\mathscr{R}/\Lambda} \varphi(\mathfrak{y}+\mathfrak{x})\,\chi(\mathfrak{y})\,d\mathfrak{x} = \chi(\mathfrak{y})\,m(C).$$

Hence, on interchanging the order of integration in (15), we obtain

$$\int_{\mathscr{R}/\Lambda} m\{(C+\mathfrak{x})\cap D\}\,d\mathfrak{x} = m(C)\int_{\mathscr{R}/\Lambda} \chi(\mathfrak{y})\,d\mathfrak{y} = m(C)\,m(D).$$

Since $\mathscr{R}/\Lambda$ has measure

$$m(\mathscr{R}/\Lambda) = \int_{\mathscr{R}/\Lambda} 1\,d\mathfrak{x} = d(\Lambda),$$

the truth of Lemma 4 now follows from the continuity of $m\{(C+\mathfrak{x})\cap D\}$ and the connectedness of $\mathscr{R}/\Lambda$.

# Chapter VIII
# Successive minima

**VIII.1. Introduction.** For some purposes one requires to know not merely that a lattice $\Lambda$ has a point in a set $\mathscr{S}$, but that it has a number of linearly independent points in $\mathscr{S}$.

Let $F(\boldsymbol{x})$ be an $n$-dimensional distance function and $\Lambda$ a lattice. If for some integer $k$ in $1 \leqq k \leqq n$ and some number $\lambda$ the star-body

$$\lambda\mathscr{S}:\ F(\boldsymbol{x}) < \lambda \tag{1}$$

contains $k$ linearly independent points

$$\boldsymbol{a}_1, \ldots, \boldsymbol{a}_k \tag{2}$$

of $\Lambda$, then so does $\mu\mathscr{S}$ for any $\mu > \lambda$, since the points (2) are also in $\mu\mathscr{S}$. We define the $k$-th successive minimum $\lambda_k = \lambda_k(F, \Lambda)$ of the distance function $F$ with respect to the lattice[1] $\Lambda$ to be the lower bound of the numbers $\lambda$ such that $\lambda\mathscr{S}$ contains $k$ linearly independent lattice points. Clearly

$$\lambda_1 \leqq \lambda_2 \leqq \cdots \leqq \lambda_n. \tag{3}$$

[1] Or of the lattice with respect to the distance function.

The numbers $\lambda_1, \ldots, \lambda_n$ defined above certainly exist, since if $\boldsymbol{a}_1, \ldots, \boldsymbol{a}_n$ are any $n$ linearly independent points of $\Lambda$, then, trivially,

$$\lambda_k \leqq \lambda_n \leqq \max_{1 \leqq j \leqq n} F(\boldsymbol{a}_j).$$

In the notation of § 4 of Chapter IV we have

$$\lambda_1 = F(\Lambda) = \inf_{\substack{\boldsymbol{a} \in \Lambda \\ \neq \boldsymbol{o}}} F(\boldsymbol{a}). \tag{4}$$

Hence, by the definition of

$$\delta(F) = \sup_{\Lambda} \frac{F^n(\Lambda)}{d(\Lambda)}, \tag{5}$$

we have

$$\lambda_1^n \leqq \delta(F)\, d(\Lambda). \tag{6}$$

The remarkable inequality

$$\lambda_1 \ldots \lambda_n \leqq 2^{\frac{1}{2}(n-1)} \delta(F)\, d(\Lambda) \tag{7}$$

was discovered independently by ROGERS (1949a) and CHABAUTY (1949a); and CHABAUTY (1949a) and MAHLER (1949a) independently produced examples to show that if $\varkappa$ is any number $< 2^{\frac{1}{2}(n-1)}$ then there are distance-functions $F$ and lattices $\Lambda$ such that

$$\lambda_1 \ldots \lambda_n > \varkappa\, \delta(F)\, d(\Lambda). \tag{8}$$

We shall give the elegant proof of (7) in § 3 and give the construction of the counter-example to show that it cannot be improved in the case $n = 2$. The difficulties in extending the counter-example to $n$ dimensions are purely algebraic. It can be shown easily by means of an example that

$$\frac{\lambda_1 \ldots \lambda_n}{\delta(F)\, d(\Lambda)}$$

can be arbitrarily small, so there is no lower bound analogous to the upper bound (7) [but see (13) below for symmetric convex $F$].

The inequality (7) holds with a suitable definition of the terms not merely to star-bodies $F(\boldsymbol{x}) < 1$ but to all point sets $\mathscr{S}$ whatsoever. There have been several different definitions of the successive minima of an arbitrary set $\mathscr{S}$. We do not discuss these further, but refer the reader to the papers quoted for the extensive literature.

It was shown already by MINKOWSKI (1896a, § 51) that, when $F(\boldsymbol{x})$ is the euclidean distance $|\boldsymbol{x}|$, the inequality (7) may be replaced by

$$\lambda_1 \ldots \lambda_n \leqq \delta(F)\, d(\Lambda). \tag{9}$$

We give his proof in § 2. More generally, it has been conjectured that (9) holds for all symmetric convex distance functions. In § 4 we shall show for these $F$ that

$$\lambda_1^{n-1} \lambda_n \leqq \delta(F)\, d(\Lambda); \tag{10}$$

which is equivalent to (9) when $n=2$. The inequality (10) was apparently discovered by CHALK and ROGERS (1949a) and CHABAUTY (1949a) independently. It has been shown by WOODS (1956a and 1958b) that (9) continues to hold for $n=3$ when $F$ is symmetric and convex and for $n=2$ when $F$ is convex but not symmetric: the proof is distinctly intricate and we do not discuss it here. For general $n$ and symmetric convex $F$, RANKIN (1953a) indicates that the constant $2^{\frac{1}{2}(n-1)}$ can be replaced by a rather smaller one.

For symmetric convex functions $F$ and any $n$, there is a result going back to MINKOWSKI (1907a) which may be regarded as a substitute for the unproved conjecture that (9) holds. In our notation, MINKOWSKI's convex body Theorem II of Chapter III states that

$$\lambda_1^n V_F \leqq 2^n d(\Lambda), \tag{11}$$

where $V_F$ is the volume of $F(\boldsymbol{x})<1$; and so $\lambda_1^n V_F$ is the volume of the body $F(\boldsymbol{x})<\lambda_1$, which, by hypothesis, contains no point of $\Lambda$ except $\boldsymbol{o}$. MINKOWSKI's theorem is that in fact

$$\lambda_1 \dots \lambda_n V_F \leqq 2^n d(\Lambda). \tag{12}$$

The proof of (12) remains difficult. Simpler proofs than the original have been given by DAVENPORT (1939c) and WEYL (1942a). We follow WEYL in § 4, since the ideas introduced will be needed in Chapter XI.

For symmetric convex $F$ there is also an inequality

$$\lambda_1 \dots \lambda_n V_F \geqq \frac{2^n}{n!} d(\Lambda), \tag{13}$$

the almost trivial proof of which is also given in § 4. From (12) and (13) it follows that the product $\lambda_1 \dots \lambda_n$ is determined by $V_F$ and $d(\Lambda)$, except for a factor which is bounded in terms of $n$.

In general, it is hopeless to expect more information about successive minima than can be deduced from the formulae for the product $\lambda_1 \dots \lambda_n$. For example, let $\lambda_1, \dots, \lambda_n$ be any numbers such that

$$\lambda_1 \leqq \lambda_2 \leqq \dots \leqq \lambda_n; \quad \lambda_1 \dots \lambda_n = 1.$$

Then the lattice $\Lambda$ of points

$$(\lambda_1 u_1, \lambda_2 u_2, \dots, \lambda_n u_n) \qquad (u_1, \dots, u_n, \text{ integers})$$

has $d(\Lambda)=1$ and has successive minima $\lambda_1, \dots, \lambda_n$ with respect to the distance function

$$F(\boldsymbol{x}) = \max_{1 \leqq j \leqq n} |x_j|,$$

as is easily verified.

**VIII.1.2.** For later purposes we shall often need the following two simple lemmas.

LEMMA 1. *Let $\lambda_1, \ldots, \lambda_n$ be the successive minima of a lattice $\Lambda$ with respect to a distance function $F$ associated with a bounded star-body $F(\boldsymbol{x}) < 1$. Then there exist $n$ linearly independent points $\boldsymbol{a}_1, \ldots, \boldsymbol{a}_n \in \Lambda$ such that*

$$F(\boldsymbol{a}_j) = \lambda_j \qquad (1 \leqq j \leqq n).$$

*If $\boldsymbol{a} \in \Lambda$ and $F(\boldsymbol{a}) < \lambda_j$, then $\boldsymbol{a}$ is linearly dependent on $\boldsymbol{a}_1, \ldots, \boldsymbol{a}_{j-1}$.*

For by the definition of $\lambda_n$ there are $n$ linearly independent points of $\Lambda$ in

$$F(\boldsymbol{x}) < \lambda_n + 1. \tag{1}$$

By Lemma 2 of Chapter IV, the set (1) is bounded and so contains only a finite number of lattice points. Only these points need be considered in the definition of the $\lambda_j$. The truth of the lemma is now obvious.

LEMMA 2. *Let $\lambda_1, \ldots, \lambda_n$ be the successive minima of the distance function $F$ with respect to the lattice $\Lambda$. Then there is a basis*

$$\boldsymbol{b}_1, \ldots, \boldsymbol{b}_n$$

*of $\Lambda$ such that, for each $j = 1, 2, \ldots, n$, the inequality*

$$F(\boldsymbol{x}) < \lambda_j$$

*implies that*

$$\boldsymbol{x} = u_1 \boldsymbol{b}_1 + \cdots + u_{j-1} \boldsymbol{b}_{j-1}$$

*for integers $u_1, \ldots, u_{j-1}$.*

When $F(\boldsymbol{x}) = 0$ only for $\boldsymbol{x} = \boldsymbol{o}$, this is a trivial consequence of Lemma 1, since we may choose $\boldsymbol{b}_1, \ldots, \boldsymbol{b}_n$ so that $\boldsymbol{a}_j$ for each $j$ is dependent only on $\boldsymbol{b}_1, \ldots, \boldsymbol{b}_j$, by Theorem I of Chapter I.

Otherwise a slightly more refined argument is needed. In general, the $\lambda_j$ will not be all unequal, but there are numbers

$$\mu_1 < \mu_2 < \cdots < \mu_s,$$

for some $s$ in $1 \leqq s \leqq n$, such that

$$\lambda_k = \mu_t \quad \text{if} \quad k_{t-1} < k \leqq k_t,$$

where

$$0 = k_0 < k_1 < \cdots < k_s = n.$$

By the definition of successive minima, there is no point of $\Lambda$ with $F(\boldsymbol{a}) < \mu_1$ except, possibly[1], $\boldsymbol{o}$. Since

$$\mu_2 > \lambda_{k_1},$$

[1] For a general distance function $F(\boldsymbol{x})$ there is, of course, no reason why $\lambda_1$ should not be 0. Indeed, if $F(\boldsymbol{x}) = |x_1 \ldots x_n|^{1/n}$, we have $\lambda_1 = \cdots = \lambda_n = 0$ for the lattice $\Lambda_0$ of points with integer coordinates.

the are $k_1$ linearly independent points

$$a_1, a_2, \ldots, a_{k_1} \tag{2}$$

of $\Lambda$ in $F(x) < \mu_2$, and, since

$$\mu_2 = \lambda_{k_1+1},$$

every other point of $\Lambda$ in $F(x) < \mu_2$ is linearly dependent on them. Similarly, we may find $k_2$ linearly independent points of $\Lambda$ in $F(x) < \mu_3$ such that every other point of $\Lambda$ in $F(x) < \mu_3$ is linearly dependent on them. Since $\mu_2 < \mu_3$ we may suppose that $k_1$ of these $k_2$ points are $a_1, \ldots, a_{k_1}$ already determined. We may thus denote by

$$a_1, \ldots, a_{k_2}$$

the maximal linearly independent set of points of $\Lambda$ in $F(x) < \mu_3$ without disturbing the notation (2). And so on. In this way we obtain $k_{s-1} < n$ points

$$a_1, a_2, \ldots, a_{k_{s-1}}$$

of $\Lambda$ such that

$$F(a_j) < \mu_t \quad \text{if} \quad j \leqq k_{t-1} \qquad (t \leqq s).$$

By Theorem I of Chapter I there is a basis $b_1, \ldots, b_n$ of $\Lambda$ such that, for each $j = 1, \ldots, k_{s-1}$, the vector $a_j$ is linearly dependent on $b_1, \ldots, b_j$ only. This basis clearly has all the properties required.

**VIII.2. Spheres.** We first prove the results for spheres, since they are simplest and the treatment forms the model for what follows.

THEOREM I. *Let*

$$F_0(x) = |x| \tag{1}$$

*and let $\lambda_1, \ldots, \lambda_n$ be the successive minima of a lattice $\Lambda$ with respect to $F_0$. Then*

$$d(\Lambda) \leqq \lambda_1 \ldots \lambda_n \leqq \delta(F_0)\, d(\Lambda). \tag{2}$$

The left-hand side of (2) was substantially proved in Theorem XIII of Chapter V. We have on the one hand

$$|\det(a_1, \ldots, a_n)| = I\, d(\Lambda) \geqq d(\Lambda),$$

where $I$ is the index of $a_1, \ldots, a_n$ in $\Lambda$, and, on the other hand,

$$|\det(a_1, \ldots, a_n)| \leqq |a_1| \ldots |a_n|$$

by HADAMARD's Lemma 9 of Chapter V. If now the $a_j$ are the linearly independent vectors of $\Lambda$ with $F(a_j) = \lambda_j$ given by Lemma 1, the required inequality follows at once.

It remains to prove the second part of (2). As in the proof of Lemma 9 of Chapter V, there is a set of mutually orthogonal[1] vectors $\mathbf{c}_1, \ldots, \mathbf{c}_n$ such that

$$\mathbf{b}_j = t_{j1}\mathbf{c}_1 + \cdots + t_{jj}\mathbf{c}_j$$

for some real numbers $t_{ji}$ $(n \geqq j)$, where $\mathbf{b}_j$ is the basis given by Lemma 2. By incorporating a factor in $\mathbf{c}_i$ we may suppose, without loss of generality, that

$$|\mathbf{c}_i|^2 = 1 \qquad (1 \leqq i \leqq n).$$

Then

$$\sum_j u_j \mathbf{b}_j = \sum_i \sum_{j \geqq i} u_j t_{ji} \mathbf{c}_i;$$

and so

$$|\sum u_j \mathbf{b}_j|^2 = \sum_i \Big(\sum_{j \geqq i} u_j t_{ji}\Big)^2. \tag{3}$$

We now show that

$$\sum_i \lambda_i^{-2} \Big(\sum_{j \geqq i} u_j t_{ji}\Big)^2 \geqq 1 \tag{4}$$

for all sets of integers $\mathbf{u} \neq \mathbf{o}$. For let $u_1, \ldots, u_n$ be integers, and suppose that

$$u_J \neq 0, \qquad u_j = 0 \qquad (j > J). \tag{5}$$

Then $u_1\mathbf{b}_1 + \cdots + u_n\mathbf{b}_n$ is not dependent on $\mathbf{b}_1, \ldots, \mathbf{b}_{J-1}$; and so

$$|\sum u_j \mathbf{b}_j|^2 \geqq \lambda_J^2. \tag{5'}$$

Further, (5) implies that all the summands in (3) and (4) with $i > J$ are 0. Hence, and since $\lambda_j \leqq \lambda_J$ if $j \leqq J$, the left-hand side of (4) is

$$\sum_{i \leqq J} \lambda_i^{-2} \Big(\sum_{j \geqq i} u_j t_{ji}\Big)^2 \geqq \sum_{i \leqq J} \lambda_J^{-2} \Big(\sum_{j \geqq i} u_j t_{ji}\Big)^2 = \lambda_J^{-2} \Big|\sum_j u_j \mathbf{b}_j\Big|^2 \geqq 1,$$

by (3) and (5′). Hence if $\Lambda'$ is the lattice with basis

$$\mathbf{b}_j' = t_{j1}\lambda_1^{-1}\mathbf{c}_1 + \cdots + t_{jj}\lambda_j^{-1}\mathbf{c}_j, \qquad (1 \leqq j \leqq n),$$

we have

$$|\sum u_j \mathbf{b}_j'|^2 \geqq 1$$

for every point $\sum u_j \mathbf{b}_j' \neq \mathbf{o}$ of $\Lambda'$; that is

$$F_0(\Lambda') = |\Lambda'| \geqq 1. \tag{6}$$

On the other hand,

$$d(\Lambda') = \lambda_1^{-1} \ldots \lambda_n^{-1} d(\Lambda). \tag{7}$$

But now

$$\frac{|\Lambda'|^n}{d(\Lambda')} \leqq \sup_{\mathsf{M}} \frac{|\mathsf{M}|^n}{d(\mathsf{M})} = \delta(F_0), \tag{8}$$

[1] We say that two vectors $\mathbf{a}, \mathbf{b}$ are orthogonal if their scalar product $\mathbf{ab}$ vanishes.

by the definition of $\delta(F_0)$. The right-hand side of (2) follows now from (6), (7) and (8). This concludes the proof of Theorem I.

**VIII.2.2.** As was remarked in Chapter V, the theory of successive minima shows that the hypotheses of Theorem III and IV of Chapter V are equivalent. This we do now.

LEMMA 3. *The following two statements* A *and* B *about a set* $\mathfrak{L}$ *of n-dimensional lattices* $\Lambda$ *are equivalent, where* $\varkappa$, $K$, $\Delta_0$, $\Delta_1$ *are supposed to depend on* $\mathfrak{L}$ *but not on* $\Lambda$.

(A) *there exist* $\Delta_1 < \infty$ *and* $\varkappa > 0$ *such that* $d(\Lambda) \leqq \Delta_1$, *and* $|\Lambda| \geqq \varkappa > 0$ *for all* $\Lambda \in \mathfrak{L}$.

(B) *there exist* $\Delta_0 > 0$ *and* $K < \infty$ *such that* $d(\Lambda) \geqq \Delta_0 > 0$ *and the sphere* $|\boldsymbol{x}| \leqq K$ *contains* $n$ *linearly independent points of* $\Lambda$, *for all* $\Lambda \in \mathfrak{L}$.

If $\lambda_1, \ldots, \lambda_n$ are the successive minima of $F_0(\boldsymbol{x}) = |\boldsymbol{x}|$ with respect to $\Lambda$, then clearly (A) and (B) are equivalent to

$$\text{(A)} \quad d(\Lambda) \leqq \Delta_1, \quad \lambda_1 \geqq \varkappa > 0,$$

and

$$\text{(B)} \quad d(\Lambda) \geqq \Delta_0 > 0, \quad \lambda_n \leqq K,$$

respectively. We now use the inequality

$$d(\Lambda) \leqq \lambda_1 \ldots \lambda_n \leqq \delta(F_0)\, d(\Lambda) \tag{1}$$

of Theorem I. Suppose first that (A) holds. Then

$$d(\Lambda) \geqq \{\delta(F_0)\}^{-1} \lambda_1 \ldots \lambda_n \geqq \{\delta(F_0)\}^{-1} \varkappa^n = \Delta_0 \quad \text{(say)},$$

and

$$\lambda_n \leqq (\lambda_1 \ldots \lambda_{n-1})^{-1} \delta(F_0)\, d(\Lambda) \leqq \varkappa^{-n+1} \delta(F_0)\, \Delta_1 = K \quad \text{(say)}.$$

These are the two conditions (B).

Suppose now that (B) holds. Then

$$\lambda_1 \geqq (\lambda_n \lambda_{n-1} \ldots \lambda_2)^{-1} d(\Lambda) \geqq K^{-n+1} \Delta_0 = \varkappa \quad \text{(say)},$$

and

$$d(\Lambda) \leqq \lambda_1 \ldots \lambda_n \leqq K^n = \Delta_1 \quad \text{(say)}.$$

These are the two conditions (B).

**VIII.3. General distance-functions.** We first prove a lemma which will be required later. Just in this section we denote by $\{x\}$ the fractional part of $x$, that is, the number such that

$$0 \leqq \{x\} < 1, \quad x - \{x\} = \text{integer}.$$

LEMMA 4. *Let* $\eta_1, \ldots, \eta_n$ *be any real numbers. Then there is a number* $\eta$ *such that*

$$\sum_{1 \leqq j \leqq n} \{\eta_j - \eta\} \leqq \tfrac{1}{2}(n-1). \tag{1}$$

For any number $\xi$ we have clearly

$$\{\xi\}+\{-\xi\}=\begin{cases}0 & \text{if } \{\xi\}=0\\ 1 & \text{otherwise}\end{cases}\Bigg\}\leqq 1.$$

Hence

$$\sum_{1\leqq k\leqq n}\ \sum_{1\leqq j\leqq n}\{\eta_j-\eta_k\}=\sum_{1\leqq k<j\leqq n}(\{\eta_j-\eta_k\}+\{\eta_k-\eta_j\})\leqq \tfrac{1}{2}n(n-1).$$

Thus there is at least one $k$ such that (1) holds with $\eta=\eta_k$.

We shall require only the more specialized

COROLLARY. *Let $\mu_1, \ldots, \mu_n$ be any numbers such that*

$$0<\mu_1\leqq\mu_2\leqq\cdots\leqq\mu_n. \tag{2}$$

*Then there exists a real number $\mu>0$ and positive integers $m_1, \ldots, m_n$ such that*

$$\text{(i)}\quad m_{j+1}/m_j \quad \text{is an integer } (1\leqq j<n),$$

$$\text{(ii)}\quad \mu m_j\leqq\mu_j \qquad (1\leqq j\leqq n),$$

*and*

$$\text{(iii)}\quad \mu_1\ldots\mu_n\leqq 2^{\frac{1}{2}(n-1)}(\mu m_1)\ldots(\mu m_n).$$

We shall in fact take all the $m_j$ to be powers of 2, say

$$m_j=2^{l_j} \qquad (1\leqq j\leqq n). \tag{3}$$

Let

$$\mu_j=2^{\eta_j} \qquad (1\leqq j\leqq n) \tag{4}$$

for real numbers $\eta_j$; and let $\eta$ be the number given by Lemma 4. By subtracting an appropriate integer from $\eta$ we may suppose, by (2) and (4), that

$$\eta\leqq\eta_1\leqq\eta_2\leqq\cdots\leqq\eta_n.$$

If now $\mu=2^\eta$ and the integers $l_j$ are defined by

$$\eta_j-\eta=l_j+\{\eta_j-\eta\},$$

then the numbers $m_j$ defined by (3) clearly satisfy (i) and (ii). Further, by the lemma,

$$\prod\left(\frac{\mu_j}{\mu m_j}\right)=2^{\Sigma\{\eta_j-\eta\}}\leqq 2^{\frac{1}{2}(n-1)};$$

which is just (iii).

**VIII.3.2.** We are now in a position to prove

THEOREM II. *Let $F(\boldsymbol{x})$ be a distance-function and $\lambda_1, \ldots, \lambda_n$ its successive minima with respect to a lattice $\Lambda$. Then*

$$\lambda_1\ldots\lambda_n\leqq 2^{\frac{1}{2}(n-1)}\delta(F)\,d(\Lambda). \tag{1}$$

We denote by $\boldsymbol{b}_1, \ldots, \boldsymbol{b}_n$ the basis for $\Lambda$ given by Lemma 2. Let $\mu$ and the integers $m_j$ be given by Lemma 4, Corollary when $\mu_j = \lambda_j$ and let $\Lambda'$ be the lattice with basis

$$\boldsymbol{b}_j' = (\mu m_j)^{-1} \boldsymbol{b}_j \quad (1 \leqq j \leqq n).$$

Then

$$d(\Lambda') = \prod_j (\mu m_j)^{-1} d(\Lambda). \tag{2}$$

We now show that

$$F(\Lambda') \geqq 1. \tag{3}$$

Any point $\boldsymbol{a}$ of $\Lambda'$ other than $\boldsymbol{o}$ may be put in the shape

$$\boldsymbol{a} = u_1 \boldsymbol{b}_1' + \cdots + u_J \boldsymbol{b}_J', \quad u_J \neq 0,$$

where $u_1, \ldots, u_J$ are integers. Then

$$(\mu m_J) \boldsymbol{a} = v_1 \boldsymbol{b}_1 + \cdots + v_J \boldsymbol{b}_J$$

where

$$v_j = \frac{m_J}{m_j} u_j \quad (1 \leqq j < J), \qquad v_J = u_J \neq 0$$

are integers, since $u_j$ and $m_J/m_j$ are integers. By Lemma 2, since $v_J \neq 0$, we have

$$F(\mu m_J \boldsymbol{a}) \geqq \lambda_J.$$

Hence

$$F(\boldsymbol{a}) \geqq \frac{\lambda_J}{\mu m_J} \geqq 1.$$

This proves (3).

Finally,

$$\frac{F^n(\Lambda')}{d(\Lambda')} \leqq \delta(F), \tag{4}$$

by the definition of $\delta(F)$. The required inequality (1) now follows from (2), (3), (4) and the inequality

$$\prod_j \left(\frac{\lambda_j}{\mu m_j}\right) \leqq 2^{\frac{1}{2}(n-1)}$$

of Lemma 4, Corollary.

A rather more detailed argument shows that the sign of equality in (1) cannot hold if $F(\boldsymbol{x}) < 1$ is a bounded star-body. Then it is possible to ensure that there are not $n$ linearly independent points $\boldsymbol{a}$ of $\Lambda'$ with $F(\boldsymbol{a}) = 1$, so $\Lambda'$ cannot be critical, and there is inequality in (4). See ROGERS (1949a).

**VIII.3.3.** We now show that the constant $2^{\frac{1}{2}(n-1)}$ in Theorem II cannot be improved. For reasons of algebra we treat only the case

$$n = 2.$$

For general $n$ see MAHLER (1949a) or CHABAUTY (1949a).

We first consider a point set which is not a star-body. Denote by $\mathscr{C}'$ the set of points

$$\mathscr{C}': \quad (\pm t, 0) \qquad t \geqq 2^{\frac{1}{2}},$$

and by $\mathscr{C}''$ the set of points

$$\mathscr{C}'': \quad (s u_1, s u_2),$$

where

$$s \geqq 1; \qquad u_1, u_2, \text{ integers}, \qquad u_2 \neq 0.$$

Finally, let $\mathscr{S}$ be the set of points which belong neither to $\mathscr{C}'$ nor to $\mathscr{C}''$. Clearly $\mathscr{S}$ is open, and if any point $\boldsymbol{x}$ is in $\mathscr{S}$, then $r\boldsymbol{x}$ is in $\mathscr{S}$ for $0 \leqq |r| \leqq 1$: so $\mathscr{S}$ has some of the attributes of a star-body. We shall later modify $\mathscr{S}$ slightly to obtain a set $\mathscr{S}_\varepsilon$ which actually is a star-body.

There certainly exist $\mathscr{S}$-admissible lattices $\Lambda$, i.e. lattices having only the origin $\boldsymbol{o}$ in $\mathscr{S}$. For example the lattice $\Lambda_2$ of points

$$(2u_1, u_2),$$

where $u_1, u_2$ are integers, is $\mathscr{S}$-admissible, since if $u_2 \neq 0$ the point $(2u_1, u_2)$ is in $\mathscr{C}''$ and if $u_2 = 0$, but $u_1 \neq 0$, then $(2u_1, u_2)$ is in $\mathscr{C}'$. We shall next show that

$$\Delta(\mathscr{S}) = d(\Lambda_2) = 2: \tag{1}$$

that is that every $\mathscr{S}$-admissible lattice $\Lambda$ has determinant $d(\Lambda) \geqq 2$.

Let $\Lambda$ be any $\mathscr{S}$-admissible lattice. By MINKOWSKI's convex body Theorem II of Chapter III, there is certainly a point $\boldsymbol{x}$ other than $\boldsymbol{o}$ of $\Lambda$ in

$$|x_1| \leqq 2d(\Lambda), \qquad |x_2| \leqq \tfrac{1}{2}.$$

This point is not in $\mathscr{S}$, so must be in $\mathscr{C}'$ or $\mathscr{C}''$ and hence has the shape

$$\boldsymbol{b}_1 = (b_{11}, 0), \qquad b_{11} \neq 0.$$

We may suppose without loss of generality that $\boldsymbol{b}_1$ is primitive. There is then a vector

$$\boldsymbol{b}_2 = (b_{12}, b_{22}) \in \Lambda,$$

which, with $\boldsymbol{b}_1$, forms a basis. Hence

$$b_{11} b_{22} = \pm d(\Lambda) \neq 0.$$

Since $\boldsymbol{b}_2$ is in the $\mathscr{S}$-admissible lattice $\Lambda$, it must be in $\mathscr{C}'$ or $\mathscr{C}''$, so

$$b_{12}/b_{22} = \text{rational}.$$

Similarly $\boldsymbol{b}_1+\boldsymbol{b}_2$ is in $\mathscr{C}'$ or $\mathscr{C}''$, so $(b_{11}+b_{12})/b_{22}$ is rational; and hence

$$b_{11}/b_{22} = \text{rational}.$$

There thus exists a real number $\xi>0$ and integers $B_{11}$, $B_{12}$, $B_{22}$ such that

$$\boldsymbol{b}_1 = (\xi B_{11}, 0), \qquad \boldsymbol{b}_2 = (\xi B_{12}, \xi B_{22}).$$

Without loss of generality, $B_{11}$, $B_{12}$ and $B_{22}$ have no common divisor except $\pm 1$.

Let $v$ be the product of the primes which divide $B_{22}$ but not $B_{12}$. Put

$$B'_{12} = v B_{11} + B_{12}.$$

We wish to show that $B'_{12}$ is prime to $B_{22}$: and must distinguish two cases for the prime divisors $p$ of $B_{22}$. If $p$ does not divide $B_{12}$, then it divides $v$. If $p$ divides $B_{12}$ then it does not divide $B_{11}$, since $B_{11}$, $B_{12}$, $B_{22}$ have no non-trivial common divisor; and $p$ does not divide $v$. In both cases $p$ does not divide $B'_{12}$. Hence, on replacing $\boldsymbol{b}_2$ by $\boldsymbol{b}_2+v\boldsymbol{b}_1$, we may suppose that $B_{12}$ and $B_{22}$ have no common non-trivial divisor.

Now $\boldsymbol{b}_2$ is in the $\mathscr{S}$-admissible lattice $\Lambda$, so is in $\mathscr{C}'$ or $\mathscr{C}''$. Hence

$$|\xi| \geqq 1,$$

since $B_{12}$ and $B_{22}$ have no common factor. Similarly $\boldsymbol{b}_1$ is in $\mathscr{C}'$ or $\mathscr{C}''$, and so

$$|\xi B_{11}| \geqq 2^{\frac{1}{2}}.$$

Hence

$$\text{either } |B_{11}| = 1, \quad |\xi| \geqq 2^{\frac{1}{2}},$$
$$\text{or } |B_{11}| \geqq 2, \quad |\xi| \geqq 1.$$

In either case,

$$d(\Lambda) = |B_{11} B_{22} \xi^2| \geqq |B_{11} \xi^2| \geqq 2.$$

This concludes the proof of (1).

We denote, as usual, by $\mu\mathscr{S}$ the set of points

$$\mu\mathscr{S}: \ \mu\boldsymbol{x}, \quad \boldsymbol{x}\in\mathscr{S};$$

and by $\Lambda_0$ the lattice of points $(u_1, u_2)$ with integer $u_1$, $u_2$. Clearly if $\mu \leqq 2^{-\frac{1}{2}}$ there are no points of $\Lambda_0$ except $\boldsymbol{o}$ in $\mu\mathscr{S}$; if $2^{-\frac{1}{2}}<\mu\leqq 1$, there are only the further points $(\pm 1, 0)$ of $\Lambda_0$ in $\mu\mathscr{S}$; while if $\mu>1$, the points $(\pm 1, 0)$ and $(0, \pm 1)$ are in $\mu\mathscr{S}$. If $\mathscr{S}$ were a star-body $F(\boldsymbol{x})<1$, these statements would imply that the successive minima of $\Lambda_0$ were $\lambda_1 = 2^{-\frac{1}{2}}$, $\lambda_2 = 1$. Hence

$$\lambda_1 \lambda_2 = 2^{\frac{1}{2}} (\Delta(\mathscr{S}))^{-1} d(\Lambda_0);$$

which is the case of equality in Theorem II if $(\Delta(\mathscr{S}))^{-1}$ is written for $\delta(F)$, the two being equal for star-bodies.

It remains now to modify $\mathscr{S}$ so as to obtain a bounded star-body, in such a way that its successive minima with respect to $\Lambda_0$ remain $2^{-\frac{1}{2}}$ and 1, and so that its lattice constant is arbitrarily close to 2. We do this by replacing the lines in $\mathscr{C}'$ and $\mathscr{C}''$ by narrow wedges.

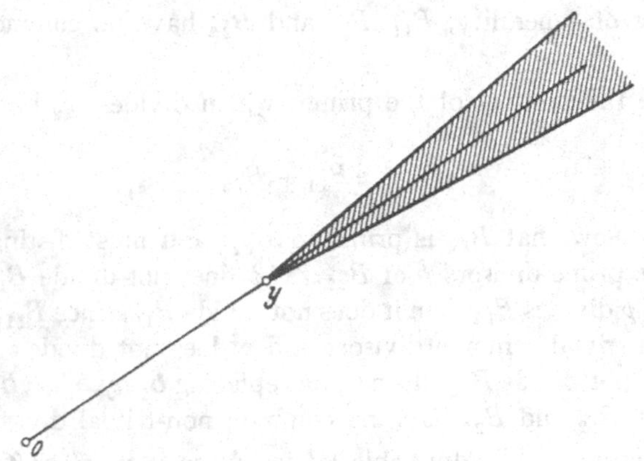


Fig. 9. The shaded portion is $\mathscr{W}_\varepsilon(\boldsymbol{y})$

Let $\varepsilon > 0$ be arbitrarily small. For any vector $\boldsymbol{y} = (y_1, y_2) \neq 0$, let $\mathscr{W}_\varepsilon(\boldsymbol{y})$ be the set of points $\boldsymbol{x}$ for which

$$\mathscr{W}_\varepsilon(y): \quad x_1 y_1 + x_2 y_2 - \varepsilon^{-1} |x_1 y_2 - x_2 y_1| \geq y_1^2 + y_2^2. \tag{2}$$

Then $\mathscr{W}_\varepsilon(\boldsymbol{y})$ is an infinite wedge having a vertex at $\boldsymbol{y}$, see Fig. 9. Its precise shape is not important. The two sides of the wedge make the small angle $\pm \operatorname{arc} \tan \varepsilon$ with the outward radius vector from $\boldsymbol{o}$ to $\boldsymbol{y}$.

Now let $\mathscr{C}'_\varepsilon$ be the set of points in $\mathscr{W}_\varepsilon(2^{\frac{1}{2}}, 0)$ and $\mathscr{W}_\varepsilon(-2^{\frac{1}{2}}, 0)$ and let $\mathscr{C}''_\varepsilon$ be the set of points in $\mathscr{W}_\varepsilon(u_1, u_2)$ for any pair of integers with $u_2 \neq 0$. Finally let $\mathscr{S}_\varepsilon$ be the set of points in

$$|\boldsymbol{x}| < \varepsilon^{-1} \tag{3}$$

which do not lie either in $\mathscr{C}'_\varepsilon$ or in $\mathscr{C}''_\varepsilon$. Clearly $\mathscr{S}_\varepsilon$ is a star-body, since there are only a finite number of the wedges composing $\mathscr{C}'_\varepsilon$ and $\mathscr{C}''_\varepsilon$ which have points in common with the disc (3). Indeed, by (2) and (3), the distance-function $F_\varepsilon(\boldsymbol{x})$ associated with $\mathscr{S}_\varepsilon$ may be written down explicitly.

Since $\mathscr{S}_\varepsilon$ is contained in $\mathscr{S}$, and since the points $(\pm 2^{\frac{1}{2}}, 0)$, $(0, \pm 1)$ are evidently still boundary points of $\mathscr{S}_\varepsilon$, at least when $\varepsilon$ is small enough, it follows that the two minima $\lambda_1$ and $\lambda_2$ of $\Lambda_0$ with respect to $F_\varepsilon(\boldsymbol{x})$ are $2^{-\frac{1}{2}}$ and 1 respectively.

Further,

$$\Delta(\mathscr{S}_\varepsilon) \leqq \Delta(\mathscr{S}) = 2.$$

Indeed

$$\lim_{\varepsilon \to 0+} \Delta(\mathscr{S}_\varepsilon) = \Delta(\mathscr{S}) = 2$$

by Theorem V of Chapter V. Hence there exist $\varepsilon$ such that

$$2^{\frac{1}{2}} \delta(F_\varepsilon)\, d(\Lambda_0) = 2^{\frac{1}{2}} (\Delta(\mathscr{S}_\varepsilon))^{-1}$$

is arbitrarily close to $\lambda_1 \lambda_2$. This shows that for $n = 2$ the constant $2^{\frac{1}{2}(n-1)} = 2^{\frac{1}{2}}$ in Theorem II cannot be improved.

**VIII.4. Convex sets.** We shall often have occasion to refer to the results of § 3.1–3.4 of Chapter IV and in particular to the properties of tac-planes.

We first need a general lemma about convex functions.

LAMMA 5. *Let $F(\boldsymbol{x})$ be a symmetric convex distance function associated with a bounded convex body $F(\boldsymbol{x}) < 1$. Let $\mathbf{c} \neq \mathbf{o}$ and let $\pi$ be the plane through the origin parallel to a tac-plane at $\mathbf{c}$ to $F(\boldsymbol{x}) < F(\mathbf{c})$. Then*

$$F(\boldsymbol{y} + \mu s \mathbf{c}) \geqq \mu F(\boldsymbol{y} + s \mathbf{c}) \tag{1}$$

*for all $\boldsymbol{y}$ in $\pi$, all real $s$, and all $\mu$ in*

$$0 < \mu < 1.$$

If $s = 0$ there is nothing to prove. Otherwise we may suppose, by homogeneity, that

$$s = 1,$$

since $s^{-1}\boldsymbol{y}$ is in $\pi$ if $\boldsymbol{y}$ is. Then

$$F(\boldsymbol{y} + \mathbf{c}) \geqq F(\mathbf{c}), \tag{2}$$

by the definition of a tac-plane. Then, by convexity,

$$\left.\begin{aligned} F(\boldsymbol{y} + \mathbf{c}) &\leqq F(\boldsymbol{y} + \mu \mathbf{c}) + F\{(1-\mu)\mathbf{c}\} \\ &= F(\boldsymbol{y} + \mu \mathbf{c}) + (1-\mu) F(\mathbf{c}). \end{aligned}\right\} \tag{3}$$

The required inequality (1) with $s = 1$ now follows from (2) and (3).

We may now prove

THEOREM III. *Let $F(\boldsymbol{x})$ be a symmetric convex distance-function associated with a bounded body $F(\boldsymbol{x}) < 1$ and let $\lambda_1, \ldots, \lambda_n$ be the successive* *minima of a lattice $\Lambda$ with respect to $F$. Then there is a lattice $\Lambda'$ with determinant*

$$d(\Lambda') = \left(\frac{\lambda_{n-1}}{\lambda_n}\right) d(\Lambda) \tag{4}$$

*and successive minima $\lambda'_j$ $(1 \leqq j \leqq n)$, where*

$$\lambda'_j = \lambda_j \qquad (1 \leqq j \leqq n-1), \qquad \lambda'_n \geqq \lambda_{n-1}. \tag{5}$$

Let $\boldsymbol{b}_1, \ldots, \boldsymbol{b}_n$ be the basis for $\Lambda$ given by Lemma 2. Let $\pm\boldsymbol{c}$ be the points on the boundary of $F(\boldsymbol{x}) < 1$ at which the tac-plane is parallel to the plane $\pi$ through $\boldsymbol{b}_1, \ldots, \boldsymbol{b}_{n-1}$ (Theorem IV of Chapter IV). Then every point in space can be uniquely put in the shape

$$\boldsymbol{x} = \boldsymbol{y} + s\boldsymbol{c}, \qquad \boldsymbol{y} \in \pi. \tag{6}$$

We put

$$\mu = \lambda_{n-1}/\lambda_n,$$

and define $\Lambda'$ to be the lattice of all points

$$\boldsymbol{y} + \mu s \boldsymbol{c}, \qquad \boldsymbol{y} + s\boldsymbol{c} \in \Lambda. \tag{7}$$

Then (4) clearly holds. If $s \neq 0$ in (7), the point $\boldsymbol{y} + s\boldsymbol{c}$ is not linearly dependent on $\boldsymbol{b}_1, \ldots, \boldsymbol{b}_{n-1}$; and so

$$F(\boldsymbol{y} + s\boldsymbol{c}) \geqq \lambda_n.$$

Hence

$$F(\boldsymbol{y} + \mu s \boldsymbol{c}) \geqq \mu \lambda_n = \lambda_{n-1} \qquad (s \neq 0) \tag{8}$$

by Lemma 5. On the other hand, the points of $\Lambda'$ with $s = 0$ are just the points of $\Lambda$ which are linearly dependent on $\boldsymbol{b}_1, \ldots, \boldsymbol{b}_{n-1}$. Hence (8) implies (5).

COROLLARY 1.

$$\lambda_1^{n-1} \lambda_n \leqq \delta(F)\, d(\Lambda).$$

For in the proof of the Theorem put

$$\mu = \lambda_1/\lambda_n$$

instead of $\lambda_{n-1}/\lambda_n$. Then (8) becomes

$$F(\boldsymbol{y} + \mu s \boldsymbol{c}) \geqq \mu \lambda_n = \lambda_1 \qquad (s \neq 0); \tag{8'}$$

so

$$F(\boldsymbol{a}') \geqq \lambda_1$$

for all $\boldsymbol{a}' \in \Lambda'$ except $\boldsymbol{o}$. That is,

$$F(\Lambda') \geqq \lambda_1.$$

Further,

$$d(\Lambda') = \mu\, d(\Lambda) = \left(\frac{\lambda_1}{\lambda_n}\right) d(\Lambda). \tag{4'}$$

But

$$F^n(\Lambda') \leqq \delta(F)\, d(\Lambda')$$

by the definition of $\delta(F)$; and then the corollary follows from (4') and (8').

COROLLARY 2.

$$\lambda_1 \ldots \lambda_n \leqq 2^{\frac{1}{2}(n-1)-\frac{1}{n}}\, \delta(F)\, d(\Lambda).$$

We only sketch the proof. By varying $\mu$ in the proof of the theorem, we may obtain a lattice $\Lambda'$ with successive minima $\lambda'_j$, where

$$\lambda'_j = \lambda_j \qquad (1 \leqq j < n), \qquad \lambda'_n = \lambda_{n-1} = \lambda'_{n-1}$$

and

$$d(\Lambda') \leqq \frac{\lambda_{n-1}}{\lambda_n}\, d(\Lambda).$$

Then

$$\frac{\lambda_1 \ldots \lambda_n}{d(\Lambda)} \leqq \frac{\lambda'_1 \ldots \lambda'_n}{d(\Lambda')}.$$

Hence it is enough to prove the corollary when $\lambda_{n-1} = \lambda_n$. But it is easy to see that if two of the numbers $\eta_j$ in Lemma 4 are equal, then the right-hand side of (1) of § 3.1 may be replaced by $\frac{1}{2}(n-1) - \frac{1}{n}$. When this improvement is inserted in the proof of Theorem II, it gives the corollary.

**VIII.4.2.** Before treating MINKOWSKI's estimates for the product of the successive minima of a bounded symmetric convex body in terms of the volume we must first prove a result, which we shall also use later, relating to convex bodies and the quotient space $\mathscr{R}/\Lambda$. We shall use the concepts and notation of Chapter VII. As was done there, we denote the points of $\mathscr{R}$ by small bold letters and those of $\mathscr{R}/\Lambda$ by small gothic letters.

THEOREM IV. *Let $F(\boldsymbol{x})$ be a convex symmetric distance-function associated with a bounded convex set*

$$\mathscr{S}: \quad F(\boldsymbol{x}) < 1 \tag{1}$$

*of volume*

$$V_F = V(\mathscr{S}). \tag{2}$$

*Let $\Lambda$ be a lattice with successive minima $\lambda_1, \ldots, \lambda_n$ with respect to $F$. For real $t > 0$ denote by $\mathsf{S}(t)$ the set of $\mathfrak{y} \in \mathscr{R}/\Lambda$ which have at least one representative $\boldsymbol{y}$ in $t\mathscr{S}$ (i.e. $F(\boldsymbol{y}) < t$). Then the measure $m\{\mathsf{S}(t)\}$ of $\mathsf{S}(t)$ satisfies the inequality*

$$m\{\mathsf{S}(t)\} \begin{cases} = t^n V_F & \text{if } t \leqq \frac{1}{2}\lambda_1 \\ \geqq (\frac{1}{2}\lambda_1) \ldots (\frac{1}{2}\lambda_J)\, t^{n-J} V_F & \text{if } \frac{1}{2}\lambda_J \leqq t \leqq \frac{1}{2}\lambda_{J+1} \\ \geqq (\frac{1}{2}\lambda_1) \ldots (\frac{1}{2}\lambda_n)\, V_F & \text{if } t \geqq \frac{1}{2}\lambda_n. \end{cases} \tag{3}$$

We first examine how the hypotheses and conclusion are affected by a homogeneous linear transformation $\boldsymbol{\tau}$. Let $\Lambda' = \boldsymbol{\tau}\Lambda$, $F'(\boldsymbol{x}) = F(\boldsymbol{\tau}^{-1}\boldsymbol{x})$. The successive minima of $\Lambda'$ with respect to $F'$ are the same as those of $\Lambda$ with respect to $F$. Clearly

$$V_{F'} = |\det(\boldsymbol{\tau})|\, V_F,$$

and by the remarks at the end of § 2.5 of Chapter VII we have

$$m'\{\boldsymbol{\tau}\, \mathsf{S}(t)\} = |\det(\boldsymbol{\tau})|\, m\{\mathsf{S}(t)\},$$

where $\boldsymbol{\tau}\mathsf{S}(t)$ is the image of $\mathsf{S}(t)$ in the natural mapping of $\mathscr{R}/\Lambda$ onto $\mathscr{R}/\boldsymbol{\tau}\Lambda$; and $m'$ is the measure in $\mathscr{R}/\boldsymbol{\tau}\Lambda$. But $\boldsymbol{\tau}\mathsf{S}(t) = \mathsf{S}'(t)$ is the set in $\mathscr{R}/\boldsymbol{\tau}\Lambda$ defined in respect of $F'$ and $\Lambda'$ as $\mathsf{S}(t)$ was defined in terms of $F$ and $\Lambda$. Hence a homogeneous linear transformation multiplies both sides of (3) by the same factor $|\det(\boldsymbol{\tau})|$.

We may therefore suppose without loss of generality that the basis $\boldsymbol{b}_1, \ldots, \boldsymbol{b}_n$ for $\Lambda$ given by Lemma 2 is just

$$\boldsymbol{b}_j = \boldsymbol{e}_j = \Bigl(\overbrace{0, \ldots, 0}^{j-1}, 1, \overbrace{0, \ldots, 0}^{n-j}\Bigr); \tag{4}$$

and that $\Lambda = \Lambda_0$ is the lattice of points with integer coordinates.

We now obtain a formula for $m\{\mathsf{S}(t)\}$ valid when

$$t \leqq \tfrac{1}{2}\lambda_{J+1}, \tag{5}$$

and $J = 1, 2, \ldots, n-1$. Let

$$\boldsymbol{x}_1 = (x_{11}, \ldots, x_{n1}), \quad \boldsymbol{x}_2 = (x_{12}, \ldots, x_{n2})$$

be two points of $F(\boldsymbol{x}) < t \leqq \frac{1}{2}\lambda_{J+1}$; and suppose that

$$\boldsymbol{x}_1 \equiv \boldsymbol{x}_2 \quad (\Lambda_0). \tag{6}$$

Then

$$F(\boldsymbol{x}_1 - \boldsymbol{x}_2) \leqq F(\boldsymbol{x}_1) + F(\boldsymbol{x}_2) < \lambda_{J+1}.$$

Since $\boldsymbol{x}_1 - \boldsymbol{x}_2 \in \Lambda_0$, we have now

$$x_{j1} = x_{j2} \quad (j > J), \tag{7}$$

by (4). Further,

$$(x_{11}, \ldots, x_{J1}) \equiv (x_{12}, \ldots, x_{J2}) \quad (\Lambda_0^J), \tag{8}$$

where $\Lambda_0^J$ is the $J$-dimensional lattice of points with integral co-ordinates. Clearly (7) and (8) together imply (6). Denote by $\mathscr{R}_J$ the $J$-dimensional euclidean space and by $m_J$ the measure in $\mathscr{R}_J/\Lambda_0^J$. For given $(n-J)$-dimensional vector $\boldsymbol{z} = (z_{J+1}, \ldots, z_n)$, denote by $\mathsf{S}_J(t, \boldsymbol{z})$ the set of points of $\mathscr{R}_J/\Lambda_0^J$ which contain representatives $(x_1, \ldots, x_J) \in \mathscr{R}_J$ such that

$$F(x_1, \ldots, x_J, z_{J+1}, \ldots, z_n) < t. \tag{9}$$

Then we assert that (5) implies

$$m\{\mathsf{S}(t)\} = \int m_J\{\mathsf{S}_J(t, \boldsymbol{z})\}\, d\boldsymbol{z} \qquad (d\boldsymbol{z} = dz_{J+1} \ldots dz_n). \tag{10}$$

In the first place, $S_J(t, \boldsymbol{z})$ certainly has a $J$-dimensional measure, since $F(\boldsymbol{x})$ is continuous by its definition as a distance function. Then, if $\boldsymbol{z}$ runs through all $(n-J)$-dimensional space and $\boldsymbol{y}=(y_1, \dots, y_J)$ runs for each $\boldsymbol{z}$ through a complete set of representatives for $S_J(t, \boldsymbol{z})$, it follows from the equivalence of (6) to (7) and (8), that

$$\boldsymbol{x}=(y_1, \dots, y_J, z_{J+1}, \dots, z_n)$$

runs through a complete set of representatives for $S(t)$. We may, for example, normalize the $\boldsymbol{y}$ by taking always $0 \leqq y_j < 1$ $(1 \leqq j \leqq J)$. This proves (10).

The next stage is to show that if $s$ is any number $\geqq 1$, so

$$0 < t \leqq st, \tag{11}$$

then

$$m_J\{S_J(st, s\boldsymbol{z})\} \geqq m_J\{S_J(t, \boldsymbol{z})\} \tag{12}$$

for any $(n-J)$-dimensional vector $\boldsymbol{z}$. This is certainly true if the right-hand side of (12) is 0. Otherwise, there is some $J$-dimensional vector $\boldsymbol{y}_0=(y_{10}, \dots, y_{J0})$ such that

$$F(\boldsymbol{y}_0, \boldsymbol{z}) < t$$

where, in an obvious notation, $(\boldsymbol{y}_0, \boldsymbol{z})=(y_{10}, \dots, y_{J0}, z_{J+1}, \dots, z_n)$: and similarly later. Let $\boldsymbol{y}$ be any $J$-dimensional vector with

$$F(\boldsymbol{y}, \boldsymbol{z}) < t.$$

Then by the convexity and homogeneity of $F(\boldsymbol{x})$, we have

$$\begin{aligned} F\{\boldsymbol{y}+(s-1)\boldsymbol{y}_0, s\boldsymbol{z}\} &= F\{(\boldsymbol{y}, \boldsymbol{z})+(s-1)(\boldsymbol{y}_0, \boldsymbol{z})\} \\ &\leqq F(\boldsymbol{y}, \boldsymbol{z})+(s-1)F(\boldsymbol{y}_0, \boldsymbol{z}) \\ &< t+(s-1)t \\ &= st. \end{aligned}$$

Hence, if $\boldsymbol{y}$ runs through a complete set of representatives for $S_J(t, \boldsymbol{z})$, then $\boldsymbol{y}+(s-1)\boldsymbol{y}_0$ runs through representatives of distinct elements[1] of $S_J(st, s\boldsymbol{z})$, when $\boldsymbol{y}_0$ is kept fixed. This proves (12).

Suppose, now, that

$$0 < t \leqq st \leqq \tfrac{1}{2}\lambda_{J+1}. \tag{13}$$

Then, by (10) and (12) we have

$$\left.\begin{aligned} m\{S(st)\} &= \int m_J\{S_J(st, \boldsymbol{z})\}\, d\boldsymbol{z} \\ &= s^{n-J} \int m_J\{S_J(st, s\boldsymbol{z})\}\, d\boldsymbol{z} \\ &\geqq s^{n-J} \int m_J\{S_J(t, \boldsymbol{z})\}\, d\boldsymbol{z} \\ &= s^{n-J}\, m\{S(t)\}, \end{aligned}\right\} \tag{14}$$

[1] Of course not every element of $S_J(st, s\boldsymbol{z})$ necessarily has a representative of the type $\boldsymbol{y}+(s-1)\boldsymbol{y}_0$. What is important, is that distinct $\boldsymbol{y}$ mod $\Lambda_0^J$ give distinct $\boldsymbol{y}+(s-1)\boldsymbol{y}_0$ mod $\Lambda_0^J$.

where in the second line we have replaced $\boldsymbol{z}$ by $s\boldsymbol{z}$ and in the third line we have used (12).

When

$$t \leqq \tfrac{1}{2}\lambda_1, \tag{15}$$

we have the simple equation

$$m\{S(t)\} = V(t\mathcal{S}) = t^n V_F, \tag{16}$$

where $t\mathcal{S}$ is the set $F(\boldsymbol{x}) < t$. Indeed, if $\boldsymbol{x}_1$ and $\boldsymbol{x}_2$ are any two points of $t\mathcal{S}$ with $\boldsymbol{x}_1 \equiv \boldsymbol{x}_2 \quad (\Lambda_0)$, we have

$$F(\boldsymbol{x}_1 - \boldsymbol{x}_2) \leqq F(\boldsymbol{x}_1) + F(\boldsymbol{x}_2) < 2t \leqq \lambda_1;$$

and so $\boldsymbol{x}_1 = \boldsymbol{x}_2$.

We may now prove (3). For $t \leqq \frac{1}{2}\lambda_1$, the truth of (3) follows from (16). Suppose that (3) is already proved for $t \leqq \frac{1}{2}\lambda_J$, where $1 \leqq J \leqq n-1$. Its truth in the range $\frac{1}{2}\lambda_J \leqq t \leqq \frac{1}{2}\lambda_{J+1}$ then follows from (13) and (14) with $t = \frac{1}{2}\lambda_J$. Finally, the truth of (2) for $t \geqq \frac{1}{2}\lambda_n$ is trivial, since $\mathcal{S}(t_1)$ includes $\mathcal{S}(t_2)$ if $t_1 \geqq t_2$: and hence $m\{S(t)\}$ increases with $t$.

**VIII.4.3.** Theorem IV provides the kernel of the proof of the following theorem of MINKOWSKI.

THEOREM V. *Let $F(\boldsymbol{x})$ be a symmetric convex distance-function associated with the bounded set $F(\boldsymbol{x}) < 1$ of volume $V_F$. Let $\lambda_1, \ldots, \lambda_n$ be the successive minima of a lattice $\Lambda$ with respect to $F$. Then*

$$\frac{2^n}{n!} d(\Lambda) \leqq \lambda_1 \ldots \lambda_n V_F \leqq 2^n d(\Lambda). \tag{1}$$

In Theorem IV the measure $m\{S(t)\}$ for any $t$ can be at most the measure of the whole space $\mathcal{R}/\Lambda$, namely $d(\Lambda)$. On applying this remark when $t = \frac{1}{2}\lambda_n$ to the inequality (3) of § 4.2 we get the right-hand side of (1) at once.

Now let $\boldsymbol{a}_1, \ldots, \boldsymbol{a}_n$ be the linearly independent points of $\Lambda$ with

$$F(\boldsymbol{a}_j) = \lambda_j$$

given by Lemma 1. By the homogeneity and convexity of $F(\boldsymbol{x})$, all points

$$\boldsymbol{x} = t_1 \boldsymbol{a}_1 + \cdots + t_n \boldsymbol{a}_n \tag{2}$$

such that

$$\sum \lambda_j |t_j| < 1 \tag{3}$$

lie in $F(\boldsymbol{x}) < 1$. Hence $V_F \geqq V'$ where $V'$ is the volume of the set of (2) subject to (3). But clearly

$$V' = \frac{2^n}{n!} |\det(\boldsymbol{a}_1, \ldots, \boldsymbol{a}_n)| = \frac{2^n I}{n!} d(\Lambda), \tag{4}$$

where $I$ is the index of $\boldsymbol{a}_1, \ldots, \boldsymbol{a}_n$ in $\Lambda$. This proves the left-hand side of (1), since $I \geqq 1$.

COROLLARY. *The index $I$ of $\boldsymbol{a}_1, \ldots, \boldsymbol{a}_n$ is at most $n!$.*

This follows from (4) and the right-hand side of (1). (Compare the proof of Theorem X of Chapter V.)

**VIII.5. Polar convex bodies.** Let $\Lambda^*$ and $F^*$ be the respective polars of the lattice $\Lambda$ (Chapter I, § 5) and the symmetric convex distance-function $F$ (Chapter IV, § 3). MAHLER (1939b) has shown that the successive minima of $\Lambda^*$ with respect to $F^*$ are determined by the successive minima of $\Lambda$ with respect to $F$ apart from factors which have bounds depending only on the dimension $n$. Thus relationship will be exploited in Chapter XI in the discussion of inhomogeneous problems and is of importance in other contexts. The theorem is, of course, closely related to Theorem VI of Chapter IV dealing with the lattice constants of mutually polar convex bodies.

THEOREM VI. *Let $\lambda_1, \ldots, \lambda_n$ be the successive minima of a lattice $\Lambda$ with respect to the symmetric convex distance-function $F$ and let $\lambda_1^*, \ldots, \lambda_n^*$ be the successive minima of the polar lattice $\Lambda^*$ with respect to the distance-function $F^*$ polar to $F$. Then*

$$1 \leqq \lambda_j \lambda_{n+1-j}^* \leqq n! \qquad (1 \leqq j \leqq n). \tag{1}$$

We attack first the left-hand inequality. By Lemma 1 there exist linearly independent vectors $\boldsymbol{a}_j, \boldsymbol{a}_j^*$ of $\Lambda$ and $\Lambda^*$ respectively such that

$$F(\boldsymbol{a}_j) = \lambda_j, \quad F^*(\boldsymbol{a}_j^*) = \lambda_j^*. \tag{2}$$

By Theorem III of Chapter IV we have

$$F(\boldsymbol{x})\, F^*(\boldsymbol{x}^*) \geqq \boldsymbol{x}\boldsymbol{x}^*$$

(scalar product) for any two vectors $\boldsymbol{x}$ and $\boldsymbol{x}^*$. On applying this $\boldsymbol{x} = \pm \boldsymbol{a}_i$, $\boldsymbol{x}^* = \pm \boldsymbol{a}_j^*$ for any pair of indices $i, j$ we have

$$\lambda_i \lambda_j^* \geqq |\boldsymbol{a}_i \boldsymbol{a}_j^*|, \tag{3}$$

since $F(\boldsymbol{x})$ and $F^*(\boldsymbol{x})$ are symmetric. But $\boldsymbol{a}_i \boldsymbol{a}_j^*$ is an integer by Lemma 5 of Chapter 1, and so

$$\text{either } \lambda_i \lambda_j^* \geqq 1 \quad \text{or} \quad \boldsymbol{a}_i \boldsymbol{a}_j^* = 0. \tag{4}$$

Let $I$ be a fixed index. The vectors $\boldsymbol{x}$ such that $\boldsymbol{x}\boldsymbol{a}_i^* = 0$ $(1 \leqq i \leqq I)$ form an $(n-I)$-dimensional subspace. Hence by the linear independence of the $\boldsymbol{a}_j$ there is some $\boldsymbol{a}_j$ with $j \leqq n+1-I$ which does not lie in this subspace; that is

$$\boldsymbol{a}_j \boldsymbol{a}_i^* \neq 0$$

for some $i, j$ with

$$i \leqq I, \quad j \leqq n+1-I.$$

Then $\lambda_i^* \leqq \lambda_I^*$, $\lambda_j \leqq \lambda_{n+1-I}$, and so, by (4),

$$\lambda_{n+1-I} \lambda_I^* \geqq \lambda_j \lambda_i^* \geqq 1.$$

Since this is true for any $I$, this gives the left-hand inequality of (1).

We now prove the right-hand inequality in the enunciation. Let $\boldsymbol{a}_j$ $(1 \leqq j \leqq n)$ be as above. Then (cf. Chapter I, § 5) there are $n$ primitive vectors $\boldsymbol{b}_j^*$ of $\Lambda^*$ such that

$$\boldsymbol{a}_i \boldsymbol{b}_j^* = 0 \qquad (i \neq j). \tag{5}$$

Since the $\boldsymbol{a}_i$ are linearly independent, the $n$ equations $\boldsymbol{a}_i \boldsymbol{x}^* = 0$ are satisfied only by $\boldsymbol{x}^* = \boldsymbol{o}$: and so

$$\boldsymbol{a}_i \boldsymbol{b}_i^* \neq 0 \qquad (1 \leqq i \leqq n). \tag{6}$$

Hence the $\boldsymbol{b}_j^*$ are linearly independent.

By Theorem III, Corollary 1 of Chapter IV, there are vectors $\boldsymbol{x}_j$ such that

$$F(\boldsymbol{x}_j) F^*(\boldsymbol{b}_j^*) = \boldsymbol{x}_j \boldsymbol{b}_j^*. \tag{7}$$

Without loss of generality

$$\boldsymbol{x}_j \boldsymbol{b}_j^* = 1 \qquad (1 \leqq j \leqq n). \tag{8}$$

The next stage is to show that for fixed $J$ the determinant $D_J$ formed from $\boldsymbol{x}_J$ and the $\boldsymbol{a}_i$ $(i \neq J)$ has absolute value at least $d(\Lambda)$. For fixed $J$, there is a basis $\boldsymbol{c}_1^*, \ldots, \boldsymbol{c}_n^*$ for $\Lambda^*$ with

$$\boldsymbol{c}_n^* = \boldsymbol{b}_J^*. \tag{9}$$

Let $\boldsymbol{c}_i$ $(1 \leqq i \leqq n)$ be the polar basis, so that, by (5) and (9),

$$\boldsymbol{a}_i = \sum_{1 \leqq j \leqq n-1} v_{ij} \boldsymbol{c}_j \qquad (i \neq J) \tag{10}$$

for some integers $v_{ij}$. Further,

$$\boldsymbol{x}_J = \pm \boldsymbol{c}_n + \sum_{1 \leqq j \leqq n-1} t_j \boldsymbol{c}_j$$

for some real numbers $t_j$, by (8) and (9). Hence

$$D_J = |\det(\boldsymbol{a}_1, \ldots, \boldsymbol{a}_{J-1}, \boldsymbol{x}_J, \boldsymbol{a}_{J+1}, \ldots, \boldsymbol{a}_n)| = |\det(v_{ij})_{\substack{i \neq J \\ j \neq n}}| \, |\det(\boldsymbol{c}_1, \ldots, \boldsymbol{c}_n)|.$$

The first factor here is a non-zero integer since the $\boldsymbol{a}_i$ are linearly independent; the second factor is just $d(\Lambda)$. Thus

$$D_J \geqq d(\Lambda), \tag{11}$$

as required.

The points

$$t_J \boldsymbol{x}_J + \sum_{i \neq J} t_i \boldsymbol{a}_i$$

with

$$|t_J| F(\boldsymbol{x}_J) + \sum_{i \neq J} |t_i| F(\boldsymbol{a}_i) < 1$$

lie all in the set $F(\boldsymbol{x}) < 1$ of volume $V_F$. This set of points has volume

$$\frac{2^n}{n!} D_J \left\{F(\boldsymbol{x}_J) \prod_{i \neq J} F(\boldsymbol{a}_i)\right\}^{-1}.$$

Hence, and by (11),

$$V_F F(\boldsymbol{x}_J) \prod_{i \neq J} F(\boldsymbol{a}_i) \geqq \frac{2^n}{n!} d(\Lambda). \tag{12}$$

But $F(\boldsymbol{a}_i) = \lambda_i$ and $V_F \prod_i \lambda_i \leqq 2^n d(\Lambda)$ by Theorem V, so

$$F(\boldsymbol{x}_J) \geqq \lambda_J / n!,$$

and finally

$$F^*(\boldsymbol{b}_J^*) \leqq n! \, \lambda_J^{-1}, \tag{13}$$

by (7), (8). The inequality (13) holds for each integer $J$ and for the independent vectors $\boldsymbol{b}_j^*$ of $\Lambda^*$.

Now $\lambda_1 \leqq \lambda_2 \leqq \cdots \leqq \lambda_n$ and so, for each integer $J$, there are the $n+1-J$ linearly independent vectors $\boldsymbol{b}^* = \boldsymbol{b}_j^*$ $(J \leqq j \leqq n)$ of $\Lambda^*$ such that $F(\boldsymbol{b}^*) \leqq n! \, \lambda_J^{-1}$. By the definition of $\lambda_{n+1-J}^*$ it follows that

$$\lambda_{n+1-J}^* \leqq n! \, \lambda_J^{-1}.$$

This is the required inequality and so concludes the proof of the theorem.

The applications of the theorem are usually only qualitative so the magnitude of the factor $n!$ on the right-hand side is usually irrelevant. MAHLER (1939b) showed that the weaker inequality

$$\lambda_J \lambda_{n+1-J}^* \leqq (n!)^2$$

can be deduced very simply from the left-hand inequalities, Theorem V and Theorem VI of Chapter IV. We have

$$V_F \lambda_1 \ldots \lambda_n \leqq 2^n d(\Lambda),$$
$$V_{F^*} \lambda_1^* \ldots \lambda_n^* \leqq 2^n d(\Lambda^*),$$

and so

$$V_F V_F^* \prod_j \lambda_j \lambda_{n+1-j}^* \leqq 2^{2n} d(\Lambda)\, d(\Lambda^*).$$

Now

$$d(\Lambda)\, d(\Lambda^*) = 1$$

by Lemma 5 of Chapter I, and

$$V_F V_{F^*} \geqq \frac{2^{2n}}{(n!)^2}$$

by Theorem VI of Chapter IV. Further,

$$\prod_j \lambda_j \lambda_{n+1-j}^* \geqq \lambda_J \lambda_{n+1-J}^*$$

for any particular $J$ by the left-hand inequality of Theorem VI. Hence $\lambda_J \lambda_{n-J}^* \leqq (n!)^2$, as required.

**VIII.5.2.** In Chapter XI we shall also need the following result of which the proof is similar to that of Theorem VI.

THEOREM VII. *Let $F(\boldsymbol{x})$ and $F^*(\boldsymbol{x})$ be polar symmetric convex distance functions. Let $\boldsymbol{b}_1, \ldots, \boldsymbol{b}_n$ be any basis of a lattice $\Lambda$ and $\boldsymbol{b}_1^*, \ldots, \boldsymbol{b}_n^*$ the polar basis of the polar lattice $\Lambda^*$. Then*

$$2^n d(\Lambda)\, F^*(\boldsymbol{b}_J^*) \leqq n!\, V_F \prod_{j \neq J} F(\boldsymbol{b}_j) \tag{1}$$

*for each integer $J = 1, 2, \ldots, n$.*

For the deduction of (12) from (5) and (6) in § 5.1 did not depend on the fact that the $\boldsymbol{a}_j$ gave the successive minima for $F$. Hence (12) of § 5.1 remains true if $\boldsymbol{b}_j$ is read for $\boldsymbol{a}_j$, where $\boldsymbol{x}_j$ is to be given by (7) and (8) of § 5.1. On substituting (7) and (8) into (12) of § 5.1 the required result follows.

COROLLARY [M. RIESZ (1936a), K. MAHLER (1939a, b)]. *Let $\lambda_1, \ldots, \lambda_n$ be the successive minima of $F$ with respect to $\Lambda$. Then the basis $\boldsymbol{b}_j$ may be chosen so that*

$$\left.\begin{aligned} F(\boldsymbol{b}_1) &= \lambda_1 \\ 2F(\boldsymbol{b}_j) &\leqq j\lambda_j \qquad (2 \leqq j \leqq n) \end{aligned}\right\} \tag{2}$$

*and*

$$F(\boldsymbol{b}_j)\, F^*(\boldsymbol{b}_j^*) \leqq (\tfrac{1}{2})^{n-1} (n!)^2. \tag{3}$$

The existence of a basis $\boldsymbol{b}_j$ satisfying (2) follows at once from Lemma 8 of Chapter V on defining $\boldsymbol{a}_j$ there to be the linearly independent points with $F(\boldsymbol{a}_j) = \lambda_j$. But now on multiplying (1) by $F(\boldsymbol{b}_J)$ and using Theorem V, we have

$$2^n d(\Lambda)\, F(\boldsymbol{b}_J)\, F^*(\boldsymbol{b}_J^*) \leqq n!\, V_F \prod_{1 \leqq j \leqq n} F(\boldsymbol{b}_j) \leqq (\tfrac{1}{2})^{n-1} (n!)^2\, V_F \prod_{1 \leqq j \leqq n} \lambda_j \leqq 2 (n!)^2 d(\Lambda).$$

# Chapter IX
# Packings

**IX.1. Introduction.** If $\mathscr{S}$ is any $n$-dimensional set and $\boldsymbol{y}$ a point, we denote by $\mathscr{S}+\boldsymbol{y}$ the set of points

$$\mathscr{S}+\boldsymbol{y}: \quad \boldsymbol{x}+\boldsymbol{y}, \quad \boldsymbol{x}\in\mathscr{S}. \tag{1}$$

By a packing of $\mathscr{S}$ in some other set $\mathscr{T}$ we shall mean a collection of sets

$$\mathscr{S}_r = \mathscr{S}+\boldsymbol{y}_r, \tag{2}$$

each of which is contained in $\mathscr{T}$, and no two of which have points in common. If $\mathscr{T}$ is the whole space $\mathscr{R}$ we speak simply of a packing of $\mathscr{S}$. If the $\boldsymbol{y}_r$ in (2) run through the points of a lattice $\Lambda$ then we say that the packing is a lattice packing. In this chapter we examine the consequences of these ideas for the geometry of numbers. This chapter may be regarded as a sequel of Chapter III but we shall also require some of the general properties of convex bodies discussed in Chapter IV. We shall find that the general theory of packings is relevant even to strictly lattice-theoretic problems.

There is an admirable account of the theory of packing in Fejes Tóth (1953a) and a conspectus of the more important results in Bambah and Rogers (1952a).

**IX.1.2.** The three following theorems show the relevance of packings to the theory of Chapter III. We give the simple proofs here

Theorem I. *A necessary and sufficient condition that the lattice $\Lambda$ give a packing of the set $\mathscr{S}$ is that no difference $\boldsymbol{x}_1-\boldsymbol{x}_2$ of two distinct points of $\mathscr{S}$ belong to $\Lambda$.*

Suppose, first, that $\boldsymbol{x}_1-\boldsymbol{x}_2=\boldsymbol{a}\in\Lambda$. Then the sets $\mathscr{S}=\mathscr{S}+\boldsymbol{o}$ and $\mathscr{S}+\boldsymbol{a}$ both contain the point $\boldsymbol{x}_1=\boldsymbol{x}_2+\boldsymbol{a}$, and so overlap. Conversely, suppose that the sets $\mathscr{S}+\boldsymbol{a}_1$ and $\mathscr{S}+\boldsymbol{a}_2$ have the point $\boldsymbol{y}$ in common where $\boldsymbol{a}_1, \boldsymbol{a}_2$ are in $\Lambda$. Then the two points $\boldsymbol{y}-\boldsymbol{a}_1=\boldsymbol{x}_1$, $\boldsymbol{y}-\boldsymbol{a}_2=\boldsymbol{x}_2$ are in $\mathscr{S}$, and their difference $\boldsymbol{a}_2-\boldsymbol{a}_1$ is in $\Lambda$.

Blichfeldt's Theorem I of Chapter III shows that

$$V(\mathscr{S}) \leqq d(\Lambda)$$

whenever $\Lambda$ packs $\mathscr{S}$. The following theorem shows when the sign of equality can occur. To avoid irrelevant topological considerations we confine attention to rather special sets $\mathscr{S}$.

Theorem II. *Let $\mathscr{S}$ be a bounded open star-body and $\Lambda$ a lattice with*

$$V(\mathscr{S}) = d(\Lambda). \tag{1}$$

(A) *If $\Lambda$ packs $\mathscr{S}$, then every point in space either belongs to precisely one set $\mathscr{S}+\boldsymbol{a}$, $\boldsymbol{a}\in\Lambda$ and is not a boundary point of any other $\mathscr{S}+\boldsymbol{a}$, or is a boundary point of at least two such sets $\mathscr{S}+\boldsymbol{a}$.*

(B) *If every point of space either belongs to or is a boundary point of at least one set $\mathscr{S}+\boldsymbol{a}$, then $\Lambda$ packs $\mathscr{S}$.*

By hypothesis, there is an $R$ such that $\mathscr{S}$ is contained in

$$|\boldsymbol{x}|<R.$$

We now prove (A). Suppose, first, that $\Lambda$ packs $\mathscr{S}$ and that there is some point $\boldsymbol{y}$ which is not in or on the boundary of any $\mathscr{S}+\boldsymbol{a}$, $\boldsymbol{a}\in\Lambda$. We may choose $\varepsilon$ in the range $0<\varepsilon<1$, so small, that the sphere $\mathscr{S}_1$ of points $\boldsymbol{x}$ with

$$\mathscr{S}_1\colon\ |\boldsymbol{x}-\boldsymbol{y}|<\varepsilon \tag{2}$$

is completely outside the finite number of bodies $\mathscr{S}+\boldsymbol{a}$ with $\boldsymbol{a}\in\Lambda$ and $|\boldsymbol{a}-\boldsymbol{y}|<R+1$. By the definition of $R$, the set $\mathscr{S}+\boldsymbol{a}$ certainly contains no points $\boldsymbol{x}$ of $\mathscr{S}_1$ if $|\boldsymbol{a}-\boldsymbol{y}|\geqq R+1$. We may suppose, further, that $\varepsilon$ is so small that the only point of $\Lambda$ in $|\boldsymbol{x}|<2\varepsilon$ is $\boldsymbol{o}$. Let

$$\mathscr{S}'=\mathscr{S}\cup\mathscr{S}_1$$

be the set of points belonging to either $\mathscr{S}$ or $\mathscr{S}_1$. Clearly, if $\boldsymbol{x}_1$ and $\boldsymbol{x}_2$ are distinct points of $\mathscr{S}'$ the difference $\boldsymbol{x}_1-\boldsymbol{x}_2$ cannot belong to $\Lambda$. Hence

$$V(\mathscr{S}')\leqq d(\Lambda)$$

by BLICHFELDT's Theorem I of Chapter III. But then $V(\mathscr{S})<V(\mathscr{S}')$, which contradicts the hypothesis. Suppose now that the hypotheses of (A) are fulfilled and that there is a point $\boldsymbol{y}$ which is on the boundary of precisely one $\mathscr{S}+\boldsymbol{a}$, $\boldsymbol{a}\in\Lambda$. Suppose, without loss of generality, that $\boldsymbol{y}$ is on the boundary of $\mathscr{S}$. As before, there is an $\varepsilon>0$ such that $\mathscr{S}_1$ defined in (2) contains no point or boundary point of any $\mathscr{S}+\boldsymbol{a}$ with $\boldsymbol{a}\in\Lambda$, $\boldsymbol{a}\neq\boldsymbol{o}$. But then the point $(1+\eta)\boldsymbol{y}$, for sufficiently small $\eta>0$, is in $\mathscr{S}_1$ and is not a point or boundary point of $\mathscr{S}$. On taking $(1+\eta)\boldsymbol{y}$ instead of $\boldsymbol{y}$, we thus have the case first considered. No point can belong to more than one $\mathscr{S}+\boldsymbol{a}$, $\boldsymbol{a}\in\Lambda$ by the definition of a packing. If $\boldsymbol{y}$ were a point of $\mathscr{S}+\boldsymbol{a}$ and a boundary point of $\mathscr{S}+\boldsymbol{b}$, where $\boldsymbol{a}$, $\boldsymbol{b}\in\Lambda$, then there would be points in the neighbourhood of $\boldsymbol{y}$ in both $\mathscr{S}+\boldsymbol{a}$ and $\mathscr{S}+\boldsymbol{b}$, since $\mathscr{S}$ is open. This completes the proof of (A).

We now prove (B). If $\mathscr{S}$ is not packed, then, by Theorem I, there are points $\boldsymbol{x}_1$ and $\boldsymbol{x}_2$ such that

$$\boldsymbol{o}\neq\boldsymbol{a}_0=\boldsymbol{x}_1-\boldsymbol{x}_2\in\Lambda.$$

Since $\mathscr{S}$ is open, by hypothesis, there is an $\varepsilon>0$ such that both spheres

$$\mathscr{S}_1: \quad |\boldsymbol{x}-\boldsymbol{x}_1|<\varepsilon,$$
$$\mathscr{S}_2: \quad |\boldsymbol{x}-\boldsymbol{x}_2|<\varepsilon$$

are contained in $\mathscr{S}$. We may suppose that $\varepsilon$ is so small that $\mathscr{S}_1$ and $\mathscr{S}_2$ have no points in common. Let $\mathscr{S}'$ be the set of points which belong to $\mathscr{S}$ but not to $\mathscr{S}_1$. Clearly every point in space is either an inner point or a boundary point of $\mathscr{S}'+\boldsymbol{a}$ for some $\boldsymbol{a}\in\Lambda$, since every point of $\mathscr{S}$ is either in $\mathscr{S}'$ or in $\mathscr{S}'+\boldsymbol{a}_0$. Let $\overline{\mathscr{S}'}$ be the closure of $\mathscr{S}'$. Since $\mathscr{S}$ is a star-body and $\mathscr{S}_1$ is a subset of $\mathscr{S}$, we have

$$V(\overline{\mathscr{S}'})=V(\mathscr{S}')<V(\mathscr{S})=d(\Lambda).$$

This is a contradiction with the Corollary to Theorem I of Chapter III since we are supposing that every point, and so every point of the fundamental parallelogram, is of the form $\boldsymbol{z}+\boldsymbol{a}$ where $\boldsymbol{z}\in\overline{\mathscr{S}'}$ and $\boldsymbol{a}\in\Lambda$. This completes the proof of Theorem II.

THEOREM III. *A necessary and sufficient condition that the convex symmetric set $\mathscr{S}$ admit the lattice $\Lambda$ is that $\Lambda$ give a lattice packing of $\frac{1}{2}\mathscr{S}$.*

This follows at once from Theorem I and Theorem II, Corollary of Chapter III.

We shall consider only packings of convex sets $\mathscr{S}$ in what follows, and we shall suppose that $\mathscr{S}$ is symmetric, whenever this gives any simplification of proofs or results.

**IX.1.3.** MINKOWSKI's convex body Theorem II of Chapter III states that if $\mathscr{S}$ is an $n$-dimensional symmetric convex body of volume $V(\mathscr{S})>2^n d(\Lambda)$, then the lattice $\Lambda$ cannot be $\mathscr{S}$-admissible. In § 2 we discuss when a lattice $\Lambda_c$ can be admissible for a convex symmetric body of volume $2^n d(\Lambda_c)$. Of course then by MINKOWSKI's convex body theorem we have

$$\Delta(\mathscr{S})=2^{-n}V(\mathscr{S}), \tag{1}$$

and the lattice $\Lambda_c$ is critical.

Even when $\mathscr{S}$ is the cube $|x_j|<1$ $(1\leqq j\leqq n)$, the critical lattices were not completely known until HAJÓS (1942) confirmed on old conjecture of MINKOWSKI. We quote the result here, but shall not prove it since it depends on considerations of group-algebra remote from the other topics in the book.

THEOREM IV. *A necessary and sufficient condition that a lattice $\Lambda$ be critical for $|x_j|<1$ $(1\leqq j\leqq n)$ is that, after a suitable permutation of* *the axes of co-ordinates, it has a basis of the shape*

$$\left.\begin{aligned} \boldsymbol{b}_1 &= (1, 0, \quad \ldots, \quad 0) \\ \boldsymbol{b}_2 &= (b_{12}, 1, 0, \quad \ldots, \quad 0) \\ \boldsymbol{b}_2 &= (b_{13}, b_{23}, 1, 0, \ldots, 0) \\ &\cdots\cdots\cdots\cdots \\ \boldsymbol{b}_n &= (b_{1n}, \ldots, b_{n-1,n}, 1). \end{aligned}\right\}$$

The reader will readily verify that a lattice of the stated kind has determinant 1 and no points other than $\boldsymbol{o}$ in $|x_j| < 1$ $(1 \leqq j \leqq n)$. For the proof of the converse the reader is referred to the original paper of HAJÓS (1942) and to RÉDEI (1955a) where there are references to the considerable amount of later literature. We proved HAJÓS' Theorem for $n=2$ incidentally as Lemma 7 of Chapter III.

MINKOWSKI (1896a) showed that any convex symmetric set $\mathscr{S}$ with $\Delta(\mathscr{S}) = 2^{-n} V(\mathscr{S})$ must have very special properties, for example that it must be a polyhedron bounded by at most $2^n - 1$ pairs of hyperplane faces. We prove this in § 2.

**IX.1.4.** VORONOÏ (1908a) suggested a simple way of finding open convex symmetric sets $\mathscr{S}$ such that

$$V(\mathscr{S}) = d(\Lambda)$$

and which are packed by a given lattice $\Lambda$. If $g(\boldsymbol{x})$ is any positive definite quadratic form, the set of points such that

$$g(\boldsymbol{x}) < \inf_{\substack{\boldsymbol{a} \in \Lambda \\ \boldsymbol{a} \neq \boldsymbol{o}}} g(\boldsymbol{x} + \boldsymbol{a})$$

has this property. The condition

$$g(\boldsymbol{x}) < g(\boldsymbol{x} + \boldsymbol{a}),$$

for any given $\boldsymbol{a}$, is linear in the coefficients $x_1, \ldots, x_n$; so $\mathscr{S}$ is convex. $\mathscr{S}$ is clearly symmetric. It is not difficult to verify that $\mathscr{S}$ is, in fact, bounded; and that then the infimum in (1) may be replaced by a minimum over a finite number of $\boldsymbol{a}$ depending on $\Lambda$ and the function $g(\boldsymbol{x})$, but not otherwise on the individual $\boldsymbol{x}$. Not every open convex symmetric $\mathscr{S}$ with $V(\mathscr{S}) = 2^n d(\Lambda)$ for which $\Lambda$ is admissible may be obtained in this way, but VORONOÏ was able to show that all, in a sense, sufficiently general such $\mathscr{S}$ could be. Unfortunately the excluded cases include some of great interest, such as those covered by HAJÓS's Theorem IV.

We do not discuss the case of general dimension $n$ in this book but deal in detail with $n=2$ in § 3. As a byproduct we obtain a result about the inhomogeneous problem for definite binary quadratics.

**IX.1.5.** Let $\mathscr{K}$ be any open 2-dimensional set and $\mathscr{S}$ the 3-dimensional set of points

$$\mathscr{C}: \quad (x_1, x_2, x_3) \quad (x_1, x_2) \in \mathscr{K} \quad |x_3| < 1; \tag{1}$$

that is, a generalized cylinder of height 2 and with cross-section $\mathscr{K}$. Then a plane section

$$x_3 = \text{constant}$$

of a lattice packing of $\mathscr{C}$ gives, in an obvious way, a packing of $\mathscr{K}$, but not necessarily a lattice packing. The idea of using non-lattice packings in this context is apparently MAHLER's (1946g). In this way we are led to consider non-lattice packings of 2-dimensional sets. This we do in § 5, after some preparatory lemmas in § 4. It turns out, as was proved independently by ROGERS (1951a) and FEJES TÓTH (1950a) [see also FEJES TÓTH (1953a)] that, in a sense which will be made precise, no packing of convex symmetric open sets is closer than the closest lattice packing. It appears unlikely that this result extends to higher dimensions. For a discussion of this point see FEJES TÓTH (1953a).

In § 6 we use the packing results to show that

$$\Delta(\mathscr{C}) = \Delta(\mathscr{K}), \tag{2}$$

when $\mathscr{K}$ is convex and symmetric and $\mathscr{C}$ is defined in (1). This result was originally proved independently by CHALK and ROGERS (1948a) and YEH (1948a). An example was given by ROGERS (1949b) which shows that (2) need not hold when $\mathscr{K}$ is a symmetric non-convex 2-dimensional star-body, and DAVENPORT and ROGERS (1950b) gave an example to show that then the ratio $\Delta(\mathscr{C})/\Delta(\mathscr{K})$ may be arbitrary small. VARNAVIDES (1948a) has shown that (2) continues to hold in one interesting non-convex case. It is trivial that $\Delta(\mathscr{C}) \leqq \Delta(\mathscr{K})$ for any $\mathscr{K}$, since if $\Lambda$ is a 2-dimensional admissible lattice for $\mathscr{C}$, the 3-dimensional lattice of points

$$(x_1, x_2, x_3) \quad (x_1, x_2) \in \Lambda, \quad x_3 = \text{integer}$$

is clearly admissible for $\mathscr{C}$ and has the same determinant as $\Lambda$.

There is an interesting unsolved problem in this connection. Let $\mathscr{K}_1$ and $\mathscr{K}_2$ be convex symmetric bodies in $n_1$ and $n_2$ dimensions respectively and let $\mathscr{C}$ be the $(n_1+n_2)$-dimensional "topological product" of $\mathscr{K}_1$ and $\mathscr{K}_2$; that is the set of points

$$\boldsymbol{x} = (\boldsymbol{y}, \boldsymbol{z}), \quad \boldsymbol{y} \in \mathscr{K}_1, \ \boldsymbol{z} \in \mathscr{K}_2.$$

The argument above shows that

$$\Delta(\mathscr{C}) \leqq \Delta(\mathscr{K}_1)\, \Delta(\mathscr{K}_2). \tag{3}$$

Can it ever happen that there is strict inequality here? The cylinder is, of course, the case $n_1=2$, $n_2=1$. WOODS (1958a) has shown that there is equality in (3) for $n_1=3$, $n_2=1$ when $\mathscr{K}_1$ is a 3-dimensional sphere.

**IX.1.6.** In §§ 7, 8 are given applications by BLICHFELDT techniques based on packing considerations, or at least BLICHFELDT's Theorem I of Chapter III, to the estimation of the lattice constants of the sets

$$x_1^2+\cdots+x_n^2<1$$

and

$$|x_1\ldots x_n|<1$$

respectively. The relationship of BLICHFELDT's results to later work will be discussed there.

**IX.2. Sets with $V(\mathscr{S})=2^n\Delta(\mathscr{S})$.** We prove here the following result of MINKOWSKI (1896a).

THEOREM V. *Let $\mathscr{S}$ be an open symmetric n-dimensional convex set which admits a lattice $\Lambda$ with $d(\Lambda)=2^{-n}V(\mathscr{S})$. Then $\mathscr{S}$ is defined by $m\leqq 2^n-1$ inequalities*[1] *of the shape*

$$\Big|\sum_j f_{ij}x_j\Big|<1. \tag{1}$$

*For each $I$ $(1\leqq I\leqq m)$ the planes*

$$\sum_j f_{Ij}x_j=\pm 1 \tag{2}$$

*give an $(n-1)$-dimensional pair of faces of $\mathscr{S}$, and each such face contains a point of $\Lambda$ as an inner point (i.e. for each $I$ there are lattice points satisfying* (2), *and* (1) *for $i\neq I$).*

By Lemma 4 of Chapter IV the set $\mathscr{S}$ is bounded since $0<V(\mathscr{S})<\infty$.

By Theorems II and III, every point either belongs to precisely one set

$$\mathscr{T}(\boldsymbol{a}):\ \tfrac{1}{2}\mathscr{S}+\boldsymbol{a},\quad \boldsymbol{a}\in\Lambda,$$

in which case it is not a boundary point of any $\mathscr{T}(\boldsymbol{b})$, $\boldsymbol{b}\in\Lambda$ or it is a boundary point of at least two $\mathscr{T}(\boldsymbol{a})$. Hence every boundary point of $\mathscr{T}(\boldsymbol{o})=\frac{1}{2}\mathscr{S}$ is also a boundary point of some $\mathscr{T}(\boldsymbol{a})$, $\boldsymbol{a}\neq\boldsymbol{o}$: and, by the boundedness of $\mathscr{S}$, only a finite number of $\boldsymbol{a}$ can occur in this way.

We note now that, for fixed $\boldsymbol{a}$, the set of points which are on the boundary of both $\mathscr{T}(\boldsymbol{o})$ and $\mathscr{T}(\boldsymbol{a})$ is convex. For if $\boldsymbol{x}, \boldsymbol{y}$ are two such points, the point

$$t\boldsymbol{x}+(1-t)\boldsymbol{y}\qquad (0<t<1) \tag{3}$$

[1] In fact there are at most $3^n-3$ faces [GROEMER, MZ **79** (1962) 364–375], and both $\mathscr{S}$ and its faces are centrally symmetric. Estimate $3^n-1$ is easy (HLAWKA, 1949a). Both GROEMER and HLAWKA give generalizations.

is certainly either a boundary point of $\mathcal{T}(\boldsymbol{a})$ or belongs to $\mathcal{T}(\boldsymbol{a})$ by convexity, and similarly for $\mathcal{T}(\boldsymbol{o})$. Hence (3) is a boundary point of both $\mathcal{T}(\boldsymbol{o})$ and $\mathcal{T}(\boldsymbol{a})$ by Theorem II.

In particular, if $\boldsymbol{z}$ is common to the boundaries of $\mathcal{T}(\boldsymbol{o})$ and $\mathcal{T}(\boldsymbol{a})$ then so is[1] $\boldsymbol{a}-\boldsymbol{z}$ by the symmetry of $\mathcal{S}$. Hence so is also

$$\tfrac{1}{2}\boldsymbol{a} = \tfrac{1}{2}\boldsymbol{z} + \tfrac{1}{2}(\boldsymbol{a}-\boldsymbol{z})$$

a common boundary point.

Denote by

$$\pm \boldsymbol{c}_k \qquad (1 \leqq k \leqq K) \tag{4}$$

the points $\boldsymbol{c}$ of $\Lambda$ such that the boundary of $\mathcal{T}(\boldsymbol{c})$ has $n$ linearly independent points in common[2] with that of $\mathcal{T}(\boldsymbol{o})$, and denote by

$$\pm \boldsymbol{b}_l \qquad (1 \leqq l \leqq L) \tag{5}$$

the remaining points $\boldsymbol{b}$ of $\Lambda$ such that the boundaries of $\mathcal{T}(\boldsymbol{b})$ and $\mathcal{T}(\boldsymbol{o})$ have points in common. From what has just been shown, the points common to the boundaries of $\mathcal{T}(\boldsymbol{o})$ and $\mathcal{T}(\boldsymbol{b}_l)$ lie in a linear subspace of dimension at most $n-2$ (not, of course, necessarily, passing through the origin. In fact, it cannot pass through the origin).

We show now that every boundary point $\boldsymbol{z}$ of $\mathcal{T}(\boldsymbol{o})$ is also a boundary point of a $\mathcal{T}(\boldsymbol{c}_k)$. The set of boundary points $\boldsymbol{x}$ of $\mathcal{T}(\boldsymbol{o})$ in any neighbourhood

$$|\boldsymbol{z}-\boldsymbol{x}| < \varepsilon \tag{6}$$

of $\boldsymbol{z}$ is $(n-1)$-dimensional, and so cannot be exhausted by the at most $(n-2)$-dimensional sets of boundary points in common with the $\mathcal{T}(\boldsymbol{b}_l)$. Hence there must be points in (6) which are common boundary points of $\mathcal{T}(\boldsymbol{o})$ and a $\mathcal{T}(\boldsymbol{c}_k)$. Thus $\boldsymbol{z}$ itself is a boundary point with a $\mathcal{T}(\boldsymbol{c}_k)$ as required, since there are only a finite number of $\boldsymbol{c}_k$.

[More precisely, let $\mathcal{S}$ be $F(\boldsymbol{x}) < 1$, where $F(\boldsymbol{x})$ is a distance-function. We may suppose, without loss of generality, that $\boldsymbol{z} = (1, 0, \ldots, 0)$. If $\boldsymbol{z}$ is common to the boundary of $\mathcal{T}(\boldsymbol{o})$ and $\mathcal{T}(\boldsymbol{b}_l)$, the common boundary points of $\mathcal{T}(\boldsymbol{o})$ and $\mathcal{T}(\boldsymbol{b}_l)$ satisfy at least two distinct equations

$$r_1(x_1 - 1) + \sum_{j \geqq 2} r_j x_j = 0,$$

and so at least one equation

$$\sum_{j \geqq 2} s_j x_j = 0.$$

There is an equation of this type for each $l$ for which $\boldsymbol{z}$ is on the boundary of $\mathcal{T}(\boldsymbol{b}_l)$. If $x_2, \ldots, x_n$ are chosen so as not to satisfy any of these conditions, and arbitrarily

---

[1] $-\boldsymbol{z}$ is on the boundary of $\mathcal{T}(\boldsymbol{o})$, by symmetry, and then $\boldsymbol{a}-\boldsymbol{z}$ is on the boundary of $\mathcal{T}(\boldsymbol{a})$.

[2] That is, the common boundary of $\mathcal{T}(\boldsymbol{o})$ and $\mathcal{T}(\boldsymbol{c})$ is a convex $(n-1)$-dimensional set with centre $\tfrac{1}{2}\boldsymbol{c}$, by what has been already proved.

small, then the point

$$\tfrac{1}{2}\{F(1, x_2, \ldots, x_n)\}^{-1}(1, x_2, \ldots, x_n)$$

is arbitrarily close to $\boldsymbol{z}$ and not on the boundary of any $\mathscr{T}(\boldsymbol{b}_l)$.]

Now we consider the boundary common to $\mathscr{T}(\boldsymbol{o})$ and $\mathscr{T}(\boldsymbol{c}_k)$. We saw already that $\frac{1}{2}\boldsymbol{c}_k$ is one point of the common boundary. Let

$$\tfrac{1}{2}\boldsymbol{c}_k, \quad \tfrac{1}{2}\boldsymbol{c}_k + \boldsymbol{y}_{kj} \qquad (1 \leqq j \leqq n-1) \tag{7}$$

be $n$ linearly independent points on the common boundary. (They exist by the definition of the $\boldsymbol{c}_k$.) Then the points

$$\tfrac{1}{2}\boldsymbol{c}_k - \boldsymbol{y}_{kj}$$

are also on the common boundary, by symmetry; and hence, by convexity[1], so are all points

$$\tfrac{1}{2}\boldsymbol{c}_k + \sum_{1 \leqq j \leqq n-1} t_j \boldsymbol{y}_{kj} \tag{8}$$

with

$$\sum |t_j| \leqq 1. \tag{8'}$$

Let $\pi_k$ be the (hyper)plane through $\frac{1}{2}\boldsymbol{c}_k$ and the $\frac{1}{2}\boldsymbol{c}_k \pm \boldsymbol{y}_{kj}$. Clearly any plane other than $\pi_k$ through $\frac{1}{2}\boldsymbol{c}_k$ contains points of $\mathscr{T}(\boldsymbol{o})$; and so $\pi_k$ must be the only tac-plane to $\mathscr{T}(\boldsymbol{o})$ at $\frac{1}{2}\boldsymbol{c}_k$. The equation of $\pi_k$ may be written in the shape

$$\sum_{1 \leqq j \leqq n} f_{kj} x_j = \tfrac{1}{2}, \tag{9}$$

since $\pi_k$ cannot pass through the inner point $\boldsymbol{o}$ of $\mathscr{T}(\boldsymbol{o})$. The corresponding tac-plane $-\pi_k$ through $-\frac{1}{2}\boldsymbol{c}_k$ is obtained by changing the sign of the $f_{kj}$ in (9). Hence every point of the open set $\mathscr{T}(\boldsymbol{o})$ satisfies the inequalities

$$\Big|\sum_j f_{kj} x_j\Big| < \tfrac{1}{2}. \tag{10}$$

Further, every point $\boldsymbol{y}$, which does not belong to $\mathscr{T}(\boldsymbol{o})$ is of the shape $\boldsymbol{y} = t\boldsymbol{y}_0$, where $t \geqq 1$ and $\boldsymbol{y}_0$ is a boundary point. We saw already that every boundary point of $\mathscr{T}(\boldsymbol{o})$ is also a boundary point to some $\mathscr{T}(\pm\boldsymbol{c}_k)$ and so satisfies

$$\pm \sum f_{kj} x_j = \tfrac{1}{2}$$

for this $k$. Hence $\boldsymbol{y}_0$, and *a fortiori* $\boldsymbol{y}$; cannot satisfy (10). Thus $\mathscr{T}(\boldsymbol{o})$ is precisely the set of $\boldsymbol{x}$ which satisfy (10). Since $\mathscr{S} = 2\mathscr{T}(\boldsymbol{o})$, the corresponding equations for $\mathscr{S}$ are (1).

[1] The point (8) is

$$t_0(\tfrac{1}{2}\boldsymbol{c}_k) + \sum_{1 \leqq j \leqq n-1} |t_j|\,(\tfrac{1}{2}\boldsymbol{c}_k \pm \boldsymbol{y}_{kj}),$$

where the $\pm$ prefixed to $\boldsymbol{y}_{kj}$ is the sign of $t_j$, and

$$t_0 + |t_1| + \cdots + |t_{n-1}| = 1.$$

Some of the inequalities (10) may be identical, since it is quite possible that the pairs of tac-planes $\pm\pi_k$ may be the same for distinct $k$. We may suppose that (10) for $1 \leqq k \leqq m$ gives a complete set of distinct inequalities, where $m \leqq K$. We saw that there is a unique tac-plane at $\boldsymbol{c}_k$, and so, since the planes $\pm\pi_l$ $(1 \leqq l \leqq m,\ l \neq k)$ are certainly tac-planes and are distinct from $\pi_k$, they cannot pass through $\boldsymbol{c}_k$. Hence $\boldsymbol{x} = \boldsymbol{c}_k$ satisfies

$$\Bigl|\sum_j f_{lj} x_j\Bigr| < \tfrac{1}{2} \qquad (1 \leqq l \leqq m,\ l \neq k), \tag{11}$$

and

$$\Bigl|\sum_j f_{kj} x_j\Bigr| = \tfrac{1}{2}. \tag{12}$$

To complete the proof of the theorem, it remains to show that $m \leqq 2^n - 1$. As in the proof of Theorem IX of Chapter V, it is enough to show that the points $\frac{1}{2}(\boldsymbol{c}_k - \boldsymbol{c}_r)$ are not in $\Lambda$ for $1 \leqq k < r \leqq m$. But from what has just been proved, the point $\frac{1}{2}(\boldsymbol{c}_k - \boldsymbol{c}_r)$ certainly satisfies $\bigl|\sum_j f_{lj}\bigr| < 1$, for $1 \leqq l \leqq m$, there being strict inequality for $l = k, r$ because then (11) holds for $\boldsymbol{x} = \frac{1}{2}\boldsymbol{c}_r, \frac{1}{2}\boldsymbol{c}_k$ respectively. Hence $\frac{1}{2}(\boldsymbol{c}_k - \boldsymbol{c}_r) \in \mathscr{S}$, so cannot be in $\Lambda$, since $\Lambda$ is $\mathscr{S}$-admissible by hypothesis.

**IX.2.2.** When $n = 2$, it is possible to specify completely the convex symmetric sets $\mathscr{S}$ with $\Delta(\mathscr{S}) = \frac{1}{4}V(\mathscr{S})$.

THEOREM VI. *A necessary and sufficient condition that the lattice $\Lambda$ be admissible for the convex open symmetric 2-dimensional set $\mathscr{S}$ with*

$$V(\mathscr{S}) = 4d(\Lambda)$$

*is that either*

*(i) $\mathscr{S}$ is a parallelogram and $\Lambda$ is generated by a mid-point of one side and a point on one of the other pair of sides or*

*(ii) $\mathscr{S}$ is a hexagon and $\Lambda$ is the lattice generated by the mid-points of any two non-opposite sides. Then $\Lambda$ contains the mid-points of all the sides.*

That $\mathscr{S}$ is a parallelogram or hexagon follows from Theorem V, since $2^n - 1 = 3$ for $n = 2$. The lattices $\Lambda$ are critical by MINKOWSKI's convex body theorem. The critical lattices of parallelograms and hexagons have already been determined in Lemma 7 of Chapter III and Lemma 13 of Chapter V respectively.

## IX.3. VORONOÏ's results.

We already saw in § 1 that if $g(\boldsymbol{x})$ is a positive definite quadratic form and $\Lambda$ a lattice, then the set of points such that

$$g(\boldsymbol{x}) < \inf_{\substack{\boldsymbol{a} \in \Lambda \\ \neq \boldsymbol{o}}} g(\boldsymbol{x} + \boldsymbol{a})$$

form a convex symmetric body $\mathscr{S}$ of volume $2^n d(\Lambda)$. We shall show that when $n=2$ every symmetric convex hexagon $\mathscr{H}$ and its unique critical lattice can be related in this way by a suitable quadratic form $g(\boldsymbol{x})$. On the other hand, if $\mathscr{S}$ is a parallelogram, then $\Lambda$ must be the particular critical lattice generated by the mid-points of the sides.

These results are clearly invariant under homogeneous linear transformation so we may suppose without loss of generality that $\Lambda=\Lambda_0$ is the lattice of points with integral co-ordinates and that

$$g(x_1, x_2) = a x_1^2 + 2h x_1 x_2 + b x_2^2$$

is reduced, in the sense that

$$0 \leqq -2h \leqq a \leqq b. \tag{1}$$

If $u_1, u_2$ are integers not both 0, the condition

$$g(x_1, x_2) < g(x_1 - u_1, x_2 - u_2) \tag{2}$$

is

$$2\{u_1(a x_1 + h x_2) + u_2(h x_1 + b x_2)\} < g(u_1, u_2).$$

Since $(-u_1, -u_2)$ occurs as well as $(u_1, u_2)$, we thus have the infinitely many conditions

$$2|u_1 X_1 + u_2 X_2| < g(u_1, u_2), \tag{3}$$

where

$$X_1 = a x_1 + h x_2, \quad X_2 = h x_1 + b x_2. \tag{4}$$

In particular,

$$\left.\begin{array}{l} 2|X_1| < a \\ 2|X_2| < b \\ 2|X_1 + X_2| < a + 2h + b = c \quad \text{(say)}, \end{array}\right\} \tag{5}$$

where

$$0 < a \leqq b \leqq c \leqq a + b. \tag{1'}$$

The set $\mathscr{H}$ defined by (5) is a proper hexagon unless $h=0$, when it degenerates into a parallelogram. The area $V(\mathscr{H})$ of $\mathscr{H}$ is readily computed from (4) and (5) to be

$$V(\mathscr{H}) = 4 = 4d(\Lambda_0).$$

But $\mathscr{S}$ is a subset of $\mathscr{H}$ and $V(\mathscr{S})=4$, by Theorem II. Hence $\mathscr{S}=\mathscr{H}$, since both are open. This implies that the infinitely many inequalities (3) all follow from (5), which the reader may verify directly with little trouble.

Further, every non-degenerate convex symmetric hexagon $\mathscr{H}$ with its critical lattice may be generated in this way, as we now show. The hexagon is given by three inequalities

$$|\boldsymbol{l}_j \boldsymbol{x}| < k_j \qquad (j = 1, 2, 3), \tag{6}$$

where

$$\boldsymbol{l}_j = (l_{1j}, l_{2j})$$

and

$$\boldsymbol{l}_j \boldsymbol{x} = l_{1j} x_1 + l_{2j} x_2$$

is the scalar product. The three 2-dimensional vectors $\boldsymbol{l}_j$ are linearly dependent and, by multiplying them by suitable factors, we may suppose without loss of generality that

$$\boldsymbol{l}_1 + \boldsymbol{l}_2 + \boldsymbol{l}_3 = 0,$$

and, on re-indexing, that

$$k_1 \leqq k_2 \leqq k_3. \tag{7}$$

On taking $X_j = \boldsymbol{l}_j \boldsymbol{x}$ $(j = 1, 2)$, the inequalities (6) become

$$\left.\begin{aligned} |X_1| &< k_1 \\ |X_2| &< k_2 \\ |X_1 + X_2| &< k_3. \end{aligned}\right\} \tag{8}$$

Further,

$$k_3 < k_1 + k_2,$$

since the hexagon $\mathscr{H}$ is not degeneratic, by hypothesis. We may identify (8) and (5) by putting

$$2k_1 = a, \quad 2k_2 = b, \quad 2k_3 = c = a + 2h + b,$$

though of course the $x_1$, $x_2$ in (4) are not necessarily to be identified with the $x_1$, $x_2$ in (6). Let $x_1'$, $x_2'$ be defined in terms of $X_1$, $X_2$ by the analogue

$$X_1 = a x_1' + h x_2', \qquad X_2 = h x_1' + b x_2'$$

of (4). On comparing with the earlier part of this section, we see that the unique critical lattice of $\mathscr{H}$ must be given by integral values of $x_1'$, $x_2'$. We may thus suppose, without loss of generality, that $(x_1', x_2')$ was in fact the original co-ordinate system $(x_1, x_2)$, and then we have the situation already discussed.

**IX.3.2.** From the results of § 3.1 we deduce the so-called "hexagon-lemma" of DIRICHLET[1] which illustrates the connection between homogeneous and inhomogeneous problems that will be discussed in more detail in Chapter XI.

[1] For an alternative derivation of the lemma and a partial generalization to $n$ dimensions, see MORDELL (1956a).

THEOREM VII. *Let*

$$g(x_1, x_2) = a x_1^2 + 2h x_1 x_2 + b x_2^2 \tag{1}$$

*be a quadratic form, reduced in the sense that*

$$0 \leqq -2h \leqq a \leqq b. \tag{2}$$

*Then to every real point* $\boldsymbol{x}_0 = (x_{10}, x_{20})$ *there is a point* $\boldsymbol{u} = (u_1, u_2)$ *with integer co-ordinates, such that*

$$4(ab - h^2)\, g(\boldsymbol{x}_0 + \boldsymbol{u}) \leqq abc, \quad c = a + 2h + b. \tag{3}$$

*The sign of equality is required when and only when*

$$2(ab - h^2)(\boldsymbol{x}_0 + \boldsymbol{v}) = \pm\{b(a+h), -a(b+h)\}, \tag{4}$$

*where* $\boldsymbol{v}$ *has integral co-ordinates.*

For by the results of § 3.1 and by Theorem II there is certainly a point $\boldsymbol{x}_0 + \boldsymbol{u}$ with integral $\boldsymbol{u}$ in the closed hexagon

$$\overline{\mathscr{H}}:\ 2|X_1| \leqq a, \quad 2|X_2| \leqq b, \quad 2|X_1 + X_2| \leqq c,$$

where

$$X_1 = a x_1 + h x_2, \quad X_2 = h x_1 + b x_2. \tag{5}$$

But the positive definite quadratic form $g(\boldsymbol{x})$ can reach its maximum in $\overline{\mathscr{H}}$ only at the vertices[1] of $\overline{\mathscr{H}}$. It is now readily verified that the vertices are of the shape (4) and that the value of $g(\boldsymbol{x})$ at all the vertices is given by the right-hand side of (3). The calculations are facilitated by the identity

$$g(X_2, -X_1) = (ab - h^2)\, g(\boldsymbol{x}),$$

where $X_1$, $X_2$ are given by (5).

Finally, the $\leqq$ in (3) cannot be replaced by $<$ if $\boldsymbol{x}_0$ is any vertex of $\overline{\mathscr{H}}$, since

$$g(\boldsymbol{x}) = \inf_{\boldsymbol{u} \in \Lambda_0} g(\boldsymbol{x} + \boldsymbol{u})$$

for the points $\boldsymbol{x}$ of $\overline{\mathscr{H}}$. This last remark also shows that it was sufficient to compute $g(\boldsymbol{x})$ at any one vertex $\boldsymbol{x}_1$ (say) since, from the nature of a critical lattice, all the other vertices are of the shape $\pm\boldsymbol{x}_1 + \boldsymbol{w}$, where $\boldsymbol{w}$ has integral co-ordinates.

**IX.3.3.** Theorem VII gives yet another proof of the result that a definite ternary quadratic form $f(\boldsymbol{x})$ represents an number $a \leqq (2D)^{\frac{1}{3}}$ for integral values of the variables not all 0, where $D$ is the determinant of $f(\boldsymbol{x})$ (§ 3.4 of Chapter II). We may suppose, without loss of generality,

[1] Perhaps the easiest way to see this is to make a homogeneous linear transformation $\boldsymbol{y} = \boldsymbol{\tau x}$ so that $g(\boldsymbol{x}) = |\boldsymbol{y}|^2$, when it is obvious.

that $f(\boldsymbol{x})$ is reduced in MINKOWSKI'S sense (cf. Chapter II, § 2.1). We have

$$\left.\begin{aligned} f(\boldsymbol{x}) &= a x_1^2 + b x_2^2 + c x_3^2 + 2h x_1 x_2 + 2g x_1 x_3 + 2f x_2 x_3 \\ &= a(x_1 + \alpha x_3)^2 + 2h(x_1 + \alpha x_3)(x_2 + \beta x_3) + b(x_2 + \beta x_3)^2 + \gamma x_3^2 \end{aligned}\right\} \quad (1)$$

for some $\alpha, \beta, \gamma$. We may suppose that $h \leqq 0$. Then

$$0 \leqq -2h \leqq a \leqq b, \quad (2)$$

and

$$f(u_1, u_2, 1) \geqq b \quad (3)$$

for all integers $u_1, u_2$, by the condition of the reduction. But now, by Theorem VII, we may choose $u_1, u_2$ so that

$$f(u_1, u_2, 1) \leqq \frac{ab(a+2h+b)}{4(ab-h^2)} + \gamma. \quad (4)$$

Hence from (1), (3), (4) we have

$$\begin{aligned} 4D = 4(ab - h^2)\gamma &\geqq 4b(ab - h^2) - ab(a + 2h + b) \\ &= -b(2h + \tfrac{1}{2}a)^2 + 3ab^2 - \tfrac{3}{4}a^2 b. \end{aligned}$$

Now

$$|2h + \tfrac{1}{2}a| \leqq \tfrac{1}{2}a,$$

by (2); and so

$$4D \geqq 3ab^2 - a^2 b \geqq 2ab^2 \geqq 2a^3,$$

by a further application of (2). This is the required result. Further, using the knowledge of the cases of equality in Theorem VII, it is easily verified that $2D = a^3$ can occur only for forms equivalent to multiples of the critical form

$$x_1^2 + x_2^2 + x_3^2 - x_1 x_2 - x_2 x_3 - x_3 x_1.$$

**IX.4. Preparatory lemmas.** In §§ 5, 6 we shall need three lemmas, each of independent interest, which it is convenient to prove first. We use the word polygon to mean indifferently a 2-dimensional set bounded by a finite number of line-segments or the boundary of such a set. Which is meant will be clear from the context. We shall say that a convex polygon is circumscribed to a convex set $\mathscr{K}$ if it contains $\mathscr{K}$ and if every side of the polygon is a tac-line[1] of $\mathscr{K}$. The first lemma is an analogue of Theorem XI of Chapter V due to REINHARDT (1934a), and found independently by MAHLER (1947c).

LEMMA 1. *Let $\mathscr{K}$ be a convex symmetric open 2-dimensional set. Then*

$$\Delta(\mathscr{S}) = \tfrac{1}{4} \inf V(\mathscr{H}), \quad (1)$$

*where $\mathscr{H}$ runs through all symmetric circumscribed hexagons and $V(\mathscr{H})$ is the area of $\mathscr{H}$.*

---

[1] We speak of a tac-line in 2-dimensions instead of a tac-plane.

Let $\mathcal{H}$ be any circumscribed hexagon and $\Lambda(\mathcal{H})$ the critical lattice of $\mathcal{H}$; so that

$$d\{\Lambda(\mathcal{H})\} = \tfrac{1}{4}V(\mathcal{H})$$

by Lemma 13 of Chapter V. But $\Lambda(\mathcal{H})$ is certainly admissible for $\mathcal{K}$, and so the left-hand side of (1) is at most equal to the right.

When $\mathcal{K}$ is a parallelogram, the lemma is trivial, so we suppose $\mathcal{K}$ is not a parallelogram. Let M be a critical lattice for $\mathcal{K}$ so that, by Theorem XI of Chapter V, it has precisely 6 points $\pm\boldsymbol{p}$, $\pm\boldsymbol{q}$, $\pm\boldsymbol{r}$ on the boundary of $\mathcal{K}$, where $\boldsymbol{p}$, $\boldsymbol{q}$ is a basis and

$$\boldsymbol{p}+\boldsymbol{q}+\boldsymbol{r}=\boldsymbol{o}.$$

Let $\mathcal{H}_0$ be the hexagon formed by tac-lines at $\pm\boldsymbol{p}$, $\pm\boldsymbol{q}$, $\pm\boldsymbol{r}$ to $\mathcal{K}$, taking the corresponding tac-line $-\pi$ at $\boldsymbol{p}$ to the tac-line $\pi$ taken at $\boldsymbol{p}$, if that is not unique, etc. Then $\mathcal{H}_0$ is a symmetric hexagon circumscribed to $\mathcal{K}$. The lattice M is admissible for $\mathcal{H}_0$ by Theorem XI of Chapter V, and so

$$\Delta(\mathcal{K}) = d(M) \geqq \Delta(\mathcal{H}_0) = \tfrac{1}{4}V(\mathcal{H}_0),$$

by Lemma 13 of Chapter V. This concludes the proof of Lemma 1.

**IX.4.2.** The following lemma due to Dowker (1944a) relates the areas of circumscribed polygons to a convex set $\mathcal{K}$, which need not be symmetric. We sketch the proof, for which see also Fejes Tóth (1953a).

Lemma 2. *Suppose that there exists a circumscribed $(n+1)$-gon $\mathcal{P}_{n+1}$ and a circumscribed $(n-1)$-gon $\mathcal{P}_{n-1}$ to a convex set $\mathcal{K}$. Then there exists a circumscribed $m$-gon with $m \leqq n$ and area*

$$\leqq \tfrac{1}{2}\{V(\mathcal{P}_{n-1}) + V(\mathcal{P}_{n+1})\}.$$

If $\boldsymbol{a}_1$, $\boldsymbol{a}_2$, $\boldsymbol{a}_3$ are three points on the boundary of $\mathcal{K}$ then in this proof we mean by

$$\boldsymbol{a}_1 < \boldsymbol{a}_2 < \boldsymbol{a}_3$$

that $\boldsymbol{a}_1$, $\boldsymbol{a}_2$, $\boldsymbol{a}_3$ occur in that order on traversing the boundary of $\mathcal{K}$ in, say, a counter-clockwise direction.

Let the sides of $\mathcal{P}_{n-1}$ be the lines $\alpha_1, \ldots, \alpha_{n-1}$. By definition, these are tac-lines to $\mathcal{K}$. Let $\boldsymbol{a}_j$ $(1 \leqq j \leqq n-1)$ be a point on the boundary of $\mathcal{K}$ at which $\alpha_j$ is a tac-line. If $\alpha_j$ is a tac-line at several points, then we choose $\boldsymbol{a}_j$ once for all. We may suppose without loss of generality that

$$\boldsymbol{a}_{n-1} < \boldsymbol{a}_1 < \boldsymbol{a}_2 < \cdots < \boldsymbol{a}_{n-1} < \boldsymbol{a}_1.$$

Similarly let $\beta_j$ and $\boldsymbol{b}_j$ be defined with respect to $\mathcal{P}_{n+1}$, where $1 \leqq j \leqq n+1$.

We distinguish two cases. Suppose, first, that three of the $\boldsymbol{b}_j$ occur between two of the $\boldsymbol{a}_j$, say,

$$\boldsymbol{a}_1 \leqq \boldsymbol{b}_1 < \boldsymbol{b}_2 < \boldsymbol{b}_3 < \boldsymbol{a}_2,$$

where the symbol between $\boldsymbol{a}_1$ and $\boldsymbol{b}_1$ means that possibly $\boldsymbol{a}_1 = \boldsymbol{b}_1$, but otherwise $\boldsymbol{a}_1 < \boldsymbol{b}_1 < \boldsymbol{b}_2$. Let $\mathscr{P}_n'$ have sides $\alpha_1, \beta_2, \alpha_2, \ldots, \alpha_n$ and $\mathscr{P}_n''$ have sides $\beta_1, \beta_3, \ldots, \beta_{n+1}$. Then

$$V(\mathscr{P}_{n+1}) + V(\mathscr{P}_{n-1}) \geqq V(\mathscr{P}_n') + V(\mathscr{P}_n''), \tag{1}$$

as is clear from Fig. 10. Indeed the difference between the two sides of (1) is the sum of the areas of the two 4-gons whose sides are formed by $\alpha_1, \beta_3, \beta_1, \beta_2$ and $\alpha_1, \alpha_2, \beta_3, \beta_2$ respectively. From (1) we have

$$\min\{V(\mathscr{P}_n'), V(\mathscr{P}_n'')\} \leqq \tfrac{1}{2}\{V(\mathscr{P}_{n-1}) + V(\mathscr{P}_{n+1})\},$$

which proves the lemma in this case.

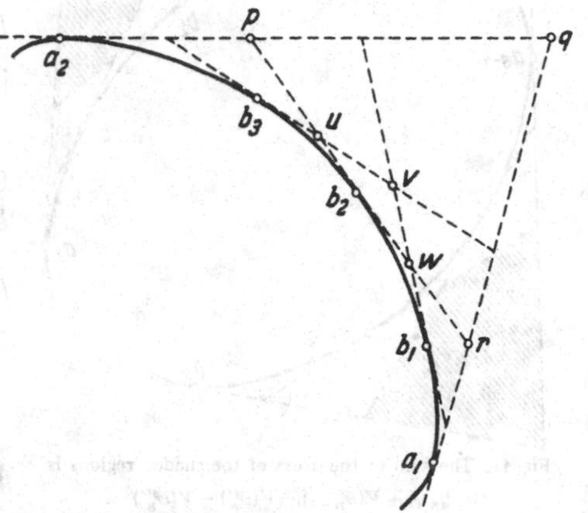


Fig. 10. From the figure,
$V(\mathscr{P}_{n-1}) - V(\mathscr{P}_n') = V(\boldsymbol{p}\boldsymbol{q}\boldsymbol{r})$,
$V(\mathscr{P}_n'') - V(\mathscr{P}_{n+1}) = V(\boldsymbol{u}\boldsymbol{v}\boldsymbol{w})$
and clearly
$V(\boldsymbol{p}\boldsymbol{q}\boldsymbol{r}) \geq V(\boldsymbol{u}\boldsymbol{v}\boldsymbol{w})$.
The point labelled $\boldsymbol{a}_3$ should be labelled $\boldsymbol{a}_2$

The polygons $\mathscr{P}_n'$, $\mathscr{P}_n''$ may have fewer than $n$ sides, since some sides of $\mathscr{P}_{n+1}$ may coincide with those of $\mathscr{P}_{n-1}$. But this possibility is covered by the enunciation of the lemma. We shall not repeat this remark which will apply at a later stage in this proof and also to the proof of Lemma 3.

If the first case does not happen, then, since there are two more $\boldsymbol{b}$'s than $\boldsymbol{a}$'s we have, on re-indexing if necessary, that

$$\boldsymbol{a}_1 \leqq \boldsymbol{b}_1 < \boldsymbol{b}_2 < \boldsymbol{a}_2 \leqq \boldsymbol{a}_{s-1} \leqq \boldsymbol{b}_s < \boldsymbol{b}_{s+1} < \boldsymbol{a}_s$$

for some $s$. Let $\mathscr{P}_n'$, $\mathscr{P}_n''$ have sides

$$\alpha_1, \beta_2, \ldots, \beta_s, \alpha_s, \ldots, \alpha_{n-1}$$

and

$$\beta_1, \alpha_2, \ldots, \alpha_{s-1}, \beta_{s+1}, \ldots, \beta_{n+1}$$

respectively. Then again

$$V(\mathscr{P}_{n+1}) + V(\mathscr{P}_{n-1}) \geqq V(\mathscr{P}_n') + V(\mathscr{P}_n''),$$

the difference being the sum of the areas of the 4-gons $\alpha_1\,\alpha_2\,\beta_1\,\beta_2$ and $\alpha_{s-1}\,\alpha_s\,\beta_s\,\beta_{s+1}$, see Fig. 11.

COROLLARY 1. *Let $U(n)$ denote the infimum of the areas of circumscribed m-gons with $m \leqq n$. Then*

$$U(n) \leqq U(n-1) \tag{2}$$

*and*

$$2U(n) \leqq U(n-1) + U(n+1). \tag{3}$$

The first inequality is a trivial consequence of the definition, the second follows at once from Lemma 2.

It is convenient to extend the definition of $U(n)$ to non-integral value of the argument. For $t \geqq 3$ put

$$U(t) = (1-l)\,U(n) + l\,U(n+1),$$

if

$$t = n + l, \quad 0 \leqq l \leqq 1.$$

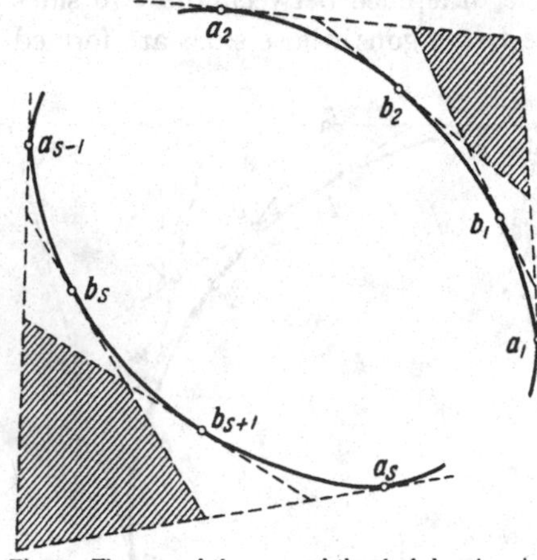

Fig. 11. The sum of the areas of the shaded regions is $V(\mathscr{P}_{n+1}) + V(\mathscr{P}_{n-1}) - V(\mathscr{P}_n') - V(\mathscr{P}_n'')$

COROLLARY 2. *Let $\mu_1, \ldots, \mu_R$ be numbers such that*

$$\mu_r \geqq 0 \quad (1 \leqq r \leqq R), \quad \sum_r \mu_r = 1.$$

*Then*

$$U\Big(\sum_r \mu_r t_r\Big) \leqq \sum_r \mu_r U(t_r), \tag{4}$$

*where $t_r$ $(1 \leqq r \leqq R)$ are any numbers with $t_r \geqq 3$.*

The inequality follows at once from Corollary 1 if $R = 2$ and then follows easily for general $R$ by induction.

By a similar argument to that used for Lemma 2 DOWKER (1944a) proved

LEMMA 3. *Suppose that $\mathscr{K}$ is symmetric as well as being convex. Let $\mathscr{P}_{2n}$ be an $2n$-gon circumscribed to $\mathscr{K}$. Then there is a symmetric $2m$-gon with $m \leqq n$, also circumscribed to $\mathscr{K}$ of area at most $V(\mathscr{P}_{2n})$.*

Let the sides

$$\alpha_1, \ldots, \alpha_{2n}$$

of $\mathscr{P}_{2n}$ be tac-lines at

$$\boldsymbol{a}_1, \ldots, \boldsymbol{a}_{2n},$$

where

$$\boldsymbol{a}_{2n} < \boldsymbol{a}_1 < \boldsymbol{a}_2 \ldots < \boldsymbol{a}_{2n} < \boldsymbol{a}_1.$$

Let

$$\beta_j = \bar{\alpha}_{j\pm n}, \quad \boldsymbol{b}_j = \bar{\boldsymbol{a}}_{j\pm n}, \tag{5}$$

where the bar denotes the image in the origin. Then, by symmetry, the $\beta_j$ are the sides of the circumscribed polygon $\overline{\mathscr{P}}_{2n}$, which is the image of $\mathscr{P}_{2n}$ in the origin. By the convexity and symmetry we have

$$\bar{a}_{j+1} < a_j < a_{j+1} < \bar{a}_j$$

for every $j$.

If $\mathscr{P}_{2n}$ is not already symmetric, we may suppose without loss of generality that $a_n \neq b_n$ and, by changing the orientation of the indexing if need be, that

$$\bar{b}_n \leqq a_n < b_n \leqq \bar{a}_n.$$

Then

$$\bar{a}_{2n} \leqq b_{2n} < a_{2n} \leqq \bar{b}_{2n},$$

by (5). There is thus a greatest $j$ in $n \leqq j < 2n$ such that

$$\bar{b}_j \leqq a_j < b_j \leqq \bar{a}_j,$$

and for this $j$ clearly

$$a_j \leqq b_j < b_{j+1} \leqq a_{j+1}.$$

It is not excluded that $b_{j+2}$ also lies between $a_j$ and $a_{j+1}$. Without loss of generality $j = n$; and then

$$a_n \leqq b_n < b_{n+1} \leqq a_{n+1}, \quad b_{2n} \leqq a_{2n} < a_1 \leqq b_1,$$

by (5).

Let $\mathscr{P}'_{2n}$, $\mathscr{P}''_{2n}$ have sides

$$\alpha_1, \ldots, \alpha_n, \beta_{n+1}, \ldots, \beta_{2n} \quad \text{and} \quad \beta_1, \ldots, \beta_n, \alpha_{n+1}, \ldots, \alpha_{2n},$$

so $\mathscr{P}'_{2n}$ and $\mathscr{P}''_{2n}$ are symmetric, by (5). Precisely as in the second case of the proof of Lemma 2 we have

$$V(\mathscr{P}'_{2n}) + V(\mathscr{P}''_{2n}) \leqq V(\mathscr{P}_{2n}) + V(\overline{\mathscr{P}}_{2n}) = 2V(\mathscr{P}_{2n});$$

and so either $\mathscr{P}'_{2n}$ or $\mathscr{P}''_{2n}$ satisfies the requirements of the lemma.

Corollary. *For convex symmetric $\mathscr{K}$,*

$$\Delta(\mathscr{K}) = \tfrac{1}{4} U(6),$$

*where $U(6)$ is the infimum of the areas of circumscribed $m$-gons with $m \leqq 6$.*

This follows at once from Lemma 1 and Lemma 3.

**IX.4.3.** We shall also need Euler's formula for convex polyhedra in a slightly unusual form (cf. Fejes Tóth 1953a). Let $v_n$ $(1 \leqq n \leqq N)$ be points in the plane (vertices). Let $\lambda_s$ $(1 \leqq s \leqq S)$ be curves joining one vertex to another vertex or, possibly coming back to the same vertex (the edges). The reader may think of the $\lambda_s$ as line-segments or composed of a finite number of line-segments. We suppose that no point of $\lambda_s$ except its ends is a $v_n$ and that no two $\lambda_s$ cross. Finally we suppose that it is possible to get from any one vertex to any other along the $\lambda_s$. Then the whole plane is dissected by the $\lambda_s$ into a number $\varphi$ of connected pieces (the "faces") one of which contains all points outside a large circle $|\boldsymbol{x}| = R$. Then EULER's formula is

LEMMA 4.
$$\varphi + N = S + 2.$$
This may be readily verified by induction on $S$.

**IX.5. FEJES TÓTH's theorem.** In this section we prove a result due to FEJES TÓTH (1950a), see also FEJES TÓTH (1953a). He proves something more general and also gives interesting related results but we give here only what is needed to treat the lattice constants of cylinders.

THEOREM VIII. *Let $\mathscr{H}$ be a convex open polygon with at most 6 sides. Let $\mathscr{K}$ be any convex open set and suppose that the sets*
$$\mathscr{K}_r = \mathscr{K} + \boldsymbol{x}_r \qquad (1 \leqq r \leqq R)$$
*are packed in $\mathscr{H}$, i.e. the $\mathscr{K}_r$ are subsets of $\mathscr{H}$ and no two have points in common. Then*
$$R\,U(6) \leqq V(\mathscr{H}),$$
*where $U(6)$ is the lower bound of the areas of m-gons circumscribed to $\mathscr{K}$ with $m \leqq 6$.*

The notation $U(6)$ is in conformity with that of Lemma 2, Corollary. FEJES TÓTH's own version of his proof is very compact, and we have found it desirable to expand it.

**IX.5.2.** The stages in the proof of Theorem VIII are enunciated for convenience as propositions.

PROPOSITION 1[1]. *Let $\mathscr{H}$ be a convex open 2-dimensional polygon and let $\mathscr{K}_r$ $(1 \leqq r \leqq R)$ be open convex sets packed in $\mathscr{H}$. Then there are open convex polygons $\mathscr{Q}_r$ $(1 \leqq r \leqq R)$ such that $\mathscr{Q}_r$ contains $\mathscr{K}_r$ and*

*(i) the $\mathscr{Q}_r$ are packed in $\mathscr{H}$,*

*(ii) if $\sigma$ is a side of $\mathscr{Q}_r$ then either,*

*(ii$_1$) $\sigma$ is part of the boundary of $\mathscr{H}$,*

*or*

*(ii$_2$) there is a subsegment $\sigma'$ of $\sigma$ containing more than a single point which is part of the boundary of a $\mathscr{Q}_s$, $(s \neq r)$, and*

*(iii) if $\sigma$ is a side of $\mathscr{H}$ then some subsegment $\sigma'$ of $\mathscr{H}$ consisting of more than a single point is part of the boundary of some $\mathscr{Q}_r$.*

Note that the $\mathscr{K}_r$ are not required to be similar to each other. We shall give two proofs of proposition 1. The first is by transfinite induc-

[1] Mr. H. L. DAVIES has pointed out that this Proposition is false as it stands by giving a counter example. The proof of Theorem VIII can, however, be salvaged.

tion (ZORN's Lemma). It involves the minimum of geometric argument, but is non-constructive. The second, which will only be sketched, gives a process for constructing the $\mathcal{Q}_r$ in a finite number of steps.

If $\{\mathcal{K}'\}$ and $\{\mathcal{K}''\}$ are two packings of $R$ open convex sets in $\mathcal{H}$, we write

$$\{\mathcal{K}'\} \prec \{\mathcal{K}''\}$$

if $\mathcal{K}_r''$ contains $\mathcal{K}_r'$ for $1 \leqq r \leqq R$, not necessarily strictly. We denote the set of all such packings by $\Pi$ and verify three statements about the symbol $\prec$.

(I) If $\{\mathcal{K}'\} \prec \{\mathcal{K}''\}$ and $\{\mathcal{K}''\} \prec \{\mathcal{K}'\}$ then $\{\mathcal{K}'\} = \{\mathcal{K}''\}$, in the sense that the sets $\mathcal{K}_r'$ and $\mathcal{K}_r''$ are identical for $1 \leqq r \leqq R$. This is trivial.

(II) If $\{\mathcal{K}'\} \prec \{\mathcal{K}''\}$ and $\{\mathcal{K}''\} \prec \{\mathcal{K}'''\}$, then $\{\mathcal{K}'\} \prec \{\mathcal{K}'''\}$. This is again trivial.

(III) Suppose that $\tilde{\Pi}$ is any subset of the set of packings $\Pi$ such that if $\{\mathcal{K}'\}$ and $\{\mathcal{K}''\}$ are in $\tilde{\Pi}$ then either $\{\mathcal{K}'\} \prec \{\mathcal{K}''\}$ or $\{\mathcal{K}''\} \prec \{\mathcal{K}'\}$. Condition (III) states that then there is some packing $\{\tilde{\mathcal{K}}\}$ in $\Pi$ (not necessarily in $\tilde{\Pi}$), such that $\{\mathcal{K}'\} \prec \{\tilde{\mathcal{K}}\}$ for all $\{\mathcal{K}'\}$ in $\tilde{\Pi}$.

To prove (III) we take for $\tilde{\mathcal{K}}_r$ the union of $\mathcal{K}_r$ for all $\{\mathcal{K}\}$ in $\tilde{\Pi}$. We must verify that $\{\tilde{\mathcal{K}}\}$ is a packing of convex open sets, and do this for the properties in turn:

First, $\tilde{\mathcal{K}}_r$ is open. For if $z_0$ is a point of $\tilde{\mathcal{K}}_r$ then it is a point of $\mathcal{K}_r'$ for some packing $\{\mathcal{K}'\}$ of $\tilde{\Pi}$. Since $\mathcal{K}_r'$ is open, a neighbourhood of $z_0$ is in $\mathcal{K}_r'$, and hence also in $\tilde{\mathcal{K}}_r$, as required.

Secondly, $\tilde{\mathcal{K}}_r$ is convex. For let $z_1, z_2$ be any points of $\tilde{\mathcal{K}}_r$, say, $z_1 \in \mathcal{K}_r'$, $z_2 \in \mathcal{K}_r''$, where $\{\mathcal{K}'\}$, $\{\mathcal{K}''\}$ are packings of $\tilde{\Pi}$. By the hypotheses of (III) we may suppose, by interchanging $z_1$ and $z_2$ if necessary, that $\{\mathcal{K}'\} \prec \{\mathcal{K}''\}$. Then $z_1 \in \mathcal{K}_r' \subset \mathcal{K}_r''$. Since $z_2 \in \mathcal{K}_r''$, the whole segment

$$t z_1 + (1-t) z_2 \qquad (0 \leqq t \leqq 1),$$

is in $\mathcal{K}_r''$; and so in $\tilde{\mathcal{K}}_r$, as required.

Thirdly, $\tilde{\mathcal{K}}_r$ and $\tilde{\mathcal{K}}_s$ have no points in common if $r \neq s$. For suppose $z_0 \in \tilde{\mathcal{K}}_r$, $z_0 \in \tilde{\mathcal{K}}_s$. Then $z_0 \in \mathcal{K}_r'$, $z_0 \in \mathcal{K}_s''$ for some packings $\{\mathcal{K}'\}$, $\{\mathcal{K}''\}$ in $\tilde{\Pi}$, where again without loss of generality $\{\mathcal{K}'\} \prec \{\mathcal{K}''\}$. Then $z_0 \in \mathcal{K}_r' \subset \mathcal{K}_r''$, so $z_0$ is common to $\mathcal{K}_r''$ and $\mathcal{K}_s''$, contrary to the hypothesis that $\{\mathcal{K}''\}$ is a packing. This concludes the verification of (I), (II) and (III).

We say that a packing $\{\mathcal{K}^\mu\}$ is maximal if

$$\{\mathcal{K}^\mu\} \prec \{\mathcal{K}'\}$$

implies $\{\mathscr{K}^\mu\} = \{\mathscr{K}'\}$. By ZORN's Lemma, since (I), (II), (III) are satisfied, to any packing $\{\mathscr{K}\}$ there is at least one maximal packing $\{\mathscr{K}^\mu\}$ such that

$$\{\mathscr{K}\} < \{\mathscr{K}^\mu\}.$$

But it is easy to see that in a maximal packing $\{\mathscr{K}^\mu\}$ the sets $\mathscr{K}_r^\mu$ must be polygons $\mathscr{Q}_r$ which satisfy the conditions (i), (ii) and (iii) of Proposition 1. Since this will be clear from the constructive proof which we give later, we do not give the detailed argument here. This concludes the first proof of Proposition 1.

We now sketch a second, constructive, proof of Proposition 1. The fundamental process is this. If $\mathscr{K}$ is any open convex bounded set and

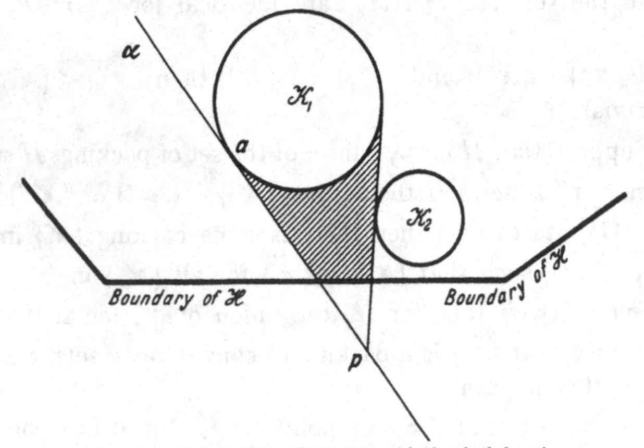


Fig. 12. $\mathscr{K}_1'$ consists of $\mathscr{K}_1$ together with the shaded region

$\boldsymbol{p}$ is any point not in $\mathscr{K}$, then the open convex cover of $\mathscr{K}$ and $\boldsymbol{p}$ is the least convex set which contains $\mathscr{K}$ and has $\boldsymbol{p}$ as a boundary point: that is, $\mathscr{K}$ is the set of

$$t\boldsymbol{p} + (1-t)\boldsymbol{q}, \quad \boldsymbol{q} \in \mathscr{K}, \quad 0 \leqq t < 1.$$

If $\boldsymbol{p}$ is on the boundary of $\mathscr{K}$, then the open convex cover of $\boldsymbol{q}$ and $\mathscr{K}$ is just $\mathscr{K}$. Otherwise the convex cover has as boundary the two tac-lines from $\boldsymbol{p}$ to $\mathscr{K}$ together with a portion of the boundary of $\mathscr{K}$.

If now $\mathscr{K}_1, \ldots, \mathscr{K}_R$ are the sets of Proposition 1, we form the polygons $\mathscr{Q}_r$ by successively taking the convex covers of the sets $\mathscr{K}_r$ and suitably chosen points. Let $\boldsymbol{a}$ be any point on the boundary of $\mathscr{K}_1$ and $\alpha$ a tac-line at $\boldsymbol{a}$. Consider points $\boldsymbol{q}$ on $\alpha$ along one direction, say, to the right of $\boldsymbol{a}$ (see Fig. 12). If $\boldsymbol{q}_2$ is to the right of $\boldsymbol{q}_1$, then the open convex cover of $\boldsymbol{q}_2$ and $\mathscr{K}_1$ contains that of $\boldsymbol{q}_1$ and $\mathscr{K}_1$. For some $\boldsymbol{q}$ to the right of $\boldsymbol{a}$ on $\alpha$ it is possible that the open convex cover of $\mathscr{K}_1$ and $\boldsymbol{q}$ overlaps some other body $\mathscr{K}_2$ of the original packing. Since the $\mathscr{K}_r$ are open, there is then a $\boldsymbol{p}$ fartherst to the right along $\alpha$ such that the open convex cover of $\boldsymbol{p}$ and $\mathscr{K}_1$ contains no points of any $\mathscr{K}_r$ $(r \neq 1)$. It is possible that $\boldsymbol{p} = \boldsymbol{a}$. We then get a new packing $\{\mathscr{K}'\}$ on replacing $\mathscr{K}_1$ by the portion of open convex cover of $\mathscr{K}_1$ and $\boldsymbol{p}$ which is in[1] $\mathscr{H}$. If the open convex cover of $\boldsymbol{p}$ and $\mathscr{K}_1$ does not meet any $\mathscr{K}_r$ $(r \neq 1)$ for all $\boldsymbol{q}$ to the right of $\boldsymbol{a}$, then $\mathscr{K}_1'$ is to be the set of points in $\mathscr{H}$ which are in the convex cover of $\mathscr{K}_1$ and any point $\boldsymbol{q}$ to the right of $\boldsymbol{a}$ on $\alpha$. Similarly one may consider points to the left of $\boldsymbol{a}$ along $\alpha$.

We may repeat the process on the new sets $\{\mathscr{K}'\}$ and will indicate how after a finite number of steps it must come to an end with polygons $\mathscr{Q}_r$ have the properties (i), (ii), (iii) of Proposition 1. We denote the sets at the $j$-th stage by $\{\mathscr{K}^j\}$, so $\{\mathscr{K}^{j-1}\} < \{\mathscr{K}^j\}$. Suppose first that there is a pair of indices $r, s$ such that $\mathscr{K}_r^j$ and $\mathscr{K}_s^j$ have a boundary point $\boldsymbol{a}$ in common. Then $\mathscr{K}_r^{j+1}$, $\mathscr{K}_s^{j+1}$ are obtained from $\mathscr{K}_r^j$, $\mathscr{K}_s^j$ by taking $\alpha$ to be a common tac-line (Chapter IV, Lemma 6) to $\mathscr{K}_r^j$, $\mathscr{K}_s^j$ at $\boldsymbol{a}$ and by applying the above process both to $\mathscr{K}_r^j$ and $\mathscr{K}_s^j$ and both to right and left along $\alpha$. Once this has been done for a pair of indices $r, s$ at the $j$-th stage we do not do it again for the same pair of indices at a later stage. If there is no pair $r, s$ of indices for which $\mathscr{K}_r^j$, $\mathscr{K}_s^j$ have a common boundary point and which have not already been treated, then there may be a body $\mathscr{K}_r^j$ with a boundary point $\boldsymbol{a}$ on the boundary of $\mathscr{H}$. If so, we take $\alpha$ to be the side of $\mathscr{H}$ on which $\boldsymbol{a}$ lies (both sides in turn if $\boldsymbol{a}$ is a vertex of $\mathscr{H}$) and apply the process. Again, once this has been done for $\mathscr{K}_r^j$ and a side of $\mathscr{H}$ we do not do it again for the same $r$ and the same side of $\mathscr{H}$. Neither of the first two steps may be allowable. Suppose that one of the $\mathscr{K}_r^j$ is not a polygon. Then $\boldsymbol{a}$ is taken to be any point on the boundary of $\mathscr{K}_r^j$ which is not in a line-segment forming part of the boundary of $\mathscr{K}_r^s$ nor on the boundary of $\mathscr{K}_s^j$ $(s \neq r)$. Finally, if all the $\mathscr{K}_r^j$ are polygons and the first two stages are impossible, then $\boldsymbol{a}$ is taken to be any vertex of a $\mathscr{K}_r^j$ at which at least one of the two sides is not also a tac-line to some $\mathscr{K}_s^j$ $(s \neq r)$.

It is clear that the steps outlined above will come to an end. And the final set of $\mathscr{K}_r^J$ is clearly a set of polygons $\mathscr{Q}_r$ having the properties (i), (ii), (iii) of the enunciation.

**IX.5.3.** The next stage is an application of EULER's formula (Lemma 4) to the configuration of Proposition 1.

PROPOSITION 2. *Let $\mathscr{Q}_r$ of Proposition 1 have $q_r$ sides* $(1 \leq r \leq R)$. *Then*[2]

$$\sum q_r \leq 6R.$$

In the application of EULER's formula, the faces will be the polygons $\mathscr{Q}_r$ together with $\mathscr{Q}_0$, the set of points not in or on the boundary of $\mathscr{H}$.

---

[1] The reader is reminded that $\mathscr{H}$ is the set in which the $\mathscr{K}_r$ are packed.
[2] The proof assumes facitly that every vertex of $\mathscr{H}$ is a vertex of a $\mathscr{Q}_r$.

Lemma 4 is not immediately applicable, since not every point is in or on the boundary of a $\mathcal{Q}_r$. The set of points which do not enjoy this property is clearly open and so consists of a finite number $\mathcal{L}_1, \ldots, \mathcal{L}_L$ of connected open sets. By (ii) and (iii) of Proposition 1, any one of these sets, say, $\mathcal{L}_l$ cannot contain the whole of a side $\sigma$ of a $\mathcal{Q}_r$. We now apply Lemma 4 where the "vertices" are of the following kinds

(α) the sets $\mathcal{L}_l$ $(1 \leqq l \leqq L)$,

(β) points not on the boundary of an $\mathcal{L}_l$ but on the boundary of at least three $\mathcal{Q}_r$ $(0 \leqq r \leqq R)$,

(γ) vertices of $\mathcal{H}$.

The "edges", for the purpose of Lemma 4, are the segments of the sides of the $\mathcal{Q}_r$ joining the "vertices". Then every side of $\mathcal{Q}_r$ gives rise to at least 1 but possibly more "edges". Let $q_r'$ be the number of "edges" surrounding $\mathcal{Q}_r$, so

$$q_r' \geqq q_r. \tag{1}$$

Since every "edge" belongs to precisely two $\mathcal{Q}_r$ $(0 \leqq r \leqq R)$, the number of "edges" is

$$S = \tfrac{1}{2} \sum_{0 \leqq r \leqq R} q_r'. \tag{2}$$

Let $\mathcal{H}$ have precisely $h$ sides, so

$$h \leqq 6. \tag{3}$$

Every vertex of type (α) or (β) above belongs to at least three $\mathcal{Q}_r$ $(0 \leqq r \leqq R)$ and there are at most $h$ vertices of type (γ). Vertices of type (γ) are on the boundary of $\mathcal{Q}_0$ and at least one $\mathcal{Q}_r$ $(r \neq 0)$. Hence the total number of "vertices" $N$ satisfies

$$3N \leqq h + \sum_{0 \leqq r \leqq R} q_r'. \tag{4}$$

Finally, the number of faces $\varphi$ is

$$\varphi = R + 1. \tag{5}$$

From (1), (3), (4) and EULER'S

$$\varphi + N = S + 2$$

(Lemma 4), we get

$$\sum_{0 \leqq r \leqq R} q_r' \leqq 6R - 6 + 2h.$$

But clearly $q_0' \geqq q_0 = h$, by (1), and so, by (1), (3),

$$\sum_{1 \leqq r \leqq R} q_r \leqq \sum_{1 \leqq r \leqq R} q_r' \leqq 6R.$$

This concludes the proof of Proposition 2.

**IX.5.4.** The proof of Theorem VIII is now comparatively rapid. Let $U(t)$ and $\mathcal{Q}_r, q_r$ have the meanings they had in Lemma 2, Corollaries 1, 2 and Propositions 1, 2. Clearly

$$V(\mathcal{Q}_r) \geqq U(q_r) \qquad (1 \leqq r \leqq R);$$

and so

$$V(\mathcal{H}) \geqq \sum V(\mathcal{Q}_r) \geqq \sum U(q_r),$$

since the $\mathcal{Q}_r$ are packed in $\mathcal{H}$.

Hence by Corollaries 1, 2 to Lemma 2 and by Proposition 2 we have

$$R^{-1} V(\mathcal{H}) \geqq \sum_{1 \leqq r \leqq R} R^{-1} U(q_r) \geqq U\left\{R^{-1} \sum_{1 \leqq r \leqq R} q_r\right\} \geqq U(6).$$

This is just the assertion of Theorem VIII, and so concludes the proof.

**IX.6. Cylinders.** We now make the application of Theorem VIII to the lattice constants of cylinders adumbrated in § 1.5.

THEOREM IX. *Let $\mathcal{K}$ be a convex symmetric 2-dimensional star-body and $\mathcal{C}$ the set of points*

$$\mathcal{C}: (x_1, x_2, x_3) \qquad (x_1, x_2) \in \mathcal{K}, \qquad |x_3| < 1.$$

*Then*

$$\Delta(\mathcal{C}) = \Delta(\mathcal{K}).$$

We may suppose without loss of generality that $\mathcal{K}$, and so $\mathcal{C}$, is open since the presence or absence of boundary points does not affect the value of the lattice constants $\Delta(\mathcal{C}), \Delta(\mathcal{K})$. It was shown already that

$$\Delta(\mathcal{C}) \leqq \Delta(\mathcal{K})$$

whether or not $\mathcal{K}$ is convex, so it remains only to show that

$$d(\Lambda) \geqq \Delta(\mathcal{K}) \tag{1}$$

for any $\mathcal{C}$-admissible lattice $\Lambda$.

We prove (1) by computing in two ways the number $N = N(X)$ of points of $\Lambda$ in a large cube

$$|x_j| < X \qquad (1 \leqq j \leqq 3).$$

In the first place we have the trivial estimate

$$d(\Lambda) N = 2^3 X^3 + O(X^2) \tag{2}$$

as $X \to \infty$.

By Theorem III, since $\Lambda$ is $\mathcal{C}$-admissible, it gives a packing of $\frac{1}{2}\mathcal{C}$. Let $\mathsf{C}$ be the set of $N$ cylinders

$$\tfrac{1}{2}\mathcal{C} + \boldsymbol{a}, \tag{3}$$

where

$$\boldsymbol{a} \in \Lambda, \quad \max_j |a_j| < X. \tag{4}$$

These cylinders are all contained in the cube

$$\max_j |x_j| < X + R, \tag{5}$$

if $\frac{1}{2}\mathscr{C}$ is contained in $|\boldsymbol{x}| < R$. We consider only the packing of the cylinders $\mathsf{C}$ in (5).

For $|y| < X + R$, let $L(y)$ be the number of cylinders of $\mathsf{C}$ which meet the plane $x_3 = y$, that is the number of $\boldsymbol{a} \in \Lambda$ satisfying (4) for which

$$|a_3 - y| < \tfrac{1}{2}.$$

These $L(y)$ cylinders give rise to a packing in the square

$$|x_j| < X + R \quad (j = 1, 2)$$

of $L(y)$ sets similar and similarly situated to $\frac{1}{2}\mathscr{K}$. Hence

$$L(y)\, U'(6) < 4(X + R)^2 \tag{6}$$

by Theorem VIII, where $U'(6)$ is infimum of the areas of circumscribed $m$-gons to $\frac{1}{2}\mathscr{K}$ with $m \leq 6$.

But clearly

$$\int_{-X-R}^{X+R} L(y)\, dy = N$$

from the definition of $L(y)$. Hence

$$U'(6)\, N < 8(X + R)^3, \tag{7}$$

by (6).

Since $R$ and $U'(6)$ are independent of $X$, the comparison of (1) and (7) as $X \to \infty$ gives

$$d(\Lambda) \geq U'(6).$$

But

$$U'(6) = 4\Delta(\tfrac{1}{2}\mathscr{K}) = \Delta(\mathscr{K})$$

by Lemma 3, Corollary. This completes the proof of (1), and so of the theorem.

**IX.7. Packing of spheres.** The unit sphere

$$\mathscr{D}_n: \quad |\boldsymbol{x}| < 1$$

in $n$ dimensions has volume

$$V_n = V(\mathscr{D}_n) = \frac{\pi^{\frac{1}{2}n}}{\Gamma\left(1 + \frac{n}{2}\right)}, \tag{1}$$

where $\Gamma\left(1+\frac{n}{2}\right)$ is the usual gamma function. In this section we estimate the lattice constant

$$\Gamma_n = \Delta(\mathcal{D}_n); \tag{2}$$

and are primarily interested in the behaviour of $\Gamma_n$ when $n$ is large.

In the literature it is customary to use $\gamma_n$ defined as the lower bound of the constants $\gamma_n'$ such that every positive definite quadratic form $\sum f_{ij} x_i x_j$ in $n$ variables represents a number $\leqq \gamma_n' \,|\det(f_{ij})|^{1/n}$ („HERMITE's Constant"). By the arguments of Chapter I, § 3 we have

$$\gamma_n^n = \Gamma_n^{-2}. \tag{3}$$

We shall need to know the asymptotic behaviour of the volume $V_n$. From STIRLING's formula[1] we have

$$\lim_{n\to\infty} n V_n^{2/n} = 2\pi e, \tag{4}$$

where

$$e = \sum (r!)^{-1}.$$

From MINKOWSKI's convex body Theorem and the Minkowski-Hlawka Theorem we have

$$\{2\zeta(n)\}^{-1} V_n \geqq \Gamma_n \geqq 2^{-n} V_n, \tag{5}$$

where $\zeta(n)$ is RIEMANN's function. These inequalities lead by (3) and (4) to

$$\limsup_{n\to\infty} n\gamma_n^{-1} \leqq 2\pi e \tag{6}$$

and

$$\liminf_{n\to\infty} n\gamma_n^{-1} \geqq \tfrac{1}{2}\pi e. \tag{7}$$

Of course the factor $2\zeta(n)$ in (5) has no effect in (6) and might as well have been replaced by 1. Indeed none of the improvements of the Minkowski-Hlawka Theorem discussed in Chapter VI affect the constant on the right-hand side of (6). On the other hand, BLICHFELDT (1929a) has improved (7) to

$$\liminf n\gamma_n^{-1} \geqq \pi e, \tag{8}$$

which appears to be the best asymptotic form to date[2]. The argument is a purely packing one and makes no use of the fact that only lattice packings are relevant to (8). BLICHFELDT's results have been improved by RANKIN (1947a), and yet further, by a more perspicuous argument,

[1] See any reputable text book on analysis, for example WHITTAKER and WATSON (1902a) Chapter XII.

[2] The improvement in (8) announced by CHABAUTY (1952a) is not correct, see the review by RANKIN in Maths. Reviews 14, 541.

by ROGERS (1958c). Their methods yield considerable improvements for small values of $n$, but do not improve the constant in (8).

BLICHFELDT'S methods may be applied to other sets than spheres, see RANKIN (1949a, b, c) and 1955a and the literature cited there.

There is a detailed discussion of the non-lattice packings of 3-dimensional spheres in FEJES TÓTH (1953a), see also S. MELMORE (1947a).

I have been helped by my recollection of a seminar talk by Professor RANKIN in Cambridge in the late 1940s on BLICHFELDT'S method.

**IX.7.2.** We observe first that BLICHFELDT'S Theorem I of Chapter III may be generalized to packings and indeed takes a quite simple shape. Let $\mathscr{S}$ be any bounded $n$-dimensional set and suppose that the sets

$$\mathscr{S}_r = \mathscr{S} + \boldsymbol{x}_r \qquad (1 \leqq r \leqq R) \tag{1}$$

are packed in some set $\mathscr{T}$. Then trivially

$$V(\mathscr{T}) \geqq R\, V(\mathscr{S}). \tag{2}$$

Suppose now that there is some function $\varphi(\boldsymbol{x})$ of the vector variable $\boldsymbol{x}$ such that

(i) $\varphi(\boldsymbol{x}) = 0$ if $|\boldsymbol{x}| \geqq \varrho$ for some $\varrho$

and

(ii) $\psi(\boldsymbol{x}) = \sum \varphi(\boldsymbol{x} - \boldsymbol{x}_r) \leqq 1$ for all $\boldsymbol{x}$,

whenever (1) is a packing of $\mathscr{S}$.

Let $\mathscr{T}(\varrho)$ be the set of points at a distance $\leqq \varrho$ from $\mathscr{T}$, including the points of $\mathscr{T}$ itself. Then, in the first place,

$$\int_{\mathscr{T}(\varrho)} \psi(\boldsymbol{x})\, d\boldsymbol{x} \leqq V\{\mathscr{T}(\varrho)\} \qquad (d\boldsymbol{x} = d x_1 \ldots d x_n) \tag{3}$$

by (ii) and, on the other hand, by (i)

$$\int_{\mathscr{T}(\varrho)} \psi(\boldsymbol{x})\, d\boldsymbol{x} = \sum_r \int \varphi(\boldsymbol{x} - \boldsymbol{x}_r)\, d\boldsymbol{x} = R \int \varphi(\boldsymbol{x})\, d\boldsymbol{x} = R\, V(\varphi) \quad \text{(say)}, \tag{4}$$

since all points with $\varphi(\boldsymbol{x} - \boldsymbol{x}_r) \neq 0$ lie in $\mathscr{T}(\varrho)$. The comparison of (3) and (4) gives

$$R \leqq V\{\mathscr{T}(\varrho)\}/V(\varphi). \tag{5}$$

Of course the characteristic function of $\mathscr{S}$, which is 1 on $\mathscr{S}$ and 0 elsewhere, has the properties (i) and (ii). With this as the function $\varphi$, the inequality (5) is rather weaker than (2), because we have replaced $V(\mathscr{T})$ by $V\{\mathscr{T}(\varrho)\}$: though of course this can be avoided by a refinement of the argument. BLICHFELDT observed that there are sometimes functions $\varphi$ which give a better estimate than the characteristic function.

For example, if $\mathscr{S} = \mathscr{D}_n$ is the sphere of unit radius, the necessary and sufficient condition that open spheres of radius 1 and centres $\boldsymbol{x}_1, \boldsymbol{x}_2$ shall not overlap is $|\boldsymbol{x}_1 - \boldsymbol{x}_2| \geqq 2$. The following lemma may be regarded heuristically as showing that, in a packing of spheres, a point may be on the boundary of two spheres but cannot be too near the boundaries of more than two spheres simultaneously.

LEMMA 4. *Put*

$$\varphi(\boldsymbol{x}) = \max\{0, 1 - \tfrac{1}{2}|\boldsymbol{x}|^2\}. \tag{6}$$

*Suppose that* $\boldsymbol{x}_r$ $(1 \leqq r \leqq R)$ *are any points such that*

$$|\boldsymbol{x}_r - \boldsymbol{x}_s| \geqq 2 \qquad (1 \leqq r < s \leqq R). \tag{7}$$

*Then*

$$\sum_{1 \leqq r \leqq R} \varphi(\boldsymbol{x} - \boldsymbol{x}_r) \leqq 1 \tag{8}$$

*for all points* $\boldsymbol{x}$.

We may clearly suppose without loss of generality that

$$0 < \varphi(\boldsymbol{x} - \boldsymbol{x}_r) = 1 - \tfrac{1}{2}|\boldsymbol{x} - \boldsymbol{x}_r|^2$$

for $1 \leqq r \leqq R$.

If $y_1, \ldots, y_R$ and $y$ are any real numbers, we have

$$R \sum_r (y - y_r)^2 = \sum_{r<s} (y_r - y_s)^2 + \Big(R y - \sum_r y_r\Big)^2 \geqq \sum_{r<s} (y_r - y_s)^2.$$

Hence, on applying this to the individual co-ordinates, since $|\boldsymbol{x}|^2 = x_1^2 + \cdots + x_n^2$, we have

$$R \sum_{1 \leqq r \leqq R} |\boldsymbol{x} - \boldsymbol{x}_r|^2 \geqq \sum_{r<s} |\boldsymbol{x}_r - \boldsymbol{x}_s|^2 \geqq 2R(R-1),$$

by (2). But this is just the same as (8).

From this we have almost immediately

THEOREM X. *Let* $\boldsymbol{x}_r$ $(1 \leqq r \leqq R)$ *be points in the n-dimensional sphere*

$$|\boldsymbol{x}| < X, \tag{9}$$

*and let*

$$|\boldsymbol{x}_r - \boldsymbol{x}_s| \geqq 2 \qquad (1 \leqq r < s \leqq R).$$

*Then*

$$R \leqq 2^{-n/2}\left(1 + \frac{n}{2}\right)(X + 2^{\frac{1}{2}})^n. \tag{10}$$

If $\varphi(\boldsymbol{x})$ is as in Lemma 4, we have

$$V(\varphi) = \int \varphi(\boldsymbol{x})\, d\boldsymbol{x} = \int_{|\boldsymbol{x}|^2 < 2} \left(1 - \frac{1}{2}|\boldsymbol{x}|^2\right) d\boldsymbol{x} = 2^{n/2}\left(1 + \frac{n}{2}\right)^{-1} V_n,$$

where $V_n$ is the volume of the unit sphere. The result now follows from (5), since $\mathcal{T}(\varrho)$ is now the sphere

$$|\boldsymbol{x}| < X + 2^{\frac{1}{2}},$$

which has volume $(X+2^{\frac{1}{2}})^n V_n$.

COROLLARY 1. *The lattice constant* $\Gamma_n$ *and volume* $V_n$ *of the unit sphere* $|\boldsymbol{x}|<1$ *satisfy*

$$\Gamma_n \geqq 2^{-n/2}\left(1+\frac{n}{2}\right)^{-1} V_n.$$

If $\Lambda$ is admissible for $|\boldsymbol{x}|<1$, then the points $\boldsymbol{x}_r$ of $2\Lambda$ satisfy the conditions of the theorem. The number of points of $2\Lambda$ in

$$\mathcal{T}: \ |\boldsymbol{x}|<X$$

is

$$\{d(2\Lambda)\}^{-1} V(\mathcal{T}) + O(X^{n-1}) = 2^{-n}\{d(\Lambda)\}^{-1} X^n V_n + O(X^{n-1}).$$

On comparing with the theorem and letting $X\to\infty$, we obtain the required inequality.

COROLLARY 2.

$$\liminf n\gamma_n^{-1} \geqq \pi e,$$

*where* $\gamma_n^n = \Gamma_n^{-2}$.

This follows from Corollary 1 and (4) of § 7.1.

**IX.8. The product of $n$ linear forms.** Denote by $\mathcal{N}_n$ the $n$-dimensional set

$$\mathcal{N}_n: \ |x_1 \dots x_n| < 1,$$

and let

$$\Delta(\mathcal{N}_n) = \nu_n^n.$$

The set $\mathcal{N}_n$ plays an important part in algebraic number theory (see Chapter X), but the only precise values of $\nu_n$ known are

$$\nu_2^2 = 5^{\frac{1}{2}}, \quad \nu_3^3 = 7$$

given by Chapter II, Theorem IV and Chapter X, Theorem V respectively. Here we shall be concerned with estimates for $\nu_n$ when $n$ is large rather as in § 7. For information about what is known for $n=4$ or 5 see Chapter II, § 6.4.

In Chapter III, § 5.3 we already gave MINKOWSKI's estimate

$$\Delta(\mathcal{N}_n) \geqq \frac{n^n}{n!}$$

which by STIRLING's Formula, gives

$$\liminf_{n\to\infty} \nu_n \geqq e = 2\cdot 71828\dots.$$

BLICHFELDT has given an elegant proof that

$$\liminf_{n\to\infty} \nu_n \geqq (2\pi)^{\frac{1}{2}} e^{\frac{3}{4}} = 5 \cdot 30653\ldots: \tag{1}$$

and this we obtain in this section.

The estimate (1) is not the best yet found. ROGERS (1950a) has shown indeed that

$$\liminf_{n\to\infty} \nu_n \geqq 4\pi^{-1} e^{\frac{3}{2}} = 5 \cdot 70626\ldots.$$

His intricate and laborious proof may be regarded as an elaboration of BLICHFELDT'S.

Since $\mathcal{N}_n$ has infinite volume, there is no estimate of $\Delta(\mathcal{N}_n)$ above from the Minkowski-Hlawka Theorem. Indeed, work of SCHOLZ (1938a) on the discriminants of algebraic number fields gives some reason to suspect that $\limsup_{n\to\infty} \nu_n = \infty$.

In § 8.2 and 8.3 we give two lemmas and then in § 8.4 BLICHFELDT'S proof of (1).

**IX.8.2.** The following Lemma of SCHUR (1918a) also occurs in the theory of the "transfinite diameter" in analysis.

LEMMA 6. *Let $\xi_1, \ldots, \xi_n$ be real numbers. Then*

$$\prod_{i<j} (\xi_i - \xi_j)^2 \leqq \vartheta_m \Big(\sum_i \xi_i^2\Big)^{\frac{1}{2}m(m-1)}, \tag{1}$$

*where*

$$\vartheta_m = \{m(m-1)\}^{-\frac{1}{2}m(m-1)} \cdot 1^2 \cdot 2^2 \cdot \ldots \cdot m^m. \tag{2}$$

The continuous function $\prod_{i<j} (\xi_i - \xi_j)^2$ of the $m$ variables $\xi_i$ attains its maximum, $\vartheta$ say, on $\sum \xi_i^2 = 1$, say at $\xi_i = \eta_i$ $(1 \leqq i \leqq m)$. Then, by homogeneity,

$$\Big(\sum_i \xi_i^2\Big)^{-\frac{1}{2}m(m-1)} \prod_{i<j} (\xi_i - \xi_j)^2 \leqq \vartheta \tag{3}$$

for all $\xi_i$, with equality when $(\xi_i) = (\eta_i)$. The derivative of the logarithm of the left-hand side of (3) with respect to each variable must vanish at the maximum $(\xi_i) = (\eta_i)$; and so

$$\sum_{j\neq i} \frac{1}{\eta_i - \eta_j} = \frac{m(m-1)\eta_i}{2} \quad (1 \leqq i \leqq m), \tag{4}$$

since $\sum \eta_i^2 = 1$. Let

$$f(\eta) = \prod_i (\eta - \eta_i) \tag{5}$$

be a polynomial in the variable $\eta$. Then (4) is

$$\frac{f''(\eta_i)}{2f'(\eta_i)} = \frac{m(m-1)\eta_i}{2}. \tag{6}$$

The polynomial

$$h(\eta) = f''(\eta) - m(m-1)\eta f'(\eta) + m^2(m-1) f(\eta)$$

is of degree at most $m-1$, since the coefficients of $\eta^m$ vanishes. By (5) and (6) we have $h(\eta_i) = 0$ $(1 \leqq i \leqq m)$; and so $h(\eta)$ vanishes identically:

$$f''(\eta) - m(m-1)\eta f'(\eta) + m^2(m-1) f(\eta) = 0. \tag{7}$$

The differential equation (7) determines $f(\eta)$ completely in terms of, say, $f(0)$ and $f'(0)$. Hence we may determine the symmetric functions $\sum \eta_i^2$ and $\prod (\eta_i - \eta_j)^2$ in terms of $f(0)$ and $f'(0)$. Since $\sum \eta_j^2 = 1$ and the coefficient of $\eta^m$ in $f(\eta)$ is 1, this determines $f(\eta)$ completely, and so also

$$\prod (\eta_i - \eta_j)^2 = \vartheta. \tag{8}$$

It is simpler, however, to use a more indirect approach which will now be described.

The resultant of two polynomials, say,

$$\varphi(\eta) = \prod_{1 \leqq i \leqq I} (\eta - \alpha_i), \quad \psi(\eta) = \prod_{1 \leqq j \leqq J} (\eta - \beta_i)$$

with highest coefficient 1 is defined to be

$$R(\varphi, \psi) = \prod_{i,j} (\alpha_i - \beta_j) \tag{9$_1$}$$

$$= \prod_i \psi(\alpha_i) \tag{9$_2$}$$

$$= (-1)^{IJ} \prod_j \varphi(\beta_j) \tag{9$_3$}$$

$$= (-1)^{IJ} R(\psi, \varphi). \tag{9$_4$}$$

If

$$\omega(\eta) = \prod_{1 \leqq k \leqq K} (\eta - \gamma_k)$$

is a third polynomial with highest coefficient 1, and if

$$\omega(\eta) = \lambda \psi(\eta) + \chi(\eta) \varphi(\eta)$$

identically for some number $\lambda$ and polynomial $\chi(\eta)$, then

$$R(\varphi, \omega) = \lambda^I R(\varphi, \psi), \tag{10}$$

by (9$_2$).

In particular, if $f(\eta)$ is defined by (5), we have

$$\left.\begin{aligned} \vartheta = \prod_{1 \leqq i < j \leqq m} (\eta_i - \eta_j)^2 &= (-1)^{\frac{1}{2}m(m-1)} \prod_{1 \leqq i \leqq m} f'(\eta_i) \\ &= (-1)^{\frac{1}{2}m(m-1)} m^m R(f, f_1), \end{aligned}\right\} \tag{11}$$

where

$$f_1(\eta) = m^{-1} f'(\eta)$$

has highest coefficient 1. More generally, put

$$f_k(\eta) = \frac{(m-k)!}{m!} f^{(k)}(\eta);$$

so that $f_k(\eta)$ has highest coefficient 1. Then on differentiating (7) $k$ times one readily obtains

$$(m-k-1) f_{k+2}(\eta) - m(m-1)\,\eta\, f_{k+1}(\eta) + m(m-1) f_k(\eta) = 0. \qquad (12)$$

Hence

$$R(f_k, f_{k+1}) = R(f_{k+1}, f_k) = \left\{\frac{-(m-k-1)}{m(m-1)}\right\}^{m-k-1} R(f_{k+1}, f_{k+2}), \qquad (13)$$

on using the rules of operation $(9_4)$ and (10). But $f_m(\eta) = 1$ and $f_{m-1}(\eta) = \eta + \gamma$ for some number $\gamma$ (in fact $\gamma = 0$); so

$$R(f_{m-1}, f_m) = 1 \qquad (14)$$

by $(9_2)$. The required value (2) for $\vartheta$ follows now from (11), (13) and (14).

**IX.8.3.** We also require an estimate of the number $\vartheta_m$ occurring in the last lemma.

LEMMA 7. *If*

$$G_m = 1 \cdot 2^2 \cdot \dots \cdot m^m,$$

*then*

$$\limsup_{m\to\infty} \{m^{-2} \log G_m - \tfrac{1}{2} \log m\} \leqq -\tfrac{1}{4}.$$

Put

$$g(x) = x \log x \qquad (x > 0).$$

Then

$$g'(x) = \log x + 1$$

increases with $x$; and so

$$g(x+t) + g(x-t) \geqq 2g(x) \qquad (1)$$

for any $t$, since if $t > 0$ we have

$$g(x+t) - g(x) = t g'(\xi_1)$$
$$g(x) - g(x-t) = t g'(\xi_2),$$

where $\xi_2 < x < \xi_1$, so $g'(\xi_2) < g'(\xi_1)$.

In particular,

$$\int_{l-\frac{1}{2}}^{l+\frac{1}{2}} g(x)\,dx = \int_0^{\frac{1}{2}} \{g(l+t) + g(l-t)\}\,dt \geqq g(l), \qquad (2)$$

for any integer $l$. Thus

$$\log G_m = \sum_{2 \leq l \leq m} g(l) \leq \int_{\frac{3}{2}}^{m+\frac{1}{2}} g(x)\, dx = \tfrac{1}{2}(m+\tfrac{1}{2})^2 \log(m+\tfrac{1}{2}) - \tfrac{1}{4}(m+\tfrac{1}{2})^2 + \gamma,$$

where $\gamma$ is independent of $m$. The required estimate now follows at once.

COROLLARY. *If* $\vartheta_m = \{m(m-1)\}^{-\frac{1}{2}m(m-1)} G_m$ *is the number defined in Lemma 6, then*

$$\limsup_{m\to\infty} \{m^{-2} \log \vartheta_m + \tfrac{1}{2} \log m\} \leq -\tfrac{1}{4}.$$

This is immediate. It is not difficult to see that "lim sup" may be replaced by "lim", but we do not need this.

**IX.8.4.** We can now prove BLICHFELDT's Theorem on the product of linear forms discussed in § 8.1.

THEOREM XI. *Let* $\nu_n^n$ *be the lattice constant of the set*

$$\mathcal{N}: \quad |x_1 \ldots x_n| < 1.$$

*Then*

$$\liminf \nu_n \geq (2\pi)^{\frac{1}{2}} e^{\frac{3}{2}}. \tag{1}$$

Let $\Lambda$ be a lattice which is admissible for $\mathcal{N}$, and let $m$ be an integer whose value will be settled later.

Consider the sphere

$$\mathcal{D}: \quad |\boldsymbol{x}| < \varrho,$$

where $\varrho$ is chosen so that

$$V(\mathcal{D}) = m\, d(\Lambda);$$

that is

$$\varrho^n V_n = m\, d(\Lambda), \tag{2}$$

where $V_n$ is the volume of $|\boldsymbol{x}|<1$. By BLICHFELDT's Theorem I of Chapter III, there are $m$ points $\boldsymbol{x}_1, \ldots, \boldsymbol{x}_m$ in $\mathcal{D}$ whose differences $\boldsymbol{x}_i - \boldsymbol{x}_j$ all lie in $\Lambda$. Put

$$\boldsymbol{x}_i = (x_{1i}, \ldots, x_{ni}) \qquad (1 \leq i \leq m),$$

and write

$$S_k = \sum_{1 \leq i \leq m} x_{ki}^2 \qquad (1 \leq k \leq n).$$

Then

$$\sum_{1 \leq k \leq n} S_k = \sum_{1 \leq i \leq m} |\boldsymbol{x}_i|^2 \leq m \varrho^2;$$

and so

$$\prod_{1 \leq k \leq n} S_k \leq \left(\frac{m}{n} \varrho^2\right)^n, \tag{3}$$

by the inequality of the arithmetic and geometric means.

Now let

$$P_k = \prod_{1 \leq i < j \leq m} (x_{ki} - x_{kj})^2.$$

Then, on the one hand

$$P_k \leq \vartheta_m S_k^{\frac{1}{2}m(m-1)} \tag{4}$$

by Lemma 6, where $\vartheta_m$ is the number defined there. On the other hand,

$$\prod_{1 \leq k \leq n} P_k = \prod_{1 \leq i < j \leq m} f^2(\boldsymbol{x}_i - \boldsymbol{x}_j),$$

where

$$f(\boldsymbol{x}) = x_1 \dots x_n.$$

The points $\boldsymbol{x}_i - \boldsymbol{x}_j$ belong to $\Lambda$, which is $\mathcal{N}$-admissible; and so

$$|f(\boldsymbol{x}_i - \boldsymbol{x}_j)| \geq 1 \qquad (i \neq j).$$

Hence

$$\prod_{1 \leq k \leq n} P_k \geq 1. \tag{5}$$

On eliminating $P_k$, $S_k$ from (3), (4) and (5) we get

$$1 \leq \vartheta_m^n \left(\frac{m}{n}\,\varrho^2\right)^{\frac{1}{2}nm(m-1)}. \tag{6}$$

Hence, on eliminating $\varrho$ between (2) and (6),

$$\{d(\Lambda)\}^{1/n} \geq \chi_1 \chi_2 \chi_3, \tag{7}$$

where

$$\chi_1 = m^{-\frac{1}{2}} \vartheta_m^{-1/m(m-1)},$$

$$\chi_2 = n^{\frac{1}{2}} V_n^{1/n},$$

and

$$\chi_3 = m^{-1/n}.$$

Now $\chi_1$ is independent of $n$ and

$$\liminf_{m \to \infty} \chi_1 \geq e^{\frac{1}{2}}, \tag{8}$$

by Lemma 7 Corollary. Further, $\chi_2$ is independent of $m$, and

$$\lim_{n \to \infty} \chi_2 = (2\pi e)^{\frac{1}{2}} \tag{9}$$

by (4) of § 7.1. Finally,

$$\lim \chi_3 = 1 \tag{10}$$

if, say, $m = n \to \infty$.

Since $v_n$ is the infimum of $\{d(\Lambda)\}^{1/n}$ over $\mathcal{N}$-admissible lattices, and since the product of the right-hand sides of (8), (9) and (10) is the right-hand side of (1), this proves the theorem, by (7).

Chapter X

# Automorphs

**X.1. Introduction.** A homogeneous linear transformation $\boldsymbol{\omega}$ is said to be an automorph of a point set $\mathscr{S}$ if $\mathscr{S}$ is just the set of points $\boldsymbol{\omega x}$, $\boldsymbol{x} \in \mathscr{S}$. The automorphs of a set $\mathscr{S}$ evidently form a group. Many of the point sets of interest in the geometry of numbers, or which occur naturally in problems arising in other branches of number-theory, have a rich group of automorphs which is reflected in the set of $\mathscr{S}$-admissible lattices. Already in the work in which he introduced the notion of limit of a sequence of lattices, MAHLER (1946d, e) laid the foundations for future work and indicated some fundamental theorems. Since then much has been done but some challenging and natural questions remain unanswered.

MAHLER (1946d, e) considers star-bodies with groups of automorphisms having special properties which he calls automorphic star-bodies. In this account we prefer in each case to state the properties of the group of automorphs which are required to hold.

We shall say that a homogeneous linear transformation $\boldsymbol{\omega}$ is an automorph of a lattice $\Lambda$ if $\boldsymbol{\omega}\Lambda = \Lambda$, that is if $\Lambda$ is precisely the set of $\boldsymbol{\omega a}$, $\boldsymbol{a} \in \Lambda$. This is really a special case of the definition at the beginning of the chapter since $\Lambda$ is a point set. Since

$$d(\boldsymbol{\omega}\Lambda) = |\det(\boldsymbol{\omega})|\, d(\Lambda),$$

we must have

$$\det(\boldsymbol{\omega}) = \pm 1.$$

We say that $\boldsymbol{\omega}$ is an automorph of a function $f(\boldsymbol{x})$ of the vector $\boldsymbol{x}$ if

$$f(\boldsymbol{\omega x}) = f(\boldsymbol{x}),$$

for all $\boldsymbol{x}$. In particular, $\boldsymbol{\omega}$ is an automorph of the distance-function $F(\boldsymbol{x})$ if and only if it is an automorph of the star-body

$$\mathscr{S}:\quad F(\boldsymbol{x}) < 1,$$

since $\mathscr{S}$ and $F(\boldsymbol{x})$ determine each other uniquely. Clearly

$$F(\boldsymbol{\omega}\Lambda) = F(\Lambda)$$

for a lattice $\Lambda$ if $\boldsymbol{\omega}$ is an automorph of the distance-function $F(\boldsymbol{x})$, since

$$F(\Lambda) = \inf_{\substack{\boldsymbol{a} \in \Lambda \\ \neq \boldsymbol{o}}} F(\boldsymbol{a}),$$

by definition.

If $\mathscr{S}$ is any point set and $\boldsymbol{\tau}$ a non-singular homogeneous linear transformation, then

$$\Delta(\boldsymbol{\tau}\mathscr{S}) = |\det(\boldsymbol{\tau})|\, \Delta(\mathscr{S}),$$

since a lattice $\Lambda$ is admissible for $\mathscr{S}$ if and only if $\boldsymbol{\tau}\Lambda$ is admissible for $\boldsymbol{\tau}\mathscr{S}$.

In this chapter we shall make great use of the properties of homogeneous linear transformation expounded in Chapter V, § 2. In particular we write

$$\boldsymbol{\varphi} = \boldsymbol{\rho} + \boldsymbol{\sigma}, \qquad \boldsymbol{\psi} = \boldsymbol{\rho}\boldsymbol{\sigma}$$

if

$$\boldsymbol{\varphi}\boldsymbol{x} = \boldsymbol{\rho}\boldsymbol{x} + \boldsymbol{\sigma}\boldsymbol{x}, \quad \boldsymbol{\psi}\boldsymbol{x} = \boldsymbol{\rho}(\boldsymbol{\sigma}\boldsymbol{x}),$$

respectively for all $\boldsymbol{x}$.

**X.1.2.** We first give three theorems which are already in MAHLER (1946a, b) but not all as formulated here. We give also the proofs: their brevity shows the power of MAHLER's techniques, particularly in the striking Theorem III.

THEOREM I. *Let $F(\boldsymbol{x})$ be a distance-function with an automorph $\boldsymbol{\omega}$ such that*

$$\det(\boldsymbol{\omega}) \neq \pm 1.$$

*Then $F(\Lambda) = 0$ for all lattices $\Lambda$.*

By taking $\boldsymbol{\omega}^{-1}$ instead of $\boldsymbol{\omega}$ if need be, we may suppose that

$$|\det(\boldsymbol{\omega})| < 1.$$

If there is a lattice $\Lambda$ with $F(\Lambda) \neq 0$, then there is a critical lattice $\mathsf{M}$ for $F(\boldsymbol{x}) < 1$, by Theorem VI of Chapter V. But then

$$F(\boldsymbol{\omega}\mathsf{M}) = F(\mathsf{M}) = 1,$$

and

$$d(\boldsymbol{\omega}\mathsf{M}) = |\det(\boldsymbol{\omega})|\, d(\mathsf{M}) < d(\mathsf{M}),$$

in contradiction to the definition of a critical lattice.

For example, Theorem I shows that

$$|x_1^2 x_2| < 1$$

is of infinite type since it has the automorphs $x_1 \to \frac{1}{2}x$, $x_2 \to 4x$ of determinant 2. This was our example of a star-body of infinite type in § 5 of Chapter V.

THEOREM II. *Let $F(\boldsymbol{x})$ be a distance-function. Suppose that every point $\boldsymbol{x}_0$ with $F(\boldsymbol{x}_0) = 1$ is of the shape*

$$\boldsymbol{x}_0 = \boldsymbol{\omega}\boldsymbol{c}_0, \tag{1}$$

*where* $\boldsymbol{\omega}$ *is an automorph of* $F$, *and* $\mathbf{c}_0$ *is in a compact set* $\mathscr{C}$. *Then for every lattice* $\Lambda$ *with* $F(\Lambda)=1$ *there exists a lattice* $\mathsf{M}$ *with*

$$F(\mathsf{M})=1, \quad d(\Lambda)=d(\mathsf{M})$$

*having a point in* $\mathscr{C}$.

Since $F(\boldsymbol{x})$ is continuous, the set $\mathscr{C}'$ of points $\mathbf{c}\in\mathscr{C}$ with $F(\mathbf{c})=1$ is compact if $\mathscr{C}$ is compact. Since $\mathbf{c}_0$ as defined in (1) has $F(\mathbf{c}_0)=F(\boldsymbol{x}_0)=1$, we may suppose without loss of generality that

$$F(\mathbf{c})=1 \qquad (\mathbf{c}\in\mathscr{C}). \tag{2}$$

Since $F(\Lambda)=1$, there is a sequence of points $\boldsymbol{a}_r\in\Lambda$, not necessarily distinct, such that

$$F(\boldsymbol{a}_r)\geqq 1: \quad F(\boldsymbol{a}_r)\to 1 \qquad (r\to\infty).$$

Then $\boldsymbol{b}_r=\{F(\boldsymbol{a}_r)\}^{-1}\boldsymbol{a}_r$ satisfies $F(\boldsymbol{b}_r)=1$; and so

$$\boldsymbol{b}_r=\boldsymbol{\omega}_r\boldsymbol{c}_r$$

for some automorph $\boldsymbol{\omega}_r$ of $F$ and some $\mathbf{c}_r\in\mathscr{C}$. Since $\mathscr{C}$ is compact, we may suppose, after extracting a subsequence and re-indexing, that

$$\mathbf{c}_r\to\mathbf{c}'\in\mathscr{C} \qquad (r\to\infty).$$

Let

$$\Lambda=\boldsymbol{\omega}_r\Lambda_r.$$

Then, since $|\det(\boldsymbol{\omega}_r)|=1$ by Theorem I, we have

$$F(\Lambda_r)=F(\Lambda)=1, \quad d(\Lambda_r)=d(\Lambda)$$

and

$$F(\boldsymbol{a}_r)\,\mathbf{c}_r\in\Lambda_r.$$

By Theorem IV, Corollary of Chapter V, the sequence $\Lambda_r$ contains a convergent subsequence, and so, without loss of generality,

$$\Lambda_r\to\mathsf{M}$$

for some lattice $\mathsf{M}$. Then

$$d(\mathsf{M})=\lim_{r\to\infty} d(\Lambda_r)=d(\Lambda)$$

and

$$F(\mathsf{M})\geqq\limsup_{r\to\infty} F(\Lambda_r)=F(\Lambda)=1 \tag{3}$$

by Theorem II of Chapter V. Further, $\mathsf{M}$ contains

$$\boldsymbol{c}'=\lim_{r\to\infty} F(\boldsymbol{a}_r)\,\mathbf{c}_r,$$

so

$$F(\mathsf{M})\leqq F(\mathbf{c}')=1, \tag{4}$$

by (2). From (3) and (4) we have $F(\mathsf{M})=1$. This concludes the proof.

COROLLARY. *There is a critical lattice for $F(\mathbf{x})<1$ having a point $\mathbf{c}$ in $\mathscr{C}$ with $F(\mathbf{c})=1$.*

For if $\Lambda$ is critical so is M. This corollary is in contrast with the example given in § 5.2 of Chapter V of a star-body no critical lattice of which has points on the boundary. Note that the corollary does not affirm that every critical lattice of $F(\mathbf{x})<1$ has points on $F(\mathbf{x})=1$; the author [CASSELS (1948a)] has given a rather artificial[1] counter-example of a body $F(\mathbf{x})<1$ satisfying the hypotheses of Theorem II and having critical lattices with no point on $F(\mathbf{x})=1$.

As an example of Theorem II consider the body $\mathscr{N}$: $|x_1 x_2 x_3|<1$ with its distance-function $|x_1 x_2 x_3|^{\frac{1}{3}}$. Here $\mathscr{C}$ may be taken to be the single point $\mathbf{c}=(1, 1, 1)$; since every point $\mathbf{x}_0=(x_{10}, x_{20}, x_{30})$ with $|x_{10} x_{20} x_{30}|=1$ is of the shape

$$\mathbf{x}_0=\boldsymbol{\omega}\,\mathbf{c},$$

where $\boldsymbol{\omega}$ is the automorph

$$x_j \rightarrow x_{j0}\, x_j \qquad (1 \leqq j \leqq 3)$$

of $\mathscr{N}$. Hence there are critical lattices for $\mathscr{N}$ with a point at $(1, 1, 1)$. If one is concerned only with the evaluation of $\Delta(\mathscr{N})$ and not with the enumeration of all the critical lattices, it is enough to consider critical lattices with a point at $(1, 1, 1)$.

THEOREM III. *Let the point-set $\mathscr{T}$ be a subset of the star-body $\mathscr{S}$ with $\Delta(\mathscr{S})<\infty$. Suppose that for every $r$ there is an automorph $\boldsymbol{\omega}_r$ of $\mathscr{S}$ such that $\boldsymbol{\omega}_r \mathscr{T}$ contains every point of $\mathscr{S}$ which is in $|\mathbf{x}|<r$. Then*

$$\Delta(\mathscr{T})=\Delta(\mathscr{S}).$$

Clearly

$$\Delta(\mathscr{T}) \leqq \Delta(\mathscr{S}).$$

By Theorem I we have $\det(\boldsymbol{\omega}_r)=\pm 1$, and so

$$\Delta(\mathscr{T})=\Delta(\boldsymbol{\omega}_r \mathscr{T}) \geqq \Delta(\mathscr{S}_r),$$

where $\mathscr{S}_r$ is the set of points of $\mathscr{S}$ in $|\mathbf{x}|<r$. But

$$\lim_{r\to\infty} \Delta(\mathscr{S}_r)=\Delta(\mathscr{S})$$

by Theorem V of Chapter V, so $\Delta(\mathscr{T})=\Delta(\mathscr{S})$, as asserted.

Clearly one may formulate theorems similar to but more general than Theorem III by making use of the full force of Theorems II and V of Chapter V. The argument used in the proof of Theorem III was already used in the proof of Theorem XV of Chapter V.

[1] As Professor ROGERS remarks, it is quite likely that the 3-dimensional body $|x_1| \max(x_2^2, x_3^2)<1$ furnishes a natural example.

As an example of Theorem III one may take for $\mathscr{S}$, $\mathscr{T}$ respectively the sets

$$\mathscr{S}: \quad |x_1 x_2 x_3| < 1$$

and

$$\mathscr{T}: \quad |x_1 x_2 x_3| < 1, \quad |x_2| < \varepsilon, \quad |x_3| < \varepsilon,$$

where $\varepsilon$ is any fixed positive number. Then the automorphism $\boldsymbol{\omega}_r$ may be taken to be

$$X_1 = r^{-2} \varepsilon^2 x_1, \quad X_2 = r \varepsilon^{-1} x_2, \quad X_3 = r \varepsilon^{-1} x_3,$$

where $\boldsymbol{X} = \boldsymbol{\omega}_r \boldsymbol{x}$. In this example one may deduce that a lattice $\Lambda$ with $d(\Lambda) < \Delta(\mathscr{S})$ has infinitely many points in $\mathscr{S}$. For $\Lambda$ must have a point in $\mathscr{T}$ for every $\varepsilon > 0$. If $\Lambda$ has no point $\boldsymbol{a} \neq \boldsymbol{o}$ with $a_2 = a_3 = 0$, this implies that $\Lambda$ has infinitely many points in $\mathscr{S}$; and on the other hand, if $\boldsymbol{a} = (a_1, 0, 0)$ is in $\Lambda$, then all the points $m\boldsymbol{a}$ $(m = 1, 2, \ldots)$ are in $\Lambda$, so there are still infinitely many points of $\Lambda$ in $\mathscr{S}$. Indeed the argument shows that for any $\varepsilon > 0$ there are infinitely many points of $\Lambda$ in $\mathscr{T}$. This sort of argument was already used for Lemma 12 of Chapter V about the existence of infinitely many points in $-1 < x_1 x_2 < k$. There we could prove rather more since this set was shown to be boundedly reducible. In § 7 we shall make a systematic study of when there are infinitely many points of a lattice in a star-body following DAVENPORT and ROGERS (1950a).

**X.1.3.** The point sets with a large group of automorphisms with which we shall be concerned will be mainly constructed simply from an algebraic form $\varphi(\boldsymbol{x})$. For example $\varphi(\boldsymbol{x})$ may be $x_1 x_2$, $x_1 x_2 x_3$, $x_1(x_2^2 + x_3^2)$ or $x_1^2 + x_2^2 - x_3^2$, and the set $\mathscr{S}$ may be defined by

$$|\varphi(\boldsymbol{x})| < 1 \tag{1}$$

or

$$0 \leqq \varphi(\boldsymbol{x}) < 1 \tag{2}$$

or

$$0 < \varphi(\boldsymbol{x}) < 1 \tag{3}$$

or

$$-k < \varphi(\boldsymbol{x}) < l, \tag{4}$$

where $k$ and $l$ are positive numbers. Of course (2) and (3) are not star-bodies. Apart from sets especially constructed from sets of the type (1)—(4) to act as counter-examples, other sets with large groups of automorphisms have proved intractable. For example the lattice constant of

$$|x_1| \max(x_2^2, x_3^2) < 1$$

is not known, though it would be of some interest in the theory of simultaneous approximation and the problem has had considerable attention [see DAVENPORT (1952a) and CASSELS (1955a) and the references given there].

We shall make continual use in this chapter of the results of Chapter I, § 4 about the relationship of lattices to forms.

A particular kind of lattice plays a special rôle in connection with sets of the type (1)—(4), where $\varphi(\boldsymbol{x})$ is an algebraic form. It is useful to introduce some new terminology. If $\varphi(\boldsymbol{a})$ is an integer for all $\boldsymbol{a}\in\Lambda$ we say that $\varphi$ is integral on $\Lambda$. If, further, $\varphi(\boldsymbol{a})=0$ for $\boldsymbol{a}\in\Lambda$ only when $\boldsymbol{a}=\boldsymbol{o}$, we will say that $\varphi$ is non-null on $\Lambda$ (the trivial zero at $\boldsymbol{o}$ being disregarded). Finally, if there is some number $t\neq 0$ such that $t\varphi$ is integral on $\Lambda$ we say that $\varphi$ is proportional to integral on $\Lambda$. Then $\varphi$ is integral on $|t|^{1/m}\Lambda$, where $m$ is the degree of $\varphi$.

In many, if not all, cases where the form $\varphi$ has infinitely many automorphs and the critical lattices $\Lambda_c$ for one of the sets (1)—(4) are known, it turns out that $\varphi$ is proportional to integral on $\Lambda_c$. Indeed in some cases $\varphi$ is proportional to integral on every known admissible lattice, and it is suspected, but not proved, that no other admissible lattices exist. In other cases, there certainly do exist admissible lattices on which $\varphi$ is not proportional to integral, but the critical lattices are not amongst them.

Before discussing the general properties of a lattice $\Lambda$ on which a form[1] $\varphi$ is proportional to integral and illustrating it with concrete examples, it is convenient to prove a simple lemma.

LEMMA 1. *Let $r>0$ and $m>0$ be integers and let*

$$\gamma(u_1, \ldots, u_r)$$

*be arbitrarily given numbers for integers $u_\varrho$ in*

$$0\leqq u_\varrho\leqq m \qquad (1\leqq\varrho\leqq r). \tag{5}$$

*Then there is a uniquely defined polynomial $f(\boldsymbol{u})$ of degree $m$ in the variables $u_1, \ldots, u_r$ such that*

$$f(\boldsymbol{u})=\gamma(\boldsymbol{u}) \tag{6}$$

*for all integers $\boldsymbol{u}=(u_1, \ldots, u_r)$ in* (5).

This is certainly true when $r=1$. For $r>1$ we use induction on $r$. We may write

$$f(\boldsymbol{u})=\sum_{0\leqq\mu\leqq m} u_r^\mu\, g_\mu(u_1, \ldots, u_{r-1}), \tag{7}$$

where the $g_\mu$ are polynomials to be determined. For any fixed values of $u_1, \ldots, u_{r-1}$, the equations (6) determine uniquely the values that must be taken by $g_\mu(u_1, \ldots, u_{r-1})$ in (5); and then there are uniquely

[1] We recollect that the word "form" implies homogeneity.

determined polynomials taking these values, since we assume that the lemma has already been proved with $r-1$ for $r$. Alternatively one could observe that the determinant of the $(m+1)^r$ equations for the $(m+1)^r$ coefficients in $f(u)$ have determinant

$$\prod_{0\leq u<v\leq m} (v-u)^{2m} \neq 0.$$

COROLLARY. *If the* $\gamma(u_1, \ldots, u_r)$ *are rational, so are the coefficients in* $f$.

This follows at once from the proof.

Now let $\varphi$ be a form which is integral on the lattice $\Lambda$ with basis $\boldsymbol{b}_1, \ldots, \boldsymbol{b}_n$. Put

$$f(\boldsymbol{u}) = f(u_1, \ldots, u_n) = \varphi\Big(\sum_j u_j \boldsymbol{b}_j\Big). \tag{8}$$

By Lemma 1, Corollary, the coefficients in the form $f(\boldsymbol{u})$ are rational. Conversely if the coefficients in $f(\boldsymbol{u})$ are rational, then $\varphi$ is proportional to integral on $\Lambda$.

We shall now describe in some detail what happens in some special cases which have been extensively investigated.

Suppose, for example, that

$$\varphi(\boldsymbol{x}) = x_1^2 + x_2^2 - x_3^2,$$

so that $f(\boldsymbol{u})$ in (8) is any indefinite ternary quadratic form of signature (2, 1) (cf. § 4 of Chapter I). No-one has yet been able to construct a ternary quadratic form which can be shown not to take arbitrarily small values for integral $\boldsymbol{u}$, apart from the multiples of forms with integral coefficients. OPPENHEIM (1953b, c) has shown[1] that an indefinite quadratic form which takes arbitrarily small values of one sign also takes arbitrarily small values of the other. Such a form then takes values in every interval, since

$$f(r\boldsymbol{u}) = r^2 f(\boldsymbol{u})$$

and $f(\boldsymbol{u})$ may be taken arbitrarily small of either sign.

The situation is much the same when

$$\varphi(\boldsymbol{x}) = x_1 x_2 x_3.$$

Then the function $f(\boldsymbol{u})$ given by (8) is the product of three real linear forms:

$$f(\boldsymbol{u}) = \prod_{1\leq j\leq 3} (b_{j1}u_1 + b_{j2}u_2 + b_{j3}u_3); \tag{9}$$

[1] He also shows that if an indefinite quadratic form is not a multiple of a form with integral coefficients and takes the value 0 then it also takes arbitrarily small non-zero values for integer values of the variables if the number of variables is greater than 5.

and conversely every product of three linear forms (9) with

$$\det(b_{jk}) \neq 0$$

gives rise in this way to a lattice $\Lambda$. A classical theorem which we shall prove in § 4 states that if the coefficients in $f(\boldsymbol{u})$ are rational and $f(\boldsymbol{u})$ may be expressed as the product of three real linear forms and if, further, $f(\boldsymbol{u}) \neq 0$ for integral $\boldsymbol{u} \neq \boldsymbol{o}$, then

$$f(\boldsymbol{u}) = t \prod_{1 \leq j \leq 3} (\beta_{j1} u_1 + \beta_{j2} u_2 + \beta_{j3} u_3),$$

where $\beta_{11}, \beta_{12}, \beta_{13}$ are numbers in a totally real cubic field $\mathfrak{K}_1$ and $\beta_{jk}$ is the conjugate of $\beta_{1k}$ in the conjugate field $\mathfrak{K}_j$.

On the other hand, there are certainly lattices $\Lambda$ which are admissible for

$$|x_1 x_2| < 1$$

and on which $x_1 x_2$ is not proportional to integral. This follows at once from the theory of continued fractions: alternatively it is not difficult to modify the proof of Theorem VIII of Chapter VI.

A rather more interesting case is

$$\varphi(\boldsymbol{x}) = x_1(x_2^2 + x_3^2). \tag{10}$$

Since

$$\varphi(\boldsymbol{x}) = x_1(x_2 + i x_3)(x_2 - i x_3),$$

where $i^2 = -1$, there is a connection with the cubic fields that are not totally real, similar to that of $x_1 x_2 x_3$ with totally real fields: it is classical, and will be proved in § 4.4 that if $x_1(x_2^2 + x_3^2)$ is proportional to integral and non-zero on $\Lambda$, then $\Lambda$ arises from a cubic field. But there certainly are other admissible lattices for

$$|x_1(x_2^2 + x_3^2)| < 1. \tag{11}$$

Let $\boldsymbol{\tau}$ be any transformation $\boldsymbol{X} = \boldsymbol{\tau x}$ of the special type

$$\left.\begin{aligned} X_1 &= \tau_{11} x_1 \\ X_2 &= \qquad \tau_{22} x_2 + \tau_{23} x_3 \\ X_3 &= \qquad \tau_{32} x_2 + \tau_{33} x_3, \end{aligned}\right\}$$

where

$$\tau_{11} \neq 0 \qquad \tau_{22}\tau_{33} - \tau_{23}\tau_{32} \neq 0.$$

Then there are clearly constants $C, c$ depending on $\boldsymbol{\tau}$, such that

$$\infty > C \geq \frac{|\varphi(\boldsymbol{\tau x})|}{|\varphi(\boldsymbol{x})|} \geq c > 0$$

for all $\boldsymbol{x}$. Hence if $\Lambda$ is admissible for $|\varphi(\boldsymbol{x})|<1$, so is $t\boldsymbol{\tau}\Lambda$ for some number $t$; and in general $\varphi(\boldsymbol{x})$ will not be proportional to integral on $\boldsymbol{\tau}\Lambda$ if it is on $\Lambda$.

This does not exhaust the admissible lattices for $x_1(x_2^2+x_3^2)<1$. One way to show this is to use the arithmetic-geometric mean inequality in the shape

$$|x_1(x_2^2+x_3^2)| \geqq 2|x_1 x_2 x_3|.$$

Hence any lattice admissible for $|x_1 x_2 x_3|<\frac{1}{2}$ is also admissible for $|x_1(x_2^2+x_3^2)|<1$; for example $2^{-\frac{1}{3}}\mathsf{M}$ has this property if $x_1 x_2 x_3$ is integral and non-null on $\mathsf{M}$ (i.e. when $\mathsf{M}$ arises from a totally real cubic field); and it is easy to see that $x_1(x_2^2+x_3^2)$ cannot be proportional to integral on $\mathsf{M}$. [In fact the $x_1$-co-ordinates of $\mathsf{M}$ for a lattice on which $x_1 x_2 x_3$ or $x_1(x_2^2+x_3^2)$ is non-null and proportional to integral determine the relevant cubic field completely and it cannot be both totally real and not totally real.] More generally, one can construct admissible lattices by the methods of Chapter VI, Theorem VIII, compare CASSELS (1955b) for a closely related problem.

It is an interesting problem to decide for any given form $\varphi(\boldsymbol{x})$ if there exist admissible lattices for a set $|\varphi(\boldsymbol{x})|<1$ on which $\varphi(\boldsymbol{x})$ is not proportional to integral. CASSELS and SWINNERTON-DYER (1955a) have considered the special cases $\varphi(\boldsymbol{x})=x_1 x_2 x_3$ and $x_1^2+x_2^2-x_3^2$, but they only transform the problems into another one. For another line of attack, see ROGERS (1953b). It is reasonable to think that essentially new ideas will be required even to cope with $x_1 x_2 x_3$ or $x_1^2+x_2^2-x_3^2$.

**X.1.4.** An important part in the theory is played by so-called isolation theorems. Their importance was first apparently recognised by DAVENPORT and ROGERS (1950a) though there are foreshadowings in MAHLER (1946e) and indeed in REMAK (1925a). A new type of isolation theorem is proved and exploited in CASSELS and SWINNERTON-DYER (1955a).

The phenomenon of isolation takes various forms all of which state, roughly speaking, that lattices in the neighbourhood of a given lattice $\mathsf{M}$, with certain exceptions, are much worse behaved than $\mathsf{M}$ itself. Thus one result we shall prove is that if $x_1 x_2 x_3$ is integral and non-null on a lattice $\mathsf{M}$, then to every $\varepsilon>0$ there is a neighbourhood $\mathfrak{L}$ of $\mathsf{M}$ in the sense of § 3.2 of Chapter V, depending on $\varepsilon$, such that

$$\inf_{\substack{\boldsymbol{x}\in\Lambda\\ \neq\boldsymbol{o}}} |x_1 x_2 x_3| < \varepsilon$$

for all $\Lambda\in\mathfrak{L}$ except the $\Lambda$ of the shape $t\mathsf{M}$, for a number $t$. This is a particularly sweeping result. Perhaps more typical is the isolation theorem for $x_1(x_2^2+x_3^2)$. This states that if

$$\inf_{\substack{\boldsymbol{x}\in\mathsf{M}\\ \boldsymbol{x}\neq\boldsymbol{o}}} |x_1(x_2^2+x_3^2)| = 1,$$

and if $x_1(x_2^2+x_3^2)$ is proportional to integral on $\mathsf{M}$, then there exists an $\eta_0>0$ and a neighbourhood $\mathfrak{L}$ of $\mathsf{M}$, such that

$$\inf_{\substack{\boldsymbol{x}\in\Lambda \\ \neq \boldsymbol{o}}} |x_1(x_2^2+x_3^2)| < 1-\eta_0$$

for all $\Lambda\in\mathfrak{L}$ except those of the type $\boldsymbol{\tau}\mathsf{M}$, where $\boldsymbol{\tau}$ is of the special type with $\tau_{12}=\tau_{13}=\tau_{21}=\tau_{31}=0$ already discussed in § 1.3. Note that for $x_1x_2x_3$, the number $\varepsilon$ could be chosen at will, whereas for $x_1(x_2^2+x_3^2)$ both $\eta_0$ and $\mathfrak{L}$ are fixed by the lattice $\mathsf{M}$.

All isolation theorems have the same general type of proof. In the first place, it is shown that if the form $\varphi(\boldsymbol{x})$ is, say, integral or integral and non-null on a lattice $\mathsf{M}$, then $\varphi(\boldsymbol{x})$ and $\mathsf{M}$ have a group $\boldsymbol{\Omega}_{\mathsf{M}}$ of automorphs $\boldsymbol{\omega}$ in common; that is

$$\varphi(\boldsymbol{\omega}\boldsymbol{x}) = \varphi(\boldsymbol{x}), \quad \boldsymbol{\omega}\mathsf{M}=\mathsf{M}.$$

For the special forms $x_1x_2$, $x_1x_2x_3$ and $x_1(x_2^2+x_3^2)$ these automorphs are given by the theory of units in algebraic number fields, and for $x_1^2+x_2^2-x_3^2$ by the theory of indefinite ternary quadratic forms; but we shall, in fact, find it easy to handle the group $\boldsymbol{\Omega}_{\mathsf{M}}$ without these theories and using only MAHLER's compactness theorem[1]. A lattice $\Lambda$ near $\mathsf{M}$, in the sense of MAHLER, is one of the shape

$$\Lambda=\boldsymbol{\tau}\mathsf{M},$$

where $\boldsymbol{\tau}$ is near the identity transformation. Suppose that there is an $\boldsymbol{a}_0\in\mathsf{M}$ such that $\varphi(\boldsymbol{a}_0)$ takes some interesting value $\alpha$. Then

$$\varphi(\boldsymbol{\omega}\boldsymbol{a}_0)=\varphi(\boldsymbol{a}_0)=\alpha, \quad \boldsymbol{\omega}\in\boldsymbol{\Omega}_{\mathsf{M}}.$$

Then $\Lambda$ contains the point $\boldsymbol{\tau}\boldsymbol{\omega}\boldsymbol{a}_0$. Although $|\boldsymbol{\tau}\boldsymbol{a}_0-\boldsymbol{a}_0|$ is small when $\boldsymbol{\tau}$ is near the identity, it does not follow that $|\boldsymbol{\tau}\boldsymbol{\omega}\boldsymbol{a}_0-\boldsymbol{\omega}\boldsymbol{a}_0|$ is uniformly small for all $\boldsymbol{\omega}$, since in general $\boldsymbol{\omega}$ may be chosen so that $\boldsymbol{\omega}\boldsymbol{a}_0$ is arbitrarily large. By suitable choice of $\boldsymbol{\omega}$ in $\boldsymbol{\Omega}_{\mathsf{M}}$ one may then show the existence of a point $\boldsymbol{\tau}\boldsymbol{\omega}\boldsymbol{a}_0$ in $\Lambda=\boldsymbol{\tau}\mathsf{M}$ having the properties desired in the problem in question, unless the transformation $\boldsymbol{\tau}$ satisfies certain conditions. Sometimes one must start not with one point $\boldsymbol{a}_0$, but with several, $\boldsymbol{a}_1,\ldots,\boldsymbol{a}_r$, so as to eliminate $\boldsymbol{\tau}$ of different kinds. This general attack will be clearer from the examples in § 5. Isolation theorems may be used to discuss the existence of infinitely many lattice points in regions, as will be shown, following DAVENPORT and ROGERS (1950a), in § 7.

**X.1.5.** Before going on to the main subject matter of the chapter we shall discuss in § 2 certain special forms and their groups of automorphs. In § 3 we shall then discuss a method of MORDELL which shows

[1] One of MINKOWSKI's first applications of the geometry of numbers was in fact to the theory of units in algebraic fields.

how a bound for the lattice-constant of an $n$-dimensional body may be obtained from a bound for that of a related $(n-1)$-dimensional body. When the original $n$-dimensional body is of a special type having many automorphs, MORDELL showed the argument can be carried a stage further. In particular it gives the lattice constants of the 3-dimensional sets $|x_1 x_2 x_3| < 1$ and $|x_1(x_2^2+x_3^2)| < 1$. In § 8 we discuss briefly the relevance of continued fractions to forms and bodies with automorphs and the possibility of generalisation.

**X.2. Special forms.** We discuss first the automorphs of the form

$$\varphi(\boldsymbol{x}) = \Big\{\prod_{1\leqq j\leqq r} x_j\Big\}\Big\{\prod_{1\leqq k\leqq s}(x_{r+k}^2 + x_{r+s+k}^2)\Big\}: \quad n = r+2s, \tag{1}$$

where both the possibilities $r=0$ and $s=0$ are permitted. We may write

$$\varphi(\boldsymbol{x}) = \psi(\boldsymbol{z}) = \prod_{1\leqq l\leqq n} z_l, \tag{2}$$

where

$$\left.\begin{aligned} z_j &= x_j \qquad (1\leqq j\leqq r) \\ \left.\begin{aligned} z_{r+k} &= x_{r+k} + i\,x_{r+s+k} \\ z_{r+s+k} &= x_{r+k} - i\,x_{r+s+k} \end{aligned}\right\} &\ (1\leqq k\leqq s), \end{aligned}\right\} \tag{3}$$

and $i^2 = -1$. If the $x_l$ are all real, then $z_j$ is real for $1\leqq j\leqq r$ and $z_{r+k}$ and $z_{r+s+k}$ are conjugate complex numbers for $1\leqq k\leqq s$; and conversely, if the $z_l$ $(1\leqq l\leqq n)$ are of this shape then the $x_l$ are real. Let now $\boldsymbol{\omega}$ be a real automorph of $\varphi(\boldsymbol{x})$. In the obvious way it gives rise to an automorph $\tilde{\boldsymbol{\omega}}$ of $\psi(\boldsymbol{z})$. Let $\boldsymbol{Z} = \tilde{\boldsymbol{\omega}}\boldsymbol{z}$. Then

$$\prod z_l = \prod Z_l \tag{4}$$

identically in $z_1, \ldots, z_n$, where the $Z_l$ are linear forms in $z_1, \ldots, z_n$. The only possibility is that $Z_L = \lambda_l z_l$ where $L = L(l)$ is a permutation of $1, \ldots, n$ and $\lambda_1, \ldots, \lambda_n$ are real or complex numbers. For our purposes, it is enough to consider the automorphs

$$Z_l = \lambda_l z_l \qquad (1\leqq l\leqq n), \tag{5}$$

where

$$\prod_{1\leqq l\leqq n} \lambda_l = 1, \tag{6}$$

by (4). But the transformation $\boldsymbol{\omega}$ transforms the real point $\boldsymbol{x}$ into the real point $\boldsymbol{X} = \boldsymbol{\omega}\boldsymbol{x}$. Hence $Z_1, \ldots, Z_r$ are real and $Z_{r+k}, Z_{r+s+k}$ are conjugate complex, and so

$$\left.\begin{aligned} \lambda_j &= \text{real} \qquad (1\leqq j\leqq r) \\ \lambda_{r+k},\ \lambda_{r+s+k} &\ \text{ conjugate complex } (1\leqq k\leqq s). \end{aligned}\right\} \tag{7}$$

Conversely, if the numbers $\lambda_l$ satisfy (6) and (7), then (5) defines a real automorph $\boldsymbol{\omega}$ of $\varphi(\boldsymbol{x})$.

We shall also need the transformation $\boldsymbol{\omega}^*$ polar to $\boldsymbol{\omega}$, that is the transformation such that identically

$$\sum_{1\leq l\leq n} x_l y_l = \sum_{1\leq l\leq n} X_l Y_l,$$

when

$$\boldsymbol{X} = \boldsymbol{\omega}\boldsymbol{x}, \quad \boldsymbol{Y} = \boldsymbol{\omega}^*\boldsymbol{y}.$$

Now

$$\sum_l x_l y_l = \sum_l z_l w_l,$$

where $z_l$ is given by (3) and

$$w_j = y_j \qquad (1 \leq j \leq r)$$

$$\left.\begin{aligned} 2w_{r+k} &= y_{r+k} - i\, y_{r+s+k} \\ 2w_{r+s+k} &= y_{r+k} + i\, y_{r+s+k} \end{aligned}\right\} (1 \leq k \leq s).$$

Hence if $\boldsymbol{\omega}^*$ induces the transformation $\tilde{\boldsymbol{\omega}}^*$ in the $\boldsymbol{w}$-co-ordinates, we must have

$$W_l = \lambda_l^{-1} w_l,$$

where $\boldsymbol{W} = \tilde{\boldsymbol{\omega}}^* \boldsymbol{w}$. In particular, the transformation $\boldsymbol{\omega}^*$ is also an automorph of $\varphi(\boldsymbol{x})$.

**X.2.2.** We shall require also to know something of the automorphs of the form

$$\varphi(\boldsymbol{x}) = x_1^2 + \cdots + x_r^2 - x_{r+1}^2 - \cdots - x_n^2, \tag{1}$$

where possibly $r = n$, so there are no negative terms. For completeness we prove the well-known

LEMMA 2. *If $\varphi(\boldsymbol{x})$ is defined by (1) and $\boldsymbol{x}_0$ is any point, then there is an automorph $\boldsymbol{\omega}$ of $\varphi(\boldsymbol{x})$ such that, for some number $t$,*

$$\boldsymbol{\omega}\boldsymbol{x}_0 = (t, 0, \ldots, 0)$$

*or*

$$\boldsymbol{\omega}\boldsymbol{x}_0 = (0, \ldots, 0, t)$$

*or*

$$\boldsymbol{\omega}\boldsymbol{x}_0 = (t, 0, \ldots, 0, t)$$

*according as $\varphi(\boldsymbol{x}_0) > 0$, $\varphi(\boldsymbol{x}_0) < 0$ or $\varphi(\boldsymbol{x}_0) = 0$.*

This is certainly true for $n = 2$, since then there are the well-known automorphs $\boldsymbol{X} = \boldsymbol{\omega}\boldsymbol{x}$ given by

$$X_1 = x_1 \cos\vartheta + x_2 \sin\vartheta, \qquad X_2 = -x_1 \sin\vartheta + x_2 \cos\vartheta \tag{2}$$

for any real $\vartheta$ when $r=n=2$, and by

$$X_1+X_2=k(x_1+x_2), \quad X_1-X_2=k^{-1}(x_1-x_2) \tag{3}$$

when $r=1$, $n=2$ and $k$ may take any values except $k=0$.

Next, the lemma is true when $r=n$. For we may suppose it proved for $n-1$. There is then an automorph $\boldsymbol{\omega}_1$ acting only on the first $n-1$ co-ordinates such that

$$\boldsymbol{x}_1=\boldsymbol{\omega}_1\boldsymbol{x}_0=(u,0,\ldots,0,x_{n0})$$

for some $u$. Then an automorph $\boldsymbol{\omega}_2$ acting only on the first and last co-ordinates makes

$$\boldsymbol{\omega}_2\boldsymbol{x}_1=(t,0,\ldots,0)$$

for some $t$. Then $\boldsymbol{\omega}=\boldsymbol{\omega}_2\boldsymbol{\omega}_1$ does what is required.

Finally, the lemma is true in general. For we may find in succession automorphs $\boldsymbol{\omega}_1, \boldsymbol{\omega}_2, \boldsymbol{\omega}_3$ such that for some numbers $u, v$ we have

$$\boldsymbol{x}_1=\boldsymbol{\omega}_1\boldsymbol{x}_0=(u,0,\ldots,0,x_{r+1,0},\ldots,x_{n0}),$$
$$\boldsymbol{x}_2=\boldsymbol{\omega}_2\boldsymbol{x}_1=(u,0,\ldots,0,0,\ldots,0,v);$$

and then

$$\boldsymbol{x}_3=\boldsymbol{\omega}_3\boldsymbol{x}_2=(t,0,\ldots,0) \quad \text{or} \quad (0,\ldots,0,t) \quad \text{or} \quad (t,0,\ldots,0,t).$$

COROLLARY. *If $\boldsymbol{\omega}$ is the automorph constructed above, then the polar $\boldsymbol{\omega}^*$ is also an automorph.*

It is readily verified that the polars of the special transformation (2) and (3) are automorphs of $\varphi(\boldsymbol{x})$. The required result now follows by induction.

[It is in fact true that if $\boldsymbol{\omega}$ is any automorph of $\varphi(\boldsymbol{x})$ then its polar is also an automorph. This is most easily proved using matrix theory. Let $\boldsymbol{\omega}$ for the nonce denote the matrix whose elements are the coefficients in the transformation $\boldsymbol{\omega}$ and let $\boldsymbol{\epsilon}$ be the matrix with 1 in the first $r$ places on the diagonal, $-1$ on the remaining diagonal places, and 0 elsewhere. The fact that $\boldsymbol{\omega}$ is an automorph is expressed by

$$\boldsymbol{\omega}'\boldsymbol{\epsilon}\boldsymbol{\omega}=\boldsymbol{\epsilon}, \tag{4}$$

where the dash (′) denotes the transposed. On taking the reciprocal of (4) we obtain

$$\boldsymbol{\omega}^{-1}\boldsymbol{\epsilon}^{-1}\boldsymbol{\omega}'^{-1}=\boldsymbol{\epsilon}^{-1}. \tag{5}$$

But the polar $\boldsymbol{\omega}^*$ of $\boldsymbol{\omega}$ is clearly $\boldsymbol{\omega}^*=\boldsymbol{\omega}'^{-1}$; and so $\boldsymbol{\omega}^*$ is an automorph of $\varphi$ by (5), since $\boldsymbol{\epsilon}^{-1}=\boldsymbol{\epsilon}$.]

**X.3. A method of Mordell.** In this section we discuss a method of MORDELL for estimating the lattice constant of an $n$-dimensional set by reducing the problem to an $(n-1)$-dimensional one.

Let $\mathscr{S}$ be any $n$-dimensional set, not for the moment necessarily endowed with any automorphs and $\Lambda$ a lattice. In Chapter I, § 5 it was shown that if $\boldsymbol{b}$ is any point of the polar lattice $\Lambda^*$, then there are $n-1$ linearly independent points of $\Lambda$ on the plane

$$\pi_{\boldsymbol{b}}: \quad \boldsymbol{x}\boldsymbol{b}=0$$

(scalar product). The plane $\pi_{\boldsymbol{b}}$ cuts $\mathscr{S}$ in an $(n-1)$-dimensional set $\mathscr{S}_{\boldsymbol{b}}$. In an obvious sense, there is an $(n-1)$-dimensional lattice $\Lambda_{\boldsymbol{b}}$ consisting of the points of $\Lambda$ in $\pi_{\boldsymbol{b}}$. Hence, if we can show that there is a point other than $\boldsymbol{o}$ of the $(n-1)$-dimensional lattice $\Lambda_{\boldsymbol{b}}$ in $\mathscr{S}_{\boldsymbol{b}}$, then there is certainly a point other than $\boldsymbol{o}$ of $\Lambda$ in $\mathscr{S}$. If $b_n \neq 0$, for example, one could project $\mathscr{S}_{\boldsymbol{b}}$ on to the hyperplane $x_n=0$ and use Lemma 6 Corollary of Chapter I. For this procedure to be effective, the vector $\boldsymbol{b}\in\Lambda^*$ must be chosen so as to give a good $(n-1)$-dimensional problem in $\pi_{\boldsymbol{b}}$; and so in general we have replaced one $n$-dimensional problem by another, rather vaguer, one for the polar lattice, together with an $(n-1)$-dimensional problem.

In this shape the technique has been applied by MULLENDER (1950a) and DAVENPORT (1952a) to the enigmatic 3-dimensional starbody

$$|x_1| \max(x_2^2, x_3^2) < 1.$$

Making use of the known (cf. § 3.3) lattice constant of the set

$$|x_1|(x_2^2+x_3^2)<1,$$

they select a point $\boldsymbol{b}$ of $\Lambda^*$ for which $b_1(b_2^2+b_3^2)$ is small and then treat the 2-dimensional problem in $\pi_{\boldsymbol{b}}$.

MORDELL (1942a, 1943a, 1944b) observed that it is sometimes possible to make the $n$-dimensional problem for the polar lattice the same as the original problem; and then the $n$-dimensional problem is reduced entirely to one or more $(n-1)$-dimensional problems without the need to solve an $n$-dimensional auxiliary problem. The sets $\mathscr{S}$ for which this procedure is feasible are those with a large group of automorphs, so it is appropriate to discuss them in this chapter. From one point of view it may be regarded as based on a generalization to non-convex bodies of the results in Chapter VIII about polar convex bodies.

**X.3.2.** We first consider quadratic forms, for which OPPENHEIM (1946a) has given a neat treatment following MORDELL (1944b).

THEOREM IV. *Let $\Gamma_{r,s}=\Delta(\mathscr{D}_{r,s})$ be the lattice constant of the $(r+s)$-dimensional star-body*

$$\mathscr{D}_{r,s} \quad |x_1^2+\cdots+x_r^2-x_{r+1}^2-\cdots-x_{r+s}^2|<1 \tag{1}$$

*for $r\geqq 0$, $s\geqq 0$. Then*

$$\Gamma_{r,s}^{r+s-2} \geqq \min(\Gamma_{r-1,s}^{r+s}, \Gamma_{r,s-1}^{r+s}) \tag{2}$$

*where the first or second term is omitted if $r=0$ or $s=0$ respectively.*

Write

$$\varphi(\boldsymbol{x}) = \varphi_{r,s}(\boldsymbol{x}) = x_1^2 + \cdots + x_r^2 - x_{r+1}^2 - \cdots - x_{r+s}^2, \tag{3}$$

and

$$|\varphi|(\Lambda) = \inf_{\substack{\boldsymbol{a}\in\Lambda \\ \neq \boldsymbol{o}}} |\varphi(\boldsymbol{a})| \tag{4}$$

for any lattice $\Lambda$. Then, by homogeneity,

$$\Gamma_{r,s}^{-2} = \sup_{\Lambda} \frac{\{|\varphi|(\Lambda)\}^{r+s}}{d^2(\Lambda)} \tag{5}$$

over all lattices $\Lambda$, with the natural convention that if $|\varphi|(\Lambda)=0$ for all $\Lambda$, then $\Gamma_{r,s}=\infty$; as most probably happens when $r>0$, $s>0$, $r+s\geqq 5$ (see appendix A).

We show first that

$$\{|\varphi|(\Lambda)\}^{r+s-1} \leqq \zeta^{-2}\{|\varphi|(\Lambda^*)\}\, d^2(\Lambda), \tag{6}$$

where $\Lambda^*$ is the polar lattice of $\Lambda$ and

$$\zeta = \min(\Gamma_{r-1,s}, \Gamma_{r,s-1}). \tag{7}$$

It is enough to show that

$$\{|\varphi|(\Lambda)\}^{r+s-1} \leqq \zeta^{-2}|\varphi(\boldsymbol{b})|\, d^2(\Lambda), \tag{8}$$

where $\boldsymbol{b}$ is any primitive point of $\Lambda^*$. After Lemma 2 we may suppose that $\boldsymbol{b}$ is one of the points

$$\boldsymbol{b}_1 = (t, 0, \ldots, 0), \quad \boldsymbol{b}_2 = (0, \ldots, 0, t), \quad \boldsymbol{b}_3 = (t, 0, \ldots, 0, t), \tag{9}$$

where $\boldsymbol{b}_1, \boldsymbol{b}_2, \boldsymbol{b}_3$ occur only if $r>0$, $s>0$ and both $r>0$, $s>0$, respectively.

Consider first $\boldsymbol{b}=\boldsymbol{b}_1$, where

$$\varphi(\boldsymbol{b}_1) = t^2. \tag{10}$$

By the results of § 5 of Chapter 1 there is a basis $\boldsymbol{a}_1, \ldots, \boldsymbol{a}_n$ for $\Lambda$ such that

$$\boldsymbol{b}_1\boldsymbol{a}_1 = 1, \quad \boldsymbol{b}_1\boldsymbol{a}_j = 0 \quad (2\leqq j\leqq n):$$

so that $\boldsymbol{a}_1=(t^{-1}, \boldsymbol{a}_1')$ and $\boldsymbol{a}_j=(0,\boldsymbol{a}_j')$ for $j\neq 1$, where $\boldsymbol{a}_j'$ is an $(n-1)$-dimensional vector. Hence the points of $\Lambda$ in $x_1=0$ form an $(n-1)$-dimensional lattice $\mathsf{M}$ in the space with co-ordinates $x_2, \ldots, x_n$ with basis $\boldsymbol{a}_j'$ $(2\leqq j\leqq n)$. Further,

$$d(\Lambda) = |\det(\boldsymbol{a}_1, \ldots, \boldsymbol{a}_n)| = |t^{-1}|\,|\det(\boldsymbol{a}_2', \ldots, \boldsymbol{a}_n')| = |t|^{-1} d(\mathsf{M}). \tag{11}$$

But now by (5) with $r-1, s$ for $r, s$ we have

$$\{|\varphi_{r-1,s}|(\mathsf{M})\}^{r+s-1} \leqq \Gamma_{r-1,s}^{-2}\, d^2(\mathsf{M}) = \Gamma_{r-1,s}^{-2}\,|\varphi(\boldsymbol{b}_1)|\, d^2(\Lambda) \tag{12}$$

by (10) and (11). This proves (8) in the case $\boldsymbol{b} = \boldsymbol{b}_1$ since the left-hand side of (12) is not less than $\{|\varphi|(\Lambda)\}^{r+s-1}$. The proof of (8) in the second case, when $\boldsymbol{b} = \boldsymbol{b}_2$ in (9), is similar except that the rôles of $r$ and $s$ are interchanged.

It remains to consider the case

$$\boldsymbol{b} = \boldsymbol{b}_3 = (t, 0, \ldots, 0, t),$$

which occurs only when $r > 0$, $s > 0$, so

$$\varphi(\boldsymbol{b}) = 0.$$

There then exist a basis $\boldsymbol{a}_1, \ldots, \boldsymbol{a}_n$ of $\Lambda$ such that

$$\boldsymbol{b}\,\boldsymbol{a}_j = 0 \quad (2 \leq j \leq n).$$

Introduce new co-ordinates $x'_j$ by

$$x'_1 = x_1 + x_n, \quad x'_n = x_1 - x_n, \quad x'_j = x_j \quad (j \neq 1, n),$$

so that

$$\varphi(\boldsymbol{x}) = x'_1\, x'_n + x'^2_2 + \cdots + x'^2_r - x'^2_{r+1} - \cdots - x'^2_{r+s-1},$$

and the points $\boldsymbol{a}_2, \ldots, \boldsymbol{a}_n$ lie on $x'_1 = 0$. The points of $\Lambda$ on $x'_1 = 0$ form an $(n-1)$-dimensional lattice $\mathsf{M}$, and $\varphi(\boldsymbol{x})$ with $x'_1 = 0$ depends only on the $n-2$ variables $x_2, \ldots, x_{n-1}$. Hence $|\varphi(\boldsymbol{x})|$ takes arbitrarily small values on $\mathsf{M}$; for example, by the degenerate case of MINKOWSKI'S convex body Theorem, there are points of $\mathsf{M}$ other than $\boldsymbol{o}$ with

$$x'_1 = 0, \quad |x'_j| < \varepsilon \quad (2 \leq j \leq n-1),$$

where $\varepsilon > 0$ is arbitrarily small, since this set has infinite $(n-1)$-dimensional volume. Hence (8) holds also when $\boldsymbol{b} = \boldsymbol{b}_3$; and so generally. This concludes the proof of (6).

We may also apply (6) to the lattice $\Lambda^*$ with its determinant $d(\Lambda^*) = d^{-1}(\Lambda)$ and its polar lattice $\Lambda$:

$$\{|\varphi|(\Lambda^*)\}^{r+s-1} \leq \zeta^{-2}\{|\varphi|(\Lambda)\}\, d^{-2}(\Lambda). \tag{6'}$$

On eliminating $|\varphi|(\Lambda^*)$ between (6) and (6'), we obtain

$$\{|\varphi|(\Lambda)\}^{(r+s)(r+s-2)} \leq \zeta^{-2(r+s)}\{d(\Lambda)\}^{2(r+s-2)}.$$

This implies the required result (2) on using (5) and (7).

In general there is no reason to expect there to be equality in (2), but this sometimes happens, as in the following

COROLLARY.

$$\Gamma_{4,0} = \tfrac{1}{2}, \quad \Gamma_{2,2} = \tfrac{3}{2}.$$

By Theorems III and VII of Chapter II, we have

$$\Gamma_{3,0} = 2^{-\frac{1}{2}}, \quad \Gamma_{2,1} = \Gamma_{1,2} = (\tfrac{3}{2})^{\frac{1}{2}}.$$

Hence the theorem shows that $\Gamma_{4,0}$ and $\Gamma_{2,2}$ have at least the values specified. The forms

$$\tfrac{1}{2}\{(x_2+x_3+x_4)^2+(x_2+x_1)^2+(x_3+x_1)^2+x_4^2\}$$

and

$$x_1^2-x_2^2-x_3^2-x_4^2+x_4x_1+x_4x_2+x_4x_3+2x_1x_3+2x_1x_2$$

have signature (4, 0), (2, 2), and determinants $\frac{1}{4}$, $\frac{9}{4}$ respectively and do not represent members less than 1 in absolute value for integer value of the variables not all 0, as is easily verified. This proves the corollary on making use of the relationship between forms and lattices of Chapter I, § 4 (especially Lemma 4).

Again, as MORDELL observed, Theorem IV gives $\Gamma_{8,0}$ once $\Gamma_{7,0}$ is known. Again, the method of proof of Theorem IV gives the lattice constant [$\frac{1}{4}$, see OPPENHEIM (1953b)] of

$$0 < x_1^2 + x_2^2 - x_3^2 - x_4^2 < 1,$$

once that [$\frac{1}{2}$, see DAVENPORT (1949a)] of

$$0 < x_1^2 + x_2^2 - x_3^2 < 1$$

is known. These sets are not star-bodies. It is necessary to choose the point $\boldsymbol{b}$ of $\Lambda^*$ so that $b_1^2+b_2^2-b_3^2-b_4^2$ is numerically small and negative. It is possible to use MORDELL's method to obtain information about the critical lattices when there is equality in (2). We do not do this here since we shall do something similar for products of linear forms in § 3.3.

**X.3.3.** Before applying MORDELL's method to ternary cubics we must translate Theorem VIII of Chapter II out of the language of forms into that of lattices.

LEMMA 3. *The lattice-constant of the 2-dimensional set*

$$\mathscr{T}: \ |\psi(\boldsymbol{x})| < 1, \tag{1}$$

*where*

$$\psi(\boldsymbol{x}) = x_1 x_2 (x_1 + x_2), \tag{2}$$

*is $\Delta(\mathscr{T}) = 7^{\frac{1}{2}}$. There are precisely two critical lattices, $\mathsf{M}_1$ and $\mathsf{M}_2$. These lattices have only $\boldsymbol{o}$ in common.*

*Let $\vartheta_1, \vartheta_2, \vartheta_3$ be the roots of*

$$\vartheta^3 + \vartheta^2 - 2\vartheta - 1 = 0 \tag{3}$$

*in some order. Then the lattice $\mathsf{M}(\vartheta_1, \vartheta_2, \vartheta_3)$ with basis $\boldsymbol{a}_1, \boldsymbol{a}_2$ defined by*

$$7^{\frac{1}{3}}\boldsymbol{a}_1 = (\vartheta_2-\vartheta_3, \vartheta_3-\vartheta_1), \quad 7^{\frac{1}{3}}\boldsymbol{a}_2 = \{\vartheta_1(\vartheta_2-\vartheta_3), \vartheta_2(\vartheta_3-\vartheta_1)\} \tag{4}$$

*is one of the two critical lattices. If $\vartheta_1', \vartheta_2', \vartheta_3'$ is a permutation of $\vartheta_1, \vartheta_2, \vartheta_3$, then $\mathsf{M}(\vartheta_1, \vartheta_2, \vartheta_3) = \mathsf{M}(\vartheta_1', \vartheta_2', \vartheta_3')$ if and only if the permutation is an even one.*

The geometrical purport of the lemma becomes clearer if new co-ordinates $y_1, y_2$ are introduced by the equations

$$x_1 = y_1, \quad x_2 = -\frac{1}{2} y_1 + \frac{\sqrt{3}}{2} y_2,$$

so

$$-x_1 - x_2 = -\frac{1}{2} y_1 - \frac{\sqrt{3}}{2} y_2.$$

In $y_1, y_2$ co-ordinates, the region $\mathcal{T}$ has three asymptotes at an angle of $2\pi/3$ and is carried into itself by either a rotation through $2\pi/3$ round the origin or by a reflection in an asymptote. The two critical lattices given by the lemma are then each invariant under a rotation through $2\pi/3$ and each is carried into the other by a reflection in an asymptote. The reader may find it instructive to draw a figure of the critical lattices each with 6 pairs of points on the boundary. For a treatment of sets $\mathcal{T}'$ which have similar symmetry and convexity properties to $\mathcal{T}$ by the geometrical methods of Chapter III see BAMBAH (1951a).

In what follows we do not introduce $y_1, y_2$ as above but we do maintain the essential cyclic symmetry between $x_1, x_2$ and $-x_1 - x_2$.

We note that the roots of (3) are

$$\Theta_1 = 2\cos\frac{2\pi}{7}, \quad \Theta_2 = 2\cos\frac{4\pi}{7}, \quad \Theta_3 = 2\cos\frac{6\pi}{7}, \tag{5}$$

so that $\vartheta_1, \vartheta_2, \vartheta_3$ are a permutation of $\Theta_1, \Theta_2, \Theta_3$. We have the trivial identities

$$\Theta_2 = \Theta_1^2 - 2, \quad \Theta_3 = \Theta_2^2 - 2, \quad \Theta_1 = \Theta_3^2 - 2, \quad \Theta_1 = 1 - \Theta_2 - \Theta_2^2 \text{ etc.} \tag{6}$$

The value of $\Delta(\mathcal{T})$ follows at once from Theorem VIII of Chapter II, so it remains only to verify the statement about the critical lattices. By Theorem VIII of Chapter II, if $\mathsf{M}$ is critical there is certainly a basis $\boldsymbol{a}_1, \boldsymbol{a}_2$ of $\mathsf{M}$ such that

$$\psi(u_1 \boldsymbol{a}_1 + u_2 \boldsymbol{a}_2) = -f_0(u_1, u_2), \tag{7}$$

where

$$f_0(u_1, u_2) = u_1^3 - u_1^2 u_2 - 2u_1 u_2^2 + u_2^3; \tag{8}$$

for one may interchange the two elements of the base given by Theorem VIII of Chapter II or take $-\boldsymbol{a}_k$ for $\boldsymbol{a}_k$ $(k = 1, 2)$. Let

$$\boldsymbol{a}_k = (a_{1k}, a_{2k}) \quad (k = 1, 2), \tag{9}$$

and define numbers $a_{3k}$ by

$$a_{1k} + a_{2k} + a_{3k} = 0 \quad (k = 1, 2). \tag{10}$$

Then (7) becomes

$$\prod_{1 \leq j \leq 3} (a_{j1} u_1 + a_{j2} u_2) = \prod_{1 \leq j \leq 3} (u_1 + \Theta_j u_2). \tag{11}$$

Hence

$$a_{j2} = \vartheta_j a_{j1} \quad (j = 1, 2, 3), \tag{12}$$

where $\vartheta_1, \vartheta_2, \vartheta_3$ is some permutation of $\Theta_1, \Theta_2, \Theta_3$. From (10) and (12) we have

$$\left.\begin{aligned} \lambda a_{j1} &= \vartheta_{j+1} - \vartheta_{j+2} \\ \lambda a_{j2} &= \vartheta_j(\vartheta_{j+1} - \vartheta_{j+2}) \end{aligned}\right\} \quad (j = 1, 2, 3), \tag{13}$$

where $\vartheta_4 = \vartheta_1$, $\vartheta_5 = \vartheta_2$ and $\lambda$ is some number. By (11) we have

$$\prod_j a_{j1} = 1,$$

and so in fact

$$\lambda^3 = (\vartheta_1 - \vartheta_2)(\vartheta_2 - \vartheta_3)(\vartheta_3 - \vartheta_1) \tag{14}$$

$$= \pm 7, \tag{15}$$

where the value $\pm 7$ may either be checked directly from (5) or from the fact that the square of the right-hand side of (14) is the discriminant of the cubic $f_0(u_1, u_2)$ by definition (§ 5.1 of Chapter II). We note that $\vartheta_1, \vartheta_2, \vartheta_3$ determine $\boldsymbol{a}_1$ and $\boldsymbol{a}_2$ absolutely uniquely, by (14).

But now we have the identity

$$f_0(w + v, v) = f_0(-v, w).$$

Hence if the point $\boldsymbol{a}_3$ of $\mathsf{M}(\vartheta_1, \vartheta_2, \vartheta_3)$ is defined by

$$\boldsymbol{a}_1 + \boldsymbol{a}_2 + \boldsymbol{a}_3 = 0,$$

we have

$$\psi(u_1 \boldsymbol{a}_2 + u_2 \boldsymbol{a}_3) = -f_0(u_1, u_2):$$

and so $\boldsymbol{a}_2, \boldsymbol{a}_3$ must correspond to a permutation $\vartheta_1', \vartheta_2', \vartheta_3'$ of $\vartheta_1, \vartheta_2, \vartheta_3$; and it cannot be the identical permutation by the last sentence of the previous paragraph. Hence the cyclic change of bases of $\mathsf{M}(\vartheta_1, \vartheta_2, \vartheta_3)$:

$$(\boldsymbol{a}_1, \boldsymbol{a}_2) \to (\boldsymbol{a}_2, \boldsymbol{a}_3) \to (\boldsymbol{a}_3, \boldsymbol{a}_1) \to (\boldsymbol{a}_1, \boldsymbol{a}_2) \to$$

must correspond to a cyclic permutation of $\vartheta_1, \vartheta_2, \vartheta_3$. Hence there are at most two distinct lattices $\mathsf{M}(\vartheta_1, \vartheta_2, \vartheta_3)$, for the permutations $\vartheta_1, \vartheta_2, \vartheta_3$ of $\Theta_1, \Theta_2, \Theta_3$.

It remains to show that $\mathsf{M}(\vartheta_1, \vartheta_2, \vartheta_3)$ is distinct from $\mathsf{M}(\Theta_1, \Theta_2, \Theta_3)$ if $\vartheta_1, \vartheta_2, \vartheta_3$ is an odd permutation of $\Theta_1, \Theta_2, \Theta_3$. We may suppose now, without loss of generality, that

$$\vartheta_1 = \Theta_2, \quad \vartheta_2 = \Theta_1, \quad \vartheta_3 = \Theta_3.$$

From (4), (6), (13) and (15), a point $\boldsymbol{b}$ of $\mathsf{M}(\Theta_1, \Theta_2, \Theta_3)$ has

$$7^{\frac{1}{3}} b_j = P(\Theta_j) \quad (j = 1, 2, 3), \tag{16}$$

where $P(t)$ is a polynomial in the variable $t$ with (rational) integer coefficients. We may suppose, by (3), that $P(t)$ is of degree $\leqq 2$; and then it is completely determined by any one of $b_1, b_2, b_3$. If $\boldsymbol{b}$ is also in $\mathsf{M}(\Theta_2, \Theta_1, \Theta_3)$, then it is also of the shape

$$7^{\frac{1}{3}} b_1 = Q(\Theta_2), \quad 7^{\frac{1}{3}} b_2 = Q(\Theta_1), \quad 7^{\frac{1}{3}} b_3 = Q(\Theta_3),$$

for some polynomial $Q(t)$ of degree $\leqq 2$ with integer coefficients. But now $P(\Theta_3) = Q(\Theta_3)$, and so the polynomials $P(t)$ and $Q(t)$ are identical. Hence

$$P(\Theta_2) = P(\Theta_1); \tag{17}$$

and so

$$P(\Theta_3) = P(\Theta_2) = P(\Theta_1), \tag{18}$$

since $P(\Theta_j)$ $(j = 1, 2, 3)$ are conjugates[1]. Finally,

$$b_1 = b_2 = -b_1 - b_2$$

by (16) and (18), and so

$$b_1 = b_2 = 0:$$

That is, $\boldsymbol{o}$ the only point common to $\mathsf{M}(\Theta_1, \Theta_2, \Theta_3)$ and $\mathsf{M}(\Theta_2, \Theta_1, \Theta_3)$, as required.

**X.3.4.** We now apply MORDELL's method to prove results for $x_1 x_2 x_3$ and $x_1(x_2^2 + x_3^2)$. These are equivalent to weaker forms of Theorems X and XI of Chapter 2, where the relevant literature is cited. We shall later prove something rather stronger by the use of isolation, but will not prove the full force of Theorem X of Chapter 2 in this book. The methods extend to products of $n$ real or complex forms in $n$ dimensions in a way which will be obvious, but do not then give the exact lattice constants [MORDELL (1941a) and (1943a)].

THEOREM V. A. *The lattice constant of the 3-dimensional set*

$$\mathscr{N}_1: \quad |x_1 x_2 x_3| < 1 \tag{1}$$

*is* $\Delta(\mathscr{N}_1) = 7$. *Denote by* $\mathsf{N}_1$ *the lattice with basis*

$$\boldsymbol{b}_1 = (1, 1, 1), \quad \boldsymbol{b}_2 = (\vartheta_1, \vartheta_2, \vartheta_3), \quad \boldsymbol{b}_3 = (\vartheta_1^2, \vartheta_2^2, \vartheta_3^2), \tag{2}$$

*where* $\vartheta_1, \vartheta_2, \vartheta_3$ *are the roots of*

$$\vartheta^3 + \vartheta^2 - 2\vartheta - 1 = 0 \tag{3}$$

*in some order. All the critical lattices* $\Lambda$ *of* $\mathscr{N}_1$ *which have a point* $\boldsymbol{a}$ *for which*

$$|a_1 a_2 a_3| = 1 \tag{4}$$

---

[1] Alternatively, (17) means that $P(\Theta_1^2 - 2) = P(\Theta_1)$; and so the polynomial $P(t^2 - 2) - P(t)$ is divisible by $t^3 + t^2 - 2t - 1$. One may now put $t = \Theta_2$ and obtain $P(\Theta_2) = P(\Theta_3)$.

*are of the shape*

$$\Lambda = \boldsymbol{\omega}\, \mathsf{N}_1, \tag{5}$$

*where* $\boldsymbol{\omega}$ *is an automorph of* $\mathcal{N}_1$.

B. *The lattice constant of*

$$\mathcal{N}_2:\ \ |x_1|(x_2^2+x_3^2)<1 \tag{6}$$

*is* $\Delta(\mathcal{N}_2)=\frac{1}{2}(23)^{\frac{1}{2}}$. *Denote by* $\mathsf{N}_2$ *the lattice with basis*

$$\left.\begin{aligned} (1,1,1),\ \ &\left\{\vartheta_1,\ \frac{1}{2}(\vartheta_2+\vartheta_3),\ \frac{1}{2i}(\vartheta_2-\vartheta_3)\right\},\\ &\left\{\vartheta_1^2,\ \frac{1}{2}(\vartheta_2^2+\vartheta_3^2),\ \frac{1}{2i}(\vartheta_2^2-\vartheta_3^2)\right\}, \end{aligned}\right\} \tag{7}$$

*where* $i^2=-1$ *and* $\vartheta_1$ *is the real, and* $\vartheta_2, \vartheta_3$ *are the complex roots of*

$$\vartheta^3-\vartheta^2+1=0. \tag{8}$$

*Every critical lattice* $\Lambda$ *for* $\mathcal{N}_2$ *which possesses a point* $\boldsymbol{a}$ *with*

$$|a_1(a_2^2+a_3^2)|=1 \tag{9}$$

*is of the shape*

$$\Lambda = \boldsymbol{\omega}\, \mathsf{N}_2, \tag{10}$$

*where* $\boldsymbol{\omega}$ *is an automorph of* $\mathcal{N}_2$.

We first prove Theorem V. A. The lattice $\mathsf{N}_1$ given by the theorem is certainly $\mathcal{N}_1$-admissible, since a point $\boldsymbol{a}$ of $\mathsf{N}_1$ has co-ordinates

$$a_j = u_1 + u_2\vartheta_j + u_3\vartheta_j^2 \qquad (j=1,2,3), \tag{11}$$

where $u_1, u_2, u_3$ are integers. Then $a_1 a_2 a_3$ is a rational integer by its symmetry in $\vartheta_1, \vartheta_2, \vartheta_3$. If $a_1a_2a_3=0$, then one of the $a_j$ is 0, say $u_1+u_2\vartheta_1+u_3\vartheta_1^2=0$; and this is impossible unless $u_1=u_2=u_3=0$, since $\vartheta_1$ does not satisfy any equation of degree less than 3. Further,

$$d(\mathsf{N}_1)=|\det(\boldsymbol{b}_1,\boldsymbol{b}_2,\boldsymbol{b}_3)|=|(\vartheta_1-\vartheta_2)(\vartheta_2-\vartheta_3)(\vartheta_3-\vartheta_1)|=7, \tag{12}$$

as was verified already in the proof of Lemma 3. The lattices obtained by different permutations of $\vartheta_1, \vartheta_2, \vartheta_3$ in (2) all differ from each other by an automorph of $\mathcal{N}_1$, namely a permutation of the co-ordinate axes.

Write

$$\varphi(\boldsymbol{x}) = x_1 x_2 x_3$$

and, as before,

$$|\varphi|(\Lambda) = \inf_{\substack{\boldsymbol{a}\in\Lambda\\ \neq \boldsymbol{o}}} |\varphi(\boldsymbol{a})|.$$

We show first that, for any lattice $\Lambda$,

$$\{|\varphi|(\Lambda)\}^2 \leq 7^{-1}\{|\varphi|(\Lambda^*)\}\, d^3(\Lambda), \tag{13}$$

where $\Lambda^*$ is the polar lattice of $\Lambda$. The proof follows closely the pattern of the proof of Theorem IV. It is enough to show that

$$\{|\varphi|(\Lambda)\}^2 \leqq 7^{-1}|\varphi(\boldsymbol{b})|\, d^3(\Lambda), \tag{14}$$

where $\boldsymbol{b}$ is any primitive point of $\Lambda^*$.

Suppose, first, that $\varphi(\boldsymbol{b})=0$. Then, after applying a suitable automorph of $\mathcal{N}_1$ furnished by § 2.1, we may suppose without loss of generality[1] that

$$\boldsymbol{b}=(1,0,0) \tag{15}$$

or

$$\boldsymbol{b}=(1,1,0). \tag{16}$$

In the first case, (15), the plane

$$\boldsymbol{b}\boldsymbol{x}=0,$$

which must contain two linearly independent elements of $\Lambda$ is just the plane $x_1=0$, and all points on it satisfy $\varphi(\boldsymbol{x})=0$. Hence (14) certainly holds in this case. In the second case, (16), there are two linearly independent points of $\Lambda$ on

$$x_1+x_2=0. \tag{17}$$

For these points

$$\varphi(\boldsymbol{x})=x_1x_2x_3=-x_2^2x_3, \tag{18}$$

and the 2-dimensional set $|x_2^2x_3|<\varepsilon$ is of infinite type for any $\varepsilon>0$. Hence there are certainly points $\boldsymbol{a}\in\Lambda$ other than $\boldsymbol{o}$ with $|\varphi(\boldsymbol{a})|<\varepsilon$. This proves (14) in the case $\boldsymbol{b}$ is given by (16).

There remains the case when $\varphi(\boldsymbol{b})\neq 0$ and so, after the application of a suitable automorph, we may suppose that

$$\boldsymbol{b}=(t,t,t), \quad t>0, \tag{19}$$

and so

$$\varphi(\boldsymbol{b})=t^3. \tag{20}$$

We have supposed that $\boldsymbol{b}$ is primitive, and so, by Lemma 6, Corollary of Chapter 1, the 2-dimensional set of points $(x_1, x_2)$ such that

$$(x_1, x_2, -x_1-x_2)\in\Lambda$$

is a lattice M of determinant

$$d(\mathsf{M})=t\,d(\Lambda).$$

But now

$$\inf_{\substack{(a_1,a_2)\in \mathsf{M}\\ \neq \boldsymbol{o}}}|a_1a_2(a_1+a_2)| \leqq \{7^{-\frac{1}{2}}d(\mathsf{M})\}^{\frac{3}{2}} = 7^{-\frac{3}{4}}t^{\frac{3}{2}}d^{\frac{3}{2}}(\Lambda),$$

[1] For we may suppose that $b_1\neq 0$, $b_3=0$. One gets the shape (15) or (16) according as $b_2=0$ or $b_2\neq 0$.

by Lemma 3, the exponent $\frac{3}{2}$ being correct for reasons of homogeneity. *A fortiori*

$$|\varphi|(\Lambda) \leqq 7^{-\frac{1}{2}} t^{\frac{3}{2}} d^{\frac{3}{2}}(\Lambda).$$

This proves (14) when $\boldsymbol{b}$ is given by (19) and (20); and so completes the proof of (13) and (14).

On interchanging $\Lambda$ and $\Lambda^*$ in (13) and using $d(\Lambda^*) = d^{-1}(\Lambda)$, we have

$$\{|\varphi|(\Lambda^*)\}^2 \leqq 7^{-1}\{|\varphi|(\Lambda)\} d^{-3}(\Lambda). \tag{13'}$$

On eliminating $|\varphi|(\Lambda^*)$ from (13) and (13′) we obtain

$$|\varphi|(\Lambda) \leqq 7^{-1} d(\Lambda), \tag{21}$$

so $\Delta(\mathscr{N}_1) \leqq 7$, since $N_1$ is the set of $\boldsymbol{x}$ with $|\varphi(\boldsymbol{x})| < 1$; and then $\Delta(\mathscr{N}_1) = 7$ since we have already exhibited an admissible lattice $\mathsf{N}_1$, with $d(\mathsf{N}_1) = 7$.

It remains to consider the critical lattices $\Lambda_c$ with a point on the boundary, and we may suppose, after the use of a suitable automorph, that

$$(1, 1, 1) \in \Lambda_c \quad d(\Lambda_c) = 7. \tag{22}$$

Clearly then the 2-dimensional lattices considered above will turn out to be critical for the relevant 2-dimensional sets, and it is necessary only to check that this can happen only when $\Lambda_c = \mathsf{N}_1$ for a suitable choice of $\vartheta_1, \vartheta_2, \vartheta_3$, where $\mathsf{N}_1$ is defined in Theorem V. A.

We note first that

$$|\varphi|(\Lambda_c^*) = 7^{-2} \tag{23}$$

by (13) and (13′). Hence the lattice $\mathsf{M}_c'$ of points

$$(x_1, x_2) \quad \text{with} \quad (x_1, x_2, -x_1 - x_2) \in \Lambda_c^*, \tag{24}$$

which has determinant

$$d(\mathsf{M}_c') = d(\Lambda_c^*) = 7^{-1},$$

must be one of the two critical lattices for

$$|x_1 x_2 (x_1 + x_2)| < 7^{-2} \tag{25}$$

given by Lemma 3. But we have already seen that $\mathsf{N}_1$ for any choice of $\vartheta_1, \vartheta_2, \vartheta_3$ is critical, and so the lattice

$$\mathsf{M}_1' = \mathsf{M}_1'(\vartheta_1, \vartheta_2, \vartheta_3),$$

defined by putting $\mathsf{N}_1 = \mathsf{N}_1(\vartheta_1, \vartheta_2, \vartheta_3)$ for $\Lambda_c$ in (24), is also critical. Clearly, by the proof of Lemma 3, both critical lattices of (25) occur as $\mathsf{M}_1'(\vartheta_1, \vartheta_2, \vartheta_3)$ for suitable choice of $\vartheta_1, \vartheta_2, \vartheta_3$. Hence we may suppose without loss of generality, that

$$\mathsf{M}_c' = \mathsf{M}_1';$$

that is, the polar lattices $\Lambda_c^*$ and $\mathsf{N}_1^*$ are identical at least on the plane

$$x_1 + x_2 + x_3 = 0.$$

Let now $\boldsymbol{b} = (b_1, b_2, b_3)$ be any point of $\Lambda_c^*$ (and so of $\mathsf{N}_1^*$) with

$$|b_1 b_2 b_3| = 7^{-2}, \quad b_1 + b_2 + b_3 = 0. \tag{26}$$

Then the lattices $\Lambda_c^{\boldsymbol{b}}$, $\mathsf{N}_1^{\boldsymbol{b}}$ consisting of the points of $\Lambda_c$ and of $\mathsf{N}_1$ respectively in the plane

$$b_1 x_1 + b_2 x_2 + b_3 x_3 = 0 \tag{27}$$

must both be critical, in the obvious sense, for the 2-dimensional section of $|x_1 x_2 x_3| < 1$ by the hyperplane (27). By Lemma 3, there are only two critical lattices and these have only the origin in common. Hence $\Lambda_1^{\boldsymbol{b}}$ and $\Lambda_c^{\boldsymbol{b}}$ must be identical, since $(1, 1, 1)$ belong to both lattices, by (27). Thus $\Lambda_c$ and $\Lambda_1$ coincide on any hyperplane (27) such that the point $\boldsymbol{b}$ satisfies (26).

But now $\mathsf{N}_1^*$ has a basis $\boldsymbol{b}_1^*, \boldsymbol{b}_2^*, \boldsymbol{b}_3^*$ (say) such that $\boldsymbol{b} = \boldsymbol{b}_1^*, \boldsymbol{b}_2^*$ satisfies (26), for we have only to choose a suitable basis $\boldsymbol{b}_1^*, \boldsymbol{b}_2^*$ for the section of $\mathsf{N}_1^*$ by $x_1 + x_2 + x_3 = 0$ and extend it to a basis for $\mathsf{N}_1^*$. Let $\boldsymbol{b}_1, \boldsymbol{b}_2, \boldsymbol{b}_3$ be the polar basis for $\mathsf{N}_1$. Then, on putting $\boldsymbol{b} = \boldsymbol{b}_1^*, \boldsymbol{b}_2^*$ in (25) in turn, we see that $\Lambda_c$ contains all points $\boldsymbol{a}$ of $\mathsf{N}_1$ such that either

$$\boldsymbol{b}_1^* \boldsymbol{a} = 0 \quad \text{or} \quad \boldsymbol{b}_2^* \boldsymbol{a} = 0;$$

that is all points of $\mathsf{N}_1$ of the shape either

$$u_2 \boldsymbol{b}_2 + u_3 \boldsymbol{b}_3 \quad \text{or} \quad v_1 \boldsymbol{b}_1 + v_3 \boldsymbol{b}_3,$$

where $u_1, u_2, v_1, v_3$ are any integers. Hence $\Lambda_c$ must contain each point

$$u_1 \boldsymbol{b}_1 + u_2 \boldsymbol{b}_2 + u_3 \boldsymbol{b}_3 = (u_2 \boldsymbol{b}_2 + u_3 \boldsymbol{b}_3) + (u_1 \boldsymbol{b}_1 + 0\, \boldsymbol{b}_3)$$

of $\mathsf{N}_1$. Since $d(\mathsf{N}_1) = d(\Lambda_c)$; we then have $\Lambda_c = \mathsf{N}_1$, as required.

This completes the proof of Theorem V. A. That of Theorem V. B is similar except that Theorems VII and VII A of Chapter III are used instead of Lemma 3. The details may be left to the reader.

**X.4. Existence of automorphs.** In this section we prove the existence of common automorphs of a lattice $\Lambda$ and a form $\varphi(\boldsymbol{x})$ which is integral and non-null on $\Lambda$, and make deductions about the possible such $\Lambda$ in a special case.

We shall require a quantitative form of MAHLER'S compactness criterion, Theorem IV of Chapter 5.

LEMMA 4. *There is a number*

$$N_0 = N_0(n, \Delta_1, \varkappa, \varepsilon) \tag{1}$$

*depending only on the integer $n>0$ and the numbers $\varDelta_1>0$, $\varkappa>0$, $\varepsilon>0$ with the following property: amongst any $N_0$ lattices $\Lambda_j$ $(1\leqq j\leqq N)$ in $n$-dimensional space such that*

$$d(\Lambda_j)\leqq\varDelta_1 \tag{2}$$

*and*

$$|\Lambda_j|\geqq\varkappa, \tag{3}$$

*there is at least one pair, say $\Lambda_1$, $\Lambda_2$, such that*

$$\Lambda_2=\boldsymbol{\tau}\Lambda_1 \tag{4}$$

*and the linear transformation $\boldsymbol{\tau}$ satisfies*

$$\|\boldsymbol{\tau}-\boldsymbol{\iota}\|<\varepsilon \quad \|\boldsymbol{\tau}^{-1}-\boldsymbol{\iota}\|<\varepsilon, \tag{5}$$

*where $\boldsymbol{\iota}$ is the identity transformation.*

We recollect that

$$|\Lambda|=\inf_{\substack{\boldsymbol{a}\in\Lambda\\ \neq\boldsymbol{o}}}|\boldsymbol{a}|, \tag{6}$$

and that the symbol $\|\boldsymbol{\sigma}\|$ for a linear transformation $\boldsymbol{X}=\boldsymbol{\sigma}\boldsymbol{x}$ with $X_j=\sum\sigma_{jk}x_k$ is $\|\boldsymbol{\sigma}\|=n\max|\sigma_{jk}|$.

It would be possible to modify the proof of Theorem IV given in Chapter V but it is simpler to follow the alternative proof sketched in § 2.2 of Chapter VIII. We suppose we have $N_0$ lattices $\Lambda_j$, where $N_0$ will be determined later. By Lemma 3 of Chapter VIII there is a $\varDelta_0>0$ and a $K$ depending only on $\varDelta_1$ and $\varkappa$, such that any $\Lambda_j$ satisfying (2) and (3) has

$$d(\Lambda_j)\geqq\varDelta_0>0 \tag{7}$$

and has $n$ linearly independent points in the sphere

$$|\boldsymbol{x}|\leqq K.$$

By Lemma 8 of Chapter V, there is then a basis

$$\boldsymbol{b}_{1j},\ldots,\boldsymbol{b}_{nj}$$

of $\Lambda_j$ with

$$|\boldsymbol{b}_{ij}|\leqq nK \quad (1\leqq i\leqq n,\ 1\leqq j\leqq N_0). \tag{8}$$

Let $\eta>0$ be arbitrarily small, to be chosen later. Then, by (8), if $N_0$ is greater than an $N_1$ depending only on $n,\eta,\varDelta_0,K$, that is on $n,\eta,\varDelta_1,\varkappa$, there are two $\Lambda_j$ say $\Lambda_1$ and $\Lambda_2$, such that

$$|\boldsymbol{b}_{i1}-\boldsymbol{b}_{i2}|<\eta \quad (1\leqq i\leqq n). \tag{9}$$

Since the $\boldsymbol{b}_{i1}$ are linearly independent, we have

$$\boldsymbol{b}_{i2}-\boldsymbol{b}_{i1}=\sum_{j=1}^{n}\sigma_{ij}\boldsymbol{b}_{j1}$$

for some numbers $\sigma_{ij}$. But now on solving for the $\sigma_{ij}$ from (7), (8) and (9), we have

$$|\sigma_{ij}| \leqq \sigma_0 \eta \quad (1 \leqq i \leqq n, \ 1 \leqq j \leqq n),$$

where $\sigma_0$ is a number depending only on $\Delta_0 K$ and $n$; a crude estimate being

$$\sigma_0 = n! \, \Delta_0^{-1} (n K)^{n-1}$$

obtained by estimating the elements of the matrix reciprocal to the matrix with columns $\boldsymbol{b}_{i1}$ $(1 \leqq i \leqq n)$. Hence

$$\|\boldsymbol{\sigma}\| < \varepsilon$$

if $\eta$ chosen to satisfy $n \sigma_0 \eta < \varepsilon$. Hence $\boldsymbol{\tau} = \boldsymbol{\iota} + \boldsymbol{\sigma}$ has $\boldsymbol{\tau}\Lambda_1 = \Lambda_2$ and $\|\boldsymbol{\tau} - \boldsymbol{\iota}\| < \varepsilon$. Since $\Lambda_1 = \boldsymbol{\tau}^{-1}\Lambda_2$ we have also $\|\boldsymbol{\tau}^{-1} - \boldsymbol{\iota}\| < \varepsilon$, because (9) is symmetric in $\Lambda_1, \Lambda_2$. This concludes the proof.

**X.4.2.** We shall also require the following rather trivial lemma which says, roughly, that a form $\varphi(\boldsymbol{x})$ cannot be integral on too many essentially distinct lattices.

LEMMA 5. *Let $\varphi(\boldsymbol{x})$ be a form integral on a lattice $\Lambda$. Then there is an $\eta > 0$ depending only on $\varphi(\boldsymbol{x})$ and $\Lambda$ with the following property: If $\varphi(\boldsymbol{x})$ is integral on $\boldsymbol{\tau}\Lambda$ and*

$$\|\boldsymbol{\tau} - \boldsymbol{\iota}\| < \eta, \tag{1}$$

*then $\boldsymbol{\tau}$ is an automorph of $\varphi(\boldsymbol{x})$.*

Let $\varphi(\boldsymbol{x})$ be of degree $m$ and let $\boldsymbol{b}_1, \ldots, \boldsymbol{b}_n$ be a basis for $\Lambda$. If $\boldsymbol{\tau}$ satisfies (1) with sufficiently small $\eta$, we have

$$\Big|\varphi\Big(\boldsymbol{\tau} \sum_j u_j \boldsymbol{b}_j\Big) - \varphi\Big(\sum_j u_j \boldsymbol{b}_j\Big)\Big| < 1 \tag{2}$$

for all integers $u_j$ such that

$$0 \leqq u_j \leqq m \quad (1 \leqq j \leqq n). \tag{3}$$

Then (2) implies

$$\varphi\Big(\boldsymbol{\tau} \sum_j u_j \boldsymbol{b}_j\Big) = \varphi(u_j \boldsymbol{b}_j) \tag{4}$$

for the integers (3), since both sides of (4) are integers. By Lemma 1, it follows that (4) holds for all real numbers $u_j$. Since every $\boldsymbol{x}$ is of the shape $\sum u_j \boldsymbol{b}_j$ with real $u_j$, we have $\varphi(\boldsymbol{\tau x}) = \varphi(\boldsymbol{x})$ for all $\boldsymbol{x}$, as required.

COROLLARY. *Suppose, further, that $\varphi(\boldsymbol{x})$ is non-null on $\Lambda$ and that*

$$d(\Lambda) \leqq \Delta_1$$

*for some $\Delta_1$. Then $\eta$ may be chosen depending only on $\varphi$ and $\Delta_1$, but not otherwise on $\Lambda$.*

For then

$$|\varphi(\boldsymbol{a})| \geqq 1 \quad (\boldsymbol{a} \in \Lambda, \ \boldsymbol{a} \neq \boldsymbol{o});$$

and so

$$|\Lambda| \geqq c > 0$$

for some $c$ depending only on $\varphi$. Hence, as in the proof of Lemma 4, there is a basis $\boldsymbol{b}_1, \ldots, \boldsymbol{b}_n$ of $\Lambda$ with

$$|\boldsymbol{b}_j| \leqq nK \qquad (1 \leqq j \leqq n)$$

for a $K$ depending only on $\Delta_1$ and $c$, i.e. on $\Delta_1$ and $\varphi$. Hence all the points $\Sigma u_j \boldsymbol{b}_j$ subject to (3) lie in a sphere

$$|\boldsymbol{x}| \leqq n^2 m K. \tag{5}$$

Then (2) holds for small enough $\eta$ depending only on $\varphi$ and $K$, since $\varphi(\boldsymbol{x})$ is uniformly continuous in (5). Hence the corollary follows.

**X.4.3.** We are now in a position to prove the main theorem on the existence of automorphs.

THEOREM VI. *Let the form $\varphi(\boldsymbol{x})$ be integral and non-null on the lattice $\Lambda$ and let $\boldsymbol{\sigma}$ be any automorph of $\varphi(\boldsymbol{x})$. Suppose $\varepsilon > 0$ is given arbitrarily small. Then there is an automorph $\boldsymbol{\tau}$ of $\varphi(\boldsymbol{x})$ with*

$$\|\boldsymbol{\tau} - \boldsymbol{\iota}\| < \varepsilon, \tag{1}$$

*such that*

$$\boldsymbol{\omega} = \boldsymbol{\sigma}^{-u} \boldsymbol{\tau} \boldsymbol{\sigma}^{v} \tag{2}$$

*is an automorph of $\Lambda$ for certain integers $u, v$ with*

$$0 \leqq u < v. \tag{3}$$

It is not excluded, of course, that $\boldsymbol{\omega}$ may be the identical transformation.

We have

$$|\varphi|(\Lambda) = \inf_{\substack{\boldsymbol{a} \in \Lambda \\ \neq \boldsymbol{o}}} |\varphi(\boldsymbol{a})| \geqq 1, \tag{4}$$

by hypothesis, and so

$$|\varphi|(\boldsymbol{\sigma}^u \Lambda) \geqq 1 \tag{5}$$

for all integers $u$. Hence

$$|\boldsymbol{\sigma}^u \Lambda| \geqq c > 0 \tag{6}$$

for all $u$ and some constant $c > 0$. Further,

$$d(\boldsymbol{\sigma}^u \Lambda) = d(\Lambda) \tag{7}$$

for all $u$ since $\det(\boldsymbol{\sigma}) = \pm 1$ by Theorem I. By (6) and (7) we may apply Lemma 3 to the $\boldsymbol{\sigma}^u \Lambda$ $(1 \leqq u \leqq N)$, if $N$ is some large enough number, to obtain two lattices $\boldsymbol{\sigma}^u \Lambda$ and $\boldsymbol{\sigma}^v \Lambda$ such that

$$\boldsymbol{\sigma}^u \Lambda = \boldsymbol{\tau} \boldsymbol{\sigma}^v \Lambda \qquad (u < v) \tag{8}$$

and

$$\|\boldsymbol{\tau}-\boldsymbol{\iota}\|<\varepsilon, \quad \|\boldsymbol{\tau}^{-1}-\boldsymbol{\iota}\|<\varepsilon. \tag{9}$$

We may suppose, by choosing a smaller number instead of the original $\varepsilon$ if necessary, that $\varepsilon<\eta$, where $\eta$ is the number in Lemma 5, Corollary with $\Delta_1=d(\Lambda)$. We may then apply Lemma 5, Corollary with $\boldsymbol{\sigma}^v\Lambda$ instead of $\Lambda$ and deduce from (8), (9) that $\boldsymbol{\tau}$ is an automorph for $\varphi(\boldsymbol{x})$. Hence $\boldsymbol{\omega}$ defined in (2) has all the properties required.

Theorem VI becomes false if the condition that $\varphi(\boldsymbol{x})$ be non-null on $\Lambda$ is omitted, as is shown by the 2-dimensional example where $\Lambda=\Lambda_0$ is the lattice of integral vectors, $\varphi(\boldsymbol{x})=x_1x_2$, and $\boldsymbol{\sigma}$ is the automorph $x_1\to 2x_1$, $x_2\to\frac{1}{2}x_2$. But in more dimensions it is sometimes possible to use the idea behind Theorem VI to construct automorphs of $\Lambda$ even when $\varphi(\boldsymbol{x})$ may be null on $\Lambda$, for example, by restricting attention to automorphs leaving fixed an element or elements of $\Lambda$ or of the polar lattice $\Lambda^*$.

**X.4.4.** Theorem VI takes a particularly simple shape when

$$\varphi(\boldsymbol{x})=\Big\{\prod_{1\leqq j\leqq r}x_j\Big\}\Big\{\prod_{1\leqq k\leqq s}(x_{r+k}^2+x_{r+s+k}^2)\Big\}, \tag{1}$$

where $n=r+2s$, which is substantially equivalent to, but rather stronger than, DIRICHLET's theorem on the existence of units in an algebraic number field. We write as usual

$$\left.\begin{aligned} z_j&=x_j \quad (1\leqq j\leqq r)\\ \left.\begin{aligned} z_{r+k}&=x_{r+k}+i\,x_{r+s+k}\\ z_{r+s+k}&=x_{r+k}-i\,x_{r+s+k}\end{aligned}\right\}&\quad(1\leqq k\leqq s).\end{aligned}\right\} \tag{2}$$

It is convenient to work with the $z_j$ rather than the $x_j$, so we shall speak of the $z_j$ as the appropriate complex co-ordinates. We shall also say for brevity that a set of numbers $\lambda_j$ $(1\leqq j\leqq n)$ is compatible with $\varphi(\boldsymbol{x})$ if

$$\lambda_j=\text{real} \qquad (1\leqq j\leqq n)$$

$$\lambda_{r+k},\lambda_{r+s+k}\ \text{conjugate complex} \qquad (1\leqq k\leqq s).$$

THEOREM VII. *Let $\varphi(\boldsymbol{x})$ be given by* (1), *and let $\lambda_j$ $(1\leqq j\leqq n)$ be numbers compatible with $\varphi(\boldsymbol{x})$ such that*

$$\prod_{1\leqq j\leqq n}\lambda_j=1.$$

*Suppose that $\varphi(\boldsymbol{x})$ is integral on $\Lambda$ and that $\varepsilon>0$ is given arbitrarily small. Then there are numbers $\omega_j$ compatible with $\varphi(\boldsymbol{x})$ and an integer $m>0$ such that*

$$\prod_{1\leqq j\leqq n}\omega_j=1, \qquad \left|\frac{\lambda_j^m}{\omega_j}-1\right|<\varepsilon \qquad (1\leqq j\leqq n), \tag{3}$$

*and such that the transformation* $\boldsymbol{\omega}$ *given in the appropriate complex co-ordinates by*

$$Z_j = \omega_j z_j \qquad (1 \leqq j \leqq n)$$

*is an automorph of* $\Lambda$.

The automorphs of $\varphi(\boldsymbol{x})$ were discussed in § 2.1. From what is said there it is clear that if $\boldsymbol{Z} = \boldsymbol{\tau z}$ is an automorph of $\varphi$ given in the appropriate complex co-ordinates and if

$$\|\boldsymbol{\tau} - \boldsymbol{\iota}\| < n, \tag{4}$$

where $n$ is the dimension, then $\tau$ must be of the shape

$$Z_j = \tau_j z_j \qquad (1 \leqq j \leqq n); \tag{5}$$

that is, there can be no permutation of the forms on the right-hand side: indeed, if $\boldsymbol{\tau}$ is written as $Z_j = \sum_k \tau_{jk} z_k$, the inequality (4) implies

$$|\tau_{jj} - 1| < 1, \quad \text{so} \quad \tau_{jj} \neq 0 \qquad (1 \leqq j \leqq n),$$

and the only automorphs of this kind are (5). If $\boldsymbol{Z} = \boldsymbol{\lambda z}$ is given in complex co-ordinates by

$$Z_j = \lambda_j z_j \qquad (1 \leqq j \leqq n),$$

it follows now that $\boldsymbol{\lambda}$ and $\boldsymbol{\tau}$ commute. Hence applying Theorem VI with $\boldsymbol{\sigma} = \boldsymbol{\lambda}$ we have

$$\boldsymbol{\omega} = \boldsymbol{\lambda}^{-u} \boldsymbol{\tau} \boldsymbol{\lambda}^{v} = \boldsymbol{\lambda}^{m} \boldsymbol{\tau},$$

where $m = v - u$. Then $\boldsymbol{\omega}$ does what is required.

We shall later require to know slightly more about the automorphs $\boldsymbol{\omega}$ of lattices on which $\varphi(\boldsymbol{x})$ given by (1) is integral; and it is convenient to prove it here.

LEMMA 6. *Let* $\varphi(\boldsymbol{x})$ *given by* (1) *be integral on* $\Lambda$ *and let the automorph* $\boldsymbol{Z} = \boldsymbol{\omega z}$ *of* $\Lambda$ *be given in the appropriate complex co-ordinates by*

$$Z_j = \omega_j z_j \qquad (1 \leqq j \leqq n).$$

*Then the* $\omega_j$ *are algebraic units, that is they satisfy an equation of the type*

$$f(\omega_j) = 0,$$

*where*

$$f(t) = t^m + c_1 t^{m-1} + \cdots + c_{m-1} t \pm 1 = 0 \tag{6}$$

*for some* $m$ *and* $c_1, \ldots, c_{m-1}$ *are rational integers.*

Let $\boldsymbol{b}_1, \ldots, \boldsymbol{b}_n$ be a basis for $\Lambda$, so that

$$\boldsymbol{\omega} \boldsymbol{b}_j = \sum_{1 \leq k \leq n} m_{jk} \boldsymbol{b}_k \tag{7}$$

for some integers $m_{jk}$. Since the $\boldsymbol{\omega}\, \boldsymbol{b}_j$ are a basis, we have

$$\det(m_{jk}) = \pm 1.$$

Let

$$\boldsymbol{b}_k = (\beta_{1k}, \ldots, \beta_{nk}) \qquad (1 \leq k \leq n) \tag{8}$$

in the appropriate complex co-ordinates and let $\mathbf{B}$ be the matrix of which the rows are given by (8). Then (7) takes the shape

$$\mathbf{B}\boldsymbol{\omega} = \boldsymbol{m}\mathbf{B}, \tag{9}$$

where $\boldsymbol{m}$ is the matrix with elements $m_{jk}$ and $\boldsymbol{\omega}$ is the diagonal matrix with elements $\omega_1, \ldots, \omega_n$ on the diagonal. Hence

$$\boldsymbol{\omega} = \mathbf{B}^{-1}\boldsymbol{m}\mathbf{B},$$

and $\omega_1, \ldots, \omega_n$ all satisfy the equation $f(\omega_j) = 0$, where

$$f(t) = \det(t\,\boldsymbol{\iota} - \boldsymbol{m}),$$

which is of the form (6).

The two following corollaries are immediate

COROLLARY 1. *$\omega_1, \ldots, \omega_n$ satisfy the same equation of type* (6) *with $m = n$.*

COROLLARY 2. *If $\omega_j$ is rational, then $\omega_j = \pm 1$.*

Although we do not need it later it is interesting to note that Theorem VII and Lemma 6 rapidly gives a complete characterisation of the lattices $\Lambda$ on which $\varphi(\boldsymbol{x})$ is proportional to integral and non-null, at least when $r > 0$. We only sketch the proof, for details see BACHMANN (1923a) Kap. 12.

LEMMA 7. *All the lattices $\Lambda$ on which $\varphi(\boldsymbol{x})$ is proportional to integral may be obtained in the following way. Let $\mathfrak{K}_1, \ldots, \mathfrak{K}_n$ be a set of conjugate algebraic fields of degree $n$ over the field of rational numbers, where $\mathfrak{K}_1, \ldots, \mathfrak{K}_r$ are real and $\mathfrak{K}_{r+k}$, $\mathfrak{K}_{r+s+k}$ are conjugate complex $(1 \leq k \leq s)$. Let $\gamma_{11}, \ldots, \gamma_{1n}$ be linearly independent elements of $\mathfrak{K}_1$ over the rationals and let $\gamma_{lk}$ $(1 \leq l \leq n)$ be the conjugate of $\gamma_{1k}$ in $\mathfrak{K}_l$. Let $\mathsf{M}$ be the lattice with basis*

$$\boldsymbol{c}_k = (\gamma_{1k}, \ldots, \gamma_{nk}) \qquad (1 \leq k \leq n)$$

*in the appropriate complex co-ordinates. Then a necessary and sufficient condition that $\varphi(\boldsymbol{x})$ be proportional to integral and non-null on a lattice $\Lambda$ is that $\Lambda$ be of the shape*

$$\Lambda = t\boldsymbol{\tau}\mathsf{M}$$

*where $t$ is real, $\boldsymbol{\tau}$ is an automorph of $\varphi(\boldsymbol{x})$, and $\mathsf{M}$ is of the type just described.*

When $r > 0$, the proof is shorter than the enunciation. By applying Theorem VII with

$$\lambda_1 = 2^{n-1}, \quad \lambda_j = \tfrac{1}{2} \qquad (2 \leq j \leq n),$$

we deduce the existence of an automorph $\boldsymbol{\omega}$ of $\varphi(\boldsymbol{x})$ and $\Lambda$ with

$$\omega_1 > 1, \quad |\omega_j| < 1 \qquad (2 \leq j \leq n). \tag{10}$$

Since $\omega_1, \ldots, \omega_n$ all satisfy the same equation of degree $n$, they must all by (10) be precisely of degree $n$ and so conjugates. Let $\boldsymbol{b}_1, \ldots, \boldsymbol{b}_n$ be a basis for $\Lambda$ and use the notation (7), (8). Then it follows from (9) that

$$(\beta_{j1}, \ldots, \beta_{jn})$$

is an eigenvector belonging to $\omega_j$ of the matrix $\boldsymbol{m}$. But clearly $\boldsymbol{m}$ has a set of conjugate eigenvectors

$$(\gamma_{j1}, \ldots, \gamma_{jn})$$

in the fields $\mathfrak{K}_j$ generated by $\omega_j$; and if these are identified with those of the enunciation it is easy to see that the lattice $\mathsf{M}$ has the required properties.

When $r=0$, the position is more difficult since it may be impossible to achieve that the $\omega_j$ are all of degree $n$, though it is possible to make them all of degree $\frac{1}{2}n$. Let $\boldsymbol{a}=(\alpha_1, \ldots, \alpha_n)$ and $\boldsymbol{b}=(\beta_1, \ldots, \beta_n)$ be two linearly independent vectors of $\Lambda$ in the appropriate complex co-ordinate system. Then $\varphi(u\boldsymbol{a}+v\boldsymbol{b})$ is a polynomial in the variables $u$ and $v$ with coefficients proportional to integers, and it vanishes for integers $u$ and $v$ only when $u=v=0$. Hence $\alpha_1/\beta_1$ is an algebraic number. Similarly, if $\boldsymbol{c}=(\gamma_1, \ldots, \gamma_n)\in\Lambda$ is linearly independent from $\boldsymbol{a}$ and $\boldsymbol{b}$, then the ratios $\alpha_1/\gamma_1$, $\beta_1/\gamma_1$ are of degree $n$ as is also $(p\alpha_1+q\beta_1)/\gamma_1$ for any integers $p$ and $q$. It is not then difficult to deduce that $\alpha_1/\beta_1$ is in a field of degree $n$ depending only on $\Lambda$ and not on the choice of $\boldsymbol{a}$ and $\boldsymbol{b}$; and the rest follows with some little trouble. We do not go into details as we do not use the result.

**X.5. Isolation theorems.** As was stated in § 1 there is a wide variety of isolation theorems, and it hardly seems worth while to formulate theorems of great generality. We shall instead consider only three concrete cases.

We shall need the following simple Lemma which is really a simple case of KRONECKER's Theorem and belongs of right in Chapter XI.

LEMMA 8. *Let $\alpha, \beta, \gamma, \delta$ be real numbers with $\alpha\delta-\beta\gamma\neq 0$. Suppose that $\alpha/\beta$ is irrational. Then to every number $\varepsilon>0$ there is an $\eta=\eta(\alpha, \beta, \gamma, \delta, \varepsilon)$ with the following property:*

*For any numbers $\lambda, \mu$ there are integers $m, n$ such that*

$$|m\alpha+n\beta-\lambda|<\varepsilon, \quad |m\gamma+n\delta-\mu|\leqq\eta.$$

By MINKOWSKI's linear forms Theorem there are integers $(m, n)\neq(0,0)$ such that $|m\alpha+n\beta|$ is arbitrarily small; and $m\alpha+n\beta\neq 0$ since $\alpha/\beta$ is irrational. Hence there are integers $(m_1, n_1)$ and $(m_2, n_2)$ such that

$$0<|m_1\alpha+n_1\beta|<\varepsilon, \quad 0<|m_2\alpha+n_2\beta|<\varepsilon,$$

and

$$m_1 n_2\neq m_2 n_1.$$

Put

$$X_j=m_j\alpha+n_j\beta, \quad Y_j=m_j\gamma+n_j\delta \quad (j=1,2),$$

so that

$$|X_j|<\varepsilon \quad (j=1,2), \qquad X_1 Y_2\neq X_2 Y_1.$$

Let $\varrho, \sigma$ be the solution of

$$\varrho X_1+\sigma X_2=\lambda, \quad \varrho Y_1+\sigma Y_2=\mu,$$

and choose integers $a, b$ such that

$$|a-\varrho| \leqq \tfrac{1}{2}, \quad |b-\sigma| \leqq \tfrac{1}{2}.$$

Then

$$|aX_1+bX_2-\lambda| = |(a-\varrho)X_1+(b-\sigma)X_2| \leqq \tfrac{1}{2}(|X_1|+|X_2|) < \varepsilon,$$

and

$$|aY_1+bY_2-\mu| \leqq \tfrac{1}{2}(|Y_1|+|Y_2|) = \eta \text{ (say)}.$$

The lemma now follows on putting

$$m = am_1+bm_2, \quad n = an_1+bn_2.$$

**X.5.2.** Perhaps the simplest isolation theorem is that for $x_1x_2$ and is due to C. A. ROGERS [unpublished, but see CASSELS (1957a), Chapter II where an application to the "Markoff chain", due to ROGERS, is given].

THEOREM VIII. *Let $x_1x_2$ be integral and non-null on the 2-dimensional lattice $\Lambda$ and let there be $\boldsymbol{a}, \boldsymbol{b} \in \Lambda$ such that*

$$a_1a_2 = -\alpha < 0 < b_1b_2 = \beta. \tag{1}$$

*Then there are numbers $\eta_0>0$, $\eta_1>0$ with the following properties:*

*Let $\boldsymbol{\tau}$ be a linear transformation and suppose that*

$$\|\boldsymbol{\tau}-\boldsymbol{\iota}\| < \eta_1 \tag{2}$$

*and*

$$\tau_{12} \neq 0, \tag{3}$$

*where the transformation $\boldsymbol{X}=\boldsymbol{\tau x}$ is given by*

$$X_1 = \tau_{11}x_1+\tau_{12}x_2, \quad X_2 = \tau_{21}x_1+\tau_{22}x_2.$$

*Then there is a point $\boldsymbol{c} \neq \boldsymbol{o}$ of $\boldsymbol{\tau}\Lambda$ such that*

$$-\alpha(1-\eta_0) < c_1c_2 < \beta(1-\eta_0), \quad |c_1| < 1.$$

We may suppose without loss of generality that

$$a_1>0, \quad b_1>0$$

and so

$$a_2<0, \quad b_2>0.$$

By Theorem VII, there is an automorph $\boldsymbol{X}=\boldsymbol{\omega x}$ of $\Lambda$ of the shape

$$X_1=\omega_1x_1, \quad X_2=\omega_2x_2,$$

where

$$0<\omega_1<1<\omega_2, \quad \omega_1\omega_2=1.$$

Then $\Lambda$ contains all the points

$$\boldsymbol{a}_m = (\omega_2^{-m}a_1, \omega_2^{m}a_2), \quad \boldsymbol{b}_m = (\omega_2^{-m}b_1, \omega_2^{m}b_2), \tag{4}$$

where $m$ is any integer, positive negative or 0.

We must now distinguish two cases according to the sign of $\tau_{12}$. Suppose, first, that

$$\tau_{12} > 0. \tag{5}$$

Let the integer $m$ be determined by

$$\tau_{11} a_1 \omega_2^{-m} + \tau_{12} a_2 \omega_2^{m} > 0 \geqq \tau_{11} a_1 \omega_2^{-m-1} + \tau_{12} a_2 \omega_2^{m+1}, \tag{6}$$

as is possible, since $\tau_{11} a_1 > 0 > \tau_{12} a_2$. Then

$$-1 < \frac{\tau_{12} a_2}{\tau_{11} a_1} \omega_2^{2m} < -\omega_2^{-2}. \tag{6'}$$

Hence

$$\omega_2^{m} = O(\tau_{12}^{-\frac{1}{2}}), \tag{7}$$

where the constant implied by the $O$ symbol may depend on $a_1$, $a_2$, and where we assume $\eta_1$ in (2) chosen so that, say, $|\tau_{11} - 1| < \frac{1}{2}$. Put

$$\mathbf{c} = \boldsymbol{\tau} \boldsymbol{a}_m, \tag{8}$$

where $\boldsymbol{a}_m$ is given by (4). Then, in the first place, it follows from (6) and (7) that

$$c_1 = \tau_{11} a_1 \omega_2^{-m} + \tau_{12} a_2 \omega_2^{m} = O(\tau_{12}^{\frac{1}{2}}),$$

so

$$|c_1| < 1$$

if $\eta_1$ is chosen small enough. Secondly, it follows from (6) or (6') that

$$0 < \omega_2^{m} c_1 \leqq \tau_{11} a_1 (1 - \omega_2^{-2}). \tag{9}$$

But now, by (7),

$$\omega_2^{-m} c_2 = \tau_{21} a_1 \omega_2^{-2m} + \tau_{22} a_2 = \tau_{22} a_2 + O(\tau_{12}). \tag{10}$$

Put $\eta_0 = \frac{1}{2}\omega_2^{-2}$. Then since $a_2 < 0 < a_1$, we have from (7), (9) and (10), that

$$a_1 a_2 (1 - \eta_0) < c_1 c_2 < 0,$$

provided that $\tau_{12}, \tau_{21}$ are small enough and that $\tau_{11}, \tau_{22}$ are near enough to 1, which may be achieved by taking $\eta_1$ small enough in (2). This concludes the proof when $\tau_{12} > 0$. The proof when $\tau_{12} < 0$ is completely similar, except that $\boldsymbol{b}$ is used instead of $\boldsymbol{a}$.

COROLLARY. *Under the hypotheses of the theorem except* (3), *there is an* $\eta_2$ *such that if*

$$\|\boldsymbol{\tau} - \boldsymbol{\iota}\| < \eta_2,$$

*and* $\boldsymbol{\tau}$ *is not an automorph of* $x_1 x_2$, *then* $\boldsymbol{\tau}\Lambda$ *contains a point* $\mathbf{c}$ *with*

$$a_1 a_2 (1 - \eta_0) < c_1 c_2 < b_1 b_2 (1 - \eta_0).$$

For if $\boldsymbol{\tau}$ is not an automorph, then either $\tau_{12} \neq 0$ or $\tau_{21} \neq 0$. If $\tau_{12} \neq 0$, then the theorem applies; and if $\tau_{21} \neq 0$ then the theorem can be applied with the rôles of $x_1$ and $x_2$ interchanged.

Note that Theorem VIII works with the values of $x_1 x_2$ at two distinct points of $\Lambda$. This rather restricts its field of application. The other isolation theorems which we shall discuss require at most knowledge of the value of the function at only one lattice point.

**X.5.3.** Before discussing the isolation results for

$$\varphi(\boldsymbol{x}) = x_1 x_2 x_3$$

we require a simple lemma.

LEMMA 9. *Let $x_1 x_2 x_3$ be integral and non-null on $\Lambda$. To every $\varepsilon > 0$ there is an $\eta > 0$, depending on $\Lambda$, with the following property:*

*To any numbers $\varrho > 0$, $\sigma > 0$ and index $k = 1, 2$ or $3$ there is an automorph $\mathbf{X} = \boldsymbol{\omega x}$ of $\Lambda$:*

$$X_j = \omega_j x_j \qquad (1 \leqq j \leqq 3), \tag{1}$$

*with*

$$\omega_j > 0 \qquad (1 \leqq j \leqq 3), \qquad \omega_1 \omega_2 \omega_3 = 1 \tag{2}$$

*and*

$$1 - \varepsilon < \frac{\varrho \omega_1}{\omega_2} < 1 + \varepsilon, \quad \eta^{-1} < \frac{\omega_k}{\sigma} < \eta. \tag{3}$$

For by Theorem VII there are certainly automorphs $\boldsymbol{\vartheta}$ and $\boldsymbol{\psi}$ of $\Lambda$ defined by

$$X_j = \vartheta_j x_j, \quad X_j = \psi_j x_j \qquad (1 \leqq j \leqq 3)$$

respectively, with

$$\vartheta_1 > 1, \quad 0 < \vartheta_2 < 1, \quad 0 < \vartheta_3 < 1, \quad \vartheta_1 \vartheta_2 \vartheta_3 = 1,$$
$$0 < \psi_1 < 1, \quad \psi_2 > 1, \quad 0 < \psi_3 < 1, \quad \psi_1 \psi_2 \psi_3 = 1.$$

Put

$$p_j = \log \vartheta_j, \quad q_j = \log \psi_j; \tag{4}$$

so

$$p_1 + p_2 + p_3 = q_1 + q_2 + q_3 = 0 \tag{5}$$

and

$$p_1 > 0, \quad p_2 < 0, \quad p_3 < 0,$$
$$q_1 < 0, \quad q_2 > 0, \quad q_3 < 0.$$

Hence

$$p_1 q_2 - p_2 q_1 = p_2 q_3 - p_3 q_2 \neq 0. \tag{6}$$

We now show that

$$(p_1 - p_2)/(q_1 - q_2) \tag{7}$$

is irrational. If not, there would be an automorph $\boldsymbol{\lambda} = \boldsymbol{\vartheta}^u \boldsymbol{\psi}^v$ with integers $(u, v) \neq (0, 0)$ for which, in an obvious notation, $\lambda_1 = \lambda_2$. But then, $\lambda_1$ would be rational, since $\lambda_1, \lambda_2, \lambda_3$ satisfy a cubic equation with integer coefficients by Lemma 6, Corollary 1. Hence $\lambda_1 = \lambda_2 = \lambda_3 = 1$, by Lemma 6, Corollary 2; that is

$$u p_j + v q_j = 0 \qquad (1 \leqq j \leqq 3),$$

which contradicts (6). By (6) we may now apply Lemma 8, with

$$\lambda = \log \varrho, \quad \alpha = p_2 - p_1, \quad \beta = q_2 - q_1,$$
$$\mu = \log \sigma, \quad \gamma = p_k, \quad \delta = q_k,$$

and

$$\min_{\pm} |\log(1 \pm \varepsilon)|, \quad \log \eta$$

or $\varepsilon, \eta$ respectively. Then

$$\boldsymbol{\omega} = \boldsymbol{\vartheta}^m \boldsymbol{\psi}^n,$$

where $m$ and $n$ are given by Lemma 8, clearly has all the properties required.

It is now a simple matter to prove

THEOREM IX. *Let $x_1 x_2 x_3$ be integral and non-null on $\Lambda$ and let $\varepsilon_1 > 0$ be arbitrarily small. There exists an $\eta_1 > 0$, depending on $\varepsilon_1$ and $\Lambda$, such that if*

$$\|\boldsymbol{\tau} - \boldsymbol{\iota}\| < \eta_1 \tag{8}$$

*and $t\boldsymbol{\tau}$ is not an automorph of $x_1 x_2 x_3$ for any number $t$, then the lattice $\boldsymbol{\tau}\Lambda$ contains a point $\boldsymbol{c} \neq \boldsymbol{o}$ for which*

$$|c_1 c_2 c_3| < \varepsilon_1. \tag{9}$$

Let $\boldsymbol{\tau}$ be given by $X_i = \sum_j \tau_{ij} x_j$, when $\boldsymbol{X} = \boldsymbol{\tau x}$. If $\boldsymbol{\tau}$ is not an automorph, there is a $\tau_{ij} \neq 0$ $(i \neq j)$. We shall suppose that

$$\tau_{12} = \max_{i \neq j} |\tau_{ij}| > 0, \tag{9'}$$

this being one of twelve possible cases[1]. Now $\Lambda$ certainly does contain some point $\boldsymbol{a}$ with

$$a_1 > 0 > a_2.$$

We shall pick one such point and keep it fixed in all that follows, so that numbers depending only on $\boldsymbol{a}$ and $\Lambda$ will be said to depend only on $\Lambda$, etc.

By Lemma 9 with an $\varepsilon > 0$ to be chosen later and

$$\varrho = -\frac{a_1 \tau_{11}}{a_2 \tau_{12}} > 0, \quad \sigma = 1, \quad k = 3, \tag{10}$$

[1] For the maximum in (9′) may correspond to any one of the six pairs $(i, j)$ with $i \neq j$; and the maximal $\tau_{ij}$ may be either positive or negative.

there is an automorph $\boldsymbol{\omega}$ of $\Lambda$ with

$$1-\varepsilon<-\frac{a_1\tau_{11}\omega_1}{a_2\tau_{12}\omega_2}<1+\varepsilon, \quad \eta^{-1}<\omega_3<\eta, \tag{11}$$

where

$$\eta=\eta(\Lambda,\varepsilon) \tag{12}$$

is independent of the $\tau_{ij}$. Since $\tau_{11}$ is assumed near to 1, say $|\tau_{11}-1|<\frac{1}{2}$, it follows from (11) and $\omega_1\omega_2\omega_3=1$ that

$$\eta'^{-1}\tau_{12}^{\frac{1}{2}}<\omega_1<\eta'\tau_{12}^{\frac{1}{2}}, \quad \eta'^{-1}\tau_{12}^{-\frac{1}{2}}<\omega_2<\eta'\tau_{12}^{-\frac{1}{2}}, \tag{13}$$

where

$$\eta'=\eta'(\varepsilon,\Lambda) \tag{14}$$

is independent of the $\tau_{ij}$.

We put

$$\boldsymbol{c}=\boldsymbol{\tau}\boldsymbol{\omega}\boldsymbol{a}\in\boldsymbol{\tau}\Lambda.$$

Then by (9′), (11) and (13), we have

$$\begin{aligned}\omega_1^{-1}|c_1|&=\omega_1^{-1}|a_1\tau_{11}\omega_1+a_2\tau_{12}\omega_2+a_3\tau_{13}\omega_3|\\ &\leqq\omega_1^{-1}\{|a_1\tau_{11}\omega_1+a_2\tau_{12}\omega_2|+|a_3\tau_{13}\omega_3|\}<\varkappa_1\varepsilon+\xi_1\tau_{12}^{\frac{1}{2}},\end{aligned}$$

where

$$\varkappa_1=\varkappa_1(\Lambda), \quad \xi_1=\xi_1(\Lambda,\varepsilon).$$

It is important that $\varkappa_1$ is independent of $\varepsilon$. Hence

$$\omega_1^{-1}|c_1|<2\varkappa_1\varepsilon, \tag{15}$$

provided that $\tau_{12}$ is smaller than a number depending on $\varepsilon$. Similarly, but more simply, by (9′), (11) and (13),

$$\begin{aligned}\omega_2^{-1}|c_2|&<\omega_2^{-1}\omega_1|\tau_{21}a_1|+|\tau_{22}a_2|+\omega_2^{-1}\omega_3|\tau_{13}a_3|\\ &<|\tau_{22}a_2|+\xi_2\tau_{12}^{\frac{1}{2}},\end{aligned}$$

where

$$\xi_2=\xi_2(\Lambda,\varepsilon);$$

and so

$$\omega_2^{-1}|c_2|<2a_2, \tag{16}$$

provided that $\tau_{12}$ is small enough and $\tau_{22}$ is near enough to 1. Similarly

$$\omega_3^{-1}|c_3|<2|a_3| \tag{17}$$

if $\tau_{33}-1$ and $\tau_{12}$ are small enough. From (15), (16) and (17) we have

$$|c_1c_2c_3|<8|\varkappa_1a_2a_3|\,\varepsilon.$$

Since $\varepsilon$ is arbitrarily small, we may put $\varepsilon_1=8|\varkappa_1a_2a_3|\,\varepsilon$, where $\varepsilon_1$ is the number in the enunciation.

This completes the proof.

Note that we have used the full force neither of Lemma 9 nor of the inequalities (13).

The proofs of the following two corollaries may be left to the reader.

COROLLARY 1. *Theorem IX remains valid if* $|c_1 c_2 c_3| < \varepsilon_1$ *is replaced by* $0 < |c_1 c_2 c_3| < \varepsilon_1$.

COROLLARY 2. *To every* $\varepsilon_2 > 0$ *there is an* $\eta_2 > 0$ *depending only on* $\Lambda$, $\varepsilon_2$ *such that, if*

$$\|\boldsymbol{\tau} - \boldsymbol{\iota}\| < \eta_2$$

*and one of* $\tau_{12}, \tau_{13}, \tau_{21}, \tau_{23}$ *is not* 0, *then there is a* $\boldsymbol{c} \in \boldsymbol{\tau}\Lambda$ *with*

$$0 < |c_1 c_2 c_3| < \varepsilon_2, \quad |c_1| < 1, \quad |c_2| < 1.$$

Corollary 1 is proved in CASSELS and SWINNERTON-DYER (1955a). A somewhat weaker form of Corollary 2 is in DAVENPORT and ROGERS (1950a).

**X.5.4.** We now discuss

$$\varphi(\boldsymbol{x}) = x_1(x_2^2 + x_3^2).$$

As in § 4.4 it is convenient to introduce the appropriate complex co-ordinates

$$z_1 = x_1, \quad z_2 = x_2 + i x_3, \quad z_3 = x_2 - i x_3 \qquad (i^2 = -1).$$

A transformation $\boldsymbol{Z} = \boldsymbol{\tau z}$ corresponds to a real transformation for the real variables $\boldsymbol{x}$ if and only if it is of the shape

$$Z_j = \sum_k \tau_{jk} z_k, \tag{1}$$

where

$$\tau_{12} = \bar{\tau}_{13}, \quad \tau_{21} = \bar{\tau}_{31}, \quad \tau_{11} = \bar{\tau}_{11} \quad \tau_{23} = \bar{\tau}_{32}, \quad \tau_{22} = \bar{\tau}_{33} \tag{2}$$

and the bar ($^-$) denotes the complex conjugate.

THEOREM X. *Let*

$$\varphi(\boldsymbol{x}) = x_1(x_2^2 + x_3^2) \tag{3}$$

*be proportional to integral and non-null on* $\Lambda$ *and let*

$$A = |\varphi|(\Lambda) = \inf_{\substack{\boldsymbol{a} \in \Lambda \\ \neq \boldsymbol{o}}} |\varphi(\boldsymbol{a})|. \tag{4}$$

*Then there are numbers* $\eta_1 > 0$, $\eta_2 > 0$ *with the following properties:*

*Suppose that* $\boldsymbol{\tau}$ *is a homogeneous transformation in the appropriate complex co-ordinates given by* (1) *and* (2) *such that*

$$\|\boldsymbol{\tau} - \boldsymbol{\iota}\| < \eta_1. \tag{5}$$

*Then*

(i) *If* $\tau_{12}=\bar{\tau}_{13}\neq 0$, *there is a* $\mathbf{c}=(\gamma_1,\gamma_2,\gamma_3)\neq 0$, *in complex coordinates, in* $\boldsymbol{\tau}\Lambda$ *such that*

$$|\gamma_1\gamma_2\gamma_3| < A(1-\eta_2), \quad |\gamma_1| < 1. \tag{6}$$

(ii) *If* $\tau_{31}=\bar{\tau}_{21}\neq 0$, *there is a* $\mathbf{c}=(\gamma_1,\gamma_2,\gamma_3)\neq 0$ *in* $\boldsymbol{\tau}\Lambda$ *such that*

$$|\gamma_1\gamma_2\gamma_3| < A(1-\eta_2), \quad |\gamma_2| = |\gamma_3| < 1. \tag{7}$$

By Theorem VII there is an automorph $\boldsymbol{Z}=\boldsymbol{\omega z}$ in complex coordinates of the shape

$$Z_j = \omega_j z_j, \quad \omega_1\omega_2\omega_3 = 1, \quad \omega_1 > 1, \quad \omega_3 = \bar{\omega}_2.$$

Define numbers $T>0$ and $\chi$ by

$$\omega_1 = T^2, \quad \omega_2 = T^{-1}e(\chi), \quad \omega_3 = T^{-1}e(-\chi), \tag{8}$$

where

$$e(\chi) = e^{2\pi i\chi}.$$

If $\chi$ were rational, say $\chi = u/v$, the transformation $\boldsymbol{\omega}^v$ would have two equal eigenvalues $\omega_2^v$, $\omega_3^v$, which would thus be rational and so 1, contrary to hypothesis (cf. proof of Lemma 9). Hence $\chi$ is irrational. Thus by Lemma 8 with $\varepsilon=\frac{1}{8}$, there is a number $\eta_3>0$ with the following property: To every pair of numbers $\varrho>0$ and $\psi$ there are integers $u$ and $v$ such that

$$|u\chi + v - \psi| < \tfrac{1}{8} \tag{9}$$

and

$$\eta_3^{-1} < \frac{T^{3u}}{\varrho} < \eta_3. \tag{10}$$

We now prove (i). Since $\varphi(\boldsymbol{x})$ is proportional to integral on $\Lambda$, there is an $\boldsymbol{a}\in\Lambda$ of the shape

$$\boldsymbol{a} = (\alpha_1,\alpha_2,\alpha_3), \quad \alpha_2 = \zeta e(\vartheta), \quad \alpha_3 = \zeta e(-\vartheta), \tag{11}$$

where

$$\alpha_1 > 0, \quad \zeta > 0, \quad A = \alpha_1\zeta^2$$

and $A$ is defined by (4). Put

$$\tau_{12} = -\sigma e(\psi), \quad \tau_{13} = -\sigma e(-\psi), \tag{12}$$

where $\sigma>0$. Then $\sigma$ is small when $\|\boldsymbol{\tau}-\boldsymbol{\iota}\|$ is small. We now choose integers $u$ and $v$ to satisfy

$$|u\chi + v - (\psi+\vartheta)| < \tfrac{1}{8} \tag{13}$$

[cf. (9)] and (10) with

$$\varrho = \frac{\alpha_1\tau_{11}}{2\eta_3\sigma\zeta}; \tag{14}$$

so that

$$\eta_3^{-2} < \frac{2\sigma\zeta T^{3u}}{\tau_{11}\alpha_1} < 1. \tag{15}$$

Since $\tau_{11}$ is near 1, there are two constants $\eta', \eta''$, depending only on $\Lambda$ (and $\boldsymbol{a}$), such that

$$0 < \eta' \sigma^{-\frac{1}{3}} < T^u < \eta'' \sigma^{-\frac{1}{3}}. \tag{16}$$

We shall show that the point

$$\boldsymbol{c} = \boldsymbol{\tau}\boldsymbol{\omega}^{-u}\boldsymbol{a} = (\gamma_1, \gamma_2, \gamma_3) \tag{17}$$

satisfies the conditions of Theorem X in case (i). In the first place,

$$\left.\begin{aligned} T^{2u}|\gamma_1| &= |\alpha_1\tau_{11} - T^{3u}\sigma\zeta\{e(\vartheta+\psi-u\chi)+e(-\vartheta-\psi+u\chi)\}| \\ &= |\alpha_1\tau_{11} - 2T^{3u}\sigma\zeta\cos 2\pi(\vartheta+\psi-u\chi)| \\ &\leqq \alpha_1\tau_{11}(1-\tfrac{1}{2}\eta_3^{-2}) < \alpha_1(1-\tfrac{1}{4}\eta_3^{-2}) \end{aligned}\right\} \tag{18}$$

by (11), (13), (15), provided that $\|\boldsymbol{\tau}-\boldsymbol{\iota}\|$ is small enough. Further,

$$T^{-u}|\gamma_3| = T^{-u}|\gamma_2| \leqq |\tau_{21}|\,\alpha_1 T^{-3u} + \zeta|\tau_{22}| + \zeta|\tau_{23}| < \zeta(1+\varepsilon) \tag{19}$$

for any given $\varepsilon > 0$, provided that $\|\boldsymbol{\tau}-\boldsymbol{\iota}\|$, and so also $\sigma$, is small enough. From (18) and (19) we then have

$$|\gamma_1\gamma_2\gamma_3| < \alpha_1\zeta^2(1-\tfrac{1}{4}\eta_3^{-2})(1+\varepsilon)^2 < \alpha_1\zeta^2(1-\tfrac{1}{8}\eta_3^{-2}) = A(1-\tfrac{1}{8}\eta_3^{-2}),$$

if $\varepsilon$ was chosen suitably. Since (16) and (18) clearly imply $|\gamma_1| < 1$ if $\|\boldsymbol{\tau}-\boldsymbol{\iota}\|$ is small enough, this completes the proof of (i) of the theorem with $\eta_2 = \frac{1}{8}\eta_3^{-2}$.

The proof of the second part is similar on considering $\boldsymbol{\tau}\boldsymbol{\omega}^u\boldsymbol{a}$ with suitable positive integer $u$. The details may be left to the reader.

For a later application we note the

Corollary 1. *The numbers $\eta_1$ and $\eta_2$ may be chosen so that the conclusion of the theorem holds uniformly for all lattices $\Lambda = \boldsymbol{\lambda}\mathsf{M}$, where $\mathsf{M}$ is some fixed lattice on which $\varphi(\boldsymbol{x})$ is proportional to integral and non-null and $\boldsymbol{\lambda}$ runs through all automorphs of $\varphi(\boldsymbol{x})$.*

It is clearly enough to consider the case when $\boldsymbol{Z} = \boldsymbol{\lambda}\boldsymbol{z}$ is of the type $Z_j = \lambda_j z_j$. Then $\boldsymbol{\omega}$ is an automorph of $\Lambda$ if it is of $\mathsf{M}$. Hence the only non-uniformity is possibly introduced by the point $\boldsymbol{a}$. But clearly there is a number $R$ depending only on $\boldsymbol{\omega}$, and so only on $\mathsf{M}$, such that $|\boldsymbol{\omega}^k\boldsymbol{a}| < R$ for some $k$. If $\boldsymbol{\omega}^k\boldsymbol{a}$ is taken for $\boldsymbol{a}$, there is then complete uniformity in the estimates.

Corollary 2. *When $\boldsymbol{\tau}$ is any automorph of $\varphi(\boldsymbol{x})$ with*

$$0 = \tau_{12} = \tau_{13} = \tau_{21} = \tau_{31},$$

*then*

$$\inf_{\substack{a\in\Lambda\\ \neq o}} |\varphi(\boldsymbol{\tau a})| \leqq |\tau_{11}| \{|\tau_{22}| - |\tau_{23}|\}^2 A.$$

We may suppose that $A=1$ and that $\boldsymbol{a}=\boldsymbol{e}=(1, 1, 1)$. For any integer $u$ positive or negative we have

$$|\varphi(\boldsymbol{\tau\omega}^u \boldsymbol{e})| = |\tau_{11}| \, |\tau_{22} e(u\chi) + \tau_{23} e(-u\chi)|^2,$$

where $\chi$ is given by (8). By Lemma 8, we may choose $u$ so that

$$|\tau_{22} e(u\chi) + \tau_{23} e(-u\chi)|$$

is arbitrarily near to $||\tau_{22}| - |\tau_{23}||$, and the corollary follows.

Note that

$$\frac{d(\boldsymbol{\tau}\Lambda)}{d(\Lambda)} = |\tau_{11}| \, ||\tau_{22}|^2 - |\tau_{23}|^2| \geqq |\tau_{11}| \{|\tau_{22}| - |\tau_{23}|\}^2,$$

with equality only when $\tau_{22}=0$ or $\tau_{23}=0$, i.e. when $\boldsymbol{\tau}$ is an automorph of $\varphi(\boldsymbol{x})$.

**X.6. Applications of isolation.** Following DAVENPORT and ROGERS (1950a) we first use isolation to strengthen Theorem V. For $x_1(x_2^2+x_3^2)$ it gives the best result to date, but for $x_1x_2x_3$ more is known, see Theorem X of Chapter II, which is not proved in this book.

THEOREM XI. A. *There is an* $\eta_1 > 0$ *such that every lattice* $\Lambda$ *admissible for*

$$\mathscr{N}_1:\ |x_1 x_2 x_3| < 1$$

*and with*

$$d(\Lambda) < 7(1+\eta_1)$$

*is of the shape*

$$\Lambda = t\boldsymbol{\omega} \mathsf{N}_1,$$

*where* $t \geqq 1$, $\boldsymbol{\omega}$ *is an automorph of* $\mathscr{N}_1$, *and* $\mathsf{N}_1$ *is defined in Theorem V.*

B. *There is an* $\eta_2 > 0$ *such that every lattice* $\Lambda$ *admissible for*

$$\mathscr{N}_2: |x_1(x_2^2+x_3^2)| < 1$$

*and with*

$$d(\Lambda) < \tfrac{1}{2}(23)^{\frac{1}{2}}(1+\eta_2)$$

*is of the shape*

$$\Lambda = \boldsymbol{\tau\omega} \mathsf{N}_2,$$

*where* $\mathsf{N}_2$ *is defined in Theorem V B,* $\boldsymbol{\omega}$ *is an automorph of* $\mathscr{N}_2$ *and* $\boldsymbol{\tau}$ *is a transformation* $X_j = \sum_k \tau_{jk} x_k$ *with* $\tau_{12}=\tau_{13}=\tau_{21}=\tau_{31}=0$.

We first prove A by *reductio ad absurdum*. Suppose, if possible, that $\eta_1$ does not exist. Then there exists an infinite sequence of admissible lattices $\mathsf{M}_r$ $(1 \leqq r < \infty)$, none of the shape $t\boldsymbol{\omega}\mathsf{N}_1$, and such that

$$d(\mathsf{M}_r) \to 7.$$

Now

$$1 \leqq \inf_{\substack{a \in M_r \\ \neq o}} |a_1 a_2 a_3| \leqq \tfrac{1}{7} d(M_r)$$

by Theorem V, and since $M_r$ is $\mathcal{N}_1$ admissible; and so there is a sequence of points

$$a_r = (a_{1r}, a_{2r}, a_{3r}) \in M_r$$

such that

$$|a_{1r} a_{2r} a_{3r}| \to 1 \qquad (r \to \infty).$$

On replacing $M_r$ by $\omega_r M_r$, with a suitable automorph $\omega_r$ of $\mathcal{N}_1$, we may suppose that

$$a_r = (s_r, s_r, s_r), \quad s_r \to 1 \qquad (r \to \infty).$$

By MAHLER's compactness principle, there is a convergent subsequence of the $M_r$ which we may also call $M_r$, say

$$M_r \to M. \tag{1}$$

Then $d(M) = 7$ and $M$ is $\mathcal{N}_1$-admissible, so is critical. Further, $(1,1,1) \in M$ and so, by Theorem V, we have

$$M = \vartheta N_1$$

where $\vartheta$ is an automorph of $\mathcal{N}_1$. In particular, $x_1 x_2 x_3$ is integral on $M$. But now

$$M_r = \tau_r M$$

for transformations $\tau_r$ such that

$$\|\tau_r - \iota\| \to 0 \qquad (r \to \infty).$$

Since $M_r$ is $\mathcal{N}_1$-admissible, the transformation $\tau_r$ must be of the shape $\tau_r = t_r \psi_r$ for some number $t_r$ and some automorph $\psi_r$ of $\mathcal{N}_1$, by Theorem IX, provided $r$ is sufficiently large. This contradicts the definition of the $M_r$. The contradiction proves Theorem XI A.

The proof of Theorem XI B is similar but using Theorem X instead of Theorem IX. The details may be left to the reader. The only point to notice is that if $\tau$ and $\omega$ are as enunciated in the theorem, then $\tau\omega = \omega'\tau'$ for some $\omega'$, $\tau'$ with similar properties to $\omega$ and $\tau$ respectively.

COROLLARY TO THEOREM XI. B. *To every $\varepsilon > 0$ there is an $\eta = \eta_3(\varepsilon) > 0$ such that every admissible lattice $\Lambda$ for $\mathcal{N}_2$ with*

$$d(\Lambda) < \tfrac{1}{2}(23)^{\frac{1}{2}}(1 + \eta_3) \tag{2}$$

*is of the shape $\Lambda = \tau\omega N_2$, where $\tau, \omega$ are as in the theorem and*

$$\|\tau - \iota\| < \varepsilon.$$

We take $\eta_3 < \eta_2$ for the $\eta_2$ of the theorem, so that $\Lambda = \boldsymbol{\tau}\boldsymbol{\omega}\mathsf{N}_2$. We may suppose that $\tau_{11} > 0$ and then, incorporating an appropriate automorph in $\omega$, that

$$\tau_{11} = 1. \tag{3}$$

Then

$$1 + \eta_3 > \frac{d(\Lambda)}{d(\mathsf{N}_2)} = \{|\tau_{22}|^2 - |\tau_{23}|^2\} \tag{4}$$

where we use the appropriate complex co-ordinates for $\boldsymbol{\tau}$ as in § 5.4. But now

$$\{|\tau_{22}| - |\tau_{23}|\}^2 \geqq 1 \tag{5}$$

by Theorem X, Corollary 2 since $\Lambda$ is $\mathscr{N}_2$-admissible; and so, in particular,

$$\frac{|\tau_{22}|^2 - |\tau_{23}|^2}{\{|\tau_{22}| - |\tau_{23}|\}^2} < 1 + \eta_3.$$

Hence if $\eta_3$ is small, either $|\tau_{22}|/|\tau_{23}|$ or $|\tau_{23}|/|\tau_{22}|$ is small; and we may suppose the latter on incorporating in $\boldsymbol{\omega}$, if necessary, the transformation which interchanges $x_2$ and $x_3$. We may further incorporate in $\boldsymbol{\omega}$ a transformation of the type

$$x_1 \to x_1, \quad x_2 \to e(\chi)\, x_2, \quad x_3 \to e(-\chi)\, x_3,$$

where $e(\chi) = e^{2\pi i \chi}$ and $\chi$ is chosen to make $\tau_{22}$ real and positive. Then from (4) and (5) we see that $\tau_{22} - 1$ and $\tau_{32}$ are small if $\eta_3$ is small. Since $\tau_{33} = \bar{\tau}_{22}$ and $\tau_{23} = \bar{\tau}_{32}$, this proves the corollary by (3), and since the remaining terms $\tau_{jk}$ are 0.

**X.6.2.** The following interesting result about $x_1 x_2 x_3$ has no analogue for $x_1(x_2^2 + x_3^2)$, since it depends on the fact that $\varepsilon$ in Theorem IX may be chosen arbitrarily. There is, however, a corresponding result for $x_1^2 + x_2^2 - x_3^2$, see CASSELS and SWINNERTON-DYER (1955a).

THEOREM XII. *Suppose that for some number $D$ there are infinitely many lattices $\mathsf{M}_r$ $(1 \leqq r < \infty)$, admissible for*

$$\mathscr{N}_1: \ |x_1 x_2 x_3| < 1,$$

*with $d(\mathsf{M}_r) \leqq D$; and such that no two, $\mathsf{M}', \mathsf{M}''$, say, are of the shape $\mathsf{M}'' = t\boldsymbol{\omega}\mathsf{M}'$, where $t$ is a number and $\boldsymbol{\omega}$ an automorph of $\mathscr{N}_1$. Then there is a lattice $\Lambda$ admissible for $\mathscr{N}_1$ with $d(\Lambda) \leqq D$ on which $x_1 x_2 x_3$ is not proportional to integral.*

For the lattices $\mathsf{M}_r$ have a convergent subsequence, say, without loss of generality

$$\mathsf{M}_r \to \Lambda \qquad (r \to \infty).$$

If $x_1 x_2 x_3$ were proportional to integral on $\Lambda$, then by Theorem IX and since $\mathsf{M}_r$ is $\mathscr{N}_1$-admissible, we should have for all sufficiently large $r$

$$\mathsf{M}_r = t_r \boldsymbol{\omega}_r \Lambda$$

for some numbers $t_r$ and some automorphs $\omega_r$ of $\mathcal{N}_1$. This clearly contradicts the hypotheses of the theorem.

As stated in § 1, it is unknown whether such a $D$ or such a $\Lambda$ exists.

**X.7. An infinity of solutions.** We now prove some results of DAVENPORT and ROGERS (1950a) about the existence of infinitely many points of a lattice in certain point-sets with groups of automorphisms. They prove more than we do here; the reader is referred to their interesting memoire for the details.

The following trivial lemma gives almost all we need for the first type of result.

LEMMA 10. *Let $\boldsymbol{\Omega}$ be some group of homogeneous linear transformations $\boldsymbol{\omega}$. Suppose that for every $\boldsymbol{x} \neq \boldsymbol{o}$ and every number $r$ there is an $\boldsymbol{\omega} \in \boldsymbol{\Omega}$ such that*

$$|\boldsymbol{\omega x}| > r.$$

*Then for every pair of numbers $c$, $C$ with*

$$0 < c < C < \infty \tag{1}$$

*and every number $r$ there is a finite set of elements $\boldsymbol{\omega}_1, \ldots, \boldsymbol{\omega}_m$ of $\boldsymbol{\Omega}$ such that*

$$\max_{1 \leq j \leq m} |\boldsymbol{\omega}_j \boldsymbol{x}| > r \tag{2}$$

*for all $\boldsymbol{x}$ in*

$$c \leqq |\boldsymbol{x}| \leqq C. \tag{3}$$

This is a simple application of the HEINE-BOREL covering theorem. The infinitely many open sets $\mathcal{T}_r(\boldsymbol{\omega})$ of points $\boldsymbol{x}$ such that $|\boldsymbol{\omega x}| > r$ cover the compact set (3). Hence a finite covering may be selected from the $\mathcal{T}_r(\boldsymbol{\omega})$.

THEOREM XIII. *Let the boundedly reducible[1] star-body $\mathcal{S}$ have a group $\boldsymbol{\Omega}$ of automorphisms $\boldsymbol{\omega}$ such that to every $\boldsymbol{x} \neq \boldsymbol{o}$ and every $r$ there is an $\boldsymbol{\omega} \in \boldsymbol{\Omega}$ such that $|\boldsymbol{\omega x}| > r$. Then to every integer $k > 0$ there is a bounded set $\mathcal{S}_k$ contained in $\mathcal{S}$ such that every lattice $\Lambda$ with $d(\Lambda) < \Delta(\mathcal{S})$ has at least $k$ points in $\mathcal{S}_k$ other than $\boldsymbol{o}$.*

That $\mathcal{S}_1$ exists is equivalent to the statement that $\mathcal{S}$ is boundedly reducible. We suppose $\mathcal{S}_k$ has been found and deduce the existence of $\mathcal{S}_{k+1}$. We may suppose without loss of generality that $\mathcal{S}_k$ is the set of points of $\mathcal{S}$ in some sphere

$$|\boldsymbol{x}| \leqq C = C_k.$$

Further, there is a positive number $c_k < C$ such that the entire sphere

$$|\boldsymbol{x}| \leqq (k+1)\, c_k \tag{4}$$

[1] For definition, see Chapter V, § 7.2.

is contained in $\mathscr{S}$. We denote by $\mathscr{S}_{k+1}$ the set of points of $\mathscr{S}$ in

$$|\boldsymbol{x}| \leqq C_{k+1}, \tag{5}$$

where

$$C_{k+1} > \max\{C_k, (k+1)\, c_k\}$$

is so large that (5) contains all the sets $\boldsymbol{\omega}_j^{-1}\mathscr{S}_k (1 \leqq j \leqq m)$, where the $\boldsymbol{\omega}_j$ are given by Lemma 10 with $c = c_k$ and $r = C = C_k$. We must verify that $\mathscr{S}_{k+1}$ has the required properties.

By hypothesis, if $d(\Lambda) < \Delta(\mathscr{S})$ there are $k$ points of $\Lambda$ in $\mathscr{S}_k$ other than $\boldsymbol{o}$. If one of them, say $\boldsymbol{a}$, is in $|\boldsymbol{x}| < c_k$, then all the points

$$l\boldsymbol{a} \qquad (1 \leqq l \leqq k+1)$$

are in $|\boldsymbol{x}| \leqq C_{k+1}$ and in $\mathscr{S}$, so in $\mathscr{S}_{k+1}$, as required. Otherwise, there is a point $\boldsymbol{b}$ of $\Lambda$ in $\mathscr{S}_{k+1}$ for which

$$c = c_k \leqq |\boldsymbol{b}| \leqq C = C_k.$$

Hence there is an automorph $\boldsymbol{\omega}_j$ of the set $\boldsymbol{\omega}_1, \ldots, \boldsymbol{\omega}_m$ such that $|\boldsymbol{\omega}_j \boldsymbol{b}| > C$. Hence $\boldsymbol{b} \notin \boldsymbol{\omega}_j^{-1}\mathscr{S}_k$. But now, since $\boldsymbol{\omega}_j$ is an automorph, we have

$$|\det \boldsymbol{\omega}_j| = 1,$$

and so

$$d(\boldsymbol{\omega}_j \Lambda) = d(\Lambda) < \Delta(\mathscr{S}).$$

Hence by the defining property of $\mathscr{S}_k$ there are $k$ points of $\boldsymbol{\omega}_j \Lambda$ in $\mathscr{S}_k$, that is there are $k$ points of $\Lambda$ in $\boldsymbol{\omega}_j^{-1}\mathscr{S}_k$. These together with $\boldsymbol{b}$ give $k+1$ points of $\Lambda$ in $\mathscr{S}_{k+1}$, as required.

COROLLARY. *When $\mathscr{S}$ is fully reducible*[1]*, the conclusions of Theorem XIII continue to hold when $d(\Lambda) = \Delta(\mathscr{S})$, provided that $\Lambda$ is not a critical lattice of $\mathscr{S}$.*

For the existence of $\mathscr{S}_1$ is equivalent to the statement that $\mathscr{S}$ is fully reducible, and the induction now goes as before.

When the star-body $\mathscr{S}$ is not boundedly reducible only slightly less than Theorem XIII is true.

THEOREM XIV. *Let $\mathscr{S}$ be a star-body and $\Delta_1$ any number in*

$$0 < \Delta_1 < \Delta(\mathscr{S}).$$

*Then to every integer $k$ there is a bounded star-body $\mathscr{S}_k$ (depending also on $\Delta_1$) such that every lattice with $d(\Lambda) \leqq \Delta_1$ has at least $k$ points other than $\boldsymbol{o}$ in $\mathscr{S}_k$.*

We may suppose that $\mathscr{S}$ is open. Suppose, if possible, that for every integer $r$ there is a lattice $\Lambda_r$ with $d(\Lambda_r) \leqq \Delta_1$ which contains no

[1] For definition, see Chapter V, § 7.2.

point other than $\boldsymbol{o}$ of $\mathscr{S}$ in $|\boldsymbol{x}|\leqq r$. Then MAHLER's compactness theorem applies, and there is a lattice $\Lambda'$ which is the limit of a convergent subsequence of $\Lambda_r$. Since $d(\Lambda')\leqq\Delta_1$ and $\Lambda'$ is $\mathscr{S}$-admissible, this contradicts the definition of $\Delta(\mathscr{S})$. The contradiction shows that $\mathscr{S}_1$ exists. The induction from $\mathscr{S}_k$ to $\mathscr{S}_{k+1}$ then goes exactly as for Theorem XIII.

**X.7.2.** Where they apply, isolation theorems may give stronger results than those § 7.1, as the following example shows.

THEOREM XV. *Put*

$$\varphi(\boldsymbol{x}) = x_1(x_2^2 + x_3^2). \tag{1}$$

*There is a number $\eta_0>0$ such that every lattice $\Lambda$ has one of the following two properties.*

*(i) there is a number $t$ such that the set of $x_1$-co-ordinates of $t\Lambda$ is identical with the set of $x_1$-co-ordinates of the critical lattice $\mathsf{N}_2$ of $|\varphi(\boldsymbol{x})|<1$ occurring in the enunciation of Theorem V B, or (ii) for every $\varepsilon>0$ there is a point $\boldsymbol{a}\neq\boldsymbol{o}$ of $\Lambda$ such that*

$$|\varphi(\boldsymbol{a})|\leqq\frac{2}{(23)^{\frac{1}{2}}}(1-\eta_0)\,d(\Lambda), \quad |a_1|<\varepsilon. \tag{2}$$

We will choose $\eta_0$ later in the course of the proof. Suppose that (ii) is false for some particular $\Lambda$ and $\varepsilon$. For integers $r=1, 2, \ldots$, let $\Lambda_r$ be the set of points $(r^2x_1, r^{-1}x_2, r^{-1}x_3)$, $(x_1, x_2, x_3)\in\Lambda$. Then there is a convergent subsequence

$$\mathsf{M}_k=\Lambda_{r_k}\to\mathsf{M}, \tag{3}$$

and $\mathsf{M}$ is admissible for

$$|x_1(x_2^2+x_3^2)|<\frac{2}{(23)^{\frac{1}{2}}}(1-\eta_0)\,d(\Lambda).$$

Hence by Theorem XI B, Corollary for any given $\varepsilon_0$ we may choose $\eta_0=\eta_0(\varepsilon_0)$ so small that

$$\mathsf{M}=t\boldsymbol{\tau}\boldsymbol{\omega}\mathsf{N}_2, \quad \|\boldsymbol{\tau}-\boldsymbol{\iota}\|<\varepsilon_0, \tag{4}$$

where $\boldsymbol{\tau}$, $\boldsymbol{\omega}$ are as in Theorem XI B and $t$ is some number. We take for $\varepsilon_0$ the number $\eta_1$ which occurs in the enunciation of Theorem X and its Corollary when $\mathsf{M}=\mathsf{N}_2$. By (3) and (4) we now have

$$\mathsf{M}_k=t\boldsymbol{\sigma}_k\boldsymbol{\omega}\Lambda_2 \tag{5}$$

for some $\boldsymbol{\sigma}_k$ such that

$$\|\boldsymbol{\sigma}_k-\boldsymbol{\iota}\|<\varepsilon_0,$$

for all sufficiently large $k$. Clearly $\mathsf{M}_r$ does not contain any points $c=(\gamma_1,\gamma_2,\gamma_3)$ with $|\gamma_1|<1$ and $|\gamma_1\gamma_2\gamma_3|<\frac{2}{(23)^{\frac{1}{2}}}(1-\eta_0)\,d(\Lambda)$ if $r$ is sufficiently large. Hence, by Theorem XI and its Corollary, if $\eta_0$ is small enough, there is a $\sigma=\sigma_k$ such that $\sigma_{12}=\sigma_{13}=0$ in the obvious notation; indeed this happens for all sufficiently large $k$. But then, by (5) this implies that (i) holds. This concludes the proof of the Theorem.

There is a similar result where $|a_1|<\varepsilon$ in (2) is replaced by $a_2^2+a_3^2<\varepsilon$, cf. DAVENPORT and ROGERS (1950a).

**X.8. Local methods.** For many questions concerning indefinite quadratic forms the appropriate tool is the theory of continued fractions. We only mention the topic briefly here since the application to specific problems not infrequently involves detailed calculation. Continued fractions appear very naturally from the point of view of the geometry of numbers. We sketch the connection here and refer the reader to the author's Cambridge Tract [CASSELS (1957a)], where they are introduced in a similar spirit[1] in a slightly different context, for a fuller treatment and references. There a knowledge of the geometry of numbers could not be assumed. For another account of the relationship of continued fractions to quadratic forms see, for example, DICKSON (1929a).

Characteristic applications of local methods are MARKOFF'S original treatment of his chain theorem (MARKOFF 1879a), [there is an account in DICKSON (1930a); compare Chapter II, § 4)], the paper of BLANEY (1957a) that will be discussed in Chapter XI, § 4, and the paper of BARNES (1951a). But applications are almost everywhere dense in the literature.

Let us suppose for convenience that the 2-dimensional lattice $\Lambda$ has no point except $\boldsymbol{o}$ on either axis. Then no two distinct points of $\Lambda$ have the same $x_1$-co-ordinate or the same $x_2$-co-ordinate. There certainly exist points $\boldsymbol{x}_0=(x_{10}, x_{20})$ of $\Lambda$ such that $\boldsymbol{o}$ is the only point of $\Lambda$ in

$$|x_1|<|x_{10}|, \quad |x_2|<|x_{20}|.$$

Let $\pm\boldsymbol{x}_1=\pm(x_{11}, x_{21})\neq\boldsymbol{o}$ be the points in $|x_1|<|x_{10}|$ for which $|x_2|$ is least. Then there is no point except $\boldsymbol{o}$ in

$$|x_1|<|x_{10}|, \quad |x_2|<|x_{21}|, \tag{1'}$$

and *a fortiori* in

$$|x_1|<|x_{11}|, \quad |x_2|<|x_{21}|.$$

We may then repeat the process with $\boldsymbol{x}_1$ instead of $\boldsymbol{x}_0$ to obtain a sequence of points $\boldsymbol{x}_1, \boldsymbol{x}_2, \ldots$. Similarly we may start with $\boldsymbol{x}_0$ and interchange the rôles of $x_1$ and $x_2$ to obtain a sequence of points $\boldsymbol{x}_{-1}, \boldsymbol{x}_{-2}, \ldots$. There is thus a sequence of

$$\boldsymbol{x}_j=(x_{1j}, x_{2j}) \qquad (-\infty<j<\infty)$$

[1] Which goes back to FELIX KLEIN (1895a and 1896a).

such that there is no point of $\Lambda$ except $\boldsymbol{o}$ in

$$|x_1| < |x_{1j}|, \quad |x_2| < |x_{2,j+1}|. \tag{1}$$

Clearly a necessary and sufficient condition that a point $\boldsymbol{y} \in \Lambda$ should occur as $\pm \boldsymbol{x}_j$ for some $j$ is that there should be no point of $\Lambda$ except $\boldsymbol{o}$ in $|x_1| < |y_1|$, $|x_2| < |y_2|$. Hence the sequence of pairs $\pm \boldsymbol{x}_j$ is completely determined by $\Lambda$, although the particular pair chosen to be $\pm \boldsymbol{x}_0$ is, of course, arbitrary. If $\boldsymbol{\omega}$ is any automorph of $x_1 x_2$ then the sequence of pairs for $\boldsymbol{\omega}\Lambda$ is either $\pm \boldsymbol{\omega x}_j$, if $\boldsymbol{\omega}$ does not interchange the axes of co-ordinates, or $\pm \boldsymbol{\omega x}_{-j}$ (i.e. in the reverse order) if it does.

Since there is no point of $\Lambda$ in (1) except $\boldsymbol{o}$, there is no point of $\Lambda$ in the closed triangle with vertices $\boldsymbol{o}, \boldsymbol{x}_j, \boldsymbol{x}_{j+1}$ except the vertices; and so $\boldsymbol{x}_j, \boldsymbol{x}_{j+1}$ is a basis of $\Lambda$ for each $j$, by Lemma 6 of Chapter III. We must now introduce an asymmetry between the $x_1$- and $x_2$-axes to study the relationship between the various bases $\boldsymbol{x}_j, \boldsymbol{x}_{j+1}$. We choose $\boldsymbol{x}_j$ to be that point of the pair $\pm \boldsymbol{x}_j$ for which

$$x_{2j} > 0 \quad (\text{all } j). \tag{2}$$

Then

$$x_{1j}\, x_{1,j+1} < 0, \tag{3}$$

since otherwise $\boldsymbol{x}_{j+1} - \boldsymbol{x}_j$ would lie in (1). Since both $\boldsymbol{x}_{j-1}, \boldsymbol{x}_j$ and $\boldsymbol{x}_j, \boldsymbol{x}_{j+1}$ are bases, we must have

$$\boldsymbol{x}_{j+1} \pm \boldsymbol{x}_{j-1} = a_j \boldsymbol{x}_j \tag{4}$$

for some integer $a_j$. Since

$$x_{2,j+1} > x_{2,j} > x_{2,j-1},$$

we must have

$$a_j > 0.$$

Then we must have the $-$ sign in (4), since

$$|x_{1,j+1}| < |x_{1j}| < |x_{1,j-1}|,$$

and (3) holds for every $j$. Hence there is a sequence of integers $a_j > 0$ such that

$$\boldsymbol{x}_{j+1} - \boldsymbol{x}_{j-1} = a_j \boldsymbol{x}_j.$$

It may be shown that if two lattices have the same sequence of integers $a_j$ then they are identical up to a transformation of the type

$$x_1 \to \omega_1 x_1, \quad x_2 \to \omega_2 x_2.$$

Further, to every sequence of positive integers $a_j$ there is a lattice.

Hence it is natural in 2-dimensional lattice problems about $x_1 x_2$ to consider not the lattice $\Lambda$ itself simply, but the sequence $a_j$. It turns out that the behaviour of any particular basis, say $\boldsymbol{x}_J, \boldsymbol{x}_{J+1}$, of $\Lambda$ is influenced very strongly by the value of $a_j$ for $j$ near to $J$ but only very weakly by $a_j$ for $j$ remote from $J$. In many problems it is possible to study the behaviour of only a few $a_j$ at a time. Hence the name "local methods".

It would be interesting if local methods could be successfully extended to problems in more than 2 dimensions, for example to problems relating to $x_1 \max(x_2^2, x_3^2)$, $x_1(x_2^2+x_3^2)$, $x_1^2+x_2^2-x_2^3$ or $x_1x_2x_3$. The difficulty is not to find the analogues of the $\boldsymbol{x}_j$ but to devise techniques to cope with their interrelations. Continued fractions have however been generalized to 2-dimensional lattices over a complex quadratic field, i.e. substantially to certain special 4-dimensional lattices, see POITOU (1953a) and the references there given.

## Chapter XI

# Inhomogeneous problems

**XI.1. Introduction.** As previously, we say that points $\boldsymbol{x}_1$ and $\boldsymbol{x}_2$ are congruent modulo $\Lambda$, written

$$\boldsymbol{x}_1 \equiv \boldsymbol{x}_2 \qquad (\Lambda),$$

where $\Lambda$ is a lattice, to mean that $\boldsymbol{x}_1 - \boldsymbol{x}_2 \in \Lambda$. The set of points $\boldsymbol{x}$ congruent to a given point $\boldsymbol{x}_0$ modulo $\Lambda$ is called a grid[1] $\mathfrak{G}$: $\Lambda$ will be called the lattice of the grid and we shall call

$$d(\mathfrak{G}) = d(\Lambda)$$

the determinant of the grid. The characteristic inhomogeneous problem of the geometry of numbers is to find conditions under which a grid has a point in a given set $\mathscr{S}$.

There is a wide variety of different problems. Thus one may be concerned with all grids of given determinant $d(\mathfrak{G})$ or one may have information about the lattice $\Lambda$. Many of the fundamental techniques for inhomogeneous problems are natural extension of those for lattices [compactness theorems and so on; for bodies with automorphs see SWINNERTON-DYER (1954a)]. For some specialized problems some extremely powerful and delicate techniques have been developed which would take too much space to discuss properly. Hence this last chapter will have more the character of a report and less that of a detailed exposition.

[1] Other terms are inhomogeneous lattice or non-homogeneous lattice.

**XI.1.2.** The following simple result due to MACBEATH (1951a) helps to fix ideas.

THEOREM I. *Let the set $\mathscr{S}$ have finite volume $V(\mathscr{S})$ and let $\varepsilon>0$ be given arbitrarily small. Then there are grids $\mathfrak{G}$ with $d(\mathfrak{G})=\varepsilon$ having no point in $\mathscr{S}$.*

We may choose $R$ so large that the portion of $\mathscr{S}$ in $|\boldsymbol{x}|\geqq R$ has volume $<\frac{1}{4}\varepsilon$. Let $\Lambda$ be the lattice with basis

$$\left.\begin{aligned} \boldsymbol{b}_1 &= (4R, 0, 0, \ldots, 0) \\ \boldsymbol{b}_2 &= (0, \eta, 0, 0, \ldots, 0) \\ \boldsymbol{b}_3 &= (0, 0, \eta, 0, \ldots, 0) \\ &\cdots\cdots\cdots\cdots \\ \boldsymbol{b}_n &= (0, 0, 0, \ldots, 0, \eta), \end{aligned}\right\} \tag{1}$$

where

$$4R\eta^{n-1} = \varepsilon. \tag{2}$$

Every point $\boldsymbol{x}_1$ of space is congruent modulo $\Lambda$ to precisely one point of the parallelopiped

$$\mathscr{P}: \quad \left\{\begin{array}{c} y_1\boldsymbol{b}_1 + \cdots + y_n\boldsymbol{b}_n \\ (-\frac{1}{2} \leqq y_j < \frac{1}{2}). \end{array}\right\} \tag{3}$$

The volume of $\mathscr{P}$ is $V(\mathscr{P})=d(\Lambda)=\varepsilon$, by (2). If a point $\boldsymbol{x}'=\sum y_j'\boldsymbol{b}_j$ of $\mathscr{P}$ is congruent modulo $\Lambda$ to a point $\boldsymbol{x}_1$ in $|\boldsymbol{x}_1|\leqq R$, then clearly $|y_1'|\leqq\frac{1}{4}$. Hence the set of points of $\mathscr{P}$ with this property has measure at most $\frac{1}{2}\varepsilon$. But now the set of points $\boldsymbol{x}_2$ of $\mathscr{S}$ with $|\boldsymbol{x}_2|>R$ has volume at most $\frac{1}{4}\varepsilon$ by construction; and hence so has the set of points $\boldsymbol{x}''$ of $\mathscr{P}$ which are congruent to at least such one point (compare the proof of Theorem I of Chapter III). Thus the set of points of $\mathscr{P}$ congruent to a point of $\mathscr{S}$ has measure at most $\frac{1}{2}\varepsilon+\frac{1}{4}\varepsilon<\varepsilon=V(\mathscr{P})$. There is thus a point $\boldsymbol{x}_0\in\mathscr{P}$ which is not congruent to any point of $\mathscr{S}$. The grid $\mathfrak{G}$ of all points congruent to $\boldsymbol{x}_0$ modulo $\Lambda$ clearly has all the properties required.

**XI.1.3.** We shall mainly be concerned with star-bodies $\mathscr{S}$ defined by a distance-function,

$$\mathscr{S}: \quad F(\boldsymbol{x}) < 1. \tag{1}$$

For any lattice $\Lambda$ and any point $\boldsymbol{x}_0$ we write[1]

$$m(\boldsymbol{x}_0) = m(\boldsymbol{x}_0, \Lambda) = \inf_{\boldsymbol{x}\equiv\boldsymbol{x}_0(\Lambda)} F(\boldsymbol{x}), \tag{2}$$

[1] So $m(\boldsymbol{x}_0)=F(\mathfrak{x}_0)$ in the notation of Chapter VII § 2.2, where $\mathfrak{x}_0$ is the element of the quotient space to which $\boldsymbol{x}_0$ belongs.

and
$$\mu(\Lambda) = \sup_{\boldsymbol{x}_0} m(\boldsymbol{x}_0, \Lambda). \tag{3}$$
Clearly
$$\mu(t\Lambda) = |t|\,\mu(\Lambda) \tag{4}$$
for any $t \neq 0$.

The infimum in (2) need not be attained, though it clearly is attained when the set $F(\boldsymbol{x}) < 1$ is bounded. The function $m(\boldsymbol{x}_0)$ need not be continuous, but it is semi-continuous:
$$\limsup_{\boldsymbol{x} \to \boldsymbol{x}_0} m(\boldsymbol{x}) \leqq m(\boldsymbol{x}_0). \tag{5}$$
Indeed given any $\varepsilon > 0$ there is a point $\boldsymbol{a} \in \Lambda$ such that
$$F(\boldsymbol{x}_0 + \boldsymbol{a}) < m(\boldsymbol{x}_0) + \varepsilon,$$
and then
$$F(\boldsymbol{x} + \boldsymbol{a}) < m(\boldsymbol{x}_0) + \varepsilon$$
for all $\boldsymbol{x}$ in a neighbourhood of $\boldsymbol{x}_0$, by the continuity of $F(\boldsymbol{x})$; so $m(\boldsymbol{x}) < m(\boldsymbol{x}_0) + \varepsilon$ in this neighbourhood. Again, when $F(\boldsymbol{x}) < 1$ is bounded, the function $m(\boldsymbol{x})$ is readily seen to be continuous. The reader will be able to supply the proofs of the positive statements just made on the lines of the proof of the semi-continuity of the function $F(\Lambda)$ in Chapter V, § 3.3. Examples to show that the infimum in (2) need not be attained and that $m(\boldsymbol{x})$ need not be continuous are provided in 2 dimensions for certain lattices $\Lambda$ when $F(\boldsymbol{x}) = |x_1 x_2|^{\frac{1}{2}}$. This case has implications in the theory of algebraic numbers and has been extensively investigated both because of this and because of its intrinsic interest; see BARNES and SWINNERTON-DYER (1952a, b and 1954a) and BARNES (1954a), where there are extensive references to earlier work. There is some work on similar lines for $|x_1 x_2 x_3|^{\frac{1}{3}}$ $(n = 3)$, but it has not been carried so far, see DAVENPORT (1947c), CLARKE (1951a) and SAMET (1954a, b).

From the definition (2) it follows that $m(\boldsymbol{x})$ may be regarded as defined on the quotient space $\mathscr{R}/\Lambda$ (compare Chapter VII). Since this is compact, it follows from (5) that the supremum in (3) is always attained; that is, there is an $\boldsymbol{x}_1$ such that
$$\mu(\Lambda) = m(\boldsymbol{x}_1, \Lambda).$$
Of course the infimum in (2) need not then be attained for $\boldsymbol{x}_1 = \boldsymbol{x}_0$. With unbounded sets $F(\boldsymbol{x}) < 1$ there may be again a phenomenon of successive minima; that is, it may happen that
$$\sup_{m(\boldsymbol{x}_0) \neq \mu(\Lambda)} m(\boldsymbol{x}_0) < \mu(\Lambda).$$

Indeed some rather elaborate patterns of successive minima have been found, see the papers of BARNES and SWINNERTON-DYER just quoted.

The quotient

$$\frac{\{\mu(\Lambda)\}^n}{d(\Lambda)} \tag{6}$$

is unchanged on replacing $\Lambda$ by $t\Lambda$, by (4). We shall write

$$\mathfrak{d}(F) = \inf_{\Lambda} \frac{\{\mu(\Lambda)\}^n}{d(\Lambda)}, \tag{7}$$

where possibly $\mathfrak{d}(F) = 0$. If the set $F(\boldsymbol{x}) < 1$ has finite volume $V_F$, we now show that

$$\mathfrak{d}(F) \geqq V_F^{-1}. \tag{8}$$

Let $\Lambda$ be some lattice and $\varepsilon > 0$ be arbitrarily small. There is a point $\boldsymbol{x}_1$ congruent to any given point $\boldsymbol{x}_0$ and satisfying

$$F(\boldsymbol{x}_1) < \mu(\Lambda) + \varepsilon. \tag{9}$$

Hence the set (9) must have volume at least $d(\Lambda)$. Since the volume of the set of points $\boldsymbol{x}_1$ satisfying (9) is

$$\{\mu(\Lambda) + \varepsilon\}^n V_F,$$

the required result (8) follows.

We shall show in § 3 that if the body $F(\boldsymbol{x}) < 1$ is bounded, the infimum in (7) is attained; that is there is a lattice M such that

$$\{\mu(\mathsf{M})\}^n = \mathfrak{d}(F)\, d(\mathsf{M}).$$

We shall treat the estimation of $\mathfrak{d}(F)$ for convex distance-functions $F$ in § 2 where the relevant literature will also be discussed.

When $V_F = \infty$ it is, of course, still possible that $\mathfrak{d}(F) > 0$. In particular, DAVENPORT (1951 a) showed this to be the case for the 2-dimensional distance-function

$$F(\boldsymbol{x}) = |x_1 x_2|^{\frac{1}{2}}. \tag{10}$$

His estimate,

$$\mathfrak{d}(F) \geqq \tfrac{1}{128},$$

was improved to

$$\mathfrak{d}(F) \geqq \tfrac{1}{45 \cdot 2}$$

by the author [CASSELS (1952a)], with a probably simpler proof. This has recently been improved by ENNOLA (1958a) to

$$\mathfrak{d}(F) \geqq (16 + 6^{\frac{3}{2}})^{-1} = \tfrac{1}{30 \cdot 69 \ldots},$$

by a modification of DAVENPORT'S original method. On the other hand, Miss PITMAN (1958a) has shown that

$$\mathfrak{d}(F) \leqq \tfrac{1}{12}$$

More recently[1], she has obtained an even smaller upper bound for $\mathfrak{d}(F)$.

The problem of determining $\mathfrak{d}(F)$ for $F$ given by (10) is closely related to the problem of determining the real quadratic numberfields with a Euclidean algorithm. DAVENPORT extended his work to number-fields of two other types corresponding to

$$F^3 = x_1(x_2^2 + x_3^2) \quad \text{and} \quad F^4 = (x_1^2 + x_2^2)(x_3^2 + x_4^2).$$

These results were proved by the author [CASSELS (1952a)] much more simply and with a better estimate of $\mathfrak{d}(F)$.

HLAWKA (1954c) has generalized these results to any distance-function $F(\boldsymbol{x})$ in $n$ variables which may be put in the shape

$$\{F(\boldsymbol{x})\}^n = \{F_r(x_1, \ldots, x_r)\}^r \{F_{n-r}(x_{r+1}, \ldots, x_n)\}^{n-r},$$

where $F_r, F_{n-r}$ are $r$- and $(n-r)$-dimensional distance-functions such that the star-bodies $F_r(\boldsymbol{x}) < 1$ and $F_{n-r}(\boldsymbol{x}) < 1$ are bounded. We do not prove these results here. A closely related problem is treated in the author's tract [CASSELS (1957a) Chapter V, § 6], where there are further references.

In general it appears to be a difficult problem to decide whether $\mathfrak{d}(F) = 0$. Thus it does not appear to be known whether this happens for[2]

$$F(\boldsymbol{x}) = |x_1^2 + x_2^2 - x_3^2|^{\frac{1}{2}} \qquad n = 3$$

or

$$F(\boldsymbol{x}) = |x_1 x_2 x_3|^{\frac{1}{3}} \qquad n = 3.$$

**XI.1.4.** It follows at once from MACBEATH'S Theorem I that

$$\mathfrak{D}(F) = \sup_{\Lambda} \frac{\{\mu(\Lambda)\}^n}{d(\Lambda)}$$

is $\infty$ whenever $V_F < \infty$. In § 4 we shall be concerned with $\mathfrak{D}(F)$ for

$$F = |x_1 \cdots x_n|^{1/n}.$$

It was conjectured by MINKOWSKI that $\mathfrak{D}(F) = 2^{-n}$, but this has been proved only for $n = 2, 3, 4$. We shall give references and a further discussion in § 4. We shall also give a result of CHALK about the set

$$x_1 x_2 \ldots x_n \leqq 1 \qquad x_j > 0 \qquad (1 \leqq j \leqq n)$$

(not a star-body!) and quote other work about sets defined in term of $x_1 \ldots x_n$.

[1] I am grateful to Miss PITMAN for allowing me to refer to this unpublished work, now published. Acta Arithmetica **6** (1960), 37–46.

[2] The first case has been settled by E. S. BARNES [J. Austral. Math. Soc. **2** (1961/62) 9–10], who shows that $\mathfrak{d}(F) = 0$.

The value of $\mathfrak{D}(F)$ for

$$F(\boldsymbol{x}) = |x_1^2 + x_2^2 - x_3^2|^{\frac{1}{2}} \qquad (n=3)$$

has been found by DAVENPORT (1948a) who showed it to be isolated and the investigation of the successive minima was carried further by BARNES (1956a). More recently BIRCH (1958a) has found $\mathfrak{D}(F)$ for

$$F(\boldsymbol{x}) = |x_1^2 + \cdots + x_r^2 - x_{r+1}^2 - \cdots - x_{2r}^2|^{\frac{1}{2}} \qquad (n = 2r),$$

for all $r \geqq 2$. Estimates for

$$F(\boldsymbol{x}) = |x_1^2 + \cdots + x_r^2 - x_{r+1}^2 - \cdots - x_n^2|^{\frac{1}{2}}$$

with $r>0$, $n-r>0$ have been given by BLANEY (1948a) and improved by ROGERS (1952a) and Miss FOSTER (1956a). All the work just described is of course equivalent to finding the best possible constant $\eta_{r,s}$ such that

$$\sup_{\boldsymbol{u}_0 \text{ real}} \inf_{\boldsymbol{u} \text{ integral}} |f(\boldsymbol{u}+\boldsymbol{u}_0)| \leqq \eta_{r,s} |D|^{1/n}$$

for all indefinite quadratic forms $f$ of signature $(r, s)$ with $r+s=n$ and with determinant $D$. We shall not discuss this work further in this book but refer the reader to the original memoires.

**XI.1.5.** For some functions $F(\boldsymbol{x})$ there are inequalities, valid for all $\Lambda$, connecting

$$\mu(\Lambda) = \sup_{\boldsymbol{x}_0} \inf_{\boldsymbol{x} \equiv \boldsymbol{x}_0 (\Lambda)} F(\boldsymbol{x})$$

and

$$F(\Lambda) = \inf_{\substack{\boldsymbol{x} \in \Lambda \\ \neq \boldsymbol{o}}} F(\boldsymbol{x})$$

or, more generally connecting $\mu(\Lambda)$ and the successive minima of $F(\boldsymbol{x})$ with respect to $\Lambda$. When $F(\boldsymbol{x})$ is convex, there are further relations with the corresponding quantities for the polar distance-function $F^*(\boldsymbol{x})$ and the polar lattice $\Lambda^*$. These relations go under the general name of transference theorems[1] (Übertragungssätze). Thus DIRICHLET'S hexagon Theorem VII of Chapter IX may be regarded as a very precise transference theorem for $|x_1^2 + x_2^2|^{\frac{1}{2}}$. We shall discuss transference theorems for convex functions $F(\boldsymbol{x})$ in § 3. Much interesting work has been done on transference theorems for the non-convex $F(\boldsymbol{x})$ defined by

$$\{F(\boldsymbol{x})\}^n = |x_1 \ldots x_r| \prod_{1 \leqq k \leqq s} (x_{r+k}^2 + x_{r+s+k}^2),$$

where $n = r+2s$, but here we can only refer the reader to the paper of DAVENPORT and SWINNERTON-DYER (1955a), where references are given to earlier work. There is a striking related result in SWINNERTON-DYER (1954a).

[1] Presumably because information is transferred from one problem to another.

There is a further type of result which may most appropriately be mentioned here since they are transference theorems of a sort. BARNES (1950a) showed that if

$$F(\boldsymbol{x}) = |x_1 x_2|^{\frac{1}{2}}$$

and if $\Lambda$ has the basis $\boldsymbol{a}_1, \boldsymbol{a}_2$ then

$$2\mu(\Lambda) \leqq \max\{F(\boldsymbol{a}_1), F(\boldsymbol{a}_2), \min_{\pm} F(\boldsymbol{a}_1 \pm \boldsymbol{a}_2)\}.$$

Other results of this general kind are known, see BAMBAH and K. ROGERS (1955a) and the references given there. In particular, K. ROGERS (1953a) showed that BARNES' result is true for all distance-functions $F(\boldsymbol{x})$ such that $F(\boldsymbol{x}) < 1$ has the same general appearance as $|x_1 x_2| < 1$. The proofs are all elementary and tend to involve a tedious splitting of cases. We do not discuss them further in this book.

**XI.2. Convex sets.** In 2 dimensions the problem of finding $\mathfrak{d}(F)$ in the notation of (7) of § 1.3 for convex functions $F$ is completely solved by the following result [BAMBAH and ROGERS (1952a)].

THEOREM II. *Let $\mathscr{S}$ be a closed 2-dimensional convex set and $\Delta_1$ some number. A necessary and sufficient condition that there exist a lattice $\Lambda$ with $d(\Lambda) = \Delta_1$ such that every point is congruent modulo $\Lambda$ to a point of $\mathscr{S}$ is that there exist a convex hexagon*[1] *$\mathscr{H}$ inscribed in $\mathscr{S}$, which is symmetrical about some point and has an area $V(\mathscr{H}) = \Delta_1$.*

Note that $\mathscr{S}$ is not required to be symmetrical about any point.

Suppose, first, that $\mathscr{H}$ exists. We may take the centre of $\mathscr{H}$ as origin $\boldsymbol{o}$. Let $\Lambda$ be a critical lattice for $2\mathscr{H}$. Then $d(\Lambda) = V(\mathscr{H}) = \Delta_1$, by Lemma 13 of Chapter V. Hence by Theorems II, III of Chapter IX applied to $2\mathscr{H}$, and since $\mathscr{H}$ is closed, every point is congruent modulo $\Lambda$ to a point of $\mathscr{H}$, and so of $\mathscr{S}$.

Suppose now that there exists a $\Lambda$ such that every point is congruent modulo $\Lambda$ to some point of $\mathscr{S}$. If $\mathscr{S}$ is unbounded, there is clearly nothing to prove, so we may suppose without loss of generality that $\mathscr{S}$ is bounded. We shall construct the hexagon $\mathscr{H}$ in stages. Suppose, first, that there is an $\boldsymbol{a} \neq \boldsymbol{o} \in \Lambda$ such that $\mathscr{S}$ and $\mathscr{S} + \boldsymbol{a}$ have inner points in common. By taking $2^s\boldsymbol{a}$ with suitable integer $s \geqq 0$ instead of $\boldsymbol{a}$, we may suppose without loss of generality that $\mathscr{S} + 2\boldsymbol{a}$ and $\mathscr{S}$ have no inner points in common. Then there exist points $\boldsymbol{c}$ and $\boldsymbol{d}$ on the boundary both of $\mathscr{S}$ and $\mathscr{S} + \boldsymbol{a}$ such that the portion of the boundary of $\mathscr{S}$ between $\boldsymbol{c}$ and $\boldsymbol{d}$ (taken in an anti-clockwise direction, say) lies in $\mathscr{S} + \boldsymbol{a}$ and the portion of the boundary of $\mathscr{S} + \boldsymbol{a}$ between $\boldsymbol{d}$ and $\boldsymbol{c}$ lies in $\mathscr{S}$. Then $\boldsymbol{c} - \boldsymbol{a}$ and $\boldsymbol{d} - \boldsymbol{a}$ are common to the boundaries of $\mathscr{S}$ and $\mathscr{S} - \boldsymbol{a}$. Let

[1] A parallelogram being allowed as a degenerate hexagon.

$\mathscr{S}_1$ be the portion of $\mathscr{S}$ lying between the line joining $\boldsymbol{c}$ and $\boldsymbol{d}$ and the line joining $\boldsymbol{c}-\boldsymbol{a}$ and $\boldsymbol{d}-\boldsymbol{a}$, and taken closed; i.e. including the points of $\mathscr{S}$ on those lines. Then clearly $\mathscr{S}_1$ is convex and every point of the plane is congruent modulo $\Lambda$ to a point of $\mathscr{S}_1$. After a finite number of steps (since $\mathscr{S}$ is bounded) we obtain a closed convex set $\mathscr{T} \subset \mathscr{S}$ such that every point is congruent modulo $\Lambda$ to a point of $\mathscr{T}$ but no two sets $\mathscr{T}$ and $\mathscr{T}+\boldsymbol{a}$, $\boldsymbol{a}\in\Lambda$ have inner points in common. Then every boundary point of $\mathscr{T}$ is also a boundary point of $\mathscr{T}+\boldsymbol{a}$ for some $\boldsymbol{a}\neq\boldsymbol{o}$ in $\Lambda$. Since $\mathscr{T}$ and $\mathscr{T}+\boldsymbol{a}$ are convex, this common boundary is either a point or a line-segment. Since $\mathscr{T}$ is bounded, only a finite number of $\boldsymbol{a}$ come into consideration, and so $\mathscr{T}$ is a convex polygon. We must now show that it is symmetric about some point. Let the vertices of $\mathscr{T}$ be $\boldsymbol{c}_1, \ldots, \boldsymbol{c}_m$, where the line segment $\boldsymbol{c}_j\boldsymbol{c}_{j+1}$ is the common boundary of $\mathscr{T}$ and $\mathscr{T}+\boldsymbol{a}_j$, $\boldsymbol{a}_j\in\Lambda$. Then the line-segment $(\boldsymbol{c}_j-\boldsymbol{a}_j)(\boldsymbol{c}_{j+1}-\boldsymbol{a}_j)$ is the common boundary of $\mathscr{T}$ and $\mathscr{T}-\boldsymbol{a}_j$. Hence $m$ is even, $m=2l$, and

$$\boldsymbol{a}_{j\pm l} = -\boldsymbol{a}_j,$$

$$\boldsymbol{c}_{j+l} = \boldsymbol{c}_{j+1} - \boldsymbol{a}_j, \quad \boldsymbol{c}_{j+l+1} = \boldsymbol{c}_j - \boldsymbol{a}_j.$$

Hence

$$\tfrac{1}{2}(\boldsymbol{c}_j + \boldsymbol{c}_{j+l}) = \tfrac{1}{2}(\boldsymbol{c}_{j+1} + \boldsymbol{c}_{j+1+l})$$

for each $j$, so $\boldsymbol{e}=\frac{1}{2}(\boldsymbol{c}_j+\boldsymbol{c}_{j+l})$ is independent of $j$. Clearly $\mathscr{T}$ is symmetric about $\boldsymbol{e}$.

We may suppose without loss of generality that $\boldsymbol{e}=\boldsymbol{o}$. Then $\Lambda$ gives a lattice packing of $\mathscr{T}$ (or, more precisely, of the interior of $\mathscr{T}$) and every point is congruent to some point of $\mathscr{T}$ modulo $\Lambda$. Hence $\mathscr{T}$ is a hexagon by Theorems II and VI of Chapter IX. This concludes the proof of Theorem II.

Using known results about hexagons inscribed in convex sets, Bambah and Rogers (1952a) deduce in our notation (§ 1.3) that

$$1 \leqq V_F\,\mathfrak{d}(F) \leqq \tfrac{3}{2}$$

for a convex 2-dimensional distance-function $F$ inequality and the stronger inequality

$$1 \leqq V_F\,\mathfrak{d}(F) \leqq \frac{2\pi}{3\sqrt{3}}$$

f $F$ is symmetric. The equalities on the right-hand side are attained when $F(\boldsymbol{x})<1$ is a triangle and a circle respectively. The left-hand inequality, which is valid whether $F$ is convex or not, was obtained in § 1.3.

There is a theory of lattice coverings and non-lattice coverings which is closely analogous to the theory of packings discussed in Chapter IX.

For details in 2 dimensions see FEJES TÓTH (1950a and 1953a) and BAMBAH and ROGERS (1952a).

Not much is known about $\mathfrak{d}(F)$ in more than 2 dimensions. When $F(\boldsymbol{x})<1$ is the unit 3-dimensional sphere, the precise value has been found by BAMBAH (1954b), and other proofs have been given by BARNES (1956b) and FEW (1956a); but all proofs are fairly complicated. The 4-dimensional sphere has been considered by BAMBAH (1954a), who obtains an estimate for $\mathfrak{d}(F)$ and gives a conjecture for the correct value. Estimates for $\mathfrak{d}(F)$ above and below and also for the corresponding number for non lattice coverings have been obtained for $n$-dimensional spheres, see BAMBAH and DAVENPORT (1952a), DAVENPORT (1952b) and WATSON (1956a) for the lattice case, and ERDÖS and ROGERS (1953a) and ROGERS (1957a) for the non-lattice case, the last treating general convex sets. Very recently ROGERS (1959a) has obtained much stronger results by more powerful methods.

**XI.2.2.**[1] ROGERS (1950b) has given an elegant proof of the following result relating $\mathfrak{d}(F)$ to the function

$$\delta(F)=\sup_{\Lambda}\frac{\{F(\Lambda)\}^n}{d(\Lambda)}$$

introduced in § 4 of Chapter IV.

THEOREM III.

$$\mathfrak{d}(F)\leqq 2^{-n}3^{n-1}\delta(F)$$

*for all symmetric convex n-dimensional distance-functions which vanish only at the origin.*

ROGERS (1950b) also proved a similar result for non-lattice packings and coverings, and indeed with the smaller constant $2^{-1}$ instead of $2^{-n}3^{n-1}$. Before proving Theorem III we note the following

COROLLARY.

$$V_F\,\mathfrak{d}(F)\leqq 3^{n-1},$$

*where $V_F$ is the volume of $F(\boldsymbol{x})<1$.*

For $V_F\delta(F)\leqq 2^n$ by MINKOWSKI'S convex body theorem.

ROGERS proves Theorem III by considering a critical lattice $\mathsf{M}$ for $F$, that is

$$F(\mathsf{M})=1,\quad d(\mathsf{M})=\{\delta(F)\}^{-1}. \tag{1}$$

We use the notation of § 1.3; in particular

$$m(\boldsymbol{x}_0)=\inf_{\boldsymbol{x}\equiv\boldsymbol{x}_0(\mathsf{M})}F(\boldsymbol{x}).$$

---

[1] When $n$ is at all large, the results of this section are superseded by ROGERS (1959a).

As was shown in § 1.3, there is then a point $\boldsymbol{x}_1$ such that

$$\left.\begin{aligned} m(\boldsymbol{x}_1) &= \sup_{\boldsymbol{x}_0} m(\boldsymbol{x}_0) \\ &= \mu(\mathsf{M}) \\ &= \mu \text{ (say)}. \end{aligned}\right\} \tag{2}$$

Then

$$m(3\boldsymbol{x}_1) \leqq \mu,$$

and so, since $F(\boldsymbol{x}) < 1$ is bounded, there is an $\boldsymbol{a} \in \mathsf{M}$ such that

$$F(3\boldsymbol{x}_1 - \boldsymbol{a}) = m(3\boldsymbol{x}_1) \leqq \mu.$$

Then

$$F(\boldsymbol{x}_1 - \tfrac{1}{3}\boldsymbol{a}) \leqq \tfrac{1}{3}\mu < \mu, \tag{3}$$

and so $\frac{1}{3}\boldsymbol{a}$ is not in M.

Let $\Lambda$ be the lattice of points

$$\boldsymbol{b} + \frac{r}{3}\boldsymbol{a}, \quad \boldsymbol{b} \in \mathsf{M}, \quad r = \text{integer},$$

so

$$d(\Lambda) = \tfrac{1}{3} d(\mathsf{M}).$$

Hence

$$\{F(\Lambda)\}^n \leqq \delta(F)\, d(\Lambda) = \tfrac{1}{3}\delta(F)\, d(\mathsf{M})$$

by the definition of $\delta(F)$; that is, there exists a point $\boldsymbol{b} + \frac{r}{3}\boldsymbol{a} \neq \boldsymbol{o}$ of $\Lambda$ such that

$$\left\{F\left(\boldsymbol{b} + \frac{r}{3}\boldsymbol{a}\right)\right\}^n \leqq \frac{1}{3}\delta(F)\, d(\mathsf{M}). \tag{4}$$

We may suppose without loss of generality that $r = 0$ or $\pm 1$. If $r = 0$, we have $\boldsymbol{b} \neq \boldsymbol{o}$, and so

$$F(\boldsymbol{b}) \geqq F(\mathsf{M}) = 1,$$

and (1) and (4) are in contradiction. Hence $r = \pm 1$, and

$$\left.\begin{aligned} F(\boldsymbol{b} \pm \tfrac{1}{3}\boldsymbol{a}) &= F\{\boldsymbol{b} \pm \boldsymbol{x}_1 \mp (\boldsymbol{x}_1 - \tfrac{1}{3}\boldsymbol{a})\} \\ &\geqq F(\boldsymbol{b} \pm \boldsymbol{x}_1) - F(\boldsymbol{x}_1 - \tfrac{1}{3}\boldsymbol{a}) \\ &\geqq \mu - \tfrac{1}{3}\mu \\ &= \tfrac{2}{3}\mu, \end{aligned}\right\} \tag{5}$$

by (2) and (3).

On substituting (5) in (4) we obtain

$$\frac{\mu^n}{d(\mathsf{M})} \leqq 2^{-n} 3^{n-1} \delta(F). \tag{6}$$

Since the left-hand side of (6) is at most $\mathfrak{d}(F)$, by the definition of $\mathfrak{d}(F)$ as an infimum (§ 1.3), the theorem follows.

**XI.3. Transference theorems for convex sets.** In this section we consider for a symmetric convex $n$-dimensional distance function $F$ which vanishes only at $\boldsymbol{o}$ the relationships between the function

$$\mu = \mu(\Lambda) = \sup_{\boldsymbol{x}_0} \inf_{\boldsymbol{x} \equiv \boldsymbol{x}_0(\Lambda)} F(\boldsymbol{x}) \tag{1}$$

discussed in § 1 and the successive minima $\lambda_1, \ldots, \lambda_n$ of $F$ with respect to $\Lambda$ which were discussed in Chapter VIII.

We first prove the inequality

$$\lambda_n \leqq 2\mu \leqq \lambda_1 + \cdots + \lambda_n. \tag{2}$$

Let $\boldsymbol{b}_1, \ldots, \boldsymbol{b}_n$ be any basis for $\Lambda$. Then by the definition of $\mu$ and the fact that $F(\boldsymbol{x}) < 1$ is bounded, there are vectors $\boldsymbol{c}_j \in \Lambda$ such that

$$F(\tfrac{1}{2}\boldsymbol{b}_j - \boldsymbol{c}_j) \leqq \mu.$$

Hence the vectors $\boldsymbol{d}_j = \boldsymbol{b}_j - 2\boldsymbol{c}_j$ all satisfy

$$F(\boldsymbol{d}_j) \leqq 2\mu.$$

Since the $\boldsymbol{d}_j$ are linearly independent, as is easily seen[1] by considering congruences modulo 2, the left-hand side of (2) follows.

We now prove the right-hand side of (2). There are linearly independent vectors $\boldsymbol{a}_j$ of $\Lambda$ such that

$$F(\boldsymbol{a}_j) = \lambda_j.$$

Every vector $\boldsymbol{x}_0$ is thus of the shape

$$\boldsymbol{x}_0 = \xi_1 \boldsymbol{a}_1 + \cdots + \xi_n \boldsymbol{a}_n$$

for some real numbers $\xi_1, \ldots, \xi_n$. Put

$$\boldsymbol{a} = u_1 \boldsymbol{a}_1 + \cdots + u_n \boldsymbol{a}_n,$$

where

$$|u_j - \xi_j| \leqq \tfrac{1}{2},$$

and $u_1, \ldots, u_n$ are integers. Then, clearly,

$$\begin{aligned} F(\boldsymbol{x}_0 - \boldsymbol{a}) &= F\Big\{\sum_j (\xi_j - u_j)\,\boldsymbol{a}_j\Big\} \\ &\leqq \sum_j |\xi_j - u_j|\, F(\boldsymbol{a}_j) \\ &\leqq \tfrac{1}{2} \sum F(\boldsymbol{a}_j) \\ &= \tfrac{1}{2} \sum \lambda_j. \end{aligned}$$

[1] For suppose that $\sum_j r_j \boldsymbol{d}_j = \boldsymbol{o}$, where the $r_j$ are integers which, without loss of generality, may be supposed to have no common factor. Then $\sum_j r_j \boldsymbol{b}_j = 2 \sum r_j \boldsymbol{c}_j$. Since the $\boldsymbol{b}_j$ are a basis, all the $r_j$ must be even. A contradiction!

This proves the right-hand side of (2).

On making use of the inequalities

$$\frac{2^n}{n!} d(\Lambda) \leqq V_F \lambda_1 \ldots \lambda_n \leqq 2^n d(\Lambda) \tag{3}$$

of Theorem V of Chapter VIII, we may deduce estimates for $\mu$ above and below in terms of

$$\lambda_1 = \inf_{\substack{a \neq o \\ \in \Lambda}} F(a) = F(\Lambda).$$

From the left-hand sides of (2) and (3), and since

$$\lambda_1 \leqq \lambda_2 \leqq \cdots \leqq \lambda_n, \tag{4}$$

we have

$$\frac{2}{n!} d(\Lambda) \leqq V_F \lambda_1 \mu^{n-1}. \tag{5}$$

On the other hand, the maximum of $\lambda_1 + \ldots + \lambda_n$ for given $\lambda_1$ and product $\lambda_1 \ldots \lambda_n$ is clearly attained when $\lambda_1 = \lambda_2 = \cdots = \lambda_{n-1}$. Hence, by (2) and (3),

$$V_F \lambda_1^{n-1} \{2\mu - (n-1)\lambda_1\} \leqq 2^n d(\Lambda). \tag{6}$$

Both the inequalities (5) and (6) may be improved. The problem of obtaining an estimate above for $\mu$ in terms of $\lambda_1$ is an old one which has been attacked by many methods. The latest result due to KNESER (1955a) and BIRCH (1956a) will be proved as Theorem V. The inequality (5) has attracted much less attention. We sketch a proof of an improvement due to BIRCH (1956b), as Theorem IV. BIRCH actually proves something slightly stronger than Theorem IV and gives examples to show that it cannot be much further improved.

THEOREM IV.

$$\mu^{n-1} \lambda_1 V_F \geqq \frac{2}{n} d(\Lambda)$$

*for convex symmetric n-dimensional distance-functions.*

BIRCH's proof is very simple. We may suppose after a suitable homogeneous linear transformation that $\Lambda = \Lambda_0$ is the lattice of points with integer co-ordinates, and that

$$F(0, \ldots, 0, 1) = \lambda_1.$$

Let $\mathscr{T}$ be the $(n-1)$-dimensional projection of the set

$$\mu \mathscr{S}: \quad F(\boldsymbol{x}) \leqq \mu$$

on to the hyperplane $x_n = 0$. Then every point with $x_n = 0$ is congruent modulo $\Lambda_0$ to a point of $\mathscr{T}$, so $\mathscr{T}$ has $(n-1)$-dimensional volume

$V_{n-1}(\mathcal{T}) \geqq 1$. Further, $\mu\mathcal{S}$ contains the points

$$\pm(0, \ldots, 0, \mu/\lambda_1).$$

Some elementary geometry[1] now shows that the volume of $\mu\mathcal{S}$ must be at least

$$V(\mu\mathcal{S}) \geqq \frac{2}{n} \cdot \frac{\mu}{\lambda_1} \cdot V_{n-1}(\mathcal{T}) \geqq \frac{2\mu}{n\lambda_1}.$$

Since $V(\mu\mathcal{S}) = \mu^n V_F$, and since we have assumed that $\Lambda = \Lambda_0$, so $d(\Lambda) = 1$, the truth of the theorem follows.

THEOREM V. *Let*

$$Q = \frac{2^n d(\Lambda)}{\lambda_1^n V_F} = q + \varkappa \tag{7}$$

*where $q$ is an integer and $0 \leqq \varkappa < 1$. Then*

$$\mu \leqq \tfrac{1}{2}\lambda_1(q + \varkappa^{1/n}). \tag{8}$$

*Further*

$$\mu \leqq \tfrac{1}{2}\lambda_1 Q, \tag{9}$$

*provided that $Q \geqq n$.*

Note that $q + \varkappa^{1/n} \geqq Q$ and $q \geqq 1$ by (3). The inequality (8) is KNESER'S (1955a)[2] and (9) is BIRCH'S, though the remark that (9) holds already for $Q \geqq n$ is KNESER'S [see BIRCH (1956a)]. BIRCH proves similar results involving other minima $\lambda_2, \ldots, \lambda_{n-1}$.

Before proceeding to the proof we note that (9) cannot be further improved[3]. Let

$$F(\boldsymbol{x}) = \max\{|x_1|, \ldots, |x_n|\},$$

and let $\Lambda$ be the lattice of points

$$(u_1, \ldots, u_{n-1}, Q u_n),$$

where $Q$ is any number $\geqq 1$ and $u_1, \ldots, u_n$ run through all integers. Clearly

$$\lambda_1 = 1 \qquad V_F = 2^n;$$

and so $Q$ is in fact the number given by (7). Further, $\mu = \frac{1}{2}Q$, as is seen by considering

$$\boldsymbol{x}_0 = (0, \ldots, 0, \tfrac{1}{2}Q).$$

---

[1] The details are given in the author's tract [CASSELS (1957a)] page 84 Lemma 1. The easiest way is to replace $\mathcal{S}$: $F(\boldsymbol{x}) < 1$ by a body of the same volume symmetric in $x_n = 0$, on replacing for each $(x_1, \ldots, x_{n-1})$ the segment of $x_n$ such that $(x_1, \ldots, x_n) \in \mathcal{S}$ by the one of equal length symmetric in $x_n = 0$ (STEINER symmetrization). The result is trivial for the symmetrized set.

[2] Professor KNESER tells me that he can show that $<$ can be substituted for $\leqq$ in (8) except when $Q$ is an integer.

[3] BAMBAH (1958a) shows that (8) and (9) may sometimes be improved if $\delta(F)$ is known.

It was long conjectured that (9) was valid for all $Q$, but the following example, due to KNESER and BIRCH (see BIRCH 1956a), shows that in fact the weaker inequality (8) cannot be improved for $1 \leqq Q < 2$. Let

$$F(\boldsymbol{x}) = \max\{|x_1|, \ldots, |x_n|\},$$

and let $\Lambda$ be the lattice of points

$$(u_1 - \varepsilon u_2,\ u_2 - \varepsilon u_3, \ldots,\ u_{n-1} - \varepsilon u_n,\ u_n + \varepsilon u_1)$$

where $0 \leqq \varepsilon < 1$ is fixed and $u_1, \ldots, u_n$ runs through all integers (note the change of sign in the last co-ordinate). Then

$$d(\Lambda) = 1 + \varepsilon^n, \quad \lambda_1 = \cdots = \lambda_n = 1, \quad \mu = \tfrac{1}{2}(1 + \varepsilon),$$

as is readily verified. No case appears to be known when (9) is false and $Q \geqq 2$.

Now to the proof of Theorem V. We work in the quotient space $\mathscr{R}/\Lambda$ and use the notation of Chapter VII and of Theorem IV of Chapter VIII. In particular, we denote by $\mathsf{S}(t)$ the set of points $\mathfrak{y}$ of $\mathscr{R}/\Lambda$ which have representatives $\boldsymbol{y}$ in $\mathscr{R}$ such that $F(\boldsymbol{y}) < t$. By Theorem IV of Chapter VIII the measure $m\{\mathsf{S}(t)\}$ satisfies

$$m\{\mathsf{S}(t)\} \begin{cases} = t^n V_F \quad \text{if} \quad t \leqq \tfrac{1}{2}\lambda_1. & (10) \\ \geqq t(\tfrac{1}{2}\lambda_1)^{n-1} V_F \quad \text{if} \quad \tfrac{1}{2}\lambda_1 \leqq t \leqq \tfrac{1}{2}\lambda_n. & (11) \end{cases}$$

We shall also need the inequality

$$m\{\mathsf{S}(t_1 + t_2)\} \geqq \min[m\{\mathsf{S}(t_1)\} + m\{\mathsf{S}(t_2)\},\ d(\Lambda)], \tag{12}$$

for any $t_1 \geqq 0$, $t_2 \geqq 0$. This follows at once from the "Sum Theorem", Theorem I of Chapter VII. Indeed, $\mathsf{S}(t_1 + t_2)$ contains the sum $\mathsf{S}(t_1) + \mathsf{S}(t_2)$, where addition of sets is as defined in § 3 of Chapter VII, since $F(\boldsymbol{y}_1 + \boldsymbol{y}_2) < t_1 + t_2$ if $F(\boldsymbol{y}_1) < t_1$ and $F(\boldsymbol{y}_2) < t_2$.

We also remark that $\mu$ is the lower bound of the numbers $t$ such that $m\{\mathsf{S}(t)\} = d(\Lambda)$. Clearly $m\{\mathsf{S}(t)\} = d(\Lambda)$ if every point of $\mathscr{R}$ is congruent modulo $\Lambda$ to a point $\boldsymbol{x}$ with $F(\boldsymbol{x}) < t$. Conversely, suppose that $m\{\mathsf{S}(t_0)\} = d(\Lambda)$. Let $\varepsilon > 0$ be arbitrarily small. Then $m\{\mathsf{S}(\varepsilon)\} > 0$ by (10), and so every point of $\mathscr{R}/\Lambda$ belongs to $\mathsf{S}(t_0) + \mathsf{S}(\varepsilon) \subset \mathsf{S}(t_0 + \varepsilon)$ by the first part of the "Sum Theorem" I of Chapter VII.

We now prove (8) very simply. By (10) we have

$$m\{\mathsf{S}(\tfrac{1}{2}\lambda_1)\} = (\tfrac{1}{2}\lambda_1)^n V_F = Q^{-1} d(\Lambda)$$

and

$$m\{\mathsf{S}(\tfrac{1}{2}\varkappa^{1/n}\lambda_1)\} = \varkappa(\tfrac{1}{2}\lambda_1)^n V_F = \varkappa\, Q^{-1} d(\Lambda).$$

Hence, by repeated use of (12), we have

$$m\left[S\left\{\tfrac{1}{2}\lambda_1(q+\varkappa^{1/n})\right\}\right] \geqq q\,m\{S(\tfrac{1}{2}\lambda_1)\} + m\{S(\tfrac{1}{2}\varkappa^{1/n}\lambda_1)\}$$
$$= (q+\varkappa)\,Q^{-1}d(\Lambda)$$
$$= d(\Lambda),$$

as required.

To prove (9) we need (11) as well as (10), where now

$$Q \geqq n.$$

We must distinguish two cases. Suppose first that

$$Q\lambda_1 \geqq n\lambda_n.$$

Then $2\mu \leqq \lambda_1 Q$ by (2), which proves (9) in this case. Otherwise, by (11),

$$m\left\{S\left(\frac{Q}{n}\cdot\frac{\lambda_1}{2}\right)\right\} \geqq \frac{Q}{n}(\tfrac{1}{2}\lambda_1)^n V_F = d(\Lambda)/n,$$

by the definition (7) of $Q$. Hence, by repeated use of (12), we have

$$m\{S(\tfrac{1}{2}Q\lambda_1)\} \geqq d(\Lambda),$$

which completes the proof of (9).

**XI.3.2.** We are now in a position to prove the result enunciated in § 1.3 that when the star-body $F(\boldsymbol{x})<1$ is bounded, then $\mathfrak{d}(F)$ is an attained minimum, that is, in the notation of § 1.3, there exists a lattice $\mathsf{M}$ such that

$$\frac{\{\mu(\mathsf{M})\}^n}{d(\mathsf{M})} = \mathfrak{d}(F) = \inf_{\Lambda} \frac{\{\mu(\Lambda)\}^n}{d(\Lambda)}.$$

We must use the transference theorem of § 3.1 to ensure that we may apply MAHLER's compactness criterion. Write

$$F_0(\boldsymbol{x}) = |\boldsymbol{x}|,$$

so that

$$F(\boldsymbol{x}) \geqq c\,F_0(\boldsymbol{x}), \quad c>0$$

for some $c$ and all $\boldsymbol{x}$, since $F(\boldsymbol{x})<1$ is bounded. Hence clearly

$$\mu^{(0)}(\Lambda) \leqq c^{-1}\mu(\Lambda),$$

where the superfix $^{(0)}$ indicates that the quantity is relative to $F_0$. In particular, if $\mu(\Lambda)$ is bounded above for some set $\mathfrak{L}$ of lattices $\Lambda$, then so is $\mu^{(0)}(\Lambda)$; and hence $\lambda_1^{(0)}$ is bounded below a strictly positive number by Theorem IV [or by the weaker inequality (5) of § 3.1]; that is

$$|\Lambda| = \inf_{\substack{\boldsymbol{a}\in\Lambda \\ \neq \boldsymbol{o}}} |\boldsymbol{a}|$$

is bounded below.

Now we select a sequence of lattices $\Lambda_r$ $(1 \leqq r < \infty)$ not necessarily distinct such that

$$d(\Lambda_r) = 1$$

and

$$\{\mu(\Lambda_r)\}^n \to \mathfrak{d}(F).$$

From what was proved in the last paragraph, $|\Lambda_r|$ is bounded below by a positive number. Hence by MAHLER's compactness criterion there is a convergent subsequence; without loss of generality

$$\Lambda_r \to \mathsf{M}.$$

Then $\mathsf{M}$ clearly has the properties required.

**XI.3.3.** Let $\Lambda$ and $\Lambda^*$ be polar lattices in the sense of Chapter I, § 5. It was there shown that a necessary and sufficient condition that a point $\boldsymbol{x}$ belong to $\Lambda$ is that the scalar product $\boldsymbol{x}\boldsymbol{a}^*$ be an integer for all $\boldsymbol{a}^* \in \Lambda^*$. We develop now what may be regarded as a quantitative generalization of this statement. For a real number $\xi$ we denote by $\|\xi\|$ the difference between $\xi$ and the nearest integer either above or below taken positively, that is

$$\|\xi\| = \inf_{m=0, \pm 1, \pm 2, \ldots} |\xi - m|.$$

There will be no possibility of confusion with the notation $\|\boldsymbol{\tau}\|$ where $\boldsymbol{\tau}$ is a homogeneous linear transformation.

THEOREM VI. *Let $F(\boldsymbol{x})$ be a symmetric convex n-dimensional distance function corresponding to a bounded set $F(\boldsymbol{x}) < 1$ and let $F^*(\boldsymbol{x})$ be the polar distance-function. Let $\Lambda$ and $\Lambda^*$ be polar lattices. For any point $\boldsymbol{x}_0$ write*

$$m(\boldsymbol{x}_0) = \inf_{\boldsymbol{x} \equiv \boldsymbol{x}_0 (\Lambda)} F(\boldsymbol{x}) \tag{1}$$

*and*

$$K(\boldsymbol{x}_0) = \sup_{\substack{\boldsymbol{a}^* \in \Lambda^* \\ \neq \boldsymbol{o}}} \frac{\|\boldsymbol{a}^* \boldsymbol{x}_0\|}{F^*(\boldsymbol{a}^*)}, \tag{2}$$

*where $\boldsymbol{a}^* \boldsymbol{x}_0$ denotes the scalar product. Then*

$$\left\{\frac{n}{2^{n-1}} (n!)^2\right\}^{-1} m(\boldsymbol{x}_0) \leqq K(\boldsymbol{x}_0) \leqq m(\boldsymbol{x}_0). \tag{3}$$

The precise values of the constants in (3) are immaterial: what matters is that the ratio $K(\boldsymbol{x}_0)/m(\boldsymbol{x}_0)$ lies between constants. Theorem VI goes back in essence to KHINTCHINE (1948a). KRONECKER's Theorem follows from it in a few lines [compare Chapter V, § 8 of the author's tract (CASSELS 1957a), where a less general form of Theorem VI is given].

We first prove the right-hand side of (3). Let

$$x_1 \equiv x_0 \,(\Lambda). \tag{4}$$

Then $x_1 a^*$ differs from $x_0 a^*$ by the integer $(x_1 - x_0) a^*$ and so

$$\|x_0 a^*\| = \|x_1 a^*\| \leqq |x_1 a^*|. \tag{5}$$

But now, by the definition of a polar function (Theorem III of Chapter IV), and since $F(x)$ is symmetric, we have

$$|x_1 a^*| \leqq F(x_1)\, F^*(a^*). \tag{6}$$

Hence

$$\|x_0 a^*\| \leqq F(x_1)\, F^*(a^*), \tag{7}$$

and so

$$\|x_0 a^*\| \leqq m(x_0)\, F^*(a^*), \tag{8}$$

on taking the infimum of the right-hand side of (7) over all $x_1 \equiv x_0 \,(\Lambda)$. This is just the right-hand side of (3).

To prove the left-hand side of (3) we need the dual bases $b_j$ and $b_j^*$ of Theorem VII, Corollary of Chapter VIII, for which

$$F(b_j)\, F^*(b_j^*) \leqq (\tfrac{1}{2})^{n-1} (n!)^2 \qquad (1 \leqq j \leqq n). \tag{9}$$

Let $x_0$ be any point, so that

$$x_0 = \xi_1 b_1 + \cdots + \xi_n b_n$$

for some real numbers $\xi_j$. Then, by (2),

$$\|\xi_j\| = \|b_j^* x_0\| \leqq K(x_0)\, F^*(b_j^*) \tag{10}$$

for $1 \leqq j \leqq n$. Choose integers $u_j$ so that

$$|u_j - \xi_j| = \|\xi_j\|, \tag{11}$$

and let

$$x_1 = (\xi_1 - u_1)\, b_1 + \cdots + (\xi_n - u_n)\, b_n,$$

so

$$x_1 \equiv x_0 \,(\Lambda).$$

Then by (9), (10) and (11),

$$\begin{aligned} m(x_0) &\leqq F(x_1) \\ &\leqq \sum_j |\xi_j - u_j|\, F(b_j) \\ &= \sum_j \|\xi_j\|\, F(b_j) \\ &\leqq K(x_0) \sum_j F^*(b_j^*)\, F(b_j) \\ &= \frac{n}{2^{n-1}} (n!)^2 K(x_0), \end{aligned}$$

which is the left-hand side of (3).

**XI.3.4.** In this section we prove a rather specialized transference theorem which we shall need in § 4. The proof uses the so-called technique of the additional variable which has often been used with success[1]. For example, the best result in the direction of Theorem V until the work of KNESER was proved by HLAWKA (1952a) using this technique. [It is reproduced in the author's tract (CASSELS 1957a) in a special case.]

LEMMA 1. *Let $F_0(\boldsymbol{x}) = |\boldsymbol{x}|$, where $\boldsymbol{x} = (x_1, x_2, x_3)$ is a 3-dimensional vector. Let $\lambda_1, \lambda_2, \lambda_3$ be the successive minima of a lattice $\Lambda$ with respect to $F_0$ and let*

$$\mu = \sup_{\boldsymbol{x}_0} \inf_{\boldsymbol{x} \equiv \boldsymbol{x}_0(\Lambda)} F_0(\boldsymbol{x}).$$

*Then*

$$\frac{\mu^2}{\lambda_3^2} \leqq 1 - \left(\frac{\lambda_1 \lambda_2 \lambda_3}{2d(\Lambda)}\right)^2, \tag{1}$$

*and*

$$4\mu^2 \leqq \lambda_1^2 + \lambda_2^2 + \lambda_3^2 \leqq 3\lambda_3^2. \tag{2}$$

We first prove (2). There are linearly independent points $\boldsymbol{a}_j$ of $\Lambda$ such that $|\boldsymbol{a}_j| = \lambda_j$. Let $\boldsymbol{c}_1, \boldsymbol{c}_2, \boldsymbol{c}_3$ be a set of mutually orthogonal vectors such that

$$\left.\begin{aligned} \boldsymbol{a}_1 &= \boldsymbol{c}_1 \\ \boldsymbol{a}_2 &= v_{21}\boldsymbol{c}_1 + \boldsymbol{c}_2 \\ \boldsymbol{a}_3 &= v_{31}\boldsymbol{c}_1 + v_{32}\boldsymbol{c}_2 + \boldsymbol{c}_3 \end{aligned}\right\} \tag{3}$$

for real numbers $v_{ij}$. Then

$$|\boldsymbol{c}_j|^2 \leqq |\boldsymbol{a}_j|^2 = \lambda_j^2 \qquad (1 \leqq j \leqq 3). \tag{4}$$

But now, if $\boldsymbol{x}_0$ is any point, it is possible to choose integers $u_3, u_2, u_1$ successively in that order, so that

$$\boldsymbol{x}_1 = \boldsymbol{x}_0 + u_1\boldsymbol{a}_1 + u_2\boldsymbol{a}_2 + u_3\boldsymbol{a}_3 = \xi_1\boldsymbol{c}_1 + \xi_2\boldsymbol{c}_2 + \xi_3\boldsymbol{c}_3,$$

where the numbers $\xi_j$ satisfy

$$|\xi_j| \leqq \tfrac{1}{2} \qquad (1 \leqq j \leqq 3).$$

Hence

$$|\boldsymbol{x}_1^2| = \xi_1^2|\boldsymbol{c}_1|^2 + \cdots + \xi_3^2|\boldsymbol{c}_3|^2 \leqq \tfrac{1}{4}(\lambda_1^2 + \lambda_2^2 + \lambda_3^2)$$

by (4). This establishes (2).

We now construct a 4-dimensional lattice M as follows. There is a point $\boldsymbol{x}_0$ such that

$$\mu = \mu(\Lambda) = \inf_{\boldsymbol{x} \equiv \boldsymbol{x}_0(\Lambda)} |\boldsymbol{x}|. \tag{5}$$

[1] Apparently first used by MORDELL (1937a).

Let the number $\varrho$ be defined by

$$\varrho^2+\mu^2=\lambda_3^2, \tag{6}$$

so

$$\varrho \geqq \tfrac{1}{2}\lambda_3, \tag{7}$$

by (2). Then $\mathsf{M}$ is the set of all 4-dimensional points

$$\boldsymbol{X}=(\boldsymbol{x}, \varrho u), \tag{8}$$

in an obvious notation, where $u$ runs through all integers and the vector $\boldsymbol{x}$ satisfies the congruence

$$\boldsymbol{x} \equiv u\,\boldsymbol{x}_0\,(\Lambda). \tag{9}$$

Clearly

$$d(\mathsf{M})=\varrho\, d(\Lambda).$$

If $\boldsymbol{X}\in\mathsf{M}$ and $u\neq 0$ we have

$$|\boldsymbol{X}|^2=|\boldsymbol{x}|^2+\varrho^2u^2 \geqq \lambda_3^2$$

by (5) and (6) or by (7) according as $u=\pm 1$ or $|u|>1$. The values taken by $|\boldsymbol{X}|$ with $u=0$ and $\boldsymbol{X}\in\mathsf{M}$ are precisely those taken by $|\boldsymbol{x}|$ with $\boldsymbol{x}\in\Lambda$. Hence the four successive minima of the function $|\boldsymbol{X}|$ with respect to $\mathsf{M}$ are $\lambda_1, \lambda_2, \lambda_3, \lambda_4$, where $\lambda_1, \lambda_2, \lambda_3$ are the minima of $|\boldsymbol{x}|$ with respect to $\Lambda$, as already defined, and

$$\lambda_4 \geqq \lambda_3.$$

(Indeed $\lambda_4=\lambda_3$, but we do not need that.)

By Theorem I of Chapter VIII and Theorem IV, Corollary of Chapter X, we have

$$\left.\begin{aligned}\lambda_1\lambda_2\lambda_3^2 &\leqq \lambda_1\lambda_2\lambda_3\lambda_4\\ &\leqq \Gamma_{4,0}^{-1}\, d(\mathsf{M})\\ &= 2d(\mathsf{M})\\ &= 2\varrho\, d(\Lambda),\end{aligned}\right\} \tag{10}$$

where $\Gamma_{4,0}$ is the lattice-constant of the 4-dimensional sphere $|\boldsymbol{X}|<1$. On eliminating $\varrho$ between (6) and (10), we obtain the required inequality (1).

We shall actually need Lemma 1 in the following shape:

COROLLARY. *To every point* $\boldsymbol{x}_0$ *there is a point* $\boldsymbol{x}_1\equiv\boldsymbol{x}_0\,(\Lambda)$ *such that*

$$|\boldsymbol{x}_1|^2 \leqq \frac{3}{4}\lambda_3^2\left\{\frac{d(\Lambda)}{\lambda_1\lambda_2\lambda_3}\right\}^{\frac{2}{3}}. \tag{11}$$

In the first place,

$$3e+e^{-3}=e+e+e+e^{-3}\geqq 4 \tag{12}$$

for every number $e>0$ by the inequality of the arithmetic and geometric mean. Hence it follows from (1) that $\mu^2$ is at most equal to the righthand side of (11), on using (12) with

$$e = \left\{ \frac{d(\Lambda)}{\lambda_1 \lambda_2 \lambda_3} \right\}^{\frac{1}{3}}.$$

But now, since $|\boldsymbol{x}| < 1$ is bounded, there is certainly an $\boldsymbol{x}_1$ such that

$$|\boldsymbol{x}_1| = \inf_{\boldsymbol{x} \equiv \boldsymbol{x}_0} |\boldsymbol{x}| \leqq \mu;$$

and the corollary follows.

**XI.4. Product of $n$ linear forms.** Let

$$F_1(\boldsymbol{x}) = |x_1 \dots x_n|^{1/n}. \tag{1}$$

As in § 1.4 we put

$$\mu_1(\Lambda) = \sup_{\boldsymbol{x}_0} \inf_{\boldsymbol{x} \equiv \boldsymbol{x}_0(\Lambda)} F_1(\boldsymbol{x}), \quad \mathfrak{D}_1 = \sup_{\Lambda} \frac{\{\mu_1(\Lambda)\}^n}{d(\Lambda)}. \tag{2}$$

There is a famous conjecture of MINKOWSKI that

$$\mathfrak{D}_1 = 2^{-n}. \tag{3}$$

That $\mathfrak{D}_1 \geqq 2^{-n}$ follows at once by considering the case when $\Lambda = \Lambda_0$ in (2) is the lattice of points with integer co-ordinates and $\boldsymbol{x}_0 = (\frac{1}{2}, \dots, \frac{1}{2})$. Clearly then $F_1(\boldsymbol{x}_0) \geqq \frac{1}{2}$ for all $\boldsymbol{x} \equiv \boldsymbol{x}_0(\Lambda_0)$, and $d(\Lambda_0) = 1$.

It is well known that

$$\{\mu_1(\Lambda)\}^n \leqq 2^{-n} d(\Lambda)$$

if $\Lambda$ is a sublattice of the integer lattice $\Lambda_0$. The proof is simple. The lattice $\Lambda$ has a basis

$$\boldsymbol{b}_j = (b_{1j}, \dots, b_{jj}, 0, \dots, 0),$$

where the $b_{ij}$ are integers and $b_{jj} \neq 0$, $b_{ij} = 0$ for $i > j$. For any real numbers $(x_{10}, \dots, x_{n0})$ we can thus choose integers $u_1, \dots, u_n$, in order, so that

$$|u_j b_{jj} + \cdots + u_n b_{jn} + x_{j0}| \leqq \tfrac{1}{2} |b_{jj}|.$$

For $\boldsymbol{x}_1 = u_1 \boldsymbol{b}_1 + \cdots + u_n \boldsymbol{b}_n + \boldsymbol{x}_0$, we then have

$$\{F(\boldsymbol{x}_1)\}^n \leqq \{\tfrac{1}{2}|b_{11}|\} \cdots \{\tfrac{1}{2}|b_{nn}|\} = 2^{-n} d(\Lambda),$$

as required.

The conjecture (3) has been proved only for $n = 2, 3, 4$. A great many proofs of (3) for $n = 2$ for have been given; we shall present one in § 4.2 due to SAWYER. This has the advantage that it gives naturally a result for the "asymmetric" distance function[1]

$$F_{k,l}(\boldsymbol{x}) = \begin{Bmatrix} k|x_1 x_2|^{\frac{1}{2}} & \text{if} & x_1 x_2 \geqq 0 \\ l|x_1 x_2|^{\frac{1}{2}} & \text{if} & x_1 x_2 \leqq 0 \end{Bmatrix},$$

[1] Of course $F_{k,l}(\boldsymbol{x}) < 1$ is symmetric about $\boldsymbol{o}$; but it is not symmetric in the four quadrants.

where $k$ and $l$ are positive numbers. These arise quite naturally even in originally symmetric problems; indeed the result we shall prove was first obtained by DAVENPORT (1948a) as a tool in his work on the "symmetric" problem for indefinite ternary quadratic forms. Further results about $F_{k,l}$ have been obtained, notably by BLANEY (1950a), BARNES and SWINNERTON-DYER (1954a) and, as an adjunct to another investigation, by BARNES (1956a). We refer the reader to these papers for further details.

When $n=3$, MINKOWSKI's conjecture (3) was proved by REMAK (1923a, b) and a simplified proof was given by DAVENPORT (1939a). We give DAVENPORT's proof in § 4.3, having already paved the way in § 3.4. A proof for $n=3$ using different ideas has been given by BIRCH and SWINNERTON-DYER (1956a).

When $n=4$ a proof of (3) has been given by DYSON (1948a) following the same general line as REMAK's proof. It is an extremely powerful piece of work and requires tools from topology as well as from number-theory proper.

For $n>4$ only estimates for $\mathfrak{D}_1$ are known. It was shown by TSCHEBOTAREW (1934a) that

$$\mathfrak{D}_1 \leqq 2^{-n/2},$$

and this was improved by MORDELL (1940a) and by DAVENPORT (1946a) to

$$D_1 \leqq \eta_n 2^{-n/2},$$

where $\eta_n$ s a number $<1$ such that $\eta_n \to (2e-1)^{-1}$ as $n\to\infty$. Recently WOODS (1958c) has shown that TSCHEBOTAREW's result may be improved simply by using BLICHFELDT's theorem instead of MINKOWSKI's convex body theorem. MORDELL (1959a) remarks that this improvement can be combined with the earlier techniques. In particular, DAVENPORT'S $\eta_n$ can be replaced by a number which is asymptotically $\frac{1}{2}\eta_n$ for large $\eta$. We give TSCHEBOTAREW's result with its impressively simple proof in § 4.4.

Some further results of a general nature are known about this problem. BIRCH and SWINNERTON-DYER (1956a) have shown that

$$\{\mu_1(\Lambda)\}^n \leqq 2^{-n} d(\Lambda)$$

for all lattices $\Lambda$ in a certain neighbourhood of the integer lattice $\Lambda_0$, and give some other facts relating to the general conjecture. The author (CASSELS 1952b) has shown that for any $\varepsilon>0$ and every $n$ there are infinitely many lattices $\Lambda$ such that

$$\{\mu_1(\Lambda)\}^n \geqq (2^{-n}-\varepsilon)\, d(\Lambda)$$

and such that no two lattices $\Lambda$, $\Lambda'$ of the set are of the type $\Lambda' = t\boldsymbol{\omega}\Lambda$, where $t$ is real and $\boldsymbol{\omega}$ an automorph of $F_1(\boldsymbol{x})$; so if MINKOWSKI's conjecture is true then the first minimum is certainly not isolated. ROGERS (1954c) has investigated the least number $\mu_1'(\Lambda)$ such that for every $\varepsilon > 0$ and every $\boldsymbol{x}_0$ there are *infinitely many* solutions of

$$F_1(\boldsymbol{x}) < \mu_1'(\Lambda) + \varepsilon \qquad \boldsymbol{x} \equiv \boldsymbol{x}_0 \, (\Lambda),$$

and obtained general conditions for $\Lambda$ under which $\mu_1'(\Lambda) = \mu_1(\Lambda)$.

CHALK (1947a, b) has obtained the complete answer for what may be regarded as an extreme asymmetric version of MINKOWSKI's problem. He shows namely that for any lattice $\Lambda$ and any point $\boldsymbol{x}_0$ there is an $\boldsymbol{x}_1 \equiv (x_{11}, \ldots, x_{n1}) \equiv \boldsymbol{x}_0 \, (\Lambda)$ such that

$$x_{j1} > 0 \qquad (1 \leqq j \leqq n), \tag{4}$$

$$x_{11} \ldots x_{n1} \leqq d(\Lambda). \tag{5}$$

That $\leqq$ in (5) cannot always be replaced by $<$ is shown by the simple example when $\Lambda = \Lambda_0$ is the lattice of integer vectors and $\boldsymbol{x}_0 = \boldsymbol{o}$. The case $n = 2$ was obtained by DAVENPORT and HEILBRONN (1947a). When $n = 2$, BLANEY (1957a) has given an interesting strengthened form: namely that for every $\boldsymbol{x}_0$ there is an $\boldsymbol{x}_1 = (x_{11}, x_{21}) \equiv \boldsymbol{x}_0 \, (\Lambda)$ such that

$$x_{j1} > 0 \qquad (j = 1, 2)$$

and

$$\tfrac{1}{2}(126^{\frac{1}{2}} - 11) \, d(\Lambda) \leqq x_{11} x_{21} \leqq d(\Lambda),$$

where the $\leqq$ on the left cannot be replaced by $<$ for a certain lattice $\Lambda$. The proof is a classic example of the local methods discussed in general terms in § 8 of Chapter X. COLE (1952a) has shown that to every $\boldsymbol{x}_0$ there is an $\boldsymbol{x}_1 \equiv \boldsymbol{x}_0$ such that

$$x_{j1} > 0 \qquad (1 \leqq j \leqq n - 1)$$

and

$$x_{11} \ldots x_{n-1,1} |x_{n1}| \leqq \tfrac{1}{2} d(\Lambda).$$

CHALK (1947b) discusses when for given $\boldsymbol{x}_0$ there are infinitely many $\boldsymbol{x}_1 \equiv \boldsymbol{x}_0 \, (\Lambda)$ satisfying (4) and (5). The principle behind the proof of CHALK's theorem is similar to TSCHEBOTAREFF's, and we prove it in § 4.4. The idea has been put in a much more general form by MACBEATH (1952a) and C. A. ROGERS (1954b), but we do not go into that here.

**XI.4.2.** The proof of MINKOWSKI's conjecture in 2-dimensions may be made to depend on the following lemma due to DELAUNAY (1947a). He used it as a tool to investigate $\mu_1(\Lambda)$ (in the notation of § 4.1) for individual 2-dimensional lattices $\Lambda$; and the so-called "algorithm of the divided cell" has been exploited further by BARNES and SWINNERTON-DYER (1954a), and BARNES (1954a, 1956c). It was remarked by DELAUNAY (1947a) that the lemma does not generalize to 3 or more dimensions; and the same counter-example in 3 dimensions was given by BIRCH (1957a) in ignorance of DELAUNAY's example.

LEMMA 2. *Let $\Lambda$ be a 2-dimensional lattice and let $\boldsymbol{x}_0$ be a point not congruent modulo $\Lambda$ to a point on either co-ordinate axis. Then there are 4 points $\boldsymbol{x}_1, \boldsymbol{x}_2, \boldsymbol{x}_3, \boldsymbol{x}_4$, each congruent to $\boldsymbol{x}_0$ modulo $\Lambda$, where $\boldsymbol{x}_j$ is in the $j$-th quadrant, so that*

$$\boldsymbol{x}_1 + \boldsymbol{x}_4 = \boldsymbol{x}_2 + \boldsymbol{x}_3 \tag{1}$$

*and $\boldsymbol{x}_2 - \boldsymbol{x}_1$, $\boldsymbol{x}_3 - \boldsymbol{x}_1$ is a basis for $\Lambda$.*

The four points $\boldsymbol{x}_1, \boldsymbol{x}_2, \boldsymbol{x}_3, \boldsymbol{x}_4$ forms a "divided cell" of the grid $\mathfrak{G}$ of points $\boldsymbol{x} \equiv \boldsymbol{x}_0\ (\Lambda)$. Simpler proofs of Lemma 2 have been given by BAMBAH (1955b) and RÉDEI (1959a). We follow RÉDEI.

The proof depends on the following two propositions.

PROPOSITION 1. *Let $\boldsymbol{y}_1, \boldsymbol{y}_2, \boldsymbol{y}_3, \boldsymbol{y}_4$ be four points of $\mathfrak{G}$ such that the quadrilateral $\boldsymbol{y}_1\boldsymbol{y}_2\boldsymbol{y}_3\boldsymbol{y}_4$ is convex and contains no other point of $\mathfrak{G}$ in its interior or boundary. Then $\boldsymbol{y}_1\boldsymbol{y}_2\boldsymbol{y}_3\boldsymbol{y}_4$ is a parallelogram and $\boldsymbol{y}_2 - \boldsymbol{y}_1$, $\boldsymbol{y}_3 - \boldsymbol{y}_1$ is a basis for $\Lambda$.*

This follows almost at once from Chapter III, Lemma 6.

PROPOSITION 2. *Let $\pi$ be a line containing points of $\mathfrak{G}$ in 3 quadrants. Let $\boldsymbol{y}_1$ be a point of $\mathfrak{G}$ in the remaining quadrant. Suppose that there are points $\boldsymbol{y}_2, \boldsymbol{y}_3$ of $\mathfrak{G}$ on $\pi$ such that $\boldsymbol{y}_1, \boldsymbol{y}_2, \boldsymbol{y}_3$ are the only points of the closed triangle $\boldsymbol{y}_1\boldsymbol{y}_2\boldsymbol{y}_3$ in $\mathfrak{G}$. Then Lemma 2 is true.*

For the line $\pi'$ through $\boldsymbol{y}_1$ and parallel to $\pi$ also contains points of $\mathfrak{G}$ in three quadrants. It is then easy to pick out a divided cell with a pair of opposite sides on $\pi$ and $\pi'$.

We now revert to the proof of Lemma 2. We can find 4 points $\boldsymbol{z}_1, \boldsymbol{z}_2, \boldsymbol{z}_3, \boldsymbol{z}_4$, with $\boldsymbol{z}_j$ in the $j$-th quadrant, such that the (not necesarily convex) closed quadrilateral $\boldsymbol{z}_1\boldsymbol{z}_2\boldsymbol{z}_3\boldsymbol{z}_4$ contains as few points of $\mathfrak{G}$ as possible. The following three cases are all that can occur.

(i) The quadrilateral $\boldsymbol{z}_1\boldsymbol{z}_2\boldsymbol{z}_3\boldsymbol{z}_4$ is convex. It is then a split parallelogram by Proposition 1.

(ii) Three of the points $\boldsymbol{z}_1, \boldsymbol{z}_2, \boldsymbol{z}_3, \boldsymbol{z}_4$ are collinear. If, say, $\boldsymbol{z}_2, \boldsymbol{z}_3, \boldsymbol{z}_4$ are on a line $\pi$, then Lemma 2 follows from Proposition 2 applied to $\boldsymbol{z}_1$ and $\pi$.

(iii) One point, say $\boldsymbol{z}_1$, is an inner point of the convex cover of the remaining three. By the minimal defining property of $\boldsymbol{z}_1, \boldsymbol{z}_2, \boldsymbol{z}_3, \boldsymbol{z}_4$, any point of $\mathfrak{G}$ in the closed quadrilateral $\boldsymbol{z}_1\boldsymbol{z}_2\boldsymbol{z}_3\boldsymbol{z}_4$ other than $\boldsymbol{z}_2, \boldsymbol{z}_3, \boldsymbol{z}_4$ must be in the first quadrant; and such points exist since $\boldsymbol{z}_1$ is one. We may thus choose a point $\boldsymbol{t}$ of $\mathfrak{G}$ in the first quadrant and in the triangle $\boldsymbol{z}_2 \boldsymbol{z}_3 \boldsymbol{z}_4$, such that the only points of $\mathfrak{G}$ in the closed triangle $\boldsymbol{z}_2 \boldsymbol{z}_3 \boldsymbol{t}$ are the vertices. Lemma 2 now follows from Proposition 2 on putting

$$\boldsymbol{y}_1 = \boldsymbol{z}_3, \quad \boldsymbol{y}_2 = \boldsymbol{z}_2, \quad \boldsymbol{y}_3 = \boldsymbol{t}.$$

Since we have now disposed of all three cases, this concludes the proof of Lemma 2.

COROLLARY. *If* $\boldsymbol{x}_j = (x_{1j}, x_{2j})$ *then*

$$\prod_j |x_{1j} x_{2j}| \leqq 2^{-8} \{d(\Lambda)\}^4.$$

For the area of the divided cell is $d(\Lambda)$. It is also the sum of the areas of the four triangles $\mathcal{T}_j$ with vertices $\boldsymbol{o}, \boldsymbol{x}_j, \boldsymbol{x}_{j+1}$ $(1 \leqq j \leqq 4; \boldsymbol{x}_5 = \boldsymbol{x}_1)$. But now the area of $\mathcal{T}_j$ is

$$\tfrac{1}{2}\{|x_{1j} x_{2,j+1}| + |x_{1,j+1} x_{2j}|\},$$

and so

$$2d(\Lambda) = \sum_j |x_{1j} x_{2,j+1}| + \sum_j |x_{2j} x_{1,j+1}|.$$

The required inequality now follows on applying the inequality of the arithmetic and geometric means to the 8 terms on the right-hand side.

We can now prove DAVENPORT'S generalization of MINKOWSKI'S conjecture for $n = 2$.

THEOREM VII. *Let* $\varrho, \sigma$ *be positive numbers and*

$$16 \varrho \sigma \geqq 1.$$

*Then to every 2-dimensional point* $\boldsymbol{x}_0$ *and every lattice* $\Lambda$ *there is a point* $\boldsymbol{x}' \equiv \boldsymbol{x}_0$ $(\Lambda)$ *such that*

$$-\varrho\, d(\Lambda) \leqq x_1' x_2' \leqq \sigma\, d(\Lambda). \tag{2}$$

The case $\varrho = \sigma = \frac{1}{4}$ is, of course, MINKOWSKI'S conjecture[1] for $n = 2$.

When $\boldsymbol{x}_0$ is congruent to a point on an axis modulo $\Lambda$, there is nothing to prove. Otherwise we show that one of the four points $\boldsymbol{x}_j$ $(1 \leqq j \leqq 4)$ given by Lemma 2 will do. If not, we should have

$$|x_{11} x_{21}| > \sigma\, d(\Lambda), \quad |x_{13} x_{23}| > \sigma\, d(\Lambda),$$
$$|x_{12} x_{22}| > \varrho\, d(\Lambda), \quad |x_{14} x_{24}| > \varrho\, d(\Lambda);$$

which is in contradiction with Lemma 2, Corollary.

The reader should not find it difficult to verify that when $\varrho = \sigma = \frac{1}{4}$ the only case when the equality signs are needed in (2) is when $\Lambda = t\boldsymbol{\omega}\Lambda_0$ and $\boldsymbol{x}_0 \equiv t\boldsymbol{\omega}(\frac{1}{2}, \frac{1}{2})$ $(\Lambda)$, where $t > 0$, $\boldsymbol{\omega}$ is an automorph of $x_1 x_2$ and

[1] Proved by MINKOWSKI in this case.

$\Lambda_0$ is the lattice of integers. DAVENPORT (1948a) showed that the equality signs may be needed when $\varrho \neq \sigma$. On the other hand it follows from CHALK'S Theorem of § 4.4 that something stronger is certainly true if $\varrho > 1$ or $\sigma > 1$; and BLANEY (1950a) has given stronger results which cover the cases when $\varrho$ or $\sigma$ is near 1.

**XI.4.3.** We now give the REMAK-DAVENPORT proof of MINKOWSKI'S conjecture in 3 dimensions, which depends on the following

LEMMA 3. *Let $\Lambda$ be any 3-dimensional lattice. Then there exist numbers $p_j > 0$, $(1 \leqq j \leqq 3)$ such that there are no points of $\Lambda$ other than $\boldsymbol{o}$ in the ellipsoid*

$$\mathscr{E}: \quad p_1 x_1^2 + p_2 x_2^2 + p_3 x_3^2 < 1, \tag{1}$$

*but there are three linearly independent points of $\Lambda$ on the boundary of $\mathscr{E}$.*

We call the ellipsoid $\mathscr{E}$ free if $\boldsymbol{o}$ is the only point of $\Lambda$ in it. We shall assume that a free ellipsoid cannot have three linearly independent points of $\Lambda$ on the boundary, for some particular lattice $\Lambda$, and will ultimately deduce a contradiction.

We note first that

$$p_1 p_2 p_3 \geqq \left(\frac{\pi}{6}\right)^2 \{d(\Lambda)\}^{-2} > 0 \tag{2}$$

for any free ellipsoid, by MINKOWSKI'S convex body theorem: the constant in (2) is not important; all that is important is that it is positive.

Secondly, if $\pm \boldsymbol{a}_1$, $\pm \boldsymbol{a}_2$, $\pm \boldsymbol{a}_3$ are three linearly dependent pairs of points of $\Lambda$ on the boundary of a free ellipsoid, we must have

$$\pm \boldsymbol{a}_1 \pm \boldsymbol{a}_2 \pm \boldsymbol{a}_3 = \boldsymbol{o}$$

for some choice of the three $\pm$ signs, since the $\pm \boldsymbol{a}_j$ lie on a plane through the origin and so are points of a 2-dimensional lattice on the boundary of an ellipse which contains no point of the lattice (Theorem XI of Chapter V).

Thirdly, under our hypothesis, if there are two pairs of points $\pm \boldsymbol{a}_1$ and $\pm \boldsymbol{a}_2$ of $\Lambda$ on the boundary of a free ellipsoid, they cannot both lie in the same co-ordinate plane, say, $x_1 = 0$. For then we should have

$$p_2 a_{21}^2 + p_3 a_{31}^2 = 1, \quad \boldsymbol{a}_1 = (0, a_{21}, a_{31})$$
$$p_2 a_{22}^2 + p_3 a_{32}^2 = 1, \quad \boldsymbol{a}_2 = (0, a_{22}, a_{32}).$$

If $p_1$ is decreased but $p_2$, $p_3$ kept constant, the points $\boldsymbol{a}_1$, $\boldsymbol{a}_2$ remain on the boundary and the volume of the ellipsoid increases. Ultimately there must come a third point on the boundary for some value of $p_1$, since it is impossible to decrease $p_1$ to 0 without a point of $\Lambda$ entering the ellipsoid, by (2). Hence for some $p_1$ the ellipsoid is free but there are points $\mathbf{a}_1, \mathbf{a}_2, \mathbf{a}_3$ on the boundary, where $\mathbf{a}_3$ is not on $x_1=0$. This contradicts the hypothesis whose absurdity we wished to prove.

Fourthly we show (on our hypothesis) that if there is a free ellipsoid (1) with the points $\pm\mathbf{a}_1, \pm\mathbf{a}_2 \in \Lambda$ on the boundary, then there is one with $\pm\mathbf{a}_1, \pm\mathbf{a}_2$ and $\pm(\mathbf{a}_1+\mathbf{a}_2)$ on the boundary. For put $\mathbf{a}_3=\mathbf{a}_1+\mathbf{a}_2$, $\mathbf{a}_4=\mathbf{a}_1-\mathbf{a}_2$, and write

$$\mathbf{a}_j = (a_{1j}, a_{2j}, a_{3j}) \qquad (1 \leqq j \leqq 4).$$

Then

$$p_1 a_{1j}^2 + p_2 a_{2j}^2 + p_3 a_{3j}^2 \begin{cases} = 1 & (j = 1, 2) \\ \geqq 1 & (j = 3, 4). \end{cases} \tag{3}$$

There are numbers $q_1, q_2, q_3$ not all 0 such that

$$q_1 a_{1j}^2 + q_2 a_{2j}^2 + q_3 a_{3j}^2 = 0 \qquad (j = 1, 2), \tag{4}$$

and after a change of sign, if need be, we may suppose without loss of generality that[1]

$$q_1 a_{14}^2 + q_2 a_{24}^2 + q_3 a_{34}^2 \geqq 0. \tag{5}$$

We now consider the ellipsoids

$$(p_1 + t q_1)\, x_1^2 + (p_2 + t q_2)\, x_2^2 + (p_3 + t q_3)\, x_3^2 = 1$$

where

$$t \geqq 0.$$

Since at least one of $q_1, q_2, q_3$ is negative by (4), as $t$ increases from 0 the inequality (2) with $p_j + t q_j$ for $p_j$ must fail for some $t$; so there must be some value of $t$ at which for the first time a lattice point enters the ellipse $\mathscr{E}$. This cannot be $\mathbf{a}_4$, by (5), and so must be $\pm\mathbf{a}_3 = \pm(\mathbf{a}_1+\mathbf{a}_2)$ by the second remark; which concludes the proof of the fourth remark.

We now prove the lemma. It is clear that we can obtain free ellipsoids with two pairs of points $\pm\mathbf{a}_1, \pm\mathbf{a}_2 \in \Lambda$ on the boundary by varying the parameters $p_j$ appropriately. By the fourth remark, there is then a free ellipse with $\mathbf{a}_1, \mathbf{a}_2, \mathbf{a}_1+\mathbf{a}_2$ on the boundary. Then by the fourth remark applied to $\mathbf{a}_1$ and $\mathbf{a}_1+\mathbf{a}_2$ there is a free ellipse with $\mathbf{a}_1, \mathbf{a}_1+\mathbf{a}_2$ and $2\mathbf{a}_1+\mathbf{a}_2$ on the boundary. By induction, there is an ellipsoid

$$p_1^{(n)} x_1^2 + p_2^{(n)} x_2^2 + p_3^{(n)} x_3^2 < 1$$

with $\mathbf{a}_1$, $n\,\mathbf{a}_1+\mathbf{a}_2$, $(n+1)\,\mathbf{a}_1+\mathbf{a}_2$ on the boundary. In particular,

$$p_1^{(n)} (n\, a_{11} + a_{12})^2 + p_2^{(n)} (n\, a_{21} + a_{22})^2 + p_3^{(n)} (n\, a_{31} + a_{32})^2 = 1. \tag{6}$$

[1] It is readily verified that there cannot be equality in (5), since the determinant of the three forms in $q_1, q_2, q_3$ in (4) and (5) does not vanish. But we do not need this.

We distinguish three cases. Suppose, first, that $a_{11} \neq 0$, $a_{21} \neq 0$, $a_{31} \neq 0$. Then, by (6),

$$p_j^{(n)} \to 0 \qquad (j = 1, 2, 3) \quad (n \to \infty),$$

in contradiction to (2). Suppose now that precisely one of $a_{11}, a_{21}, a_{31}$ vanishes, say, $a_{11} = 0$, $a_{21} \neq 0$, $a_{31} \neq 0$. Then by the third remark above we have $a_{12} \neq 0$, and so

$$p_1^{(n)} \leqq a_{12}^{-2} < \infty, \quad p_j^{(n)} \to 0 \qquad (j = 2, 3),$$

again in contradiction to (2). Finally, suppose that two of $a_{11}, a_{21}, a_{31}$ vanish, say, $a_{11} = a_{21} = 0 \neq a_{31}$. Then $a_{12} \neq 0 \neq a_{22}$, and so

$$p_j^{(n)} \leqq a_{j2}^{-2} \qquad (j = 1, 2), \qquad p_3^{(n)} \to 0,$$

again in contradiction with (2). Since we have reached a contradiction in every case, we have proved the absurdity of our initial hypothesis and so the lemma is true.

MINKOWSKI's conjecture for $n = 3$ now follows in a few lines from Lemma 3 and Lemma 1 Corollary.

THEOREM VIII. *Let $\Lambda$ be any 3-dimensional lattice and $\boldsymbol{x}_0$ any point. Then there is an $\boldsymbol{x}_1 = (x_{11}, x_{21}, x_{31}) \equiv \boldsymbol{x}_0$ ($\Lambda$) such that*

$$|x_{11} x_{21} x_{31}| \leqq \tfrac{1}{8} d(\Lambda). \tag{7}$$

Let $p_1, p_2, p_3$ be the numbers given by Lemma 3, so that $\Lambda$ has no point in $p_1 x_1^2 + p_2 x_2^2 + p_3 x_3^2 < 1$, but three linearly independent points on the boundary. Hence the three successive minima of $\Lambda$ with respect to the distance-function

$$F(\boldsymbol{x}) = (p_1 x_1^2 + p_2 x_2^2 + p_3 x_3^2)^{\frac{1}{2}} \tag{8}$$

are

$$\lambda_1 = \lambda_2 = \lambda_3 = 1. \tag{9}$$

We may now apply Lemma 1 Corollary to the lattice M of points

$$(p_1^{\frac{1}{2}} x_1, p_2^{\frac{1}{2}} x_2, p_3^{\frac{1}{2}} x_3), \qquad (x_1, x_2, x_3) \in \Lambda,$$

with determinant

$$d(\mathsf{M}) = (p_1 p_2 p_3)^{\frac{1}{2}} d(\Lambda)$$

and with successive minima with respect to $|\boldsymbol{x}|$ given by (9). Hence to any $\boldsymbol{x}_0$ there is a congruent $\boldsymbol{x}_1$ such that

$$\left.\begin{aligned} p_1 x_{11}^2 + p_2 x_{21}^2 + p_3 x_{31}^2 &\leqq \frac{3}{4} \left\{ \frac{d(\mathsf{M})}{\lambda_1 \lambda_2 \lambda_3} \right\}^{\frac{2}{3}} \\ &= \frac{3}{4} (p_1 p_2 p_3)^{\frac{1}{3}} \{d(\Lambda)\}^{\frac{2}{3}}. \end{aligned}\right\} \tag{10}$$

The required inequality (7) now follows at once from (10) and from the inequality of the arithmetic and geometric means.

The reader should have no difficulty in showing that the sign of equality in (7) is required only when $\Lambda = t\boldsymbol{\omega}\Lambda_0$ for some number $t \neq 0$, and some automorph $\boldsymbol{\omega}$ of $x_1 x_2 x_3$, where $\Lambda_0$ is the lattice of points with integral co-ordinates; and then only for $\boldsymbol{x}_0 \equiv t\boldsymbol{\omega}_0(\frac{1}{2}, \frac{1}{2}, \frac{1}{2})$ $(\Lambda)$. Note that to have equality in (7) one must have equality in both applications of the inequality of the arithmetic and geometric means; that going from Lemma 1 to Lemma 1 Corollary and that going from (10) to (7).

**XI.4.4.** We now prove[1] the theorems of TSCHEBOTAREW and CHALK. Since CHALK's theorem is slightly simpler, we prove that first.

THEOREM IX. *Let $\Lambda$ be an n-dimensional lattice and $\boldsymbol{x}_0$ a point. Then there is an $\boldsymbol{x}_1 = (x_{11}, \ldots, x_{n1}) \equiv \boldsymbol{x}_0$ $(\Lambda)$ such that*

$$x_{j1} > 0 \qquad (1 \leqq j \leqq n), \tag{1}$$

$$x_{11} \cdots x_{n1} \leqq d(\Lambda). \tag{2}$$

There certainly is a point $\boldsymbol{x}_2 = (x_{12}, \ldots, x_{n2}) \equiv \boldsymbol{x}_0$ $(\Lambda)$ for which

$$x_{j2} > 0 \qquad (1 \leqq j \leqq n). \tag{3}$$

If $\prod x_{j2} \leqq d(\Lambda)$, then we may put $\boldsymbol{x}_1 = \boldsymbol{x}_2$. Otherwise we have

$$\prod x_{j2} > d(\Lambda), \tag{4}$$

and so, by MINKOWSKI's convex body theorem, there is a point $\boldsymbol{a} \neq \boldsymbol{o}$ of $\Lambda$ such that

$$|a_j| < |x_{j2}| \qquad (1 \leqq j \leqq n). \tag{5}$$

By considering $2^r \boldsymbol{a}$ instead of $\boldsymbol{a}$ with a suitably chosen integer $r \geqq 0$, we may suppose, further, that

$$|a_J| \geqq \tfrac{1}{2} |x_{J2}| \tag{6}$$

for at least one integer $J$. Then the two points

$$\boldsymbol{x}_2 \pm \boldsymbol{a} = \boldsymbol{x}^{\pm} = (x_1^{\pm}, \ldots, x_n^{\pm})$$

are both congruent to $\boldsymbol{x}_0$ and lie in the quadrant $x_j > 0$ $(1 \leqq j \leqq n)$. Further,

$$\frac{\prod_j x_j^+ \prod_j x_j^-}{\prod_j x_{j2}^2} = \prod_j \left( \frac{x_{j2}^2 - a_j^2}{x_{j2}^2} \right) \leqq \frac{3}{4},$$

since by (5) and (6) every factor on the right-hand side is $\leqq 1$, and one at least is $\leqq \frac{3}{4}$. Hence choosing for $\boldsymbol{x}_3$ that one of $\boldsymbol{x}^{\pm}$ for which $\prod x_j$ is least, we have

$$x_{j3} > 0 \quad (1 \leqq j \leqq n); \qquad \prod_j x_{j3} \leqq (\tfrac{3}{4})^{\frac{1}{2}} \prod_j x_{j2}.$$

---

[1] Following MACBEATH (1952a), but in our special cases the argument can be simplified.

If $\prod x_{j3} \leq d(\Lambda)$, then we may put $\boldsymbol{x}_1 = \boldsymbol{x}_3$. Otherwise we repeat the process with $\boldsymbol{x}_3$ instead of $\boldsymbol{x}_2$ and obtain an $\boldsymbol{x}_4$ with

$$x_{j4} > 0 \qquad (1 \leq j \leq n); \qquad \prod_j x_{j4} \leq (\tfrac{3}{4})^{\frac{1}{2}} \prod_j x_{j3} \leq \tfrac{3}{4} \prod_j x_{j2}.$$

And so on. Clearly an $\boldsymbol{x}_1$ is reached in a bounded number of steps, with a bound that can be given explicitly in terms of $\prod_j x_{j2}$. This concludes the proof.

A similar idea gives TSCHEBOTAREW'S

THEOREM X. *Let $\Lambda$ be any $n$-dimensional lattice, $\varepsilon$ an arbitrarily small number and $\boldsymbol{x}_0$ a point. Then there is a point $\boldsymbol{x}_1 = (x_{11}, \ldots, x_{n1}) \equiv \boldsymbol{x}_0$ $(\Lambda)$, such that*

$$|x_{11} \ldots x_{n1}| \leq (2^{-n/2} + \varepsilon)\, d(\Lambda). \tag{7}$$

Let $t$ be the number such that

$$(2^{-n/2} + \varepsilon)\, t^n = 1, \tag{8}$$

so

$$0 < t < 2^{\frac{1}{2}}. \tag{9}$$

If $\prod_j |x_{j0}| \leq (2^{-n/2} + \varepsilon)\, d(\Lambda)$, there is nothing to prove, so we may suppose that

$$\left.\begin{aligned} \prod_j |x_{j0}| &> (2^{-n/2} + \varepsilon)\, d(\Lambda) \\ &= t^{-n} d(\Lambda). \end{aligned}\right\} \tag{10}$$

By MINKOWSKI's convex body theorem, there is a point $\boldsymbol{a} \neq \boldsymbol{o}$ in $\Lambda$ for which

$$|a_j| \leq t\, |x_{j0}| \qquad (1 \leq j \leq n). \tag{11}$$

As in the proof of Theorem IX, we may suppose, on taking $2^r \boldsymbol{a}$ with suitable integer $r \geq 0$, that

$$|a_J| \geq \tfrac{1}{2} t\, |x_{J0}| \tag{12}$$

for some $J$. Put

$$\boldsymbol{x}^{\pm} = \boldsymbol{x}_0 \pm \boldsymbol{a},$$

so that

$$\frac{\prod_j x_j^+ \prod_j x_j^-}{\prod_j x_{j0}^2} = \prod \left(1 - \frac{a_j^2}{x_{j0}^2}\right). \tag{13}$$

But

$$-1 < 1 - t^2 \leq 1 - \frac{a_j^2}{x_{j0}^2} \leq 1,$$

by (9) and (11). Further,

$$1 - t^2 \leq 1 - \frac{a_J^2}{x_{J0}^2} \leq 1 - \frac{1}{4} t^2$$

by (12). Hence, on taking for $\boldsymbol{x}_2$ that one of $\boldsymbol{x}^{\pm}$ for which $\prod |x_j|$ is least, we have

$$\prod_j |x_{j2}| \leqq s \prod_j |x_{j0}|,$$

where

$$s^2 = \max\{|1-t^2|,\ |1-\tfrac{1}{4}t^2|\} < 1.$$

As in the proof of IX we reach an $\boldsymbol{x}_1$ satisfying (7) after a finite number of steps, the number of steps being bounded by a number depending only on $n$, $\prod_j |x_{j0}|$ and $\varepsilon$. This concludes the proof.

# Appendix

In this appendix we list the lattice constants of some sets connected with quadratic forms and give further references and some additional comments. We write

$$\varphi_{r,s}(\boldsymbol{x}) = x_1^2 + \cdots + x_r^2 - x_{r+1}^2 - \cdots - x_{r+s}^2,$$

and denote by $\Gamma_{r,s}$ the lattice constant of the set

$$|\varphi_{r,s}| < 1$$

in $n$-dimensional space, where

$$n = r + s.$$

Results about definite forms are usually given in terms of $\gamma_n$ where $\gamma_n^n = \Gamma_{n,0}^{-2}$. The first 8 values are known:

$$\gamma_1^1 = 1, \quad \gamma_2^2 = \tfrac{4}{3}, \quad \gamma_3^3 = 2, \quad \gamma_4^4 = 4,$$
$$\gamma_5^5 = 8, \quad \gamma_6^6 = \tfrac{64}{3}, \quad \gamma_7^7 = 64, \quad \gamma_8^8 = 2^8.$$

The value of $\gamma_1$ is trivial; the values of $\gamma_2, \gamma_3, \gamma_4$ have been found in this book (Chapter II, Theorems II, III and Chapter X, Theorem IV, Corollary). For references and a list of the corresponding critical forms see CHAUNDY (1946a), who gives proofs that $\gamma_9^9 = 2^9$, $\gamma_{10}^{10} = 2^{10}/3$; but CHAUNDY's proofs contain a lacuna. Presumably his line of argument would lead to incorrect results by $n = 12$; see COXETER and TODD (1953a) for a special form in 12 variables.

For indefinite forms we have

$$\Gamma_{1,1}^2 = \tfrac{5}{4}$$
$$\Gamma_{2,1}^2 = \Gamma_{1,2}^2 = \tfrac{3}{4}$$
$$\Gamma_{2,2}^2 = \tfrac{9}{4}$$
$$\Gamma_{3,1}^2 = \Gamma_{1,3}^2 = \tfrac{7}{4}$$

due to HURWITZ, MARKOFF, OPPENHEIM and OPPENHEIM respectively, the proofs being reproduced in DICKSON (1930a). We have proved all except the last line in the book (Chapter II, Theorems IV, VII and Chapter X, Theorem IV, Corollary). All are isolated. The successive minima of $|\varphi_{1,1}|<1$ are the MARKOFF Chain (see Chapter 2, § 4). The first 11 minima for $|\varphi_{2,1}|<1$ and the first 7 minima for $|\varphi_{2,2}|<1$ have been given by VENKOV (1945a) and OPPENHEIM (1934a) respectively. It is conjectured that $|\varphi_{r,s}|<1$ is of infinite type when $r>0$, $s>0$, $r+s\geqq 5$, see DAVENPORT (1956a)[1].

Let $B_{r,s}$ be the lattice constant of

$$0<\varphi_{r,s}<1.$$

Then

$$B_{1,1}^2=\tfrac{1}{4}$$

$$B_{2,1}^2=\tfrac{1}{4},\quad B_{1,2}^2=\tfrac{4}{27}$$

$$B_{2,2}^2=\tfrac{1}{16},\quad B_{3,1}^2=\tfrac{3}{16},\quad B_{1,3}^2=\tfrac{27}{256}.$$

The value of $B_{1,1}$ is given by Theorem V of Chapter II. The results in the second row are due to DAVENPORT (1949a); both are isolated and something is known about further minima, see OPPENHEIM (1953a). The results in the third row are due to OPPENHEIM (1953b) and again something is known about successive minima. In all cases the critical lattice has points $\boldsymbol{a}\neq\boldsymbol{o}$ at which $\varphi_{r,s}(\boldsymbol{a})=0$.

Let $A_{r,s}$ be the lattice constant of

$$0\leqq\varphi_{r,s}<1.$$

Then

$$A_{1,1}^2=\tfrac{1}{4}$$

$$A_{2,1}^2=\tfrac{3}{4},\quad A_{1,2}^2=\tfrac{5}{16}$$

$$A_{2,2}^2=\tfrac{81}{64},\quad A_{3,1}^2\geqq\tfrac{27}{32},\quad A_{1,3}^2\geqq\tfrac{27}{64}.$$

The value of $A_{1,1}$ follows at once from Theorem VI of Chapter II. The rest are due to BARNES (1955a) and BARNES and OPPENHEIM (1955a).

If a quadratic form in $n\geqq 3$ variables takes arbitrarily small non-zero values of one sign then it also takes arbitrarily small values of the other sign. If a quadratic form represents 0, has two of its coefficients in an irrational ratio and has $n\geqq 5$ variables, then it takes arbitrarily small values of both signs (OPPENHEIM 1953c, d).

---

[1] For later work on this problem, mainly due to DAVENPORT and BIRCH, see RIDOUT (1958a).

# References


BACHMANN, P. (1923a): Die Arithmetik der quadratischen Formen II. Leipzig and Berlin.

BAMBAH, R. P. (1951a): On the geometry of numbers of non-convex star-regions with hexagonal symmetry. Phil. Trans. Roy. Soc. Lond. **243**, 431–462.

— (1954a): Lattice coverings with four-dimensional spheres. Proc. Cambridge Phil. Soc. **50**, 203–208 (1954).

— (1954b): On lattice coverings by spheres. Proc. Nat. Inst. Sci. India **20**, 25–52.

— (1954c): On polar reciprocal convex domains (and addendum). Proc. Nat. Inst. Sci. India **20**, 119–120, 324–325.

— (1955a): Polar Reciprocal Convex bodies. Proc. Cambridge Phil. Soc. **51**, 377–378.

— (1955b): Divided cells. Res. Bull. Panjab Univ. **81**, 173—174.

— (1958a): Some transference theorems in the geometry of numbers. Mh. Math. **62**, 243—249.

—, and H. DAVENPORT (1952a): The covering of $n$-dimensional space by spheres. J. Lond. Math. Soc. **27**, 224–229.

—, and C. A. ROGERS (1952a): Covering the plane with convex sets. J. Lond. Math. Soc. **27**, 304–314.

—, and K. ROGERS (1955a): An inhomogeneous minimum for non-convex star regions with hexagonal symmetry. Canad. J. Math. **7**, 337–346.

BARNES, E. S. (1950a): Non-homogeneous binary quadratic forms. Quart. J. Math. Oxford (2) **1**, 199–210.

— (1951a): The minimum of the product of two values of a quadratic form I. Proc. Lond. Math. Soc. (3) **1**, 257–283.

— (1954a): The inhomogeneous minima of binary quadratic forms IV. Acta Math. **92**, 235–264.

— (1955a): The non-negative values of quadratic forms. Proc. Lond. Math. Soc. (3) **5**, 185–196.

— (1956a): The inhomogeneous minimum of a ternary quadratic form. Acta Math. **96**, 67–97.

— (1956b): The coverings of space by spheres. Canad. J. Math. **8**, 293–304.

— (1956c): On linear inhomogeneous Diophantine approximation. J. Lond. Math. Soc. **31**, 73–79.

— (1957a): On a theorem of Voronoï. Proc. Cambridge Phil. Soc. **53**, 537–539.

— (1957b): The complete enumeration of extreme senary forms. Phil. Trans. Roy. Soc. Lond. **249**, 461–506.

—, and A. OPPENHEIM (1955a): The non-negative values of a ternary quadratic form. J. Lond. Math. Soc. **30**, 429–439.

—, and H. P. F. SWINNERTON-DYER (1952a, b): The inhomogeneous minima of binary quadratic forms I, II. Acta Math. **87**, 259–323; **88**, 279–316.

— — (1954a): The inhomogeneous minima of binary quadratic forms III. Acta Math. **92**, 199–234.

BIRCH, B. J. (1956a): A transference theorem of the geometry of numbers. J. Lond. Math. Soc. **31**, 248–251.

BIRCH, B. J. (1956b): Another transference Theorem of the geometry of numbers. Proc. Cambridge Phil. Soc. **53**, 269—272.
— (1957a): A grid with no split parallelopiped. Proc. Cambridge Phil. Soc. **53**, 536.
— (1958a): The inhomogeneous minima of quadratic forms of signature 0. Acta Arithmetica **4**, 85—98.
—, and H. P. F. SWINNERTON-DYER (1956a): On the inhomogeneous minimum of the product of $n$ linear forms. Mathematika **3**, 25—39.
BLANEY, H. (1948a): Indefinite quadratic forms in $n$ variables. J. Lond. Math. Soc. **23**, 153—160.
— (1950a): Some asymmetric inequalities. Proc. Cambridge Phil. Soc. **46**, 359—376.
— (1957a): On the Davenport-Heilbronn Theorem. Mh. Math. **61**, 1—36.
BLICHFELDT, H. F. (1914a): A new principle in the geometry of numbers with some applications. Trans. Amer. Math. Soc. **15**, 227—235.
— (1929a): The minimum value of quadratic forms and the closest packing of spheres. Math. Ann. **101**, 605—608.
— (1939a): Note on the minimum value of the discriminant of an algebraic field. Mh. Math. Phys. **48**, 531—533.
BONNESEN, T., u. W. FENCHEL (1934a): Theorie der Konvexen Körper. Ergebnisse der Math. usw. **3** (1). Berlin.
BRUNNGRABER, E. (1944a): Über Punktgitter. Diss. Wien.
CASSELS, J. W. S. (1947a): On a theorem of Rado in the geometry of numbers. J. Lond. Math. Soc. **22**, 196—200.
— (1948a): On two problems of Mahler. Proc. Kon. Ned. Akad. Wet. **51**, 854—857 (= Indag Math. **10**, 282—285).
— (1952a): The inhomogeneous minimum of binary quadratic, ternary cubic and quaternary quartic forms. Proc. Cambridge Phil. Soc. **48**, 72—86, 519—520.
— (1952b): The product of $n$ inhomogeneous linear forms in $n$ variables. J. Lond. Math. Soc. **27**, 485—492.
— (1953a): A short proof of the Minkowski-Hlawka Theorem. Proc. Cambridge Phil. Soc. **49**, 165—166.
— (1955a, b): Simultaneous diophantine approximation I, II. J. Lond. Math. Soc. **30**, 119—121 and Proc. Lond. Math. Soc. (3), **5**, 435—448.
— (1956a): On a result of Marshall Hall. Mathematika **3**, 109—110.
— (1957a): An introduction to diophantine approximation. Cambridge: Cambridge University Press.
— (1958a): On subgroups of infinite abelian groups. J. Lond. Math. Soc. **33**, 281—284.
— W. LEDERMANN and K. MAHLER (1951a): Farey section in $k(i)$ and $k(\varrho)$. Phil. Trans. Roy. Soc. Lond. A **243**, 585—628.
—, and H. P. F. SWINNERTON-DYER (1955a): On the product of three homogeneous linear forms and indefinite ternary quadratic forms. Phil. Trans. Roy. Soc. Lond. **248**, 73—96.
ČERNÝ, K. (1952a): On the minima of binary biquadratic forms I (in Russian). Czechos. Math. J. **2**, 1—56.
CHABAUTY, C. (1949a): Sur les minima arithmétiques des formes. Ann. Sci. Éc. Norm. Sup., Paris (3) **66**, 367—394.
— (1950a): Limite d'ensembles et géométrie des nombres. Bull. Soc. Math. France **78**, 143—151.
— (1952a): Empilement de sphères égales dans $R^n$. etc. C. R. Acad. Sci., Paris **235**, 529—532 (but see Math. Reviews **14**, 511).

CHALK, J. H. H. (1947a, b): On the positive values of linear forms I, II. Quart. J. Math. Oxford **18**, 215–227; **19**, 67–80 (1948).
— (1949a): Reduced binary cubic forms. J. Lond. Math. Soc. **24**, 280–284.
— (1950a): On the frustrum of a sphere. Ann. of Math. (2) **52**, 199–216.
—, and C. A. ROGERS (1948a): The critical determinant of a convex cylinder. J. Lond. Math. Soc. **23**, 178–187.
— — (1949a): The successive minima of a convex cylinder. J. Lond. Math. Soc. **24**, 284–291.
— — (1951): On the product of three homogeneous linear forms. Proc. Cambridge Phil. Soc. **47**, 251–259.
CHAUNDY, T. W. (1946a): The arithmetic minima of positive quadratic forms. Quart. J. Math, Oxford **17** (67), 166–192.
CLARKE, L. E. (1951a): On the product of three non-homogeneous forms. Proc. Cambridge Phil. Soc. **47**, 260–265.
— (1958a): The critical lattices of a star-shaped octagon. Acta Math. **99**, 1–32.
COHN, H. (1953a, b): Stable lattices I, II. Canad. J. Math. **5**, 261–270; **6**, 265–273 (1954).
COLE, A. J. (1952a): On the product of *n* linear forms. Quart. J. Math. Oxford (2) **3**, 56–62.
CORPUT, J. G. VAN DER (1936a): Verallgemeinerung einer Mordellschen Beweismethode in der Geometrie der Zahlen. Acta Arithmetica **2**, 145–146.
COXETER, H. S. M., and J. A. TODD (1953a): An extreme duodenary form. Canad. J. Math. **5**, 384–392.
DAVENPORT, H. (1938a): On the product of three homogeneous linear forms I. Proc. Lond. Math. Soc. (2) **44**, 412–431.
— (1939a): A simple proof of Remak's Theorem on the product of 3 linear forms. J. Lond. Math. Soc. **14**, 47–51.
— (1939b): On the product of three homogeneous linear forms III. Proc. Lond. Math. Soc. **45**, 98–125.
— (1939c): Minkowski's inequality for the minima associated with a convex body. Quart. J. Math. Oxford **10**, 119–121.
— (1941a): Note on the product of three homogeneous linear forms. J. Lond. Math. Soc. **16**, 98–101.
— (1941b): On a conjecture of Mordell concerning binary cubic forms. Proc. Cambridge Phil. Soc. **37**, 325–330.
— (1943a): On the product of three homogeneous linear forms. IV. Proc. Cambridge Phil. Soc. **39**, 1–21.
— (1945a, b): The reduction of a binary cubic form I, II. J. Lond. Math. Soc. **20**, 14–22, 139–157.
— (1946a): On a theorem of Tschebotareff. J. Lond. Math. Soc. **21**, 28–34 and Corrigendum **24**, 316 (1949).
— (1947a): On a theorem of Markoff. J. Lond. Math. Soc. **22**, 96–99.
— (1947b): The geometry of numbers. Math. Gazette **31**, 206–207.
— (1947c): On the product of three non-homogeneous linear forms. Proc. Cambridge Phil. Soc. **43**, 137–152.
— (1948a): Non-homogeneous ternary quadratic forms. Acta Math. **80**, 65–95.
— (1949a): On indefinite ternary quadratic forms. Proc. Lond. Math. Soc. (2) **51**, 145–160.
— (1950a): Note on a binary quartic form. Quart. J. Math. Oxford (2) **1**, 253–261.
— (1951a): Indefinite quadratic forms and Euclid's algorithm in real quadratic fields. Proc. Lond. Math. Soc. (2) **53**, 65–82.

DAVENPORT, H. (1952a): Simultaneous diophantine approximation. Proc. Lond. Math Soc. (3) **2**, 406–416.
— (1952b): The covering of space by spheres. Rend. Circ. Mat. Palermo (2) **1**, 92–107.
— (1955a): On a theorem of Furtwängler. J. Lond. Math. Soc. **30**, 186–195.
— (1956a): Indefinite quadratic forms in many variables. Mathematica **3**, 81–101.
—, and M. HALL (1948a): On the equation $ax^2+by^2+cz^2=0$. Quart. J. Math. Oxford **19**, 189–192.
—, and H. HEILBRONN (1947a): Asymmetric inequalities for non-homogeneous linear forms. J. Lond. Math. Soc. **22**, 53–61.
—, and C. A. ROGERS (1947a): Hlawka's Theorem in the geometry of numbers. Duke Math. J. **14**, 367–375.
— — (1950a): Diophantine inequalities with an infinity of solutions. Phil. Trans. Roy. Soc. Lond. **242**, 311–344.
— — (1950b): On the critical determinants of cylinders. Quart. J. Math. Oxford (2) **1**, 215–218.
—, and H. P. F. SWINNERTON-DYER (1955a): Products of inhomogeneous linear forms. Proc. Lond. Math. Soc. (3) **5**, 474–499.
DAVIS, C. S. (1951a): The minimum of a binary quartic form. Acta Math. **84**, 263–298.
DELAUNAY, B. N. (1947a): An algorithm for divided cells [Russian]. Izv. Akad. Nauk SSSR. (Ser. Mat.) **11**, 505–538.
DICKSON, L. E. (1929a): Introduction to the theory of numbers. Chicago, Ill.: Chicago University Press.
— (1930a): Studies in the Theory of Numbers. Chicago, Ill.: Chicago University Press.
DOWKER, C. H. (1944a): On minimum circumscribed polygons. Bull. Amer. Math. Soc. **50**, 120–122.
DYSON, F. J. (1948a): On the product of four non-homogeneous forms. Ann. of Math. (2) **49**, 82–109.
EGGLESTON, H. G. (1958a): Convexity. Cambridge: Cambridge Univ. Press.
ENNOLA, V. (1958a): On the first inhomogeneous minimum of indefinite binary quadratic forms and Euclid's algorithm in real quadratic fields. Ann. Univ. Turkuensis (Turun Yliopiston Julkaisuja) A 1 **28**, 1–58.
ERDÖS, P., and C. A. ROGERS (1953a): The covering of $n$-dimensional space by spheres. J. Lond. Math. Soc. **28**, 287–293.
FEJES TÓTH, L. (1950a): Some packing and covering theorems. Acta Univ. Szeged, Acta Sci. Math **12**/A, 62–67.
— (1953a): Lagerungen in der Ebene, auf der Kugel und im Raum. Berlin: Springer.
FENCHEL, W. (1937a): Verallgemeinerung einiger Sätze aus der Geometrie der Zahlen. Acta Arithmetica **2**, 230–241.
FEW, L. (1956a): Covering space by spheres. Mathematika **3**, 136–139.
FOSTER, D. M. E. (1956a): Indefinite quadratic polynomials in $n$ variables. Mathematika **3**, 111–116.
GAUSS, C. F. (1831a): Besprechung des Buchs von L. A. Seeber: Untersuchungen über die Eigenschaften der positiven ternären quadratischen Formen usw. Göttingsche Gelehrte Anzeigen **1831**, Juli 9. Reprinted in Werke (1876), Vol. II, 188–196.
GODWIN, H. J. (1950a): On the product of five homogeneous linear forms. J. Lond. Math. Soc. **25**, 331–339.
HAJÓS, G. (1942a): Über einfache und mehrfache Bedeckung des $n$-dimensionalen Raumes mit einem Würfelgitter. Math. Z. **47**, 427–467.

HALL, M. (1947a): On the sum and product of continued fractions. Ann. Math. (2) **48**, 966–993.

HARDY, G. H., and E. M. WRIGHT (1938a): An introduction to the theory of numbers. Oxford.

HLAWKA, E. (1944a): Zur Geometrie der Zahlen. Math. Z. **49**, 285–312.

— (1949a): Ausfüllung und Überdeckung konvexer Körper durch konvexe Körper. Mh. Math. Phys. **53**, 81–131.

— (1952a): Zur Theorie des Figurengitters. Math. Ann. **125**, 183–207.

— (1954a): Grundbegriffe der Geometrie der Zahlen. Jber. DMV **57**, 37–55.

— (1954b): Zur Theorie der Überdeckung durch konvexe Körper. Mh. Math. Phys. **58**, 287–291.

— (1954c): Inhomogene Minima von Sternkörpern. Mh. Math. Phys. **58**, 292–305.

JOHN, F. (1948a): Extremum problems with inequalities as subsidiary conditions. Studies and essays presented to R. COURANT, p. 187–204. New York.

KELLER, O. H. (1954a): Geometrie der Zahlen. Enzyklopädie der Math. Wissenschaften, Bd. $I_2$, H. 11, Teil iii.

KHINTCHINE, A. YA (1948a): A quantitative formulation of Kronecker's Theory of approximation [in Russian]. Izv. Akad. Nauk SSSR. (Ser. Mat.) **12**, 113–122.

KLEIN, F. (1895a): Über eine geometrische Auffassung der gewöhnlichen Kettenbruchentwicklung. Nachr. Ges. Wiss. Göttingen **1895**, 357—359.

— (1896a): Ausgewählte Kapitel der Zahlentheorie. (Ausgearbeitet von A. SOMMERFELD und PH. FURTWÄNGLER.) Leipzig.

KNESER, M. (1955a): Ein Satz über abelsche Gruppen mit Anwendungen auf die Geometrie der Zahlen. Math. Z. **61**, 429–434.

— (1956a): Summenmengen in lokalkompakten abelschen Gruppen. Math. Z. **66**, 88–110.

LEKKERKERKER, C. G. (1956a): On the Minkowski-Hlawka theorem. Proc. Kon. Ned. Acad. Wet. Amsterdam A **59** (= Indag. Math. **18**), 426–434.

— (1957a): On the volume of compound convex bodies. Proc. Kon. Ned. Acad. Wet. A **60** (= Indag. Math. **19**), 284–289.

MACBEATH, A. M. (1951a): The finite volume theorem for non-homogeneous lattices. Proc. Cambridge Phil. Soc. **47**, 627–628.

— (1952a): A theorem on non-homogeneous lattices. Ann. of Math. (2) **56**, 269–293.

— (1953a): On measure of sum sets. II. The sum-theorem for the torus. Proc. Cambridge Phil. Soc. **49**, 40–43.

—, and C. A. ROGERS (1955a and 1958b): A modified form of Siegel's mean-value theorem I, II. Proc. Cambridge Phil. Soc. **51**, 565–576 and **54**, 322—325 (1958).

— — (1958a): Siegel's mean value theorem in the geometry of numbers, Proc. Cambridge Phil. Soc. **54**, 139–151.

MAHLER, K. (1938a): On Minkowski's theory of reduction of positive definite quadratic forms. Quart. J. Math. Oxford **9**, 259–263.

— (1939a): Ein Übertragungsprinzip für lineare Ungleichungen. Časopis pro pest. mat. a fys. **68**, 85–92.

— (1939b): Ein Übertragungsprinzip für konvexe Körper. Časopis pro pest. mat. a fys. **68**, 93–102.

— (1946a, b, c): Lattice points in two-dimensional star domains I, II, III. Proc. Lond. Math. Soc. (2) **49**, 128–157, 158–167, 168–183.

— (1946d, e): On lattice points in $n$-dimensional star-bodies. I. Existence theorems. Proc. Roy. Soc. Lond. A **187**, 151–187. — II. Reducibility theorems. Proc. Kon. Ned. Akad. Wet. **49**, 331–343, 444–454, 524–532, 622–631.

— (1946f): The theorem of Minkowski-Hlawka. Duke Math. J. **13**, 611–621.

MAHLER, K. (1946g): On lattice points in a cylinder. Quart. J. Math. Oxford **17**, 16–18.

— (1947a): On irreducible convex domains. Proc. Kon. Ned. Akad. Wet. **50**, 98–107 (= Indag. Math. **9**, 3–12).

— (1947b): On the area and the densest packing of convex domains. Proc. Kon. Ned. Akad. Wet. **50**, 108–118 (= Indag. Math. **9**, 14–24).

— (1947c): On the minimum determinant and the circumscribed hexagons of a convex domain. Proc. Kon. Ned. Akad. Wet. **50**, 692–703 (= Indag. Math. **9**, 326–337).

— (1948a): On lattice points in polar reciprocal convex domains. Proc. Kon. Ned. Wet. **51**, 482–485 (= Indag. Math. **10**, 176–179).

— (1949a): On the minimum determinant of a special point set. Proc. Kon. Ned. Akad. Wet. **52**, 633–642 (= Indag. Math. **11**, 195–204).

— (1949b): On the critical lattices of arbitrary point sets. Canad. J. Math. **1**, 78–87.

— (1950a): The geometry of numbers (Duplicated lectures, Boulder, Colorado, U.S.A.).

— (1955a, b): On compound convex bodies I, II. Proc. Lond. Math. Soc. (3) **5**, 358–384.

MARKOFF, A. (1879a): Sur les formes quadratiques binaires indéfinies. Math. Ann. **15**, 381–409.

MELMORE, S. (1947a): Densest packing of equal spheres. Nature, Lond. **159**, 817.

MINKOWSKI, H. (1896a): Geometrie der Zahlen. Leipzig and Berlin.

— (1904a): Dichteste gitterförmige Lagerung kongruenter Körper. Nachr. K. Ges. Wiss. Göttingen **1904**, 311–355. (Reprinted in Gesammelte Abhandlungen II, 3–42).

— (1907a): Diophantische Approximationen. Leipzig.

MORDELL, L. J. (1931a): Indefinite quadratic forms in $n$ variables. Proc. Roy. Soc. A **131**, 99–108.

— (1937a): A theorem of Khintchine on linear diophantine approximation. J. Lond. Math. Soc. **12**, 166–167.

— (1940a): Tschebotareff's theorem on the product of non-homogeneous linear forms. Vjschr. naturforsch. Ges. Zürich **85**, Beiblatt (Festschrift Rudolf Fueter) 47–50.

— (1941a): The product of homogeneous linear forms. J. Lond. Math. Soc. **16**, 4–12.

— (1942a): The product of three homogeneous linear ternary forms. J. Lond. Math. Soc. **17**, 107–115.

— (1943a): The product of $n$ homogeneous forms. Mat. Sbornik (Rec. Math.) **12** (54), 273–276.

— (1943b): On numbers represented by binary cubic forms. Proc. Lond. Math. Soc. (2) **48**, 198–228.

— (1944a): Lattice points in the region $|x^3+y^3|\leq 1$. J. Lond. Math. Soc. **19**, 92–99.

— (1944b): Observation on the minimum of a positive definite quadratic form in eight variables. J. Lond. Math. Soc. **19**, 3–6.

— (1946a): Further contributions to the geometry of numbers for non-convex regions. Trans. Amer. Math. Soc. **59**, 189–215.

— (1948a): The minimum of a definite ternary quadratic form. J. Lond. Math. Soc. **23**, 175–178.

— (1951a): On the equation $ax^2+by^2-cz^2=0$. Mh. Math. **55**, 323–327.

— (1952a): The minima of some non-homogeneous functions of two variables. Duke Math. J. **19**, 519–527.

— (1956a): The minimum of an inhomogeneous quadratic polynomial in $n$ variables. Math. Z. **63**, 525–528.

MORDELL, L. J. (1959a): Tschebotareff's theorem on the product of non-homogeneous linear forms II. (To appear.)

MULLENDER, P. (1945a): Toepassing van de meetkunde der getallen op ongelijkheden in $K(1)$ en $K(i\sqrt{m})$. Diss. Amsterdam.

— (1948a): Lattice points in non-convex regions. I, II, III. Proc. Kon. Ned. Acad. Wet. **51**, 874—884, 1251—1261; **52**, 50—60 (1949) [= Indag. Math. **10**, 302—312, 395—405; **11**, 18—28 (1949)].

— (1950a): Simultaneous approximation. Ann. of Math. **52**, 417—426.

MULLINEUX, N. (1951a): Lattice points in the star-body $|x_1^2+x_2^2-x_3^2|\leqq 1, |x_3|\leqq\sqrt{2}$. Proc. Lond. Math. Soc. (2) **54**, 1—41.

OLLERENSHAW, K. (1945a): The minima of a pair of indefinite harmonic binary quadratic forms. Proc. Cambridge Phil. Soc. **41**, 77—96.

— (1945b): The critical lattices of a circular quadrilateral formed by arcs of three circles. Quart. J. Math. **17**, 223—239.

— (1953a): An irreducible non-convex region. Proc. Cambridge Phil. Soc. **49**, 194—200.

— (1953b): Irreducible convex bodies. Oxford Quart. J. Math. (2) **4**, 293—302.

OPPENHEIM, A. (1932a): The lower bounds of indefinite Hermitian forms. Quart. J. Math. Oxford **3**, 10—14.

— (1934a): Minima of quaternary quadratic forms of signature 0. Proc. Lond. Math. Soc. (2) **37**, 63—81.

— (1936a): The lower bounds of Hermitian quadratic forms in any quadratic field. Proc. Lond. Math. Soc. (2) **40**, 541—555.

— (1946a): Remark on the minimum of quadratic forms. J. Lond. Math. Soc. **21**, 251—252.

— (1953a): One-sided inequalities for quadratic forms. (I) Ternary forms. Proc. Lond. Math. Soc. (3) **3**, 328—337.

— (1953b): One-sided inequalities for quadratic forms. (II) Quaternary forms. Proc. Lond. Math. Soc. (3) **3**, 417—429.

— (1953c, d): Value of quadratic forms I, II. Quart. J. Math. Oxford (2) **4**, 54—59, 60—66.

— (1953e): Value of quadratic forms III. Mh. Math. **57**, 97—101.

— (1953f): One-sided inequalities for hermitian quadratic forms. Mh. Math. **57**, 1—5.

PITMAN, J. (1958a): The inhomogeneous minima of a sequence of symmetric MARKOV forms. Acta Arithmetica **5**, 81—116.

POITOU, G. (1953a): Sur l'approximation des nombres complexes par les nombres des corps imaginaires quadratiques etc. Ann. Sci. Éc. Norm. Sup. Paris (3) **70**, 199—265.

PRASAD, A. V. (1949a): A non-homogeneous inequality for integers in a special cubic field. Proc. Kon. Ned. Akad. Wet. Amst. **52**, 240—250, 338—350 (= Indag. Math. **11**, 55—65, 112—124).

RADO, R. (1946a): A theorem on the geometry of numbers. J. Lond. Math. Soc. **21**, 34—47.

RANKIN, R. A. (1947a): On the closest packing of spheres in $n$ dimensions. Ann. of Math. **48**, 1062—1081.

— (1949a, b, c): On sums of powers of linear forms I, II, III. I. Ann. of Math. **50**, 691—698. — II. Ann. of Math. **50**, 699—704. — III. Proc. Kon. Ned. Akad. Wet. **51**, 846—853 (1948) (= Indag. Math. **10**, 274—281).

— (1953a): The anomaly of convex bodies. Proc. Cambridge Phil. Soc. **49**, 54—58.

— (1955a): The closest packing of spherical caps in $n$ dimensions. Proc. Glasgow Math. Assoc. **2**, 139—144.

RÉDEI, L. (1955a): Neuer Beweis des Hajósschen Satzes über die endlichen abelschen Gruppen. Acta Math. Hungarica **6**, 27–40.

— (1959a): Neuer Beweis eines Satzes von DELONE über ebene Punktgitter. J. Lond. Math. Soc. **34**, 205—207.

REINHARDT, K. (1934a): Über die dichteste gitterförmige Lagerung kongruenter Bereiche in der Ebene usw. Abh. Math. Sem. Hansische Univ. **10**, 216–230.

REMAK, R. (1923a, b): Verallgemeinerung eines Minkowskischen Satzes I, II. Math. Z. **17**, 1–34; **18**, 173–200 (1924).

— (1925a): Über die geometrische Darstellung der indefiniten binären quadratischen Formen. Jber. DMV **33**, 228–245.

— (1938a): Über die Minkowskische Reduktion. Comp. Math. **5**, 368–391.

RIDOUT, D. (1958a): Indefinite quadratic forms. Mathematika **5**, 122—124.

RIESZ, M. (1936a): Modules réciproques. Comptes Rendus, Congr. Intern. des Math., Oslo **2**, 36—37.

ROGERS, C. A. (1947a): A note on irreducible star-bodies. Proc. Kon. Ned. Akad. Wet. **50**, 868–872 (= Indag. Math. **9**, 379–383).

— (1947b): Existence theorems in the geometry of numbers. Ann. of Math. **48**, 994–1002.

— (1947c): A note on a problem of Mahler. Proc. Roy. Soc. Lond. A **191**, 503–517.

— (1949a): The product of the minima and the determinant of a set. Proc. Kon. Ned. Akad. Wet. **52**, 256–263 (= Indag. Math. **11**, 71–78).

— (1949b): On the critical determinant of a certain non-convex cylinder. Quart. J. Math. Oxford **20**, 45–47.

— (1950a): The product of $n$ real homogeneous linear forms. Acta Math. **82**, 185–208.

— (1950b): A note on coverings and packings. J. Lond. Math. Soc. **25**, 327–331.

— (1951a): The closest packing of convex two-dimensional domains. Acta Math. **86**, 309–321.

— (1951b): The number of lattice points in a star-body. J. Lond. Math. Soc. **26**, 307–310.

— (1952a): The reduction of star-sets. Phil. Trans. Roy. Soc. Lond. A **245**, 59–93.

— (1952b): Indefinite quadratic forms in $n$ variables. J. Lond. Math. Soc. **27**, 314–319.

— (1953a): Almost periodic critical lattices. Arch. der Math. **4**, 267–274.

— (1954a): The Minkowski-Hlawka Theorem. Mathematika **1**, 111–124.

— (1954b): A note on the theorem of Macbeath. J. Lond. Math. Soc. **29**, 133–143.

— (1954c): The product of $n$ non-homogeneous linear forms. Proc. Lond. Math. Soc. (3) **4**, 50–83.

— (1955a): Mean values over the space of lattices. Acta Math. **94**, 249–287.

— (1955b): The moments of the number of points of a lattice in a bounded set. Phil. Trans. Roy. Soc. Lond. A **248**, 225–251.

— (1956a): The number of lattice points in a set. Proc. Lond. Math. Soc. (3) **6**, 305–320.

— (1957a): A note on coverings. Mathematika **4**, 1–6.

— (1958a): Lattice coverings of space: the Minkowski-Hlawka theorem. Proc. Lond. Math. Soc. (3) **8**, 447–465.

— (1958b): Lattice coverings of space with convex bodies. J. Lond. Math. Soc. **33**, 208–212.

— (1958c): The packing of equal spheres. Proc. Lond. Math. Soc. (3) **8**, 609–620.

— (1959a): Lattice coverings of space. Mathematika **6**, 33—39.

ROGERS, K. (1953a): The minima of some inhomogeneous functions of two variables. J. Lond. Math. Soc. **28**, 394–402.

ROGERS, K. (1955a): On the generators of an ideal, with an application to the geometry of numbers in unitary space $U_2$. Amer. J. Math. **77**, 621–627.

— (1956a): Indefinite binary hermitian forms. Proc. Lond. Math. Soc. (3) **6**, 205–223.

—, and H. P. F. SWINNERTON-DYER (1958a): The geometry of numbers over algebraic number fields. Trans. Amer. Math. Soc. **88**, 227–242.

SAMET, P. A. (1954a, b): The product of linear non-homogeneous linear forms I, II. Proc. Cambridge Phil. Soc. **50**, 372–379, 380–390.

SAWYER, D. B. (1948a): The product of two non-homogeneous linear forms. J. Lond. Math. Soc. **23**, 250–251.

— (1950a): A note on the product of two non-homogeneous linear forms. J. Lond. Math. Soc. **25**, 239–240.

— (1953a): The minima of indefinite binary quadratic forms. J. Lond. Math. Soc. **28**, 387–394.

SCHMIDT, W. (1955a): Über höhere kritische Determinanten von Sternkörpern. Mh. Math. **59**, 274–304.

— (1956a): Eine neue Abschätzung der kritischen Determinanten von Sternkörpern. Mh. Math. **60**, 1–10.

— (1956b): Eine Verschärfung des Satzes von Minkowski-Hlawka. Mh. Math., **60**, 110–113.

— (1958a): The measure of the set of admissible lattices. Proc. Amer. Math. Soc. **9**, 390–403.

SCHOLZ, A. (1938a): Minimaldiskriminanten algebraischer Zahlkörper. Crelle **179**, 16–21.

SCHUR, I. (1918a): Über die Verteilung der Wurzeln bei gewissen algebraischen Gleichungen mit ganzzahligen Koeffizienten. Math. Z. **1**, 377–402.

— (1929a): Einige Sätze über Primzahlen mit Anwendungen auf Irreduzibilitätsfragen I, II. Sitzgsber. preuß. Akad. Wiss. **1929**, 125–136, 370–391.

SEGRE, B. (1945a): Lattice Points in infinite domains and asymmetric diophantine approximations. Duke Math. J. **12**, 337–365.

SIEGEL, C. L. (1935a): Über Gitterpunkte in konvexen Körpern und ein damit zusammenhängendes Extremalproblem. Acta Math. **65**, 309–323.

— (1940a): Einheiten quadratischer Formen. Abh. Hans. Univ. **13**, 209–239.

— (1945a): A mean value theorem in the geometry of numbers. Ann. of Math. **46**, 340–347.

SWINNERTON-DYER, H. P. F. (1953a): Extremal lattices of convex bodies. Proc. Cambridge Phil. Soc. **49**, 161–162.

— (1954a): Inhomogeneous lattices. Proc. Cambridge Phil. Soc. **50**, 20–24.

— (1954b): The inhomogeneous minima of complex cubic norm forms. Proc. Cambridge Phil. Soc. **50**, 209–219.

SYLVESTER, J. J. (1892a): On arithmetical series. Messenger of Maths. **21**, 1–19, 87–120 (= Collected works **4**, 687–731).

TORNHEIM, L. (1955a): Asymmetric minima of quadratic forms and asymmetric diophantine approximation. Duke Math. J. **22**, 287–294.

TSCHEBOTAREW, N. (1934a): Beweis des Minkowskischen Satzes über lineare inhomogene Formen [Russian]. Ucen. Zapiski Kazansk. Gos. Univ. **94**, 3–16. There is a German version in Vjschr. naturforsch. Ges. Zürich **85** (1940), Beiblatt (Festschrift Rudolf Fueter) 27–30.

VARNAVIDES, P. (1948a): On lattice points in a hyperbolic cylinder. J. Lond. Math. Soc. **23**, 195–199.

VENKOV, B. A. (1945a): On the extremal problem of Markov for indefinite ternary quadratic forms [Russian]. Izv. Akad. Nauk SSSR. (Ser. Mat.) **9**, 429–494.

VORONOI, G. (1907a): Sur quelques propriétés des formes quadratiques positives parfaites. Crelle **133**, 97–178.

— (1908a): Récherches sur les paralléloèdres primitifs I, II. Crelle **134**, 198–287; **136**, 67–181 (1909).

WAERDEN, B. L. VAN DER (1956a): Die Reduktionstheorie der positiven quadratischen Formen. Acta Math., Stockh. **96**, 265–309.

WATSON, G. L. (1956a): The covering of space by spheres. Rend. Circ. Mat. Palermo (2) **5**, 1–8.

WEIL, A. (1951a): L'intégration dans les groupes topologiques et ses applications (Actualités Sci. Ind. 869–1145). Paris: Hermann.

WEYL, H. (1940a): The theory of reduction for arithmetical equivalence. Trans. Amer. Math. Soc. **48**, 126–165; **51**, 203–231 (1942).

— (1942a): On geometry of numbers. Proc. Lond. Math. Soc. (2) **47**, 268–289.

WHITWORTH, J. V. (1948a): On the densest packing of sections of a cube. Ann. Mat. pura appl. (4) **27**, 29–37.

— (1951a): The critical lattices of the double cone. Proc. Lond. Math. Soc. (2) **53**, 422–443.

WHITTAKER, E. T., and G. N. WATSON (1902a): A course of Modern Analysis. Cambridge: Cambridge University Press.

WOLFF, K. H. (1954a): Über kritische Gitter im vierdimensionalen Raum. Mh. Math. **58**, 38–56.

WOODS, A. C. (1956a): The anomaly of convex bodies. Proc. Cambridge Phil. Soc. **52**, 406–423.

— (1958a): The critical determinant of a spherical cylinder. J. Lond. Math. Soc. **33**, 357–368.

— (1958b): On two-dimensional convex bodies. Pacific J. Maths. **8**, 635–640.

— (1958c): On a theorem of TSCHEBOTAREFF. Duke Math. J. **25**, 631–638.

YEH, Y. (1948a): Lattice points in a cylinder over a convex domain. J. Lond. Math. Soc. **23**, 188–195.

ŽILINKAS, G. (1941a): On the product of four homogeneous linear forms. J. Lond. Math. Soc. **16**, 27–37.

# Index

**Commonly used symbols are listed first, followed by words and phrases in alphabetical order**

Druck: STRAUSS OFFSETDRUCK, MÖRLENBACH
Verarbeitung: SCHÄFFER, GRÜNSTADT

CIM

**M. Aigner** Combinatorial Theory ISBN 978-3-540-61787-7
**A. L. Besse** Einstein Manifolds ISBN 978-3-540-74120-6
**N. P. Bhatia, G. P. Szegő** Stability Theory of Dynamical Systems ISBN 978-3-540-42748-3
**J. W. S. Cassels** An Introduction to the Geometry of Numbers ISBN 978-3-540-61788-4
**R. Courant, F. John** Introduction to Calculus and Analysis I ISBN 978-3-540-65058-4
**R. Courant, F. John** Introduction to Calculus and Analysis II/1 ISBN 978-3-540-66569-4
**R. Courant, F. John** Introduction to Calculus and Analysis II/2 ISBN 978-3-540-66570-0
**P. Dembowski** Finite Geometries ISBN 978-3-540-61786-0
**A. Dold** Lectures on Algebraic Topology ISBN 978-3-540-58660-9
**J. L. Doob** Classical Potential Theory and Its Probabilistic Counterpart ISBN 978-3-540-41206-9
**R. S. Ellis** Entropy, Large Deviations, and Statistical Mechanics ISBN 978-3-540-29059-9
**H. Federer** Geometric Measure Theory ISBN 978-3-540-60656-7
**S. Flügge** Practical Quantum Mechanics ISBN 978-3-540-65035-5
**L. D. Faddeev, L. A. Takhtajan** Hamiltonian Methods in the Theory of Solitons ISBN 978-3-540-69843-2
**I. I. Gikhman, A. V. Skorokhod** The Theory of Stochastic Processes I ISBN 978-3-540-20284-4
**I. I. Gikhman, A. V. Skorokhod** The Theory of Stochastic Processes II ISBN 978-3-540-20285-1
**I. I. Gikhman, A. V. Skorokhod** The Theory of Stochastic Processes III ISBN 978-3-540-49940-4
**D. Gilbarg, N. S. Trudinger** Elliptic Partial Differential Equations of Second Order ISBN 978-3-540-41160-4
**H. Grauert, R. Remmert** Theory of Stein Spaces ISBN 978-3-540-00373-1
**H. Hasse** Number Theory ISBN 978-3-540-42749-0
**F. Hirzebruch** Topological Methods in Algebraic Geometry ISBN 978-3-540-58663-0
**L. Hörmander** The Analysis of Linear Partial Differential Operators I – Distribution Theory and Fourier Analysis ISBN 978-3-540-00662-6
**L. Hörmander** The Analysis of Linear Partial Differential Operators II – Differential Operators with Constant Coefficients ISBN 978-3-540-22516-4
**L. Hörmander** The Analysis of Linear Partial Differential Operators III – Pseudo-Differential Operators ISBN 978-3-540-49937-4
**L. Hörmander** The Analysis of Linear Partial Differential Operators IV – Fourier Integral Operators ISBN 978-3-642-00117-8
**K. Itô, H. P. McKean, Jr.** Diffusion Processes and Their Sample Paths ISBN 978-3-540-60629-1
**T. Kato** Perturbation Theory for Linear Operators ISBN 978-3-540-58661-6
**S. Kobayashi** Transformation Groups in Differential Geometry ISBN 978-3-540-58659-3
**K. Kodaira** Complex Manifolds and Deformation of Complex Structures ISBN 978-3-540-22614-7
**Th. M. Liggett** Interacting Particle Systems ISBN 978-3-540-22617-8
**J. Lindenstrauss, L. Tzafriri** Classical Banach Spaces I and II ISBN 978-3-540-60628-4
**R. C. Lyndon, P. E Schupp** Combinatorial Group Theory ISBN 978-3-540-41158-1
**S. Mac Lane** Homology ISBN 978-3-540-58662-3
**C. B. Morrey Jr.** Multiple Integrals in the Calculus of Variations ISBN 978-3-540-69915-6
**D. Mumford** Algebraic Geometry I – Complex Projective Varieties ISBN 978-3-540-58657-9
**O. T. O'Meara** Introduction to Quadratic Forms ISBN 978-3-540-66564-9
**G. Pólya, G. Szegő** Problems and Theorems in Analysis I – Series. Integral Calculus. Theory of Functions ISBN 978-3-540-63640-3
**G. Pólya, G. Szegő** Problems and Theorems in Analysis II – Theory of Functions. Zeros. Polynomials. Determinants. Number Theory. Geometry ISBN 978-3-540-63686-1
**W. Rudin** Function Theory in the Unit Ball of $\mathbb{C}^n$ ISBN 978-3-540-68272-1
**S. Sakai** $C^*$-Algebras and $W^*$-Algebras ISBN 978-3-540-63633-5
**C. L. Siegel, J. K. Moser** Lectures on Celestial Mechanics ISBN 978-3-540-58656-2
**T. A. Springer** Jordan Algebras and Algebraic Groups ISBN 978-3-540-63632-8
**D. W. Stroock, S. R. S. Varadhan** Multidimensional Diffusion Processes ISBN 978-3-540-28998-2
**R. R. Switzer** Algebraic Topology: Homology and Homotopy ISBN 978-3-540-42750-6
**A. Weil** Basic Number Theory ISBN 978-3-540-58655-5
**A. Weil** Elliptic Functions According to Eisenstein and Kronecker ISBN 978-3-540-65036-2
**K. Yosida** Functional Analysis ISBN 978-3-540-58654-8
**O. Zariski** Algebraic Surfaces ISBN 978-3-540-58658-6